Quality Auditing in Construction Projects

This book provides construction professionals, designers, contractors and quality auditors involved in construction projects with the auditing skills and processes required to improve construction quality and make their projects more competitive and economical.

The processes within the book focus on auditing compliance to ISO, corporate quality management systems, project-specific quality management systems, contract management, regulatory authorities' requirements, safety and environmental considerations. The book is divided into seven chapters and each chapter is divided into numbered sections covering auditing-related topics that have importance or relevance for understanding quality auditing concepts for construction projects.

No other book covers construction quality auditing in such detail and with this level of practical application. It is an essential guide not only for construction and quality professionals, but also for students and academics interested in learning about quality auditing in construction projects.

Abdul Razzak Rumane (PhD) is a Chartered Quality Professional Fellow of The Chartered Quality Institute (UK) and a certified consultant engineer in electrical engineering. Presently he is associated with SIJJEEL Co., Kuwait, as Advisor and Director, Construction Management.

Quality Auditing in Construction Projects
A Handbook

Abdul Razzak Rumane

Routledge
Taylor & Francis Group
LONDON AND NEW YORK

First published 2019
by Routledge
2 Park Square, Milton Park, Abingdon, Oxon OX14 4RN

and by Routledge
52 Vanderbilt Avenue, New York, NY 10017

Routledge is an imprint of the Taylor & Francis Group, an informa business

© 2019 Abdul Razzak Rumane

The right of Abdul Razzak Rumane to be identified as author of this work has been asserted by him in accordance with sections 77 and 78 of the Copyright, Designs and Patents Act 1988.

All rights reserved. No part of this book may be reprinted or reproduced or utilised in any form or by any electronic, mechanical, or other means, now known or hereafter invented, including photocopying and recording, or in any information storage or retrieval system, without permission in writing from the publishers.

Trademark notice: Product or corporate names may be trademarks or registered trademarks, and are used only for identification and explanation without intent to infringe.

British Library Cataloguing-in-Publication Data
A catalogue record for this book is available from the British Library

Library of Congress Cataloging-in-Publication Data
Names: Rumane, Abdul Razzak, author.
Title: Quality auditing in construction projects : a handbook / Abdul Razzak Rumane.
Description: Abingdon, Oxon ; New York, NY : Routledge, [2019] | Includes bibliographical references and index. |
Identifiers: LCCN 2018061411 (print) | LCCN 2019000602 (ebook) | ISBN 9781351201858 (ePub) | ISBN 9781351201841 (Mobipocket) | ISBN 9781351201865 (Adobe PDF) | ISBN 9780815385318 (hardback) | ISBN 9781351201872 (ebook)
Subjects: LCSH: Building—Quality control—Handbooks, manuals, etc. | Civil engineering—Quality control—Handbooks, manuals, etc. | Auditing—Handbooks, manuals, etc.
Classification: LCC TH438.2 (ebook) | LCC TH438.2 .R856 2019 (print) | DDC 624.068/5—dc23
LC record available at https://lccn.loc.gov/2018061411

ISBN: 978-0-8153-8531-8 (hbk)
ISBN: 978-1-351-20187-2 (ebk)

Typeset in Times New Roman
by Apex CoVantage, LLC

Printed and bound in Great Britain by
TJ International Ltd, Padstow, Cornwall

To
My Parents
For their prayers and love
My prayers are always for my father who encouraged me all the times.
I wish he could have been here to see this book and give me blessings.
My prayers and love for my mother who is always inspiring me.

Contents

List of figures xii
List of tables xviii
Foreword xxiv
Acknowledgements xxvi
Preface xxviii
List of abbreviations xxxii
List of synonyms xxxiv
About the author xxxv
Contributors xxxvii

1 **Overview of construction projects** 1
   1.1 Construction projects 1
      1.1.1 Construction project life cycle 5
   1.2 Quality definition for construction projects 12
   1.3 Quality management system 14
      1.3.1 Integrated quality management system 18
   1.4 Quality management in construction projects 20
      1.4.1 Quality during the design stage 22
         1.4.1.1 Quality plan 22
         1.4.1.2 Quality assurance 22
         1.4.1.3 Quality control 24
      1.4.2 Quality during the bidding and tendering stage 24
      1.4.3 Quality during the construction stage 24

2 **ISO certification for the construction industry** 27
   2.1 Introduction 27
   2.2 Importance of standards 28
   2.3 Standards organizations 29
   2.4 QMS manuals 29
      2.4.1 Development of QMS 30
   2.5 QMS documents auditing 48

Contents

    Appendix A: Typical auditing questionnaire for quality manual documents – owner (client, project owner, developer) 49
    Appendix B: Typical auditing questionnaire for quality manual documents – consultant (design and supervision) 63
    Appendix C: Typical auditing questionnaire for quality manual documents – contractor 80
    Appendix D: Typical auditing questionnaire for quality manual documents – manufacturer 97

3 Auditing standards for construction projects   109
MUSTAFA SHRAIM

    3.1 Introduction 109
        3.1.1 Management system standards – auditing requirements 109
    3.2 ISO standards for auditing 110
        3.2.1 ISO 19011:2018 – Guidelines for Auditing Management Systems 111

4 Auditing fundamentals   120
SHIRINE L. MAFI

    4.1 Introduction 120
    4.2 Categories of auditing 121
        4.2.1 First party auditing 122
        4.2.2 Second party auditing 122
        4.2.3 Third party auditing 122
    4.3 Auditing in construction projects 123
        4.3.1 Purpose 124
    4.4 Auditing process 124
    4.5 Audit tools 126
    4.6 Audit findings 127
    4.7 Risks in auditing 127
    4.8 Training for auditing 128
    Audit exercises 132

5 Auditor/auditing team selection for construction projects   133
    5.1 Construction project auditing stakeholders 133
        5.1.1 Roles and responsibilities 138
        5.1.2 Competencies and expertise 143
    5.2 Selection of the auditor for project life cycle phases 149

       5.2.1 *Auditor for the bidding and tendering process 151*
              5.2.1.1 *Auditor for the consultant selection process 154*
              5.2.1.2 *Auditor for the designer selection process 159*
       5.2.2 *Auditor for the conceptual design phase 162*
              5.2.2.1 *Auditor for the study stage 167*
              5.2.2.2 *Auditor for the concept design 167*
       5.2.3 *Auditor for the schematic design phase 169*
       5.2.4 *Auditor for the design development phase 175*
       5.2.5 *Auditor for the construction documents phase 176*
       5.2.6 *Auditor for the bidding and tendering for the construction contractor 185*
       5.2.7 *Auditor for the construction phase 189*
       5.2.8 *Auditor for the testing, commissioning and handover phase 195*

6   Auditing processes for project life cycle phases   201
    6.1 Development of auditing processes for the construction project life cycle 201
       6.1.1 *Auditing performance criteria 201*
    6.2 Auditing process for the Design-Bid-Build type of project delivery system 208
       6.2.1 *Auditing process for bidding and tendering 208*
              6.2.1.1 *Auditing process for the selection of the consultant for the study stage 212*
              6.2.1.2 *Auditing process for the selection of the designer 219*
              6.2.1.3 *Auditing process for the designer's proposal submission process 224*
       6.2.2 *Auditing process for design phases 237*
              6.2.2.1 *Auditing process for the conceptual design phase 237*
              6.2.2.2 *Auditing process for the schematic design phase 249*
              6.2.2.3 *Auditing process for the design development/ detail design phase 258*
              6.2.2.4 *Auditing process for the construction documents phase 261*
       6.2.3 *Auditing process for the bidding and tendering phase 272*

    6.2.3.1 Auditing process for the selection of a contractor (Design-Bid-Build) *278*
    6.2.3.2 Auditing process for the contractor's tender submission process *280*
    6.2.3.3 Auditing process for the selection of the construction supervisor (consultant) *292*
    6.2.3.4 Auditing process for the construction supervisor's tender submission process *301*
  6.2.4 Auditing process for the construction phase (Design-Bid-Build project delivery system) *314*
    6.2.4.1 Auditing process for the construction phase (contractor's works) *320*
    6.2.4.2 Auditing process for the construction phase (construction supervisor's activities) *367*
  6.2.5 Auditing process for the testing, commissioning and handover phase *385*
    6.2.5.1 Contractor's work *387*
    6.2.5.2 Construction supervisor's activities *393*
6.3 Auditing process for the construction project (Design-Build type of project delivery system) phases *401*
  6.3.1 Auditing process for bidding and tendering (Design-Build project delivery system) *404*
    6.3.1.1 Auditing process for the selection of the consultant for the study stage (Design-Build) *404*
    6.3.1.2 Auditing process for the selection of the concept designer (Design-Build) *404*
    6.3.1.3 Auditing process for the Design-Build concept designer's proposal submission process *406*
  6.3.2 Auditing process for contract/construction documents *407*
  6.3.3 Auditing process for the bidding and tendering (Design-Build) contractor *414*
    6.3.3.1 Auditing process for the selection of the contractor (Design-Build) *414*
    6.3.3.2 Auditing process for the Design-Build contractor's tender submission process *416*
    6.3.3.3 Auditing process for the selection of the construction supervisor (Design-Build) *431*

   6.3.3.4 Auditing process for the Design-Build construction supervisor tender submission process 437
  6.3.4 Auditing process for construction (Design-Build) activities 448
   6.3.4.1 Contractor's work 453
   6.3.4.2 Construction supervisor's activities 455
  6.3.5 Auditing process for the testing, commissioning and handover phase 463
   6.3.5.1 Contractor's work 463
   6.3.5.2 Construction supervisor's activities 463
 6.4 Auditing process for the project manager type of project delivery system 463
  6.4.1 Auditing process for the selection of the project manager 464
  6.4.2 Auditing process for the project manager's proposal submission process 474
  6.4.3 Auditing process for the project manager's activities 479
   6.4.3.1 Selection of project teams 481
   6.4.3.2 Design management 481
   6.4.3.3 Construction supervision activities 501
 6.5 Auditing process for the construction management type of project delivery system 522
  6.5.1 Auditing process for the selection of the agency construction manager 522
  6.5.2 Auditing process for the agency construction manager's proposal submission process 526
  6.5.3 Auditing process for construction management activities 537
   6.5.3.1 Selection of project teams 544
   6.5.3.2 Design management 544
   6.5.3.3 Construction management activities 545

7 Audit reporting 547
 7.1 Audit report 547
 7.2 Corrective and preventive action 549
 7.3 Project closeout audit 552

Bibliography 553
Index 554

Figures

1.1	Types of construction projects	3
1.2	Concept of Design-Bid-Build (traditional contracting system)	4
1.3	Construction project trilogy	13
1.4	QMS documentation pyramid	16
1.5	Logic flow diagram for the development of IQMS	19
2.1	QMS documentation model	32
2.2	Black Box	32
2.3	Process-based QMS development procedure	32
2.4	Organizational structure of ISO 9001:2015 sections	33
2.5	PDCA cycle for QMS activities	34
2.6	Quality management process model	35
2.7	ISO QMS design and implementation process flow diagram	47
3.1	Process flow for managing audit programs	116
3.2	Process approach in auditing	117
4.1	Basic structure of certification	121
4.2	Relationship between system, process and output as defined by the audit process	125
5.1	Construction project stakeholders	134
5.2	Audit structure	135
5.3	Classification of audits	136
5.4	Organization structure of a first party (internal auditing) audit team	136
5.5	Organization structure of a second party audit team	137
5.6	Organization structure of a third party audit team	137
5.7	Organization structure of a project owner auditee team	138
5.8	Organization structure of a designer (consultant) auditee team	139
5.9	Organization structure of a contractor (head office) auditee team	140
5.10	Organization structure of a contractor (construction site) auditee team	141
5.11	Overview of project management process groups	150
5.12	Bidding and tendering stages for construction projects	152
5.13	Bidding and tendering process	153
5.14	Logic flow diagram for the selection of a designer (A/E)	160

5.15	Logic flow process for the conceptual design phase	163
5.16	Major activities relating to the conceptual design process	164
5.17	Logic flow process for the schematic design phase	170
5.18	Major activities relating to the schematic design process	171
5.19	Logic flow process for design development phase	176
5.20	Major activities relating to the design development phase	177
5.21	Logic flow process for the construction documents phase	181
5.22	Major activities relating to the construction documents phase	182
5.23	Logic flow process for the bidding and tendering phase	186
5.24	Major activities relating to the bidding and tendering phase	187
5.25	Logic flow process for the construction phase	190
5.26	Major activities relating to the construction phase	191
5.27	Logic flow process for testing, commissioning, and handover phase	196
5.28	Major activities relating to testing, commissioning and handover phase	197
6.1	PDCA cycle for the quality auditing process	202
6.2	Typical quality auditing process for construction projects	203
6.3	Triple constraints	204
6.4a	Design-Bid-Build (traditional contracting system) contractual relationship	206
6.4b	Design-Build type of delivery system contractual relationship	206
6.4c	Project Manager type of delivery system contractual relationship	207
6.4d	Agency construction management contractual relationship	207
6.5a	Logic flow diagram for construction projects – Design-Bid-Build project delivery system	209
6.5b	Quality auditing categories in the construction project life cycle (Design-Bid-Build) phases	210
6.6	Bidding and tendering (procurement) process	211
6.7	Quality auditing stages of the bidding and tendering (procurement) process for the selection of the consultant	213
6.8	Typical quality auditing process for the bidding and tendering procedure for the selection of the designer (A/E)	221
6.9	Bidding and tendering (procurement) process stages for the selection of the designer (A/E)	222
6.10	Typical proposal submission procedure by the designer	231
6.11	Typical quality auditing process for the designer's proposal submission process	232
6.12	Logic flow diagram for the quality auditing process for the designer's proposal submission process	236
6.13	Typical quality auditing process for the study stage of conceptual design	239
6.14	Typical quality auditing process for concept design	244

xiv *Figures*

6.15	House of Quality for the development of the concept design	250
6.16	Typical quality auditing process for the schematic design phase	252
6.17	Logic flow diagram for the quality auditing of schematic design	257
6.18	Design development stages for building construction	259
6.19	Typical quality auditing process for the design development phase	260
6.20	Framework for the quality auditing of the design development phase	262
6.21	Design review procedure	263
6.22	Logic flow diagram for the quality auditing of the design development phase	269
6.23	Typical quality auditing process for the construction documents phase	270
6.24	Logic flow diagram for the quality auditing of the construction documents phase	277
6.25	Typical Quality Auditing Process for the Bidding and Tendering Procedure for the Selection of the Contractor (Design-Bid-Build)	279
6.26	Bidding and tendering (procurement) process stages for the selection of the contractor (Design-Bid-Build)	281
6.27	Bid clarification form	284
6.28	Contractor selection criteria	285
6.29	Logic flow diagram for the quality auditing of the bidding and tendering phase	293
6.30	Typical quality auditing process for the contractor's tender submission process	294
6.31	Typical tender submission procedure by the contractor	295
6.32	Logic flow diagram for the quality auditing process for the contractor's tender submission process	298
6.33	Typical quality auditing process for the bidding and tendering procedure for the selection of the construction supervisor (consultant)	300
6.34	Typical quality auditing stages of the bidding and tendering (procurement) process for the selection of the construction supervisor (consultant)	302
6.35	Logic flow diagram for the quality auditing of the selection of the construction supervisor (consultant)	311
6.36	Typical quality auditing process for the construction supervisor's tender submission process	312
6.37	Typical tender submission procedure by the construction supervisor	313
6.38	Logic flow diagram for the quality auditing process for the construction supervisor's tender submission process	317
6.39	Typical quality auditing stages for the construction phase	318
6.40	Framework for the quality auditing of the construction phase	321

Figures xv

6.41	Typical quality auditing process for the construction phase	322
6.42	Logic flow diagram for the quality auditing process for mobilization activities during the construction phase	325
6.43	Logic flow diagram for the development of the contractor's quality plan	327
6.44	Logic flow diagram for the quality auditing process for the preparation of management plans activities during the construction phase	330
6.45	Contractor's submittal status log	331
6.46	Submittal process	332
6.47	Submittal transmittal form	333
6.48	Specification comparison statement	334
6.49	Logic flow diagram for the quality auditing process for submittal activities during the construction phase	337
6.50	Flow chart for the concrete casting process for structural work	338
6.51	Material procurement procedure	342
6.52	Logic flow diagram for the project monitoring and control process	343
6.53	Traditional monitoring system	344
6.54	Digitized monitoring system	345
6.55	Logic flow diagram for the quality auditing process for construction/execution activities during the construction phase	353
6.56	Sequence of execution of works	354
6.57	Logic flow diagram for the quality auditing process for the inspection of executed/installed activities during the construction phase	355
6.58	Typical quality auditing process for construction phase activities	357
6.59	Logic flow diagram for the quality auditing process for construction phase activities	368
6.60	Typical quality auditing process for the construction supervisor's activities related to the construction phase	370
6.61	Logic flow diagram for the quality auditing process for the construction supervisor's activities related to the construction phase	386
6.62	Typical quality auditing process for the contractor's activities related to the testing, commissioning and handover phase	388
6.63	Development of the inspection and test plan	389
6.64	Logic flow diagram for the quality auditing process for the contractor's activities related to the testing, commissioning and handover phase	394
6.65	Typical quality auditing process for the construction supervisor's activities related to the testing, commissioning and handover phase	395
6.66	Handing over certificate	397
6.67	Logic flow diagram for the quality auditing process for the construction supervisor's activities related to the testing, commissioning and handover phase	400

xvi *Figures*

6.68a	Logic flow diagram for construction projects – Design-Build project delivery system	402
6.68b	Quality auditing categories in the construction project life cycle (Design-Build) phases	403
6.69	Typical quality auditing process for construction documents for Design-Build projects	409
6.70	Logic flow diagram for the quality auditing of construction documents for Design-Build projects	413
6.71	Typical quality auditing process for the bidding and tendering procedure for the selection of the contractor (Design-Build projects)	415
6.72	Logic flow diagram for the quality auditing of the bidding and tendering for the selection of the contractor (Design-Build projects)	426
6.73	Typical quality auditing process for the contractor's (Design-Build) tender submission process	427
6.74	Typical tender submission procedure by the contractor (Design-Build)	428
6.75	Logic flow diagram for the quality auditing process for the contractor's (Design-Build) tender submission process	432
6.76	Typical quality auditing process for the bidding and tendering procedure for the selection of the construction supervisor (Design-Build)	435
6.77	Typical quality auditing stages of the bidding and tendering process for the selection of the construction supervisor (Design-Build)	436
6.78	Logic flow diagram for the quality auditing of the selection of the construction supervisor (Design-Build)	449
6.79	Typical quality auditing process for the construction supervisor's (Design-Build) tender submission process	450
6.80	Typical tender submission procedure by the construction supervisor (Design-Build)	451
6.81	Logic flow diagram for the quality auditing process for the construction supervisor's (Design-Build) tender submission process	452
6.82	Typical quality auditing stages for the construction work (phase) of the Design-Build project	455
6.83	Typical quality auditing process for the construction supervisor's activities related to the design management stages of the Design-Build project	457
6.84	Logic flow diagram for the quality auditing process for the construction supervisor's activities related to the design management stages of the Design-Build project	462
6.85	Project manager type delivery system contractual relationship	464
6.86	Typical quality auditing process for the bidding and tendering procedure for the selection of the project manager	466

Figures xvii

6.87	Logic flow diagram for the quality auditing of the selection of the project manager	475
6.88	Typical quality auditing process for the project manager's proposal submission process	476
6.89	Logic flow diagram for the quality auditing process for the project manager's proposal submission process	480
6.90	Typical quality auditing process for concept design	482
6.91	Logic flow diagram for the quality auditing of the concept design	486
6.92	Typical quality auditing process for the schematic design phase	487
6.93	Logic flow diagram for the quality auditing of schematic design	492
6.94	Typical quality auditing process for the design development phase	493
6.95	Logic flow diagram for the quality auditing of the design development phase	498
6.96	Typical quality auditing process for the construction documents phase	499
6.97	Logic flow diagram for the quality auditing of the construction documents phase	505
6.98	Typical quality auditing process for the project manager's activities related to the construction phase	506
6.99	Logic flow diagram for the quality auditing process for the project manager's activities related to the construction phase	516
6.100	Typical quality auditing process for the project manager's activities related to the testing, commissioning and handover phase	518
6.101	Logic flow diagram for the quality auditing process for the project manager's activities related to the testing, commissioning and handover phase	521
6.102	Construction manager contractual relationships (agency CM)	523
6.103	Construction manager contractual relationships (at risk CM)	523
6.104	Sequential activities of agency CM – Design-Bid-Build	524
6.105	Typical quality auditing process for the bidding and tendering procedure for the selection of the agency construction manager	525
6.106	Logic flow diagram for the quality auditing of the selection of the agency construction manager	535
6.107	Typical quality auditing process for the agency construction manager's proposal submission process	536
6.108	Logic flow diagram for the quality auditing process for the agency construction manager's proposal submission process	540

Tables

1.1	Construction project life cycle (Design-Bid-Build) phases	10
1.2	Principles of quality in construction projects	15
1.3	Major quality planning activities during the design stage	22
1.4	Major quality assurance activities during the design stage	23
1.5	Major quality control activities during the design stage	25
1.6	Contents of the contractor's quality control plan	25
2.1	ISO 9001:2015 sections (clauses)	31
2.2	List of quality manual documents – owner (client, project owner, developer)	36
2.3	List of quality manual documents – consultant (design and supervision)	39
2.4	List of quality manual documents – contractor	41
2.5	List of quality manual documents – manufacturer	44
3.1	Typical procedure for QMS audits	114
3.2	Checklist for audit preparation	116
3.3	Auditor evaluation methods	118
4.1	Audit schedule	125
4.2	List of certification for quality audits	129
4.3	An example of an internal audit schedule	129
4.4	An example of audit findings	130
4.5	An example of a nonconformance report	130
4.6	Audit report example (area for improvement)	131
5.1	Client's roles and responsibilities	142
5.2	Internal auditor's roles and responsibilities	143
5.3	Lead auditor's roles and responsibilities	144
5.4	Auditor's roles and responsibilities	144
5.5	Auditee's roles and responsibilities	145
5.6	Audit program manager's roles and responsibilities	145
5.7	Internal auditor's competencies and expertise	146
5.8	Lead auditor's competencies and expertise	147
5.9	Auditor's competencies and expertise	148
5.10	Subject matter expert auditor's competencies and expertise	149
5.11	Major elements of the study stage	157

5.12	Designer's (A/E) selection criteria	160
5.13	Consultant's qualification for feasibility study	162
5.14	Auditor selection criteria for study stage audit	167
5.15	Auditor selection criteria for the concept design phase audit	168
5.16	Auditor selection criteria for the schematic design phase audit	174
5.17	Auditor selection criteria for the design development phase audit	180
5.18	Auditor selection criteria for the construction documents phase audit	184
5.19	Auditor selection criteria for the bidding and tendering phase audit	189
5.20	Auditor selection criteria for the construction phase audit	195
5.21	Auditor selection criteria for the testing, commissioning and handover phase audit	199
6.1	Contents of an audit plan	204
6.2	Categories of project delivery systems	205
6.3	Audit methodology for the assessment of the bidding and tender process for the selection of a consultant – Stage I (tender documents)	213
6.4	Audit methodology for the assessment of the bidding and tender process for the selection of a consultant – Stage II (contract bid solicitation)	215
6.5	Audit methodology for the assessment of the bidding and tender process for the selection of a consultant – Stage III (contract award)	217
6.6	Prequalification questionnaires (PQQ) for the registration of a designer (A/E)	223
6.7	Audit methodology for the assessment of the bidding and tender process for the selection of a designer (A/E) – Stage I (shortlisting/registration of designers)	225
6.8	Audit methodology for the assessment of the bidding and tender process for the selection of a designer (A/E) – Stage II (proposal documents)	226
6.9	Audit methodology for the assessment of the bidding and tender process for the selection of a designer (A/E) – Stage III (contract bid solicitation)	228
6.10	Audit methodology for the assessment of the bidding and tender process for the selection of a designer (A/E) – Stage IV (contract award)	229
6.11	Contents of a Request for Proposal (RFP) for the designer (A/E)	233
6.12	Audit methodology for the assessment of the designer's bid (proposal) submission process	234
6.13	Major considerations for the need analysis of a construction project	240
6.14	Audit methodology for the assessment of the study stage of the conceptual phase	241

xx Tables

6.15	Typical contents of Terms of Reference (TOR) documents	245
6.16	Audit methodology for assessment of the concept design	247
6.17	Schematic design deliverables	253
6.18	Audit methodology for the assessment of the schematic design	254
6.19	Mistake proofing for eliminating design errors	264
6.20	Audit methodology for the assessment of the design development phase	264
6.21	Construction documents deliverables	271
6.22	Audit methodology for assessment of construction documents	273
6.23	Prequalification questionnaires (PQQ) for selecting contractor (Design-Bid-Build)	282
6.24	Audit methodology for the assessment of the bidding and tender process for the selection of a contractor (Design-Bid-Build) – Stage I (shortlisting/registration of contractors)	286
6.25	Audit methodology for the assessment of the bidding and tender process for the selection of the contractor (Design-Bid-Build) – Stage II (bidding and tender documents)	287
6.26	Audit methodology for the assessment of the bidding and tender process for the selection of the contractor (Design-Bid-Build) – Stage III (contract bid solicitation)	288
6.27	Audit methodology for the assessment of the bidding and tender process for the selection of the contractor (Design-Bid-Build) – Stage IV (contract award)	290
6.28	Audit methodology for the assessment of the contractor's tender submission process	296
6.29	Responsibilities of the construction supervisor (consultant)	299
6.30	Prequalification questionnaires for the registration of the construction supervisor (consultant)	303
6.31	Audit methodology for the assessment of the bidding and tender process for the selection of the construction supervisor (consultant) – Stage I (shortlisting/registration of construction supervisor)	304
6.32	Audit methodology for the assessment of the bidding and tender process for the selection of the construction supervisor (consultant) – Stage II (bidding and tender documents)	305
6.33	Audit methodology for the assessment of the bidding and tender process for the selection of the construction supervisor (consultant) – Stage III (contract bid solicitation)	307
6.34	Audit methodology for the assessment of the bidding and tender process for the selection of the construction supervisor (consultant) – Stage IV (Contract Award)	309
6.35	Audit methodology for the assessment of the construction supervisor's tender submission process	315
6.36	Audit methodology for the assessment of mobilization activities during the construction phase	323

6.37	Risk register	328
6.38	Audit methodology for the assessment of the preparation of management plans during the construction phase	329
6.39	Audit methodology for the assessment of submittal activities during the construction phase	335
6.40	Contractor's responsibilities to manage construction quality	339
6.41	Audit methodology for the assessment of the construction/execution activities/processes during the construction phase	345
6.42	Audit methodology for the assessment of inspection activities related to execution/installation works during the construction phase	354
6.43	Audit methodology for the assessment of construction phase activities/processes after completion of the construction	358
6.44	Responsibilities of the construction supervisor (consultant)	369
6.45	Consultant's checklist for the smooth functioning of the project	371
6.46	List of project control documents	374
6.47	List of logs	376
6.48	Audit methodology for the assessment of the construction supervisor's activities related to the construction phase	377
6.49	Project closeout documents	390
6.50	Audit methodology for the assessment of the contractor's activities related to the testing, commissioning and handover phase	392
6.51	Typical responsibilities of the construction supervisor during the testing, commissioning and handover phase	395
6.52	Audit methodology for the assessment of the construction supervisor's activities related to the testing, commissioning and handover phase	398
6.53	Prequalification questionnaires (PQQ) for the registration of the designer (A/E) for Design-Build projects	405
6.54	Construction documents deliverables for Design-Build types of project	407
6.55	Audit methodology for the assessment of construction documents (Design-Build)	410
6.56	Prequalification questionnaires (PQQ) for selecting the Design-Build contractor	417
6.57	Audit methodology for the assessment of the bidding and tender process for the selection of contractor (Design-Build) – Stage I (shortlisting/registration of contractors)	418
6.58	Audit methodology for the assessment of the bidding and tender process for the selection of the contractor (Design-Build) – Stage II (bidding and tender documents)	420
6.59	Audit methodology for the assessment of the bidding and tender process for the selection of the contractor (Design-Build) – Stage III (contract bid solicitation)	422

6.60	Audit methodology for the assessment of the bidding and tender process for the selection of the contractor (Design-Build) – Stage IV (contract award)	424
6.61	Audit methodology for the assessment of the contractor's (Design-Build) tender submission process	429
6.62	Responsibilities of the construction supervisor (Design-Build)	433
6.63	Prequalification questionnaires for the registration of the construction supervisor (Design-Build)	438
6.64	Audit methodology for the assessment of the bidding and tender process for the selection of the construction supervisor (Design-Build) – Stage I (shortlisting/registration of construction supervisor)	439
6.65	Audit methodology for the assessment of the bidding and tender process for the selection of the construction supervisor (Design-Build) – Stage II (bidding and tender documents)	440
6.66	Audit methodology for the assessment of the bidding and tender process for the selection of the construction supervisor (Design-Build) - Stage III (contract bid solicitation)	442
6.67	Audit methodology for the assessment of the bidding and tender process for the selection of the construction supervisor (Design-Build) – Stage IV (contract award)	444
6.68	Audit methodology for the assessment of the construction supervisor's (Design-Build) tender submission process	446
6.69	Audit methodology for the assessment of the construction supervisor's activities related to the design development stages of the Design-Build project	459
6.70	Audit methodology for the assessment of the bidding and tender process for the selection of the project manager firm – Stage I (shortlisting/registration of designers)	467
6.71	Audit methodology for the assessment of the bidding and tendering process for the selection of the project manager firm – Stage II (proposal documents)	468
6.72	Audit methodology for the assessment of the bidding and tendering process for the selection of the project manager firm – Stage III (contract bid solicitation)	470
6.73	Audit methodology for the assessment of the bidding and tendering process for the selection of the project manager firm – Stage IV (contract award)	472
6.74	Audit methodology for the assessment of the project manager's bid (proposal) submission process	477
6.75	Audit methodology for the assessment of the concept design	484
6.76	Audit methodology for the assessment of the schematic design	489
6.77	Audit methodology for the assessment of the design development phase	494
6.78	Audit methodology for the assessment of construction documents	501

6.79	Audit methodology for the assessment of the project manager's activities related to the construction phase	508
6.80	Typical responsibilities of the project manager during the testing, commissioning and handover phase	519
6.81	Audit methodology for the assessment of the project manager's activities related to the testing, commissioning and handover phase	520
6.82	Prequalification questionnaires (PQQ) for selecting the construction manager	527
6.83	Audit methodology for the assessment of the bidding and tender process for the selection of the agency construction manager firm – Stage I (shortlisting/registration of designers)	528
6.84	Audit methodology for the assessment of the bidding and tendering process for the selection of the agency construction manager firm – Stage II (proposal documents)	529
6.85	Audit methodology for the assessment of the bidding and tendering process for the selection of the agency construction manager firm – Stage III (contract bid solicitation)	531
6.86	Audit methodology for the assessment of the bidding and tendering process for the selection of the agency construction manager firm – Stage IV (contract award)	533
6.87	Audit methodology for the assessment of the agency construction manager's bid (proposal) submission process	538
7.1	Typical contents of an audit report	548
7.2	Typical audit findings report	548
7.3	Typical nonconformance report	549
7.4	Typical areas for improvement report	550
7.5	Typical corrective action plan	550
7.6	Process improvement tools	550
7.7	Typical corrective action taken report	551

Foreword

I first met Dr. Rumane at the ASQ World Quality and Improvement Conference, when I was Chair of the ASQ Design and Construction Division and he, as a member of the division, was one of the presenters at the conference. I was impressed by his passion for construction management and quality and his approach to the issue of quality management in construction as indicated in his previous books. He is considered an expert in quality assurance processes, specifically to construction projects, and, from his extensive background and education, provided a practical approach to construction practitioners in the industry.

This book provides a widely useful compilation of ideas, cases, innovative approaches and practical strategies for the quality auditing of construction projects. By taking a diversified look at construction project auditing, Dr. Abdul Razzak Rumane identifies a substantial approach in the overview of quality in construction projects, ISO certification for the construction industry and the important effort to prepare, perform and participate in the quality audits of construction projects. In this book he provides an enormously useful range of approaches for designing, planning, implementing, performing and evaluating construction project audits. This work would be an important resource if it only highlighted the ways construction quality audit programs can improve overall construction project quality, but the book goes beyond the basics for quality auditing and making us aware of this sometimes underutilized management tool. It covers and describes all the major factors in building a successful construction project quality audit program that enhances construction project success. This volume is an important resource for project construction directors, project managers, quality managers, auditors and third-party agencies that provide quality audit programs, and other organizations that use this service. First, it provides a first-hand perspective on construction auditing programs, showing how they can be an important source for both project process improvement and construction project success. Audit programs are found across the United States serving many projects. Understanding how they can more effectively serve the quality needs of construction projects adds a major cache of time for helping all projects achieve success. Second, the book provides a relevant and constructive set of approaches and concepts for making these programs effective. Numerous research studies, Internet websites and publications extend these ideas. This book should be read by anyone involved in construction, and construction quality programs, because it offers practical ideas for auditing these programs. All the key processes necessary for revising an existing or implementing a new program are found in this volume.

Any leader, or construction project management team, who wants to develop effective construction audit programs that foster employee learning, process improvement and overall construction project quality will gain much from this book. For example, the reader is offered an overview of quality in a construction project as well as auditing fundamentals, auditor selection, the audit process. These topics indicate what a good construction quality audit program should contain, how it is managed and its key features. In addition, the book provides practical steps for an overall program. Dr. Rumane lists an important set of topics to consider when designing a construction quality audit program. Throughout the book, Dr. Rumane sets out to make the material understandable to readers interested in implementing these programs – these ideas are found in the form of suggestions, tables and figures of issues to address, and practical approaches to consider. The sections of the program are quite practical and detailed. They offer a clear description of the ways construction projects can and should evaluate construction quality audit programs. Overall, this book offers construction project top management and quality leaders a concrete, useful and in-depth look at ways to design, implement and evaluate a major resource in the audits of construction projects. Clearly written, well organized and enormously practical, it should be in every construction organization's professional library.

John F. Mascaro, ASQ Fellow – Construction Quality Consultant

Acknowledgements

Share the knowledge with others is the motto of this book.

Many colleagues and friends extended help while I was preparing the book, by arranging reference material; many thanks to all of them for their support.

I thank the publishers and authors whose writings are included in this book, for extending their support by allowing me to reprint their material.

I thank the reviewers, from various professional organizations, for their valuable input to improve my writing. I thank members of ASQ Audit Division, ASQ Design & Construction Division, The Institution of Engineers (India) and Kuwait Society of Engineers for their support to bring out this book.

I thank Edward Needle of Routledge for his support. I thank Mr. Patrick Hetherington and other Routledge staff for making this book a reality.

I thank Mr. John Mascaro, former Chair, Design and Construction Division and ASQ Audit Division, for his support and nicely worded thought-provoking Foreword.

I thank Dr. Adedeji B. Badiru, Series Editor, CRC Press and Cindy Carelli, Executive Editor, CRC Press for their best wishes all the time.

I thank Mr. Raymond R. Crawford of Parsons Brinckerhoff for his valuable and timely suggestions during my writing. I thank Dr. Ayed Alamri, President of Saudi Quality Council, for his support and good wishes. I thank Engr. Ahmed Almershed, former Undersecretary, MSNA, Kuwait, for his good wishes all the time. I thank Dr. N. N. Murthy of Jagruti Kiran Consultants for his support. I thank Mr. Bashir Ibrahim Parkar of SSH International, Engr. Ganeshan Swaminathan of SSH International, Engr. Mohammed Ramjan of SDPM, Engr. Syed Fadal of Kuwait Oil Company and Engr. D.M. Tripathi of Alghanim International for their valuable input and support. I extend my thanks to Dr. Ted Coleman for his everlasting support.

My special thanks to H.E. Sheikh Rakan Nayef Jaber Al Sabah for his support and good wishes.

I thank my well-wishers whose inspiration made me to complete this book.

Most of the data discussed in this book is from the author's practical and professional experience and are accurate to the best of the author's knowledge and ability. However, if any discrepancies are observed in the presentation, I would appreciate them being communicated to me.

The contribution of my son Ataullah, my daughter Farzeen and my daughter-in-law Masum is worth mentioning here. They encouraged me and helped me in my preparatory work to achieve the final product. I thank my mother, brothers, sisters and family members for their support, encouragement and good wishes all the time.

Finally, my special thanks to my wife, Noor Jahan, for her patience as she had to suffer a lot because of my busy schedule.

Abdul Razzak Rumane

Preface

Quality has been of great concern throughout the recorded history of human beings. Examples of specification and inspection can be found in the Bible dating to at least 500 BCE. The desire for products that do as well or better than the customer's needs and requirements has been a constant in human history, matched only by the determination of builders and makers to meet that desire.

Construction projects have the involvement of many participants comprising the owner, designer, contractor and many other professionals from construction related industries. Each of these participants is involved in implementing quality in construction projects. These participants are both influenced by and depend on each other in addition to "other players" involved in the construction process. Therefore construction projects have become more complex and technical and extensive efforts are required to reduce rework and costs associated with time, materials and engineering.

Quality is a distinguishing characteristic of products or services, which satisfy the customer. Most production or services systems are of a repetitive nature and are designed for mass production or batch (lot) production. The definition of quality for construction projects is different to that of manufacturing or services industries. In the case of mass production and batch oriented production systems, quality can be achieved by getting feedback from the process by observing the actual performance and regulating the process to meet the established standards. Whereas, because of the non-repetitive nature of construction projects, it is not possible to compare the actual performance of the project as past experience may be of limited value.

The quality management of manufactured products is performed by the manufacturer's own team which has control over all the activities of the product life cycle, whereas construction projects have a diversity of interactions and relationships between owners, architects/engineers and contractors. Construction projects differ from manufacturing or production. Construction projects are custom oriented and custom designed, having specific requirements set by the customer/owner to be completed within a finite duration and assigned budget. Every project has elements that are unique. No two projects are alike. It is always the owner's desire that their project should be unique and better. To a great extent, each project has to be designed and built to serve a specified need. Construction project is a custom than a routine and repetitive business.

Quality management in construction projects is different to that of manufacturing. Quality in construction projects is not only the quality of products and equipment used in the construction, it is the total management approach to complete the facility as per the scope of works to customer/owner satisfaction to be completed within the specified schedule and within the budget to meet the owner's defined purpose.

Quality in construction projects is achieved through the application of various quality control principles, procedures, concepts, methods, tools and techniques and their applications to various activities/components/subsystems at different phases of the life cycle of a construction project to improve the construction process to conveniently manage the project and make it more qualitative, competitive and economical.

In order to achieve a qualitative, competitive and economical project, the quality compliance at each phase of the construction project has to be assessed/measured to ensure it meets the owner needs and is mainly performed during the following phases:

1 Conceptual phase
2 Schematic phase
3 Design development phase
4 Construction documents phase
5 Bidding and tendering phase
6 Construction phase
7 Testing, commissioning and handover phase

The assessment of quality activities can be evaluated with two perspectives:

1 Assessment that focuses on customer satisfaction results but includes an evaluation of the present quality system
2 Assessment that focuses on the evaluation of present quality system with little emphasis on customer satisfaction results

In either case the assessment can be performed by the organization itself or by an external body. The assessment performed by the organization is known as 'self-assessment', whereas an assessment performed by an external body is referred to as a quality audit. An audit is a systematic, planned, independent and documented process to verify or evaluate and report the degree of compliance to the agreed-upon quality criteria, or the specification or contract requirements of the product, services or project. A quality audit is the formal or methodical examining, reviewing and investigating of the existing system to determine whether agreed-upon requirements are being met.

Audits are mainly classified as follows:

- **First Party** – Audit your own organization (internal audit)
- **Second Party** – Customer audits the supplier
- **Third Party** – Audits performed by independent audit organization.

xxx   Preface

This book is developed to provide all the related information to construction professional practitioners, designers (consultant), contractors and quality auditors involved in construction projects,and construction related industries about auditing fundamentals, processes to improve construction quality in order to make the project the most qualitative, competitive and economical. The auditing processes will focus on auditing compliance to ISO, corporate quality management systems, project-specific quality management, contract management, regulatory authorities' requirements, safety and environmental considerations.

For the benefit of audit professionals who are not familiar with construction management practices, procedures, tools, methods and principles that are applicable to various activities of different phases/stages, the book contains many valuable figures and tables elaborating major activities in construction projects, procedures, processes and logs, which are generally used by the construction industry, which can be used as reference material to conduct the audit of a particular phase/stage.

The book contains useful material and information for students interested in acquiring knowledge about quality audits in the field of construction projects. The book will also provide much information to academics about the quality auditing practices followed in construction projects.

For the sake of proper understanding, the book is divided into seven chapters and each chapter is divided into a number of sections covering quality related topics that have importance or relevance for understanding quality concepts for construction projects.

Chapter 1 is an overview of quality in construction projects. It gives a brief introduction to types of construction projects and construction project life cycle phases of major projects. It also discusses quality definition for construction projects, integrated quality management system and three elements of quality management which consists of quality plan, quality assurance and quality control.

Chapter 2 is about ISO certification for the construction industry. It discusses the importance of standards and standards organizations. It also discusses QMS manuals for the owner, designer, contractor and manufacturing organization. Typical auditing questionnaires to assess these manuals are listed in Appendix A, Appendix B, Appendix C and Appendix D.

Chapter 3 is about auditing standards for construction projects.

Chapter 4 is about auditing fundamentals. It discusses different categories of auditing such as internal auditing and external auditing. It also discusses different types of audits such as product, process and system.

Chapter 5 is about auditor/audit team selection for construction projects. It discusses the roles, responsibilities, competencies and expertise of auditing stakeholders. It elaborates the auditor selection procedure for different phases/stages of construction projects starting with the bidding and tendering processes, study and design phases, construction phase and testing, commissioning and handover phase. It also discusses the qualification and experience requirements of the auditors to perform an audit during the conceptual phase, schematic phase, design development phase, construction documents phase, bidding and tendering phase, construction phase and testing, commissioning and handover phase.

Chapter 6 details auditing processes for project life cycle phases. The processes focus on auditing compliance to ISO, corporate quality management systems, project specific quality management, contract management, regulatory authorities' requirements, safety and environmental considerations. It elaborates on the auditing/assessment of various activities/components/systems at different phases/stages of construction projects with coverage that spans from inception through to handover of the auditing process for the designer, contractor and construction supervisor to improve the quality of completed projects and make them the most qualitative, competitive and economical. The quality audit objectives of construction projects vary in line with clients' requirements and to what extent the details are to be assessed and verified. Each process discussed in this book for the different phases/stages covers the main aims and objectives, audit tools, auditee and methodology. Though there are similar activities in different phases and stages, each process is considered as an individual element for a better understanding of the audit process. This will help the professionals to perform an in-depth audit of each phase/stage. The chapter also discusses the auditing process for different types of project delivery systems.

Chapter 7 details auditing reporting.

The book, I am certain, will meet the requirements of construction professionals, students and academics and satisfy their needs.

Abbreviations

A/E	architect/engineer
AISC	American Institute of Steel Construction
AMCA	American Composite Manufacturers Association
ANSI	American National Standards Institute
ASCE	American Society of Civil Engineers
ASHRAE	American Society of Heating, Refrigeration and Air-conditioning Engineers
ASTM	American Society of Testing Materials
ASQ	American Society for Quality
BMS	Building Management System
BREEAM	Building Research Establishment Environmental Assessment
BSI	British Standard Institute
CEN	European Committee for Standardization
CHA	Certified HACCP Auditor
CQA	Certified Quality Auditor
CQI	Chartered Quality Institute
EVM	earned value management
HACCP	hazard analysis and critical control points
ICE	Institute of Civil Engineers (UK)
IEC	International Electrotechnical Commission
IEEE	Institute of Electrical and Electronics Engineers
IRCA	International Register of Certified Auditors
ISO	International Organization for Standardization
JSI	job site instruction
LEED	Leadership in Energy and Environmental Design
NCR	non-conformance report
NFPA	National Fire Protection Association
PMI	Project Management Institute
PMBOK	Project Management Book of Knowledge
PQQ	prequalification questionnaire
QMS	quality management system
RFI	request for information

RFP	request for proposal
SOW	statement of work
SWI	site work instruction
TOR	Terms of Reference
VO	variation order

Synonyms

Consultant	Architect/Engineer (A/E), Designer, Design Professionals, Consulting Engineers, Supervision Professional, Construction Supervisor
Contractor	Constructor, Builder
Engineer	Resident Project Representative
Engineer's Representative	Resident Engineer
Main Contractor	General Contractor
Owner	Client, Employer
Project Manager	Construction Manager (Agency CM)
Quantity Surveyor	Cost Estimator, Contract Attorney, Cost Engineer, Cost and Works Superintendent

About the author

Abdul Razzak Rumane (PhD) is a Chartered Quality Professional-Fellow of The Chartered Quality Institute (UK) and a certified consultant engineer in electrical engineering. He obtained a Bachelor of Engineering (Electrical) degree from Marathwada University (now Dr. Babasaheb Ambedkar Marathwada University) India in 1972, and received his PhD from Kennedy Western University, USA (now Warren National University) in 2005. His dissertation topic was "Quality Engineering Applications in Construction Projects". Dr. Rumane's professional career exceeds 46 years including 10 years in manufacturing industries and over 36 years in construction projects. Presently he is associated with SIJJEEL Co., Kuwait as Advisor and Director, Construction Management.

Dr. Rumane is associated with a number of professional organizations. He is a Chartered Quality Professional-Fellow of The Chartered Quality Institute (UK) and Fellow of The Institution of Engineers (India), and he has an Honorary Fellowship of Chartered Management Association (Hong Kong). He is also a Senior Member of the Institute of Electrical and Electronics Engineers (USA), a Senior Member of American Society for Quality, a Member of Kuwait Society of Engineers, a Member of SAVE International (The Value Society) and a Member of Project Management Institute. He is also an Associate Member of American Society of Civil Engineers, a Member of London Diplomatic Academy, Member of International Diplomatic Academy and a Member of the Board of Governors of International Benevolent Research Forum.

As an accomplished engineer, Dr. Rumane has been awarded an honorary doctorate in engineering from The Yorker International University, USA (2007). The World Quality Congress awarded him a "Global Award for Excellence in Quality Management and Leadership." The Albert Schweitzer International Foundation honored him with a gold medal for "Outstanding contribution in the field of Construction Quality Management" and "Outstanding contribution in the field of electrical engineering/consultancy in construction projects in Kuwait." He was selected as one of the Top 100 Engineers in 2009 of IBC (International Biographical Centre, Cambridge, UK). The European Academy of Informatisation honored him with the "World Order of Science-Education-Culture" and a title of "Cavalier", and The Sovereign Order of the Knights of Justice, England honored him with a Meritorious Service Medal.

Dr. Rumane has attended many international conferences and has made technical presentations at various conferences. Dr. Rumane is an author whose books include *Quality Management in Construction Projects*, first edition (2010), *Quality Tools for Managing Construction Projects* (2013), *Quality Management in Construction Projects*, second edition (2017) and the editor of *Handbook of Construction Management: Scope, Schedule, and Cost Control* (2016). All of these books are published by CRC Press (a Taylor & Francis Group Company), USA.

Dr. Rumane was Secretary of ASQ Kuwait GC since 2017. He was the honorary chairman of The Institution of Engineers (India), Kuwait chapter for the years 2016–2017, 2013–2014 and 2005–2007.

Contributors

Dr. Shirine Mafi earned her BS and MBA from Marshall University. She then joined the largest engineering design firm in the U.S., Gilbert Common Wealth in Reading, PA, as a management consultant. She ultimately settled in Columbus with her family and started her teaching career at Otterbein University. In 2000, she received her PhD in Human Resource Development from the Ohio State University. She is also a certified quality auditor (CQA) with experience in auditing both in manufacturing and service entities. She is currently a Professor of Management at Otterbein University. She has served as the MBA Director and Acting Chair for the Department at Otterbein. Her areas of interest are in operations, quality, auditing and training for performance improvement.

Mustafa Shraim is an Assistant Professor in the Department of Engineering Technology & Management at Ohio University in Athens, Ohio. He received both his BS and MS degrees from Ohio University, and a PhD in Industrial Engineering from West Virginia University. He has over 20 years of industrial experience as a quality engineer, manager and consultant in quality management systems, statistical methods and Lean/Six Sigma. In addition, he coaches and mentors Green and Black Belts on process improvement projects in both manufacturing and service. He is a Certified Quality Engineer (CQE) and a Certified Six Sigma Black Belt (CSSBB) by ASQ, and a certified QMS Principal Auditor by IRCA in London. He was elected a Fellow by ASQ in 2007.

1 Overview of construction projects

1.1 Construction projects

Construction is the translation of the owner's goals and objectives, by the contractor, to build the facility as stipulated in the contract documents, plans and specifications on schedule and within budget.

Construction has a history of several thousand years. The first shelters were built from stone or mud with materials collected from the forests to provide protection against the cold, wind, rain and snow. These buildings were primarily for residential purposes, although some may have had commercial functions.

During the New Stone Age, people introduced dried bricks, wall construction, metal working and irrigation. Gradually people developed the skills to construct villages and cities and considerable skills in building were acquired. This can be seen from the great civilizations in different parts of the world some 4000–5000 years ago. During Greek settlements, which were about 2000 BC, the buildings were made of mud using timber frames. Later temples and theatres were built from marble. Some 1500–2000 years ago Rome became the leading centre of world culture which extended to construction.

Marcus Vitruvius Pollo, a military and civil engineer in the first century BC, published books in Rome. These included the world's first major publication on architecture and construction, which dealt with building materials, the styles and design of building types, the construction process, building physics, astronomy and building machines.

During the Middle Ages (476–1492), improvements in agriculture and artisanal productivity, and exploration and consequent broadening of commerce took place and in the late Middle Ages building construction became a major industry. Craftsmen were given training and education in order to develop skills and to raise their status. At this time, Guilds were responsible for managing quality.

The fifteenth century brought a 'renaissance' or renewal in architecture, building and science. Significant changes occurred during the seventeenth century and thereafter due to the increasing transformation of construction and urban habitat.

The scientific revolution of the seventeenth and eighteenth centuries gave birth to the great Industrial Revolution of the eighteenth century. After some delay, construction followed these developments in the nineteenth century.

The first half of the twentieth century witnessed the construction industry becoming an important sector throughout the world, employing many workers. During this period skyscrapers, long-span dams, shells and bridges were developed to satisfy new requirements and marked the continuing progress of construction techniques. The provision of services such as heating, air conditioning, electrical lighting, mains water and elevators to buildings became common. The twentieth century saw the transformation of the construction and building industry into a major economic sector. During the second half of the twentieth century, the construction industry began to industrialize, introducing mechanization, prefabrication and system building. The design of building services systems changed considerably in the last twenty years of the twentieth century.

Construction projects are constantly increasing in technological complexity. In addition, the requirements of construction clients are on the increase and as a result, construction projects must meet varied performance standards. Therefore, to ensure the adequacy of the client brief, which addresses the numerous complex client/user needs, it has became the responsibility of the designer to evaluate the requirements in terms of activities and their relationship and follow health, safety and environmental regulations while designing any building.

Building and commercial, traditional architect/engineer (A/E) type of construction projects accounts for an estimated 25% of the annual construction volume. Building construction is a labor intensive endeavor. Every construction project has some elements that are unique. No two construction or R&D projects are alike. Though it is clear that projects are usually more routine than research and development projects, some degree of customization is a characteristic of the projects.

There are several types of projects. Figure 1.1 illustrates types of projects.

A project is a temporary endeavor undertaken to create a unique product or service. "Temporary" means that every project has a definite beginning and a definite end. Unique means that the product or service is different in some distinguishing way from all similar product or services. Projects are often critical components of the performing organization business strategy. Examples of projects include:

- Developing a new product or service
- Effecting a change in structure, staffing or style of an organization
- Designing a new transportation vehicle/aircraft
- Developing or acquiring a new or modified information system
- Running a campaign for political office
- Implementing a new business procedure or process
- Constructing a building or facilities

The duration of a project is finite; projects are not ongoing efforts and the project ceases when its declared objectives have been attained. Some of the characteristics of projects are:

1   Performed by people
2   Constrained by limited resources
3   Planned, executed and controlled

1	**Process type projects**		
1.1	Liquid chemical plants		
1.2	Liquid/solid plants		
1.3	Solid process plants		
1.4	Petrochemical plants		
1.5	Petroleum refineries		
2	**Non-process type projects**		
2.1	Power plants		
2.2	Manufacturing plants		
2.3	Support facilities		
2.4	Miscellaneous (R&D) projects		
2.5	Civil construction projects	Residential construction	Family homes, multi-unit town houses, garden, apartments, condominiums, high-rise apartments, villas
2.6	Commercial A/E projects	Building construction (institutional and commercial)	Schools, universities, hospitals, commercial office complexes, shopping malls, banks, theatres, stadiums, government buildings, ware houses, recreation centers, amusement parks, holiday resorts, neighborhood centers
		Industrial construction	Petroleum refineries, petroleum plants, power plants, heavy manufacturing plants, steel mills, chemical processing plants
		Heavy engineering	Dams, tunnels, bridges, highways, railways, airports, urban rapid transit system, ports, harbors, power lines and communication network
		Environmental	Water treatment and clean water distribution, sanitary and sewage system, waste management

Categories of civil construction projects and commercial A/E projects

Figure 1.1 Types of construction projects

Source: Abdul Razzak Rumane (2013). *Quality Tools for Managing Construction Projects.* Reprinted with permission of Taylor & Francis Group

Based on various definitions, the project can be defined as: "a plan or program performed by people with assigned resources to achieve an objective within a finite duration."

Construction projects comprise a cross-section of many different participants. These participants are both influenced by and depend on each other in addition to "other players" involved in the construction process. Figure 1.2 illustrates the concept of traditional construction project organization.

Traditional construction projects have the involvement of three main groups. These are:

1. Owner – A person or an organization that initiates and sanctions a project. They request the need for the facility and are responsible for arranging the financial resources for the creation of the facility.
2. Designer (A/E) – This consists of architects or engineers or consultant. They are the owner's appointed entity responsible for converting the owner's conception and need into a specific facility with detailed directions through drawings and specifications within the economic objectives. They are responsible for the design of the project and in certain cases the supervision of the construction process.
3. Contractor – A construction firm engaged by the owner to complete the specific facility by providing the necessary staff, workforce, materials, equipment, tools and other accessories to the satisfaction of the owner/end user in compliance with the contract documents. The contractor is responsible for implementing the project activities and for achieving the owner's objectives.

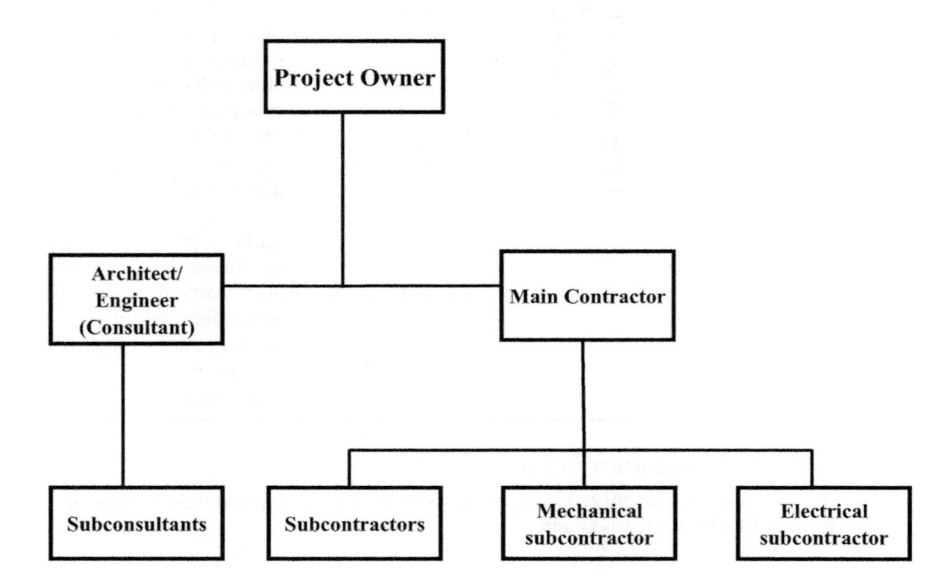

Figure 1.2 Concept of Design-Bid-Build (traditional contracting system)

Construction projects are executed based on predetermined set of goals and objectives. In traditional construction projects, the owner heads the team and designates a project manager. The project manager is a person/member of the owner's staff or independently hired person/firm who has overall or principal responsibility for the management of the project as a whole.

Complex and major construction projects face many challenges such as delays, changes, disputes, and accidents on site and therefore the projects need to be efficiently managed from the beginning to the end to meet the intended use and the owner's expectations. The owner/client may not have the necessary staff/resources in-house to manage the planning, design, and construction of the project to achieve the desired results. Therefore, in such cases, owners engage a professional construction manager, who is trained in the management of construction processes, to assist in developing bid documents, overseeing, and coordinating the project for the owner. The basic construction management concept is that the owner assigns a contract to a firm that is knowledgeable and capable of coordinating all the aspects of the project to meet the intended use of the project by the owner. In the construction management type of construction projects, the consultant (A/E) prepares the complete design drawings and contact documents, then the project is put for competitive bid and the contact is awarded to the most competitive bidder (contractor). The owner hires a third party (construction manager) to oversee and coordinate the construction.

Construction projects are mainly capital investment projects. They are customized and non-repetitive in nature. Construction projects have become more complex and technical, and the relationships and the contractual grouping of those who are involved are also more complex and contractually varied. The products used in construction projects are expensive, complex, immovable, and long-lived. Generally a construction project is composed of building materials (civil), electro-mechanical items, finishing items and equipment. These are normally produced by other construction-related industries/manufacturers. These industries produce products in line with their own quality management practices, complying with certain quality standards or against specific requirements for a particular project. The owner of the construction project or their representative has no direct control over these companies unless they/their representative/appointed contractor commit to buy the product for use in their facility. These organizations may have their own quality management program. In manufacturing or service industries the quality management of all in-house manufactured products is performed by the manufacturer's own team or under the control of the same organization having jurisdiction over their manufacturing plants at different locations. Quality management of vendor-supplied items/products is carried out as stipulated in the purchasing contract as per the quality control specifications of the buyer.

1.1.1 Construction project life cycle

Most construction projects are custom oriented, having a specific need and a customized design. It is always the owner's desire that their project should be unique and better. Further it is the owner's goal and objective that the facility is completed on time. The expected time schedule is important both from a financial viewpoint and for the acquisition of the facility by the owner/end user.

The system life cycle is fundamental to the application of systems engineering. A systems engineering approach to construction projects helps to understand the entire process of project management and to manage and control its activities at different levels of various phases to ensure timely completion of the project with economical use of resources to make the construction project the most qualitative, competitive, and economical.

Systems engineering starts from the complexity of the large-scale problem as a whole and moves towards structural analysis and the partitioning process until the questions of interest are answered. This process of decomposition is called a Work Breakdown Structure (WBS). The WBS is a hierarchical representation of system levels. Being a family tree, the WBS consists of a number of levels, starting with the complete system at level 1 at the top and progressing downward through as many levels as necessary, to obtain elements that can be conveniently managed.

Benefits of systems engineering applications are:

- Reduction in cost of system design and development, production/construction, system operation and support, system retirement and material disposal
- Reduction in system acquisition time
- More visibility and reduction in the risks associated with the design decision making process.

It is difficult to generalize project life cycle to system life cycle. However, considering that there are innumerable processes that make up the construction process, the technologies and processes applied to systems engineering can also be applied to construction projects. The number of phases will depend on the complexity of the project. Duration of each phase may vary from project to project. Generally construction projects have five most common phases, however major projects are divided into seven phases in order to conveniently manage and control the project at each stage. These are as follows:

1. Conceptual Design
2. Schematic/Preliminary Design
3. Design Development
4. Construction Documents
5. Bidding and Tendering
6. Construction
7. Testing, Commissioning and Handover.

Each phase can further be sub-divided on the WBS principle to reach a level of complexity where each element/activity can be treated as a single unit which can be conveniently managed. WBS represents a systematic and logical breakdown of the project phase into its components (activities). It is constructed by dividing the project into major elements with each of these being divided into sub-elements. This is carried out until a breakdown is done in terms of manageable units of work for which responsibility can be defined. WBS involves envisioning the project as a hierarchy of goal, objectives, activities, sub-activities and work packages. The hierarchical decomposition of activities continues until the entire project is displayed as a network of separately identified and non-overlapping activities, Each activity will be single purposed, of a specific time duration and manageable, its time and cost estimates easily derived, deliverables clearly understood and responsibility for its completion clearly assigned. The work breakdown structure helps in:

- Effective planning by dividing the work into manageable elements which can be planned, budgeted and controlled.
- Assignment of responsibility for work elements to project personnel and outside agencies.
- Development of control and information system.

WBS facilitates the planning, budgeting, scheduling and control activities for the project manager and their team. By the application of the WBS phenomenon, the construction phases are further divided into various activities. Division of these phases will improve the control and planning of the construction project at every stage before a new phase starts. The components/activities of construction project lifecycle phases divided on WBS principle are listed below:

1 Conceptual Design Phase

    i   Develop Project Charter
    ii   Develop Preliminary Project Management Plan
    iii Identify Project Stakeholders
    iv Develop Concept Design Scope
    v   Develop Concept Design
    vi Prepare Preliminary Schedule
    vii Estimate Conceptual Cost
    viii Establish Quality Requirements
    ix Estimate Resources
    x   Manage Risks
    xi Monitor Work Progress
    xii Review Concept Design
    xiii Finalize Concept Design
    xiv Submit Concept Design

2 Schematic Design Phase

    i   Develop Schematic Design Requirements
    ii   Develop Preliminary Project Management Plan

- iii Identify Project Stakeholders
- iv Develop Schematic Design Scope
- v Develop Schematic Design
- vi Prepare Preliminary Schedule
- vii Estimate Preliminary Cost
- viii Perform Value Engineering
- ix Manage Design Quality
- x Estimate Resources
- xi Manage Risks
- xii Monitor Work Progress
- xiii Review Schematic Design
- xiv Finalize Schematic Design
- xv Submit Schematic Design

3 Design Development Phase
- i Develop Design Development Requirements
- ii Develop Project Management Plan
- iii Identify Project Stakeholders
- iv Develop Design Development Scope
- v Develop Detail Design
- vi Prepare Schedule
- vii Estimate Cost
- viii Manage Design Quality
- ix Estimate Resources
- x Manage Risks
- xi Monitor Work Progress
- xii Review Detail Design
- xiii Finalize Detail Design
- xiv Submit Detail Design

4 Construction Documents Phase
- i Develop Construction Documents Requirements
- ii Identify Stakeholders
- iii Develop Construction Documents Scope
- iv Develop Construction Documents
- v Estimate Construction Schedule
- vi Estimate Construction Cost
- vii Manage Document Quality
- viii Estimate Resources
- ix Manage Risk
- x Monitor Work Progress
- xi Review Construction Documents
- xii Finalize Construction Documents
- xiii Release for Tendering

5 Bidding and Tendering Phase

   i   Organize Tendering Documents
   ii   Identify Stakeholders
   iii Identify Tendering Procedure
   iv   Identify Bidders
   v   Manage Tendering Process
   vi   Manage Risks
   vii Review Bid Documents
   viii Award Contract

6 Construction Phase

   i   Develop Project Execution Requirements
   ii   Develop Construction Management Plan
   iii Identify Stakeholders
   iv   Develop Project Execution Scope
   v   Develop Contractor's Construction Schedule
   vi   Develop Project S-Curve
   vii Develop Contractor's Quality Control Plan
   viii Develop Resource Management Plan
   ix   Develop Communication Plan
   x   Develop Risk Management Plan
   xi   Develop Health, Safety and the Environment (HSE) Plan
   xii Execute Project Works
   xiii Monitor and Control Project Works
   xiv Validate Executed Works

7 Testing, commissioning and handover phase

   i   Identify Testing and Startup Requirements
   ii   Develop Testing, Commissioning and Handover Plan
   iii Manage Stakeholders
   iv   Develop Scope of Work
   v   Establish Testing, Commissioning Quality Procedure
   vi   Execute Testing and Commissioning Works
   vii Develop Documents
   viii Monitor Work Progress
   ix   Train Owner's/End User's Personnel
   x   Handover of the Project
   xi   Close Contract
   xii Settle Payments
   xiii Settle Claims

Table 1.1 illustrates sub-divided activities/components of a major construction project life cycle.

Table 1.1 Construction project life cycle (Design-Bid-Build) phases

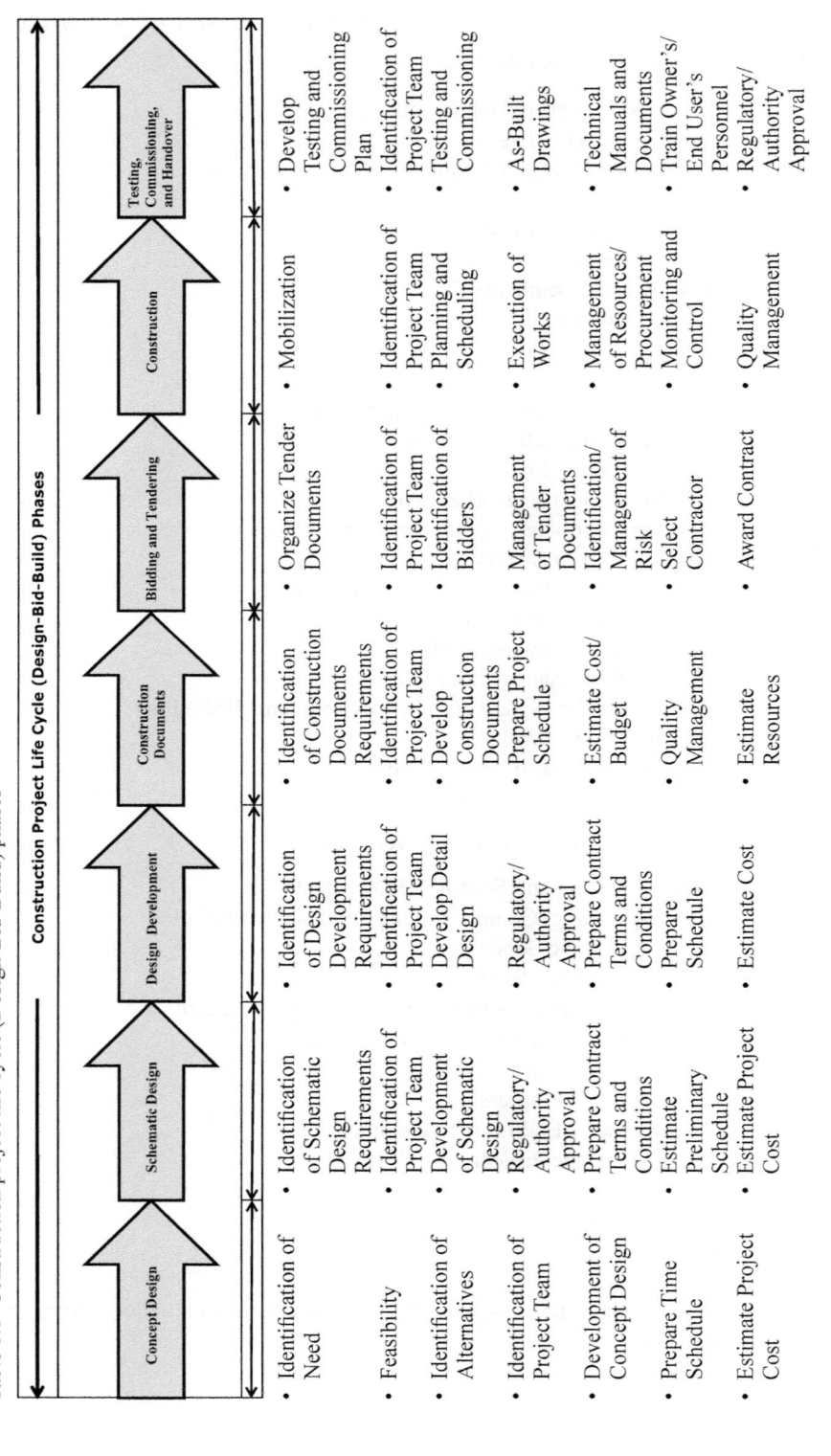

Concept Design	Schematic Design	Design Development	Construction Documents	Bidding and Tendering	Construction	Testing, Commissioning, and Handover
• Identification of Need	• Identification of Schematic Design Requirements	• Identification of Design Development Requirements	• Identification of Construction Documents Requirements	• Organize Tender Documents	• Mobilization	• Develop Testing and Commissioning Plan
• Feasibility	• Identification of Project Team	• Identification of Project Team	• Identification of Project Team	• Identification of Project Team	• Identification of Project Team	• Identification of Project Team
• Identification of Alternatives	• Development of Schematic Design	• Develop Detail Design	• Develop Construction Documents	• Identification of Bidders	• Planning and Scheduling	• Testing and Commissioning
• Identification of Project Team	• Regulatory/ Authority Approval	• Regulatory/ Authority Approval	• Prepare Project Schedule	• Management of Tender Documents	• Execution of Works	• As-Built Drawings
• Development of Concept Design	• Prepare Contract Terms and Conditions	• Prepare Contract Terms and Conditions	• Estimate Cost/ Budget	• Identification/ Management of Risk	• Management of Resources/ Procurement	• Technical Manuals and Documents
• Prepare Time Schedule	• Estimate Preliminary Schedule	• Prepare Schedule	• Quality Management	• Select Contractor	• Monitoring and Control	• Train Owner's/ End User's Personnel
• Estimate Project Cost	• Estimate Project Cost	• Estimate Cost	• Estimate Resources	• Award Contract	• Quality Management	• Regulatory/ Authority Approval

Construction Project Life Cycle (Design-Bid-Build) Phases

• Quality Management	• Quality Management	• Quality Management	• Quality Management	• Risk Management
• Estimate Resources	• Estimate Resources	• Estimate Resources	• Identification/ Management of Risk	• Contract Management
• Identification/ Management of Risk	• Identification/ Management of Risk	• Identification/ Management of Risk	• Finalize Construction Documents	• Site Safety
• Finalize Concept Design	• Perform Value Engineering Study	• Finalize Detail Design		• Move-in-Plan
	• Finalize Schematic Design			• Handover of Facility to Owner/End User
				• Substantial Completion

Source: Abdul Razzak Rumane (2016). *Handbook of Construction Management: Scope, Schedule, and Cost Control*. Reprinted with permission of Taylor & Francis Group

These activities may not be strictly sequential, however the breakdown allows implementation of project management functions more effectively at different stages.

1.2 Quality definition for construction projects

Quality has different meanings for different people. The definition of quality relating to manufacturing, processes and service industries is as follows:

- Meeting the customer's need
- Customer satisfaction
- Fitness for use
- Conforming to requirements
- Degree of excellence at an acceptable price.

The International Organization for Standardization (ISO) defines quality as "the totality of characteristics of an entity that bears on its ability to satisfy stated or implied needs."

However, the definition of quality for construction projects is different to that of manufacturing or services industries as the product is not repetitive, but an unique piece of work with specific requirements.

Quality in construction projects is not only the quality of the product and equipment used in the construction of the facility/project, it is the total management approach to complete the facility. Quality of construction depends mainly upon the control of construction, which is the primary responsibility of the contractor.

Quality in manufacturing passes through a series of processes. Material and labor are input through a series of process out of which a product is obtained. The output is monitored by inspection and testing at various stages of production. Any non-conforming product identified is repaired, reworked or scrapped, and proper steps are taken to eliminate problem causes. Statistical process control methods are used to reduce the variability and to increase the efficiency of the process. In construction projects, the scenario is not the same. If anything goes wrong, the non-conforming work is very difficult to rectify and remedial actions are sometimes not possible.

Quality management in construction projects is different to that of manufacturing. Construction quality management is the total management approach to complete the facility as per the scope of works to customer/owner satisfaction to be completed within the specified schedule and within the budget to meet the owner's defined purpose. The nature of the contracts between the parties plays a dominant part in the quality system required from the project and the responsibility for achieving them must therefore be specified in the project documents. The documents include plans, specifications, schedules, bill of quantities and so on. Quality control in construction typically involves ensuring compliance with minimum standards of material and workmanship in order to ensure the performance of the facility according to the design. These minimum standards are contained in the specification documents. For the purpose of ensuring compliance, random samples and statistical methods are commonly used as the basis for accepting or rejecting work completed and batches of materials. Rejection of a batch is based on non-conformance or violation of the relevant design specifications.

Based on the above, quality of construction projects can be defined as follows: construction project quality is the fulfillment of the owner's needs as per the defined scope of works within a specified schedule and budget to satisfy the owner's/user's requirements. The phenomenon of these three components can be called a "construction project trilogy" and is illustrated in Figure 1.3.

Thus quality of construction projects can be evolved as follows:

1. Properly defined scope of work
2. Owner, project manager, design team leader, consultant and constructor's manager are responsible for implementing the quality
3. Continuous improvement can be achieved at different levels as follows:

   a. Owner – specify the latest needs
   b. Designer – specification to include latest quality materials, products and equipment
   c. Constructor – use latest construction equipment to build the facility

4. Establishment of performance measures

   a. Owner:

      I. To review and ensure that designer has prepared the contract documents which satisfy their needs
      II. To check the progress of work to ensure compliance with the contract documents

   b. Consultant:

      I. As a consultant designer to include the owner's requirements explicitly and clearly defined in the contact documents
      II. As a supervision consultant supervise contractor's work as per contract documents and the specified standards

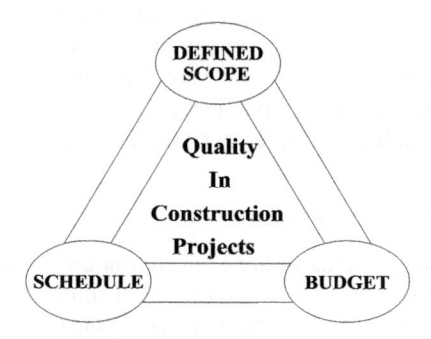

Figure 1.3 Construction project trilogy

    c   Contactor – To construct the facility as specified and use materials, products and equipment which satisfy the specified requirements
5. Team Approach – Every member of the project team should know the principles of Total Quality Management (TQM), understanding that TQM is a collaborative and collective effort and everybody should participate in all the functional areas to improve the quality of project works.
6. Training and Education – Consultant and contractor should have customized training plans for their management, engineers, supervisors, office staff, technicians and labor (workers).
7. Establish Leadership – Organizational leadership should be established to achieve the specified quality. Encourage and help the staff and workers to understand the quality to be achieved for the project.

These definitions when applied to construction projects relate to the contract specifications or owner/end user requirements to be constructed in such a way that the construction of the facility is suitable for the owner's use or it meets the owner's requirements. Quality in construction is achieved through complex interaction of many participants in the facilities development process.

The quality plan for construction projects is part of the overall project documentation consisting of:

1. Well-defined specification for all the materials, products, components and equipment to be used to construct the facility.
2. Detailed construction drawings.
3. Detailed work procedure.
4. Details of the quality standards and codes to be complied.
5. Cost of the project.
6. Manpower and other resources to be used for the project.
7. Project completion schedule.

Participation involvement of all three parties at different levels of the construction phases is required to develop a quality system and application of quality tools and techniques. With the application of various quality principles, tools and methods by all the participants at different stages of the construction project, rework can be reduced resulting in savings in the project cost and making the project qualitative and economical. This will ensure the completion of the construction and making the project as qualitative, competitive and economical as possible. Table 1.2 illustrates quality principles of construction projects.

1.3 Quality management system

ISO 9000 quality system standards are a tested framework for taking a systematic approach to the business process so that the organizations turn out products or services conforming to customers' satisfaction. The typical ISO quality management

Table 1.2 Principles of quality in construction projects

Principle	Construction Projects' Quality Principles
Principle 1	Owner, consultant, contractor are fully responsible for application of quality management system to meet defined scope of work in the contract documents
Principle 2	Consultant is responsible for providing owner's requirements explicitly and clearly in the contract documents
Principle 3	Contractor should study all the documents during tendering/bidding stage and submit their proposal taking into consideration all the requirements specified in the contract documents
Principle 4	Method of payments (work progress, material, equipment, etc.) to be clearly defined in the contract documents. Rate analysis of Bill of Quantities (BOQ) or Bill of Materials (BOM) items to be agreed before signing of contract
Principle 5	Contract documents should include a clause to settle disputes arising during construction stage
Principle 6	Contractor shall follow an agreed-upon quality assurance and quality control plan. Consultant shall be responsible for overseeing the compliance with contract documents and specified standards
Principle 7	Contractor shall follow the submittal procedure specified in the contract documents
Principle 8	Contractor is responsible for constructing the facility as specified and using the material, products, equipment and methods which satisfy the specified requirements
Principle 9	Each member of the project team should participate in all the functional areas to continuously improve the quality of the project
Principle 10	Contractor is responsible for providing all the resources, manpower, material, equipment, etc. to build the facility as per specifications
Principle 11	Contractor to build the facility as stipulated in the contract documents, plan and specifications within budget and on schedule to meet owner's objectives
Principle 12	Contractor should perform the works as per agreed-upon construction program and hand over the project as per contracted schedule

Source: Abdul Razzak Rumane (2013) *Quality Tools for Managing Construction Projects*. Reprinted with permission of Taylor & Francis Group

system is structured on four levels, usually portrayed as a pyramid as shown in Figure 1.4.

At the top of the pyramid is the quality policy, which sets out what management requires its staff to do in order to ensure the quality management system. Underneath policy is quality manual which details the work is to be done. Beneath quality manual are work instructions, procedures and records. The number of manuals containing work instructions or procedures is determined based on the size and complexity of the organization. The procedures mainly discuss the following:

- What is to be done?
- How is it done?

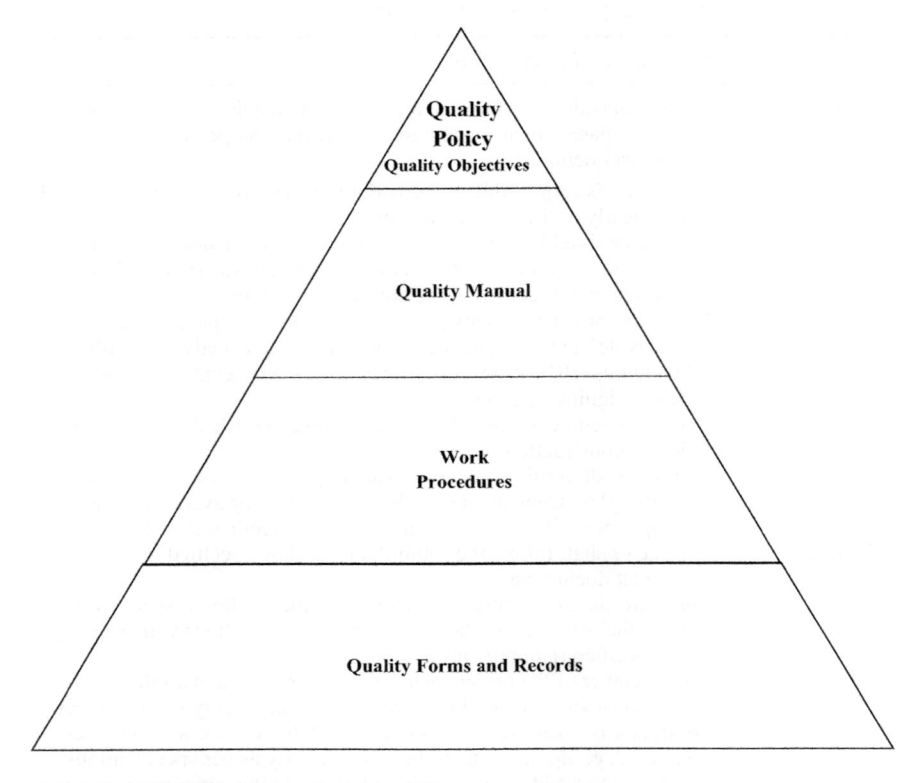

```
                    /\
                   /  \
                  /Quality\
                 / Policy  \
                /Quality Objectives\
               /_____\
              /              \
             /                \
            /   Quality Manual  \
           /_____\
          /                      \
         /     Work Procedures    \
        /_____\
       /                            \
      /   Quality Forms and Records   \
     /_____\
```

Figure 1.4 QMS documentation pyramid

- How does one know that it has been done properly (for example, by inspecting, testing or measuring)?
- What is to be done if there are problems (for example, failure)?

The bottom level of hierarchy contains forms and records that are used to capture the history of routine events and activities.

The ISO 9000 quality management system requires documentation that includes a quality manual and quality procedures as well as work instructions and quality records. All documentation (including quality records) must be controlled according to a document control procedure. The structure of the quality management system depends largely on the management structure in the organization.

ISO 9001:2008 identifies certain minimum requirements that all quality management system must meet to ensure customer satisfaction. ISO 9001:2008 specifies requirements for quality management system when an organization:

- Needs to demonstrate its ability to consistently provide a product that meets customer and applicable regulatory requirements, and

- Aims to enhance customer satisfaction through the effective application of the system, including processes for continual improvement of the system and the assurance of conformity to customer and applicable regulatory requirements.

A quality system has to cover all the activities leading to the turning out of the final product or service. The quality system depends entirely on the scope of operation of the organization and particular circumstances such as the number of employees, type of organization and physical size of the premises of the organization. The quality manual is the document that identifies and describes the quality management system. The quality manual:

1. Identifies the process (activities and necessary elements) needed for quality management system
2. Determines the sequence and interaction of these processes and how they fit together to accomplish quality goals
3. Determines how these processes are effectively operated and controlled
4. Measures, monitors and analyzes these processes and implement the action necessary to correct the process and achieve continual requirement
5. Ensures that all information is available to support the operation and monitoring of the process
6. Displays the most options, thus helping make the right management system

ISO 9001:2015 requirements fall into the following sections (clauses):

1. Scope
2. Normative References
3. Terms and Definitions
4. Context of Organization/Quality Management System
5. Leadership
6. Planning for Quality Management System
7. Support
8. Operation
9. Performance Evaluation
10. Improvement

ISO 9000:2015 has 10 clauses compared to 8 clauses in ISO 9000:2008. The most important is that the revised quality management system (QMS) focuses on risk based thinking which has to be considered from the beginning and throughout the life cycle of the project.

In the construction industry, a contractor may be working at any one time on a number of projects of a varied nature. These projects have their own contract documents to implement project quality which require the contractor to submit the contractor's quality control plan to ensure that specific requirements of the project are considered to meet the client's requirements. Therefore, while preparing a quality management system at a corporate level, the organization has to take into account tailor-made requirements for the projects and the manual should be prepared accordingly.

1.3.1 Integrated quality management system

Integrated quality management system (IQMS) is the integration and proper coordination of functional elements of quality to achieve efficiency and effectiveness in implementation and maintaining an organization's quality management system to meet customer requirements and satisfaction. IQMS consists of any element or activity which has an effect on quality. Customer satisfaction is the goal of quality objectives.

During the past three decades, many programs have been implemented for organizational improvements. In 1980s programs such as statistical process control, various quality tools and total quality management (TQM) were implemented. In the 1990s the most popular ISO 9000 came into being, which resulted in improved productivity, cost reduction and improved time, quality and customer satisfaction.

With globalization and competition, it became necessary for organizations to improve continuously to achieve highest performance and competitive advantage.

In the 1980s, the major challenge facing most organizations was to improve quality. In 1990s, it was to improve faster by restructuring and reengineering all operations.

In today's global competitive environment, organizations are facing many challenges due to the increase in customer demand for higher performance requirements at a competitive cost. They are finding that their survival in the competitive market is increasingly in doubt. To achieve competitive advantage, effective quality improvement is critical.

Processes and systems are essential for the performance and expansion of any organization. ISO 9000 is an excellent tool to develop a strong foundation for good processes and systems. The ISO 9000 quality management system is accepted worldwide and ISO 9000 certification has global recognition.

IQMS is developed by merging recommendations and specifications from ISO 9000 (Quality Management System), ISO 14000 (Environmental Management System), OHSAS 18000 (Occupational Health and Safety Management), together with other contract documents. If an organization has a certified Quality Management System (ISO 9000), it can build an IQMS by adding environmental, health, safety and other requirements of management system standards.

Benefits of implementing an IQMS are:

- Reduced duplication and therefore cost
- Improved resource allocation
- Standardized process
- Elimination of conflicting responsibilities and relationship
- Create consistency
- Improved communication

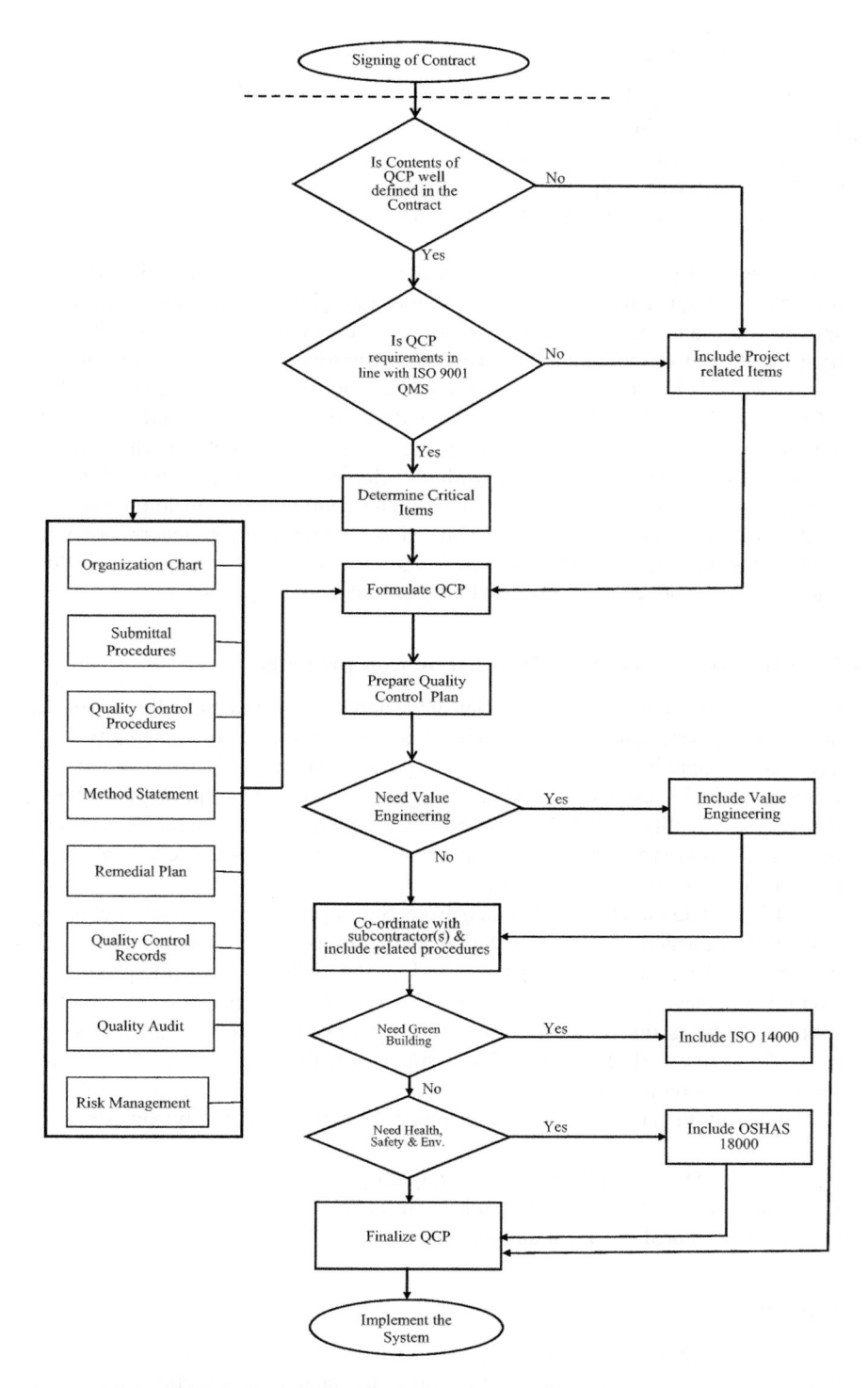

Figure 1.5 Logic flow diagram for the development of IQMS

- Reduced risk and increase profitability
- Facilitate training development
- Simplify document maintenance
- Reduced record keeping
- Ease of managing legal and other requirements

Construction projects are unique and non-repetitive in nature and have their own quality requirements which can be developed by the integration of project specifications and the organization's quality management system. Normally quality management system manuals consist of procedures to develop a project quality control plan taking into consideration contract specifications. This plan is called a Contractor's Quality Control Plan (CQCP). Certain projects specify for value engineering studies to be undertaken during the construction phase. The contractor is required to include the same while developing the CQCP. This plan can be termed as IQMS for construction projects. The contractor has to implement the quality system to ensure that the construction is carried out in accordance with the specification details and approved COQP. Figure 1.5 illustrates a logic flow diagram for the development of IQMS for construction projects.

1.4 Quality management in construction projects

Quality management is an organization-wide approach to understanding customer needs and delivering the solutions to fulfill and satisfy the customer. Quality management is managing and implementing a quality system to achieve customer satisfaction at the lowest overall cost to the organization while continuing to improve the process. Quality system is a framework for quality management. It embraces the organization structure, policies, procedures and processes needed to implement a quality management system.

Quality management in construction projects is different to that of manufacturing.

Quality management in construction addresses both the management of the project and the product of the project and all the components of the product. It also involves incorporation of changes or improvements, if needed. Construction project quality is the fulfillment of the owner's needs as per the defined scope of works in line with the specified schedule and within the budget to satisfy the owner's/user's requirements.

The quality management system in construction projects mainly consists of:

- Quality management planning (Plan quality)
- Quality assurance (Perform quality assurance)
- Quality control (Perform quality control)

Plan quality

The quality plan for construction projects is part of the overall project documentation addressing and describing the procedures to manage construction quality and project deliverables. The quality plan mainly consists of:

- Stakeholders' quality requirements
- Well-defined specification for all the materials, products, components and equipment to be used to construct the facility
- Detailed construction drawings
- Detailed work procedure
- Details of the quality standards and codes to be complied
- Regulatory requirements
- Manpower and other resources to be used for the project
- Project completion schedule
- Cost of the project

Perform quality assurance

Quality assurance in construction projects covers all activities performed by the design team, contractor and quality controller/auditor (supervision staff) to meet the owner's objectives as specified and to ensure and guarantee that the project/facility is fully functional to the satisfaction of the owner/end user. Auditing is part of the quality assurance function.

Perform quality control

Quality control in construction projects is performed at every stage through use of various control charts, diagrams, checklists, etc. and can be defined as:

- Checking of executed/installed works to confirm that works have been performed/executed as specified, using specified/approved materials, installation methods and specified references, codes and standards to meet intended use
- Planning, monitoring and controlling project schedule
- Controlling budget

Construction projects have the involvement of the owner, designer (consultant) and contractor. In order to achieve project objectives, both the designer and the contractor have to develop a project quality management plan. The designer's quality management plan is based on the owner's project objectives whereas the contractor's plan takes into consideration the requirements of contract documents.

The following sections discuss in brief quality management processes (quality planning, quality assurance and quality control) and activities to be performed during the following main stages of construction projects:

1. Design stage (conceptual design, preliminary design, design development, construction documents)
2. Bidding & Tendering stage
3. Construction stage (construction, testing, commissioning and handover)

1.4.1 Quality during the design stage

1.4.1.1 Quality plan

Table 1.3 lists the major quality planning activities to be performed during the design stage.

1.4.1.2 Quality assurance

Table 1.4 lists the major quality assurance activities to be performed during the design stage.

Table 1.3 Major quality planning activities during the design stage

Serial Number	Activities
1	**Conceptual design**
	1.1 Owner's requirements
	1.2 Quality standards and codes to be complied
	1.3 Regulatory requirements
	1.4 Design review procedure
	1.5 Drawings review procedure
	1.6 Document review procedure
	1.7 Quality management during all the phases of project life cycle
2	**Schematic design**
	2.1 Establish owner's requirements
	2.2 Determine number of drawings to be produced
	2.3 Establish scope of work
	2.4 Identify quality standards and codes to be complied
	2.5 Establish design criteria
	2.6 Identify regulatory requirements
	2.7 Identify requirements listed in terms of reference (TOR)
	2.8 Establish quality organization with responsibility matrix
	2.9 Develop design (drawings and documents) review procedure
	2.10 Establish submittal plan
	2.11 Establish design review procedure
3	**Design development**
	3.1 Review comments on schematic design
	3.2 Determine number of drawings to be produced
	3.3 Establish scope of work for preparation of detail design
	3.4 Identify requirements listed in TOR
	3.5 Identify quality standards and codes to be complied
	3.6 Establish design criteria
	3.7 Identify regulatory requirements
	3.8 Identify environmental requirements
	3.9 Establish quality organization with responsibility matrix
	3.10 Develop design (drawings and documents) review procedure
	3.11 Establish submittal plan
	3.12 Establish design review procedure

Serial Number	Activities
4	**Construction documents** 4.1 Review comments on design development package 4.2 Determine number of drawings to be produced 4.3 Establish scope of work for preparation of construction documents 4.4 Identify requirements listed in TOR 4.5 Identify quality standards and codes to be complied 4.6 Identify regulatory requirements 4.7 Identify environmental requirements 4.8 Establish quality organization with responsibility matrix 4.9 Develop review procedure for the produced working drawings 4.10 Develop review procedure for the specifications and contract documents 4.11 Establish submittal plan for construction documents

Table 1.4 Major quality assurance activities during the design stage

Serial Number	Activities
1	**Conceptual design** 1.1 Prepare concept design 1.2 Reports 1.3 Model 1.4 Project schedule 1.5 Project cost
2	**Schematic design** 2.1 Collect data 2.2 Investigate site conditions 2.3 Prepare preliminary drawings 2.4 Prepare outline specifications 2.5 Ensure functional and technical compatibility 2.6 Coordinate with all disciplines 2.7 Select material to meet owner objectives
3	**Design development** 3.1 Collect data 3.2 Investigate site conditions 3.3 Prepare design drawings 3.4 Prepare detailed specifications 3.5 Prepare contract documents 3.6 Prepare Bill of Quantities (BOQ) 3.7 Ensure functional and technical compatibility 3.8 Ensure the design is constructible 3.9 Ensure operational objectives are met 3.10 Ensure drawings are fully coordinate with all disciplines 3.11 Ensure the design is cost-effective 3.12 Ensure selected/recommended material meet owner objectives 3.13 Ensure that design fully meets the owner's objectives/goals

(*Continued*)

Table 1.4 (Continued)

Serial Number	Activities
4	**Construction documents** 4.1 Prepare working drawings 4.2 Prepare detailed specifications 4.3 Prepare contract documents 4.4 Prepare Bill of Quantities and Schedule of Rates 4.5 Ensure functional and technical compatibility 4.6 Ensure the design is constructible 4.7 Ensure operational objectives are met 4.8 Ensure drawings are fully coordinate with all disciplines 4.9 Ensure the design is cost-effective 4.10 Prepare working drawings 4.11 Prepare detailed specifications 4.12 Prepare contract documents

1.4.1.3 Quality control

Table 1.5 lists the major quality control activities to be performed during the design stage.

1.4.2 Quality during the bidding and tendering stage

The following are the main quality management related activities to be considered:

- Tendering procedure/process
- Manage submittals/bids
- Manage project budget.

1.4.3 Quality during the construction stage

1 Quality plan

Develop the contractor's quality control plan (CQCP).
　Table 1.6 lists the contents of the contractor's quality control plan.

2 Quality assurance

The contractor's quality assurance activities mainly consist of the following:

1. Selecting the materials, systems fully complying with contract specifications and installing the approved material, systems only.
2. Preparing the shop drawings detailing all the requirements included in the working drawings and installing/executing the works as per approved shop drawings.

Table 1.5 Major quality control activities during the design stage

Serial Number	Activities
1	**Conceptual design** 1.1 Conformance to owner's requirements 1.2 Conformance to requirements listed in TOR 1.3 Regulatory compliance
2	**Schematic design** 2.1 Check design drawings 2.2 Check specifications/contract documents 2.3 Check for regulatory compliance 2.4 Check preliminary schedule 2.5 Check cost of project (preliminary cost)
3	**Design development** 3.1 Check quality of design drawings 3.2 Check accuracy and correctness of design 3.3 Verify Bill of Quantities for correctness as per design drawings and specifications 3.4 Check specifications 3.5 Check contract documents 3.6 Check for regulatory compliance 3.7 Check project schedule 3.8 Check project cost 3.9 Check interdisciplinary requirements 3.10 Check required number of drawings prepared drawing
4	**Construction documents** 4.1 Check quality of design drawings 4.2 Check accuracy and correctness of design 4.3 Verify Bill of Quantities for correctness as per design drawings and specifications 4.4 Check specifications 4.5 Check contract documents 4.6 Check for regulatory compliance 4.7 Check project schedule 4.8 Check project cost 4.9 Check interdisciplinary requirements 4.10 Check required number of drawings prepared drawing 4.11 Check quality of design drawings 4.12 Check accuracy and correctness of design 4.13 Verify Bill of Quantities for correctness as per design drawings and specifications

Table 1.6 Contents of the contractor's quality control plan

Serial Number	Description
1.0	Introduction
2.0	Description of project
3.0	Quality control organization

(*Continued*)

Table 1.6 (Continued)

Serial Number	Description
4.0	Qualification of quality control staff
5.0	Responsibilities of quality control personnel
6.0	Procedure for submittals
6.1	Submittals of subcontractor(s)
6.2	Submittals of shop drawings
6.3	Submittals of materials
6.4	Modification request
6.5	Construction program
7.0	Quality control procedure
7.1	Procurement
7.2	Inspection of site activities (checklists)
7.3	Inspection and testing procedure for systems
7.4	Off-site manufacturing, inspection, and testing
7.5	Procedure for laboratory testing of material
7.6	Inspection of material received on site
7.7	Protection of works
7.8	Material storage and handling
8.0	Method statement for various installation activities
9.0	Project-specific procedures
10.0	Risk management
11.0	Quality control records
12.0	Company's quality manual and procedures
13.0	Periodical testing
14.0	Quality updating program
15.0	Quality auditing program
16.0	Testing, commissioning, and handover
17.0	Health, safety and the environment

    3   Installing the works, materials and systems as per the specified method statement and as per recommendations from the manufacturer of products.

3   Quality control

The construction project quality control process is part of the contract documents which provide details about specific quality practices, resources and activities relevant to the project. On a construction site, inspection and testing is carried out at three stages during the construction period to ensure quality compliance:

1. During construction process: This is carried with the checklist request submitted by the contractor for testing of ongoing works before proceeding to the next step.
2. Receipt of material, equipment or services. This is performed by a material inspection request submitted by the contractor to the consultant upon receipt of material.
3. Before final delivery or commissioning and handover.

2 ISO certification for the construction industry

2.1 Introduction

Construction projects are mainly capital investment projects. They are customized and non-repetitive in nature. Quality management in construction projects is different from that in manufacturing. Quality in construction projects encompasses not only the quality of products and equipment used in the construction, but the total management approach to completing the facility per the scope of works to customer/owner satisfaction within the specified schedule and within the budget and in accordance with the need to meet the owner's defined purpose. Generally, a construction project comprises building materials (civil), electromechanical items, finishing items and equipment. These are normally produced by other construction-related industries/manufacturers. Quality in manufacturing passes through a series of processes. The output is monitored by inspection and testing at various stages of production. These industries/manufacturers produce products in line with their own quality management practices complying with certain quality standards or against specific requirements for a particular project.

Construction projects have the involvement of many stakeholders; however, the following three are the main participants:

1 Project owners (Developers)
2 Design firms (Consultants)
3 Contractors (Constructors)

Each of these parties is involved in maintaining the quality in a construction project. Quality in construction typically involves ensuring compliance with minimum standards of material and workmanship in order to ensure the performance of the facility according to the design/owner's requirements. These minimum standards are contained in the specification documents. The adequacy of the quality system, quality of products, services, process and projects are judged by their compliance to specified/relevant standards. In construction projects, the standards are referred in the contract documents that include plans, specifications, schedule, bill of quantities and so on.

Standards are published documents that establish specifications, guidelines and procedures designed to ensure the reliability of materials, products, systems, processes and services people use every day to satisfy their requirement. In the construction sector, the compliance with minimum standards in implementation of workmanship, materials, products, systems and processes provides a competitive advantage in terms of customer satisfaction and achievement of a qualitative, competitive and economical project. Standards have important economic, social and environmental repercussions.

2.2 Importance of standards

Standards are used to ensure that a product, material, system or service measures up to its specifications and is safe for use. Standards are the key to any conformity assessment activity. Standards form the fundamental building blocks for product/process, material and system development by establishing consistent protocols that can be universally understood and adopted. Standards are used to understand and compare competing products/processes, materials and systems. Standards help ensure requirements of interconnectivity and interoperability.

The International Organization for Standardization (ISO) has given the importance of standards as follows:

> Standards make an enormous contribution to most aspects of our lives. Standards ensure desirable characteristics of products and services such as quality, environmental fitness, safety, reliability, efficiency, and interchangeability and at an economical cost.

When products and services meet our expectations, we tend to take this for granted. And be unaware of the role of standards. However, when standards are absent, we soon notice. We soon care when products turn out to be poor quality, do not fit, are incompatible with equipment that we already have, are unreliable or dangerous. When products, systems, machinery and devices work well and safely, it is often because they meet standards.

Construction industry is a key sector in national and international economies. ISO has described the importance of ISO standards in the construction industry as follows:

- Standards make construction industry more efficient and effective
- Standards solve problems and provide solutions in all stages of the construction development process
- Standards provide construction industry stakeholders with the information they need to compete in global markets
- Standards provide a state-of-the-art technical base for regulators and help to drive down costs for producers, customers and consumers

2.3 Standards organizations

There are many organizations that produce standards; some of the best-known organizations in the quality field are:

- International Organization for Standardization (ISO)
- International Electrotechnical Commission (IEC)
- American Society for Quality (ASQ)
- American National Standards Institute (ANSI)
- American Society for Testing and Materials (ASTM)
- American Society of Mechanical Engineers (ASME)
- American Society for Heating, Refrigerating, and Air-Conditioning Engineers (ASHRAE)
- National Fire Protection Association (NFPA)
- Institute of Electrical and Electronic Engineers (IEEE)
- European Committee for Standardization (CEN)
- European Committee for Electrotechnical Standardization (CENELEC)
- British Standards Institution (BSI)

Standards produced by these organizations/institutes are recognized worldwide. These standards are referred in contract documents by designers to specify products or systems or services to be used in a project. They are also used to specify the installation method to be followed or the fabrication works to be performed during the construction process.

Apart from these there have been many other national and international quality system standards. These various standards have commonalities and historical linkage. However, in order to facilitate international trade, delegates from 25 countries met in London in 1946 to create a new international organization. The objective of this organization was to facilitate international coordination and unification of industrial standards. The new organization, the International Organization for Standardization (ISO), officially began operation on February 23, 1947. ISO is the world's largest developer and publisher of international standards. It is a non-governmental organization that forms a bridge between the public and private sectors. ISO has more than 16500 international standards. Of all the standards produced by ISO the ones that are most widely known are those of the ISO 9000 and ISO 14000 series. ISO is a network of national standards institutes of 162 countries (as of March 2017), on the basis of one member per country, with a Central Secretariat in Geneva, Switzerland, that co-ordinates the system.

2.4 QMS manuals

A quality management system (QMS) is a set of coordinated documentation that includes a quality manual, quality procedures and processes as well as work instructions (details of works to be done), quality forms and records that are developed to meet customer and regulatory requirements taking into consideration the organization's quality policies and business objectives and improving the effectiveness and efficiency on a continuous basis.

ISO 9001 is the international standard that is most recognized and used as an international reference for the development of an effective quality management system.

ISO 9001 contains 8 quality management principles (CLIPSCFM). These are:

- Principle 1 – Customer focus
- Principle 2 – Leadership
- Principle 3 – Involvement of people
- Principle 4 – Process approach
- Principle 5 – System approach to management
- Principle 6 – Continual improvement
- Principle 7 – Factual approach to design making
- Principle 8 – Mutual beneficial supplier relationship

2.4.1 Development of QMS

The ISO 9000 family is primarily concerned with "quality management.. This means what the organization does to:

- Fulfill customers' quality requirements
- Meet applicable regulatory requirements, while aiming to enhance customer satisfaction
- Achieve continual improvement of its performance in pursuit of the objectives

Although any ISO 9000 quality management system should be created to address organization's business objectives, needs and customer satisfaction, there are some general elements that are common in all ISO compliant quality management systems. These elements have importance as per the levels mentioned in Figure 1.4 (please refer to Chapter 1, section 1.3) known as QMS Documentation Pyramid that includes quality policy and objectives, a quality manual, quality procedures as well as work instructions and quality records. The manual is developed taking into consideration the following:

1. Eight principles (CLIPSCFM) of QMS as defined by the ISO Technical Committee, TC 176
2. All the related and applicable documents produced taking into consideration 10 sections/clauses listed in ISO 9001:2015 to ensure that the manual is in compliance with ISO 9001:2015. Table 2.1 illustrates these sections/clauses.

Figure 2.1 illustrates the relationship between QMS principles and ISO sections.

Table 2.1 ISO 9001:2015 sections (clauses)

Section (Clause) No.	Relevant Clause in 9001:2015	Description
1	**Scope**	
2	**Normative References**	
3	**Terms and References**	
4	**Context of the Organization**	
	4.1	Understanding the organization and its context
	4.2	Understanding the needs and expectations of interested parties
	4.3	Determining the scope of quality management system
	4.4	Quality management system and its processes
5	**Leadership**	
	5.1	Leadership and commitment
	5.2	Policy
	5.3	Organizational roles, responsibilities and authorities
6	**Planning**	
	6.1	Actions to address risks and opportunities
	6.2	Quality objectives and planning to achieve them
	6.3	Planning of changes
7	**Support**	
	7.1	Resources
	7.2	Competence
	7.3	Awareness
	7.4	Communication
	7.5	Documented information
8	**Operation**	
	8.1	Operational planning and control
	8.2	Requirements for product and services
	8.3	Design and development of products and services
	8.4	Control of externally provided processes, products and services
	8.5	Production and service provision
	8.6	Release of products and services
	8.7	Control of nonconforming outputs
9	**Performance evaluation**	
	9.1	Monitoring, measurement, analysis and evaluation
	9.2	Internal audit
	9.3	Management review
10	**Improvement**	
	10.1	General
	10.2	Nonconformity and corrective action
	10.3	Continual improvement

32 Abdul Razzak Rumane

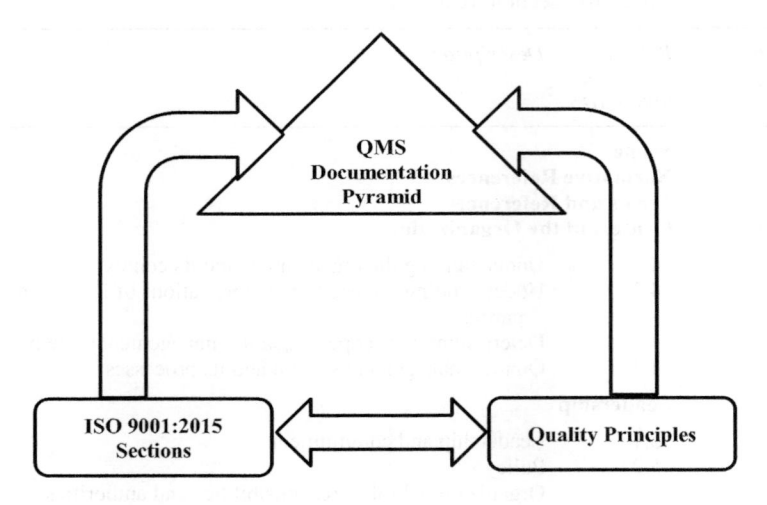

Figure 2.1 QMS documentation model

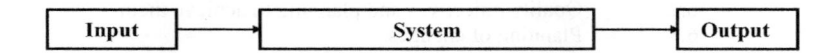

Figure 2.2 Black Box

Figure 2.3 Process-based QMS development procedure

A QMS is made up of elements (components) having a functional relationship to achieve a common objective (customer satisfaction). These elements need to be coordinated taking a systematic approach to improve and sustain the overall performance of the products and services. Figure 2.2 illustrates a simple behavioral approach to the system and is generally known as Black Box.

Systems engineering or a process based approach can be used to develop a QMS. Figure 2.3 illustrates how the concept of system phenomenon can be applied to develop a QMS.

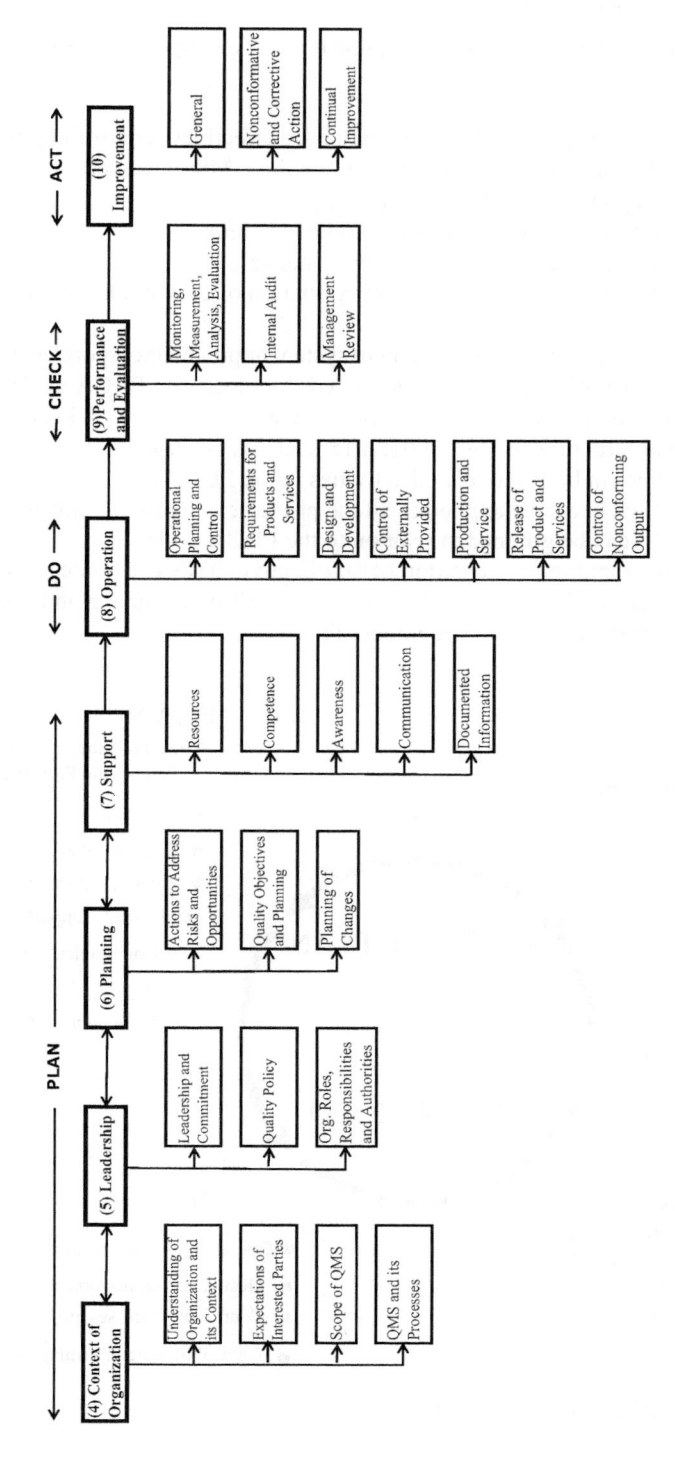

Figure 2.4 Organizational structure of ISO 9001:2015 sections

The application of the process approach in a quality management system enhances:

- Overall performance of the organization by effectively controlling the interrelationship and interdependencies among the QMS processes
- Consistency in meeting the customer requirements
- Consideration of processes in terms of added value
- Achievement of effective process performance
- Improvement of processes based on evaluation of data and information

The process based quality management system approach incorporates the Plan-Do-Check-Act (PDCA) cycle and risk based thinking. The PDCA cycle can be applied to all processes and to the QMS as a whole. Figure 2.4 illustrates the organizational structure of ISO 9001:2015 sections (clauses) 4 to 10 and Figure 2.5 describes in brief the related activities in the PDCA cycle.

The PDCA cycle can be applied to all the processes. Figure 2.6 illustrates how clauses 4 to 10 can be grouped in relation to the PDCA cycle.

Development and implementation of an effective QMS is a strategic decision of the organization that helps to improve its overall performance and provide a sound basis for sustainable development initiatives.

ACT
- Nonconformity and Corrective action
- Continual improvement

PLAN
- Establish the objectives (System and Processes)
- Resources
- Quality Policy
- Roles and Responsibilities
- Risk and Opportunities
- Communication
- Changes
- Documentation

CHECK
- Monitoring, measurement, analysis and evaluation
- Internal Audit
- Management review

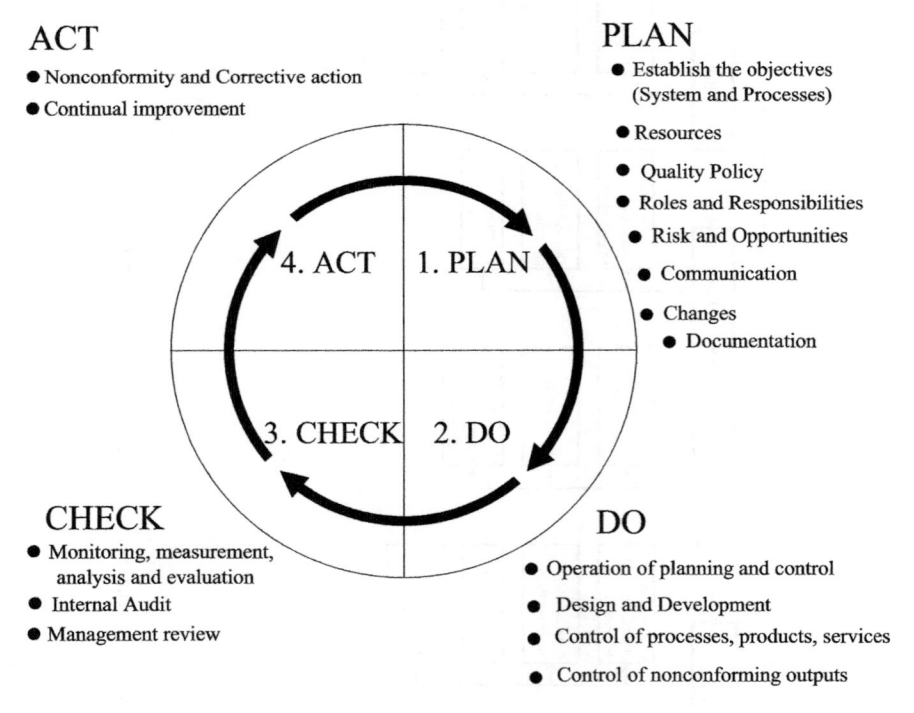

DO
- Operation of planning and control
- Design and Development
- Control of processes, products, services
- Control of nonconforming outputs

Figure 2.5 PDCA cycle for QMS activities

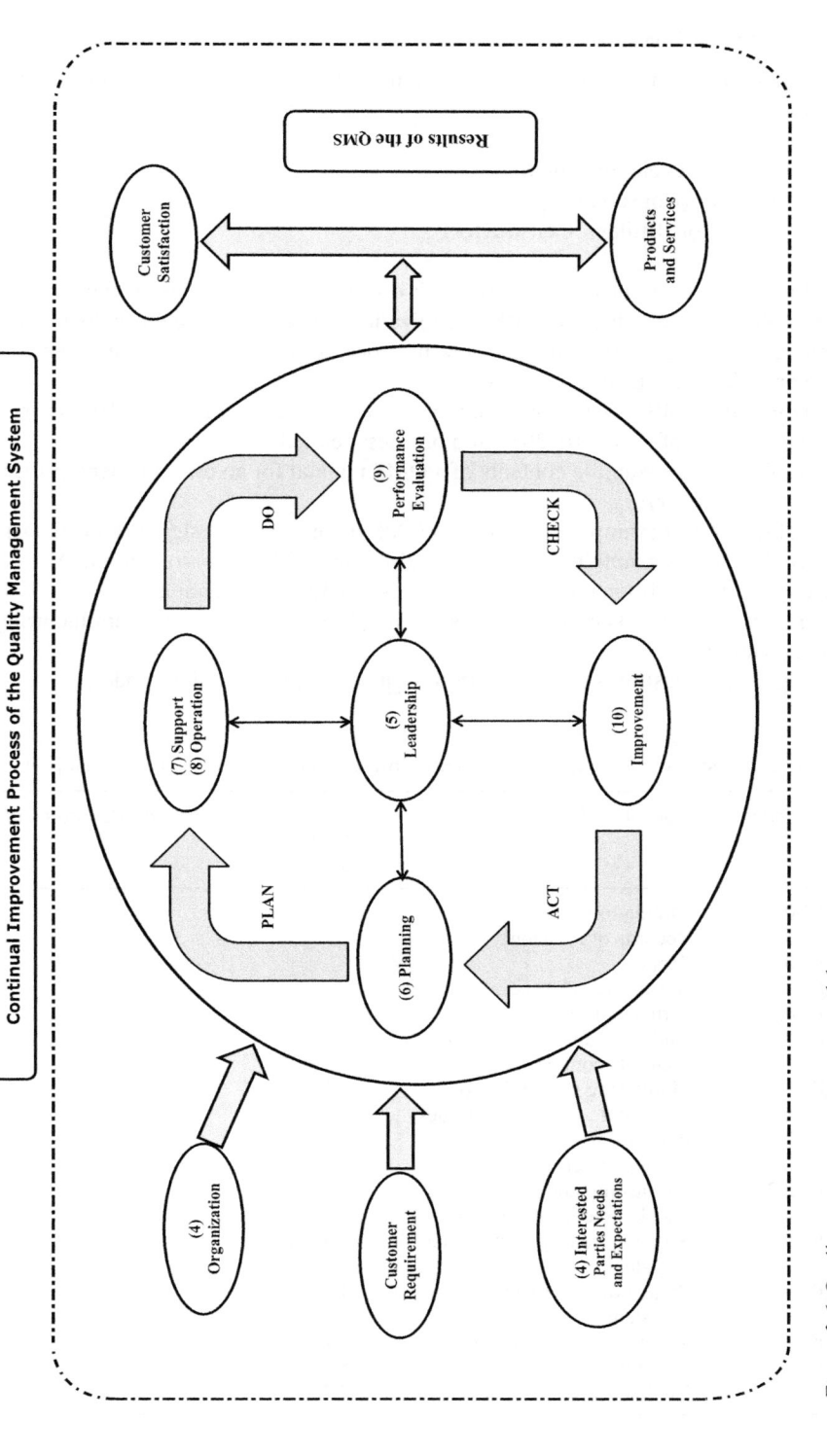

Figure 2.6 Quality management process model

Traditional construction projects have the involvement of three main groups. These are:

1. Owner (Developer, Client)
2. Designer (Consultant)
3. Contractor (Builder, Constructor)

Each of these groups should have their own QMS to meet their business objectives. While developing the QMS, organizations have to include those documents that are required to perform the relevant processes and specific requirements of the organization's business interests.

Normally QMS consists of documents produced taking into consideration relevant sections of ISO 9001:2015 and business related activities.

Table 2.2 lists example contents of a QMS manual for an owner (client, project owner, developer).

Table 2.3 lists example contents of a QMS manual for a designer (consultant).

Table 2.4 lists example contents of a QMS manual for a contractor and Annexure A 2.4 lists additional documents for Design-Build contracting.

Table 2.5 lists example contents of a QMS manual for a manufacturing organization.

Figure 2.7 illustrates an ISO QMS design and implementation model.

Table 2.2 List of quality manual documents – owner (client, project owner, developer)

Document No.	Document Title	Relevant Clause in 9001:2015	Version/Revision Date
QC-0	Circulation list		
QC-00	Records of revision		
QC-01	Scope		
QC-02	Normative references		
QC-03	Terms and references		
QC-1.1	Understanding the organization and its context	4.1	
QC-1.2	Monitoring and review of internal and external issues	4.1	
QC-2.1	Relevant requirements of stakeholders	4.2	
QC-2.2	Monitoring and review of stakeholder information	4.2	
QC-3	Scope of quality management system	4.3/4.4	
QC-4	Project quality management system	4.4	
QC-5	Management responsibilities	5.1	
QC-6	Customer focus	5.1.2	
QC-7.1	Quality policy (organization)	5.2	

Document No.	Document Title	Relevant Clause in 9001:2015	Version/Revision Date
QC -7.2	Quality policy (project)	5.2.2	
QC-8	Organizational roles, responsibilities, and authorities (organization chart)	5.3	
QC-9	Preparation and control of project quality plan	6.0	
QC-10.1	Project risk (study stage)	6.1	
QC-10.2	Project risk (during design)	6.1	
QC-10.3	Project risk (during bidding & tendering)	6.1	
QC-10.4	Project risk (during construction)	6.1	
QC-10.5	Project risk (selection of designer, contractor)	6.1	
QC-11	Project quality objective	6.2	
QC-12.1	Change management (during design phase)	6.3	
QC-12.2	Change management (during construction phase)	6.3	
QC-12.3	Change management (owner initiated)	6.3	
QC-13.1	Office resources (office equipment)	7.1	
QC-13–2	Human resources (project management, construction management)	7.1.2	
QC-13.3	Human resources (office staff)	7.1.2	
QC-14.1	Infrastructure	7.1.3	
QC-14.2	Work environment	7.1.4	
QC-15	Monitoring and measuring office equipment	7.1.5	
QC-16	Organizational knowledge	7.1.6	
QC-17.1	Competence	7.2	
QC-17.2	Training in quality system	7.2	
QC-17.3	Training in quality auditing	7.2	
QC-17.4	Training in supervision/ technical skills	7.2	
QC-17.5	Records of training, skills, experience and qualifications	7.2	
QC-18	Communication internal and external	7.4	
QC-19.1	Control of documents for general application	7.5.2/3	
QC-19.2	Control of documents for specific projects	7.5.2/3	
QC-20	Records updates	7.5.2	

(*Continued*)

Table 2.2 (Continued)

Document No.	Document Title	Relevant Clause in 9001:2015	Version/Revision Date
QC-21	Control of quality records	7.5.3	
QC-22.1	Planning of project/facility and quality plan	8.1/8.3/ 8.3.1/2/3	
QC-22.2	Project study and feasibility	8.1/8.3/ 8.3.1/2/3	
QC-22.3	Project brief	8.1/8.3/ 8.3.1/2/3	
QC-22.4	Records of feasibility inputs	8.3.3	
QC-22.5	Records of terms of reference	8.3.3	
QC-22.6	Record of changes	8.3.6	
QC-23.1	Selection and evaluation of designer, project manager, construction manager, contractor	8.4	
QC-23.2	Control of design, contracting services	8.4	
QC-23.3	Communication with designer, contractor	8.4.3	
QC-24.1	Designer selection procedure	8.5	
QC-24.2	Contractor selection procedure	8.5	
QC-25.1	Design review procedure	8.5	
QC-25.2	Construction supervision procedure	8.5	
QC-25.3	Project management procedure	8.5	
QC-25.4	Construction management procedure	8.5	
QC-26.1	Release of project documents for tender	8.6	
QC-26.2	Release of project documents for construction	8.6	
QC-27	Control of nonconforming work (project)	8.7	
QC-28.1	Project document review (during design)	9.1	
QC-28.2	Project document review (during construction)	9.1	
QC-28.3	Project manager/construction manager comments	9.1.2	
QC-29	Internal quality audits	9.2	
QC-30	Management review	9.3	
QC-31	Nonconformity and corrective action	10.2.2	
QC-32	Control of client complaints	10.2.2	
QC-33	Continual improvement	10.3	

Table 2.3 List of quality manual documents – consultant (design and supervision)

Document No.	Document Title	Relevant Clause in 9001:2015	Version/Revision Date
QC-0	Circulation list		
QC-00	Records of revision		
QC-01	Scope		
QC-02	Normative references		
QC-03	Terms and references		
QC-1.1	Understanding the organization and its context	4.1	
QC-1.2	Monitoring and review of internal and external issues	4.1	
QC-2.1	Relevant requirements of stakeholders	4.2	
QC-2.2	Monitoring and review of stakeholder information	4.2	
QC-3	Scope of quality management system	4.3/4.4	
QC-4	Project quality management system	4.4	
QC-5	Management responsibilities	5.1	
QC-6	Customer focus	5.1.2	
QC-7.1	Quality policy (organization)	5.2	
QC-7.2	Quality policy (project)	5.2.2	
QC-8	Organizational roles, responsibilities, and authorities (organization chart)	5.3	
QC-9	Preparation and control of project quality plan	6.0	
QC-10.1	Project risk (design proposal)	6.1	
QC-10.2	Project risk (supervision proposal)	6.1	
QC-10.3	Project risk (during design)	6.1	
QC-10.4	Project risk (during construction)	6.1	
QC-11	Project quality objective	6.2	
QC-12.1	Change management (during design phase)	6.3	
QC-12.2	Change management (during construction phase)	6.3	
QC-13.1	Office resources (office equipment, design software)	7.1	

(*Continued*)

Table 2.3 (Continued)

Document No.	Document Title	Relevant Clause in 9001:2015	Version/Revision Date
QC-13–2	Human resources (design team, supervision team)	7.1.2	
QC-13.3	Human resources (office staff)	7.1.2	
QC-14.1	Infrastructure	7.1.3	
QC-14.2	Work environment	7.1.4	
QC-15	Monitoring and measuring office equipment	7.1.5	
QC-16	Organizational knowledge	7.1.6	
QC-17.1	Competence	7.2	
QC-17.2	Training in quality system	7.2	
QC-17.3	Training in quality auditing	7.2	
QC-17.4	Training in operational/technical skills	7.2	
QC-17.5	Record of training, skills, experience and qualifications	7.2	
QC-18	Communication internal and external	7.4	
QC-19.1	Control of documents for general application	7.5.2/3	
QC-19.2	Control of documents for specific projects	7.5.2/3	
QC-20	Records updates	7.5.2	
QC-21	Control of quality records	7.5.3	
QC-22.1	Proposal (design, supervision) documents	8,2	
QC-22.2	Proposal review	8.2	
QC-22.3	Records of proposal changes	8.2	
QC-23.1	Planning of engineering design and quality plan	8.1/8.3/ 8.3.1/2/3	
QC-23.2	Design development (Design-Bid-Build)	8.1/8.3/ 8.3.1/2/3	
QC-23.3	Design development (Design-Build)	8.1/8.3/ 8.3.1/2/3	
QC-23.4	Record of design development inputs	8.3.3	
QC-23.5	Records of design development control	8.3.4	
QC-23.6	Records of design development outputs	8.3.5	
QC-23.7	Records of design development changes	8.3.6	
QC-24.1	Selection and evaluation of subconsultant	8.4	
QC-24.2	Control of subconsultant services	8.4	

Document No.	Document Title	Relevant Clause in 9001:2015	Version/Revision Date
QC-24.3	Communication with subconsultant	8.4.3	
QC-24.4	Evaluation and selection of equipment, provisions	8.4	
QC-25	Engineering design procedure	8.5	
QC-25.1	Design review procedure	8.5	
QC-26.1	Construction supervision procedure	8.5	
QC-26.2	Project management procedure	8.5	
QC-26.3	Construction management procedure	8.5	
QC-27	Release of project documents	8.6	
QC-28	Control of nonconforming work (design errors)	8.7	
QC-29.1	Project document review (management and control)	9.1	
QC-29.2	Client/owner comments	9.1.2	
QC-30	Internal quality audits	9.2	
QC-31	Management review	9.3	
QC-32.1	Nonconformity and corrective action	10.2.2	
QC-32.2	Preventive action	10.2.2	
QC-33	Control of client complaints	10.2.2	
QC-34	Continual improvement	10.3	

Table 2.4 List of quality manual documents – contractor

Document No.	Document Title	Relevant Clause in 9001:2015	Version/Revision Date
QC-0	Circulation list		
QC-00	Records of revision		
QC-01	Scope		
QC-02	Normative references		
QC-03	Terms and references		
QC-1.1	Understanding the organization and its context	4.1	
QC-1.2	Monitoring and review of internal and external issues	4.1	
QC-2.1	Relevant requirements of stakeholders	4.2	
QC-2.2	Monitoring and review of stakeholder information	4.2	

(*Continued*)

Table 2.4 (Continued)

Document No.	Document Title	Relevant Clause in 9001:2015	Version/Revision Date
QC-3	Scope of quality management system	4.3/4.4	
QC-4	Project quality management system	4.4	
QC-5	Management responsibilities	5.1	
QC-6	Customer focus	5.1.2	
QC-7.1	Quality policy (organization)	5.2	
QC-7.2	Quality policy (project)	5.2.2	
QC-8	Organizational roles, responsibilities, and authorities (organization chart)	5.3	
QC-9	Preparation and control of project quality plan	6.0	
QC-10.1	Project risk management (tendering)	6.1	
QC-10.2	Project risk management (construction)	6.1	
QC-10.3	Project risk management (selection of subcontractor, supplier)	6.1	
QC-11	Project quality objectives	6.2	
QC-12.1	Change management (scope)	6.3	
QC-12.2	Change management (variation orders, site work instructions)	6.3	
QC-13.1	Office resources (office staff, office equipment)	7.1	
QC-13.2	Construction resources (human resources, equipment and machinery)	7.1	
QC-14.1	Infrastructure	7.1.3	
QC-14.2	Work environment	7.1.4	
QC-15	Control of construction, material, measuring and test equipment	7.1.5	
QC-16	Control of human resources	7.1.6	
QC-17	Organizational knowledge	7.1.6	
QC-18.1	Competence	7.2	
QC-18.2	Training and development in quality system	7.2	
QC-18.3	Training in quality auditing	7.2	
QC-18.4	Training in operational/technical skills	7.2	
QC-18.5	Records of training, skills, experience and qualifications	7.2	
QC-19	Communication internal and external	7.4	

Document No.	Document Title	Relevant Clause in 9001:2015	Version/Revision Date
QC-20.1	Control of documents for general application	7.5.2/3	
QC-20.2	Control of documents for specific projects	7.5.2/3	
QC-21	Records updates	7.5.2	
QC-22	Control of quality records	7.5.3	
QC-23	Documents control (logs)	7.5.3.2	
QC-24	Project planning and control	8.1	
QC-25	Project specific requirements	8.2	
QC-26	Project specific quality control plan (contractor's quality control plan)	8.2	
QC-27–1	Tender documents	8.2	
QC-27.2	Tender review	8.2	
QC-27.3	Contract review	8.2.1	
QC-28	Construction processes	8.2.2	
QC-29	Variation review	8.2.3	
QC-30.1	Engineering and shop drawings	8.3	
QC-30.2	Records of engineering and shop drawing input	8.3.3	
QC-31	Design developments for Design-Build projects	8.3	Refer Annexure A2.4
QC-32	Selection and evaluation of subcontractors	8.4	
QC-32.1	Selection and evaluation and of suppliers	8.4.1	
QC-33	Communication with subcontractors, material suppliers, vendors	8.4.3	
QC-34.1	Inspection of subcontracted work	8.4.2	
QC-34.2	Incoming material inspection and testing	8.4.3	
QC-35	Installation procedures	8.5	
QC-36	Product identification and traceability	8.5.2	
QC-37	Identification of inspection and test status	8.5.2	
QC-38	Control of owner supplied items	8.5.3	
QC-39	Handling and storage	8.5.4	
QC-40	Construction inspection, testing and commissioning	8.6	
QC-41	Project handover	8.6	
QC-42.1	Control of nonconforming work	8.7	
QC-42.2	Control of nonconforming subcontractor	8.7	

(Continued)

Table 2.4 (Continued)

Document No.	Document Title	Relevant Clause in 9001:2015	Version/Revision Date
QC-43.1	Project performance review	9.1.1	
QC-43.2	Project quality assessment and measurement	9.1.2	
QC-44	Internal quality audits	9.2	
QC-45	Management review	9.3	
QC-46	New technology in construction	10.1	
QC-47.1	Nonconformity and corrective action	10.2.2	
QC-47.2	Preventive action	10.3	
QC-48	Control of client complaints	10.2.2	
QC-49	Continual improvement	10.3	
Annexure A2.4 (Additional documents for Design-Build Contracting)			
QC-10.4	Project risk management (tendering)	6.1	
QC-10.5	Project risk management (design)	6.1	
QC-10.6	Project risk management (construction)	6.1	
QC-12.3	Change management (design changes)	6.3	
QC-13.3	Design resources (design team, design software)	7.1	
QC-18.4	Competence	7.2	
QC-25.1	Planning of engineering design and quality plan	8.1/8.3	
QC-25.2	Design development	8.1/8.3	
QC-25.3	Records of design development inputs	8.3.3	
QC-41.1	Release of project design	8.6	
QC-43.3	Control of nonconforming work (design errors)	8.7	
QC-44.3	Client/owner/PM/CM comments	9.1	

Table 2.5 List of quality manual documents – manufacturer

Document No.	Document Title	Relevant Clause in 9001:2015	Version/Revision Date
QC-0	Circulation list		
QC-00	Records of revision		
QC-01	Scope		
QC-02	Normative references		
QC-03	Terms and references		
QC-1.1	Understanding the organization and its context	4.1	

Document No.	Document Title	Relevant Clause in 9001:2015	Version/Revision Date
QC-1.2	Monitoring and review of internal and external issues	4.1	
QC-2.1	Relevant requirements of stakeholders	4.2	
QC-2.2	Monitoring and review of stakeholder information	4.2	
QC-3	Scope of quality management system	4.3/4.4	
QC-4	Project quality management system	4.4	
QC-5	Management responsibilities	5.1	
QC-6	Customer focus	5.1.2	
QC-7.1	Quality policy (organization)	5.2	
QC-7.2	Quality policy (procurement)	5.2.2	
QC-8	Organizational roles, responsibilities and authorities (organization chart)	5.3	
QC-9	Preparation and control of quality plan	6.0	
QC-10.1	Risk management (quotation/proposal)	6.1	
QC-10.2	Risk management (manufacturing process)	6.1	
QC-10.3	Risk management (selection of vendor, supplier)	6.1	
QC-11	Quality objectives	6.2	
QC-12.1	Change management	6.3	
QC-13.1	Office resources (office staff, office equipment)	7.1	
QC-13.2	Plant resources (human resources, equipment and machinery)	7.1	
QC-14.1	Infrastructure	7.1.3	
QC-14.2	Work environment	7.1.4	
QC-15	Control of processes, material, measuring and test equipment	7.1.5	
QC-16	Control of human resources	7.1.6	
QC-17	Organizational knowledge	7.1.6	
QC-18.1	Training and development in quality system	7.2	
QC-18.2	Training in quality auditing	7.2	
QC-18.3	Training in operational/technical skills	7.2	
QC-18.4	Records of training, skills, experience and qualifications	7.2	
QC-19	Communication internal and external	7.4	
QC-20.1	Control of documents for general application	7.5.2/3	

(*Continued*)

Table 2.5 (Continued)

Document No.	Document Title	Relevant Clause in 9001:2015	Version/Revision Date
QC-20.2	Control of documents for specific product/material/system	7.5.2/3	
QC-21	Records updates	7.5.2	
QC -22	Control of quality records	7.5.3	
QC-23	Process planning and control	8.1	
QC-24	Customer specific requirements	8.2	
QC-25	Customer specific quality control plan	8.2	
QC-26–1	Tender documents	8.2	
QC-26.2	Tender review	8.2	
QC-26.3	Contract (purchase order) review	8.2.1	
QC-27	Manufacturing processes	8.2.2	
QC-28	Engineering and shop drawings	8.3	
QC-29	Selection and evaluation of vendors, suppliers	8.4	
QC-30	Communication with material suppliers, vendors	8.4.3	
QC-31.1	Inspection of outsourced work	8.4.2	
QC-31.2	Incoming material inspection and testing	8.4.3	
QC-32	Fabrication/assembling procedures	8.5	
QC-33	Product identification and traceability	8.5.2	
QC-34	Identification of inspection and test status	8.5.2	
QC-35	Handling and storage	8.5.4	
QC-36	Product inspection and testing	8.6	
QC-37	Product packing and dispatch	8.6	
QC-38.1	Control of nonconforming work	8.7	
QC-38.2	Control of nonconforming work	8.7	
QC-39.1	Production performance review	9.1.1	
QC-39.2	Product quality assessment and measurement	9.1.2	
QC-40	Internal quality audits	9.2	
QC-41	Management review	9.3	
QC-42	New technology in manufacturing	10.1	
QC-43.1	Nonconformity and corrective action	10.2.2	
QC-43.2	Preventive action	10.3	
QC-44	Control of client/customer complaints	10.2.2	
QC-45	Continual improvement	10.3	

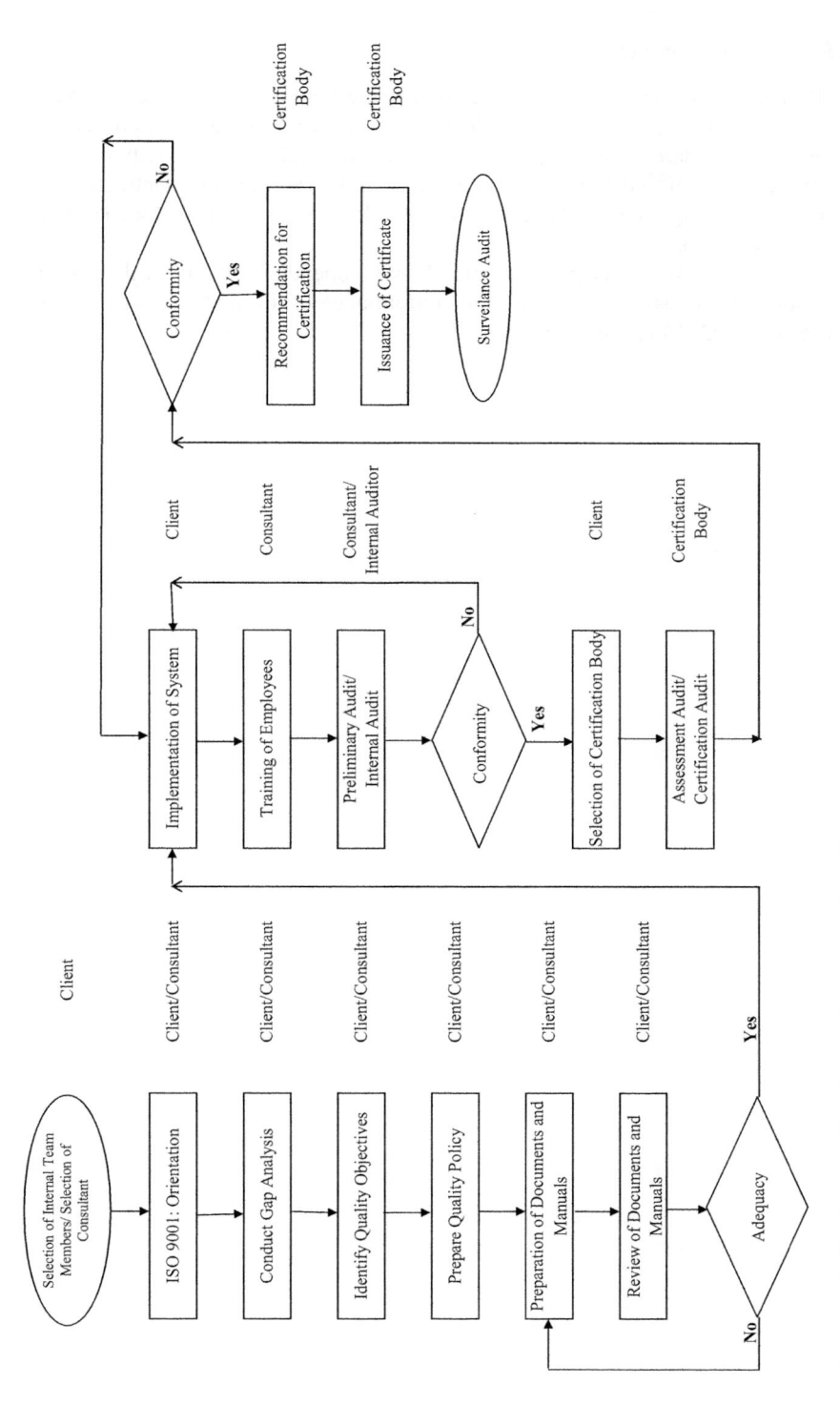

Figure 2.7 ISO design and implementation process flow diagram

2.5 QMS documents auditing

QMS has to cover all the activities leading to the final product, process, project and services. The QMS is based on the guidelines for performance improvement as per ISO standards and the quality management requirements. The tables listing the contents of QMS manual for owner, designer (consultant), contractor and manufacturing organization have already been discussed in section 2.4.1 (Development of QMS).

Appendix A, Appendix B, Appendix C, and Appendix D list typical auditing questionnaires to assess and verify the conformance of documented activities as per ISO 9001:2015 requirements.

Appendix A

Typical auditing questionnaire for quality manual documents – owner (client, project owner, developer)

Document No.	Document Title	Relevant Clause in 9001:2015	Typical Audit Questionnaire
QC-0	Circulation list		
QC-00	Records of revision and ownership of documents		
QC-01	Scope		
QC-02	Normative references		
QC-03	Terms and references		
QC-1.1	Understanding the organization and its context	4.1	• How were the strategic and business plans of organization established? • How was the mission and vision statement developed? • Is there any strategic policy register maintained?
QC-1.2	Monitoring and review of internal and external issues	4.1	• How does the organization monitor external influence on the organization's policy? • How does the organization recognize and review internal issues including that of employees?
QC-1.3	Review of QMS program	4.1	• Are there regular meetings conducted to discuss, review and analyze the organization's quality program?

(*Continued*)

Document No.	Document Title	Relevant Clause in 9001:2015	Typical Audit Questionnaire
QC-2.1	Relevant requirements of stakeholders	4.2	• Does the organization identify and maintain stakeholder lists? • How are the customer requirements communicated to the stakeholders? • How are the customer's quality requirements developed, compiled and complied with?
QC-2.2	Monitoring and review of stakeholder's information	4.2	• How are the customer's and stakeholder's satisfactions taken care of? • Is there any survey carried out periodically or at the end of each project? • How are the applicable requirements of the regulatory, statutory authority monitored and protected?
QC-3	Scope of quality management system	4.3/4.4	• How is the quality requirement of the facility/project established? • Is there a customer requirements survey carried out on a project to project basis or are general market requirements and data considered for the development of project?
QC-4	Project quality management system	4.4	• Does each of the facility/project have its own quality requirements and quality management plan?

Document No.	Document Title	Relevant Clause in 9001:2015	Typical Audit Questionnaire
QC -5	Management responsibilities	5.1	• How is the quality management system for each project documented? • Does the management regularly get involved in maintaining the project quality and its commitment towards customer satisfaction?
QC -6	Customer focus	5.1.2	• Does the organization maintain the records of customer requirements and ensure that the facility project satisfies and meets their requirements? • How are the risks related to customer quality requirements/ satisfaction addressed? • How does the organization ensure that the facility/ project complies with the regulatory authority's requirements? • How does the organization measure customer satisfaction? • What are the steps taken to enhance customer satisfaction?
QC -7.1	Quality policy (organization)	5.2	• How the management establish a quality policy that meets and satisfies the customer requirements?

(*Continued*)

Document No.	Document Title	Relevant Clause in 9001:2015	Typical Audit Questionnaire
QC -7.2	Quality policy (project)	5.2.2	• Does the policy meet statutory and regulatory requirements, legislation and applicable laws? • Is the policy committed for continual improvement and reviewed frequently? • How does the management establish a project quality policy that meets and satisfies the customer and stakeholder requirements? • Does the management establish project quality based on specific requirements of the customers? • Does the policy meet statutory and regulatory requirements, legislation and applicable laws?
QC-8	Organizational roles, responsibilities, and authorities (organization chart)	5.3	• Is there is an organization chart, with names of team members/ staff, that specifies the responsibilities and authorities of each member of the project team for general as well as specific/individual project?
QC-9	Preparation and control of project quality plan	6.0	• What quality requirements and other issues are considered while preparing project quality plan and how they are implemented?

Document No.	Document Title	Relevant Clause in 9001:2015	Typical Audit Questionnaire
QC-10.1	Project risk (study stage)	6.1	• Is there a list of major risk factors affecting the owner during the study stage and their impact on the project? • How are risk management actions planned to address the risks to avoid, eliminate or reduce?
QC-10.2	Project risk (during design)	6.1	• Is there a list of major risk factors affecting the owner during the study stage and their impact on the project? • How are risk management actions planned to address the risks to avoid, eliminate or reduce?
QC-10.3	Project risk (during bidding & tendering)	6.1	• Is there a list of major risks factors affecting the owner during the bidding & tendering phase and their impact on the project? • How are risk management actions planned to address the risks to avoid, eliminate or reduce?
QC-10.4	Project risk (during construction)	6.1	• Is there a list of major risk factors affecting the owner during the construction phase and their impact on the project? • How are risk management actions planned to address the risks to avoid, eliminate or reduce?

(Continued)

Document No.	Document Title	Relevant Clause in 9001:2015	Typical Audit Questionnaire
QC-10.5	Project risk (selection of designer, contractor)	6.1	• Is there a list of major risk factors affecting the owner during selection of the designer, contractor and their impact on the project? • How are risk management actions planned to address the risks to avoid, eliminate or reduce?
QC-11	Project quality objective	6.2	• How are the project quality objectives achieved? • Are they are in line with quality policy? • How does the organization ensure that the project quality objectives are SMART?
QC-12.1	Change management (during design phase)	6.3	• Is there any standard change management plan to approve scope change, variation/ alternatives? • Who is authorized to approve scope changes?
QC-12.2	Change management (during construction phase)	6.3	• Is there any standard change management plan to approve scope change, variation/ alternatives? • Who is authorized to approve scope changes?
QC-12.3	Change management (owner initiated)	6.3	• Who is authorized to initiate scope changes, additional works? • Is there any procedure to approve the changes?

Document No.	Document Title	Relevant Clause in 9001:2015	Typical Audit Questionnaire
QC-13.1	Office resources (office equipment)	7.1	• How are resources (office equipment) for the office as well as for the project established?
QC-13–2	Human resources (project management, construction management)	7.1.2	• How are resources for project management established? • How is the project delivery system decided? • How are additional resources for the project, if required, obtained and managed?
QC-13.3	Human resources (office staff)	7.1.2	• How are resources (office staff) requirements established? • Who is responsible to authorize hiring of new staff?
QC-14.1	Infrastructure	7.1.3	• How does the organization maintain the infrastructure for proper and smooth function of the works?
QC-14.2	Work environment	7.1.4	• How does the organization determine that adequate space is available for safe and comfortable working? • Is the work environment congenial to work?
QC-15	Monitoring and measuring office equipment	7.1.5	• How does the organization monitor proper utilization of office equipment? • How does the organization ascertain that office equipment is

(Continued)

Document No.	Document Title	Relevant Clause in 9001:2015	Typical Audit Questionnaire
			equipped with latest technology and latest software and is suitable for use?
• How does the organization ensure that the installed software is original and genuine?
QC-16	Organizational knowledge	7.1.6	• How does the organization determine that the organization has the necessary knowledge for construction of facility/project to meet and satisfy customer requirement?
QC-17.1	Competence	7.2	• Does the organization possess the necessary competency for competitive advantage?
QC-17.2	Training in quality system	7.2	• Does the organization conduct training for its staff in quality management, project management, construction management, quality management as applicable?
QC-17.3	Training in quality auditing	7.2	• Has the organization trained the auditor/lead auditor and does it provide training in auditing?
QC-17.4	Training in supervision/ technical skills	7.2	• Does the organization conduct training to achieve skills to achieve organization objectives?

Document No.	Document Title	Relevant Clause in 9001:2015	Typical Audit Questionnaire
			• Does the organization conduct training to achieve skills in design development? • Does the organization conduct training to achieve skills in supervision activities, project quality control and document review?
QC-17.5	Records of training, skills, experience and qualifications	7.2	• Does the organization maintain records of training and conduct need assessment for training?
QC-18	Communication internal and external	7.4	• How does the organization determine a communication method both internal and external that covers customers and all other stakeholders? • Are there are meetings to appraise the project progress and new developments?
QC-19.1	Control of documents for general application	7.5.2/3	• How are general documents controlled and archived?
QC-19.2	Control of documents for specific projects	7.5.2/3	• How are project specific documents controlled and archived?
QC-20	Records updates	7.5.2	• Are document records regularly updated?
QC-21	Control of quality records	7.5.3	• How does the organization control the quality records?

(*Continued*)

Document No.	Document Title	Relevant Clause in 9001:2015	Typical Audit Questionnaire
QC-22.1	Planning of project/ facility and quality plan	8.1/8.3/ 8.3.1/2/3	• How does the organization plan the quality of the planned project/ facility? • How are the requirements of customers determined?
QC-22.2	Project study and feasibility	8.1/8.3/ 8.3.1/2/3	• Is a feasibility study carried out for all projects? • Who conducts the feasibility study • How is the consultant to conduct the feasibility study selected?
QC-22.3	Project brief	8.1/8.3/ 8.3.1/2/3	• Are project needs properly identified and analyzed? • How is the project statement developed and project goals and objectives established? • How are preliminary schedule, cost and resources established? • Is project risk considered and analyzed? • How is the project delivery system identified? • How is the contracting/pricing system determined? • Who develops the project brief/project charter?
QC-22.4	Records of feasibility inputs	8.3.3	• Are all the inputs for preparation of the feasibility report considered? • Is any external agency involved for the feasibility study?

Document No.	Document Title	Relevant Clause in 9001:2015	Typical Audit Questionnaire
QC-22.5	Records of terms of reference	8.3.3	• Who prepares the terms of reference?
QC-22.6	Record of changes	8.3.6	• How are the records of approved preferred alternatives maintained?
QC-23.1	Selection and evaluation of designer, project manager, construction manager and contractor	8.4	• Does the organization have a bidding & tendering process to select the designer, project manager, construction manager and contractor? • What are the criteria for the evaluation of submitted bids?
QC-23.2	Control of design, contracting services	8.4	• Who is responsible for controlling the services of the designer and contractor? • How are the changes controlled?
QC-23.3	Communication with designer and contractor	8.4.3	• What method of communication is used among the designer, contractor and other stakeholders? • Is there an administrative matrix for communication and responsibilities?
QC-24.1	Designer selection procedure	8.5	• Does the organization have a list of shortlisted designers? • Is the designer selected on a low bid basis or quality based system? • Who is responsible for selecting the designer and awarding the contractor

(Continued)

Document No.	Document Title	Relevant Clause in 9001:2015	Typical Audit Questionnaire
QC-24.2	Contractor selection procedure	8.5	• Does the organization have a standard procedure to select the contractor? • How are the contractors shortlisted? • What are the criteria to select the contractor?
QC-25.1	Design review procedure	8.5	• Does the quality management system have a design review procedure to be followed by the construction supervisor/project manager (owner's representative)?
QC-25.2	Construction supervision procedure	8.5	• Is there any construction supervision manual/procedure?
QC-25.3	Project management procedure	8.5	• Is there any project management manual/procedure?
QC-25.4	Construction management procedure	8.5	• Is there any construction management manual procedure?
QC-26.1	Release of project documents for tender	8.6	• Who is responsible for approving construction documents? • Who is authorized to release the documents for tendering? • Is there any authorization matrix?
QC-26.2	Release of project documents for construction	8.6	• Who is authorized to issue the letter to proceed? • Who is responsible for releasing the construction documents to the construction

Document No.	Document Title	Relevant Clause in 9001:2015	Typical Audit Questionnaire
			supervisor (project manager, construction manager, quality manager as applicable)? • How are client supplied items and material controlled?
QC-27	Control of nonconforming work (project)	8.7	• How does the organization control nonconformity of construction works? • Who is responsible for issuing the nonconformance report?
QC-28.1	Project document review (during design)	9.1	• Who is responsible for reviewing design drawings, specifications and construction documents?
QC-28.2	Project document review (during construction)	9.1	• Who is responsible for reviewing and approving shop drawings, material, certificates and other submittals? • Who is responsible for approving the executed works? • Who is responsible for testing and commissioning of the project? • Who is authorized to take over the completed facility/ project?
QC-28.3	Project manager/ construction manager/quality manager comments	9.1.2	• Who is reviewing the comments by the project manager/ construction manager/quality manager on the submittals?

(*Continued*)

Document No.	Document Title	Relevant Clause in 9001:2015	Typical Audit Questionnaire
			• Who is reviewing the correspondence between project manager/ construction manager to the contractor and other stakeholders?
• Is there any agreed procedure?
QC-29	Internal quality audits	9.2	• Is an internal audit conducted at planned intervals, results analyzed and actions taken?
QC-30	Management review	9.3	• How frequently does the management review the documents to ensure quality plans are implemented at all levels?
QC-31	Nonconformity and corrective action	10.2.2	• Who is responsible for taking action on nonconformities and taking corrective actions?
QC-32	Control of client complaints	10.2.2	• Is there any register maintained for customer complaints?
• Is there any register maintained for designer and contractor complaints against project manager/ construction manager?
• Is there any register maintained for suggested improvements by the customer?
• Who is responsible for taking action against complaints and suggestions?
QC-33	Continual improvement	10.3	• Is there any procedure for continual improvements?

Appendix B

Typical auditing questionnaire for quality manual documents – consultant (design and supervision)

Document No.	Document Title	Relevant Clause in 9001:2015	Typical Audit Questionnaire
QC-0	Circulation list		
QC-00	Records of revision and ownership documents		
QC-01	Scope		
QC-02	Normative references		
QC-03	Terms and references		
QC-1.1	Understanding the organization and its context	4.1	• How were the strategic and business plans of the organization established? • How was the mission and vision statement developed? • Is there any strategic policy register maintained?
QC-1.2	Monitoring and review of internal and external issues	4.1	• How does the organization monitor external influence on organization's policy? • How does the organization recognize and review internal issues including that of employees? • Does the organization maintain a list of internal and external issues?

(*Continued*)

Document No.	Document Title	Relevant Clause in 9001:2015	Typical Audit Questionnaire
QC-1.3	Review of QMS program	4.1	• Are there regular meetings conducted to discuss, review and analyze the organization's quality management program?
QC-2.1	Relevant requirements of stakeholders	4.2	• Does the organization identify and maintain stakeholder lists? • How are the customer/project owner/ developer requirements communicated to the stakeholders? • How are the project owner's quality requirements developed, compiled and complied with?
QC-2.2	Monitoring and review of stakeholder information	4.2	• How are the project owner's and other stakeholders' satisfactions are taken consideration? • Is there any customer satisfaction survey carried out periodically or at the end of each project? • How are the applicable requirements of the regulatory, statutory authority monitored and protected?
QC-3	Scope of quality management system	4.3/4.4	• Is there is an organizational quality management system covering the quality requirement of projects in line with ISO standards?

Document No.	Document Title	Relevant Clause in 9001:2015	Typical Audit Questionnaire
QC-4	Project quality management system	4.4	• How are the quality requirements/ quality management system for each project established and documented?
QC-5	Management responsibilities	5.1	• Is the management regularly involved in maintaining the corporate quality management system? • Is the management involved in maintaining project quality and its commitment towards the project owner's satisfaction?
QC-6	Customer focus	5.1.2	• How does the organization ensure that project quality complies with project specifications and the owner's requirements? • How does the organization measure customer/ client/project owner satisfaction? • How are the risks related to customer quality requirements/ satisfaction addressed? • What are the steps taken to enhance customer satisfaction? • How does the organization ensure that the facility/ project complies with the regulatory authority's requirements?

(Continued)

Document No.	Document Title	Relevant Clause in 9001:2015	Typical Audit Questionnaire
QC -7.1	Quality policy (organization)	5.2	• How does the management establish quality policies that address the project requirements and satisfy the owner's goals and objectives? • Does the policy meet statutory and regulatory requirements, legislation and applicable laws? • Is the policy committed to continual improvement and reviewed frequently?
QC -7.2	Quality policy (project)	5.2.2	• How does the management establish a project quality policy that meets and satisfies the project specification requirements? • Does the management establish a project quality policy based on the specific requirements of the project? • Does the policy meet statutory and regulatory requirements, legislation and applicable laws? • How is the project quality policy documented and communicated to related stakeholders?

Document No.	Document Title	Relevant Clause in 9001:2015	Typical Audit Questionnaire
QC-8	Organizational roles, responsibilities and authorities (organization chart)	5.3	• Is there an organization chart, with names of project team members/project staff that specifies the responsibilities and authorities of each member of the project team?
QC-9	Preparation and control of project quality plan	6.0	• What quality requirements and other issues are considered while preparing the project quality plan and how they are implemented and controlled? • Does the quality plan include an organization chart specifying the responsibilities and authorities of each of the project team members?
QC-10.1	Project risk (design proposal)	6.1	• What are the major risks factors considered while preparing the design proposal for the project? • Is there any list of major risk factors that may affect the design proposal? • Are the identified risks reviewed qualitatively and quantified? • How are risk management actions planned to address the risks to avoid, eliminate or reduce?

(*Continued*)

Document No.	Document Title	Relevant Clause in 9001:2015	Typical Audit Questionnaire
QC-10.2	Project risk (supervision proposal)	6.1	• What are the major risk factors considered while preparing the supervision proposal for the project? • Is there any list of major risk factors that may affect the construction supervision proposal? • Is there any periodical risk review to assess the effectiveness of risk responses? • How are risk management actions planned to address the risks to avoid, eliminate or reduce?
QC-10.3	Project risk (during design)	6.1	• Is there any list of major risks factors that may affect the project or the owner due to errors/omissions in the design? • How is the risk related to conflict in the design/construction documents analyzed and resolved?
QC-10.4	Project risk (during construction)	6.1	• Is there any list of major risk factors that may affect the construction activities due to errors/omissions in the design? • How is the risk related to conflict in construction documents analyzed and resolved?

Document No.	Document Title	Relevant Clause in 9001:2015	Typical Audit Questionnaire
QC-11	Project quality objective	6.2	• How are the project quality objectives established? • How does the organization ensure that the project objectives are SMART? • At what stage is the project quality prepared? • Is there any provision in the corporate QMS about the development of the project specific quality system?
QC-12.1	Change management (during design phase)	6.3	• Is there any process to submit the variation during the design phase? • How are the changes implemented?
QC-12.2	Change management (during construction phase)	6.3	• Is there any process to submit the variation during construction phase for any changes to the contract?
QC-13.1	Office resources (office equipment, design software)	7.1	• How are the resources (office equipment) for the office as well as for the project established? • How does the organization ensure that the office equipment is equipped with the latest technological standards and installed with approved software for office use?

(*Continued*)

Document No.	Document Title	Relevant Clause in 9001:2015	Typical Audit Questionnaire
QC-13–2	Human resources (design team, supervision team)	7.1.2	• How are the resources for the design team established? • How are the resources for the construction supervision team established? • Is there any approved manpower plan/ histogram for the project? • How are additional resources for the project, if required, obtained and managed?
QC-13.3	Human resources (office and project office)	7.1.2	• How are resources (office staff) requirements established? • Who is responsible for authorizing the hiring of new office staff as well project team members for the project?
QC-14.1	Infrastructure	7.1.3	• How does the organization maintain the infrastructure for proper and smooth function of the works?
QC-14.2	Work environment	7.1.4	• How does the organization determine that adequate space is available for safe and comfortable working? • Is the work environment congenial to work?
QC-15	Monitoring and measuring office equipment	7.1.5	• How does the organization monitor the proper utilization of office equipment?

Document No.	Document Title	Relevant Clause in 9001:2015	Typical Audit Questionnaire
			• How does the organization ascertain that office equipment is equipped with the latest technology and latest software and is suitable for use? • How does the organization ensure that the installed software is original and genuine?
QC-16	Organizational knowledge	7.1.6	• How does the organization determine that the organization has the necessary knowledge for development design and construction supervision of the project(s) to meet the project specifications, project goals and objectives and owner satisfaction?
QC-17.1	Competence	7.2	• Does the organization possess the necessary competency to design and supervise the projects as per customer needs to achieve a qualitative, competitive and economical project? • Does the organization possess the competency to develop sustainable projects?

(*Continued*)

Document No.	Document Title	Relevant Clause in 9001:2015	Typical Audit Questionnaire
			• Does the organization have competent professionals having relevant certification to work as design team members and supervision team members?
QC-17.2	Training in quality and project management system	7.2	• Does the organization conduct training for its staff/ project team members in quality management, project management, construction management, quality management as applicable
QC-17.3	Training in quality auditing	7.2	• Does the organization have a trained auditor/lead auditor and provide training in auditing?
QC-17.4	Training in operational/ technical skills	7.2	• Does the organization conduct training to achieve the skills to meet the organization's objectives and competency in design and supervision?
QC-17.5	Record of training, skills, experience and qualifications	7.2	• Does the organization maintain records of training conducted and identify and assess training needs?
QC-18	Communication internal and external	7.4	• How does the organization determine communication

Document No.	Document Title	Relevant Clause in 9001:2015	Typical Audit Questionnaire
			methods both internal and external that cover project owners, developers, regulatory authorities, contractor, consultants and all other stakeholders? • Are there periodic progress and coordination meetings conducted to monitor the project progress?
QC-19.1	Control of documents for general application	7.5.2/3	• How are general documents controlled and archived? • Who is responsible for reviewing and approving the quality system documents prior to use? • How are revisions and amendments controlled?
QC-19.2	Control of documents for specific projects	7.5.2/3	• How are project specific documents controlled and archived? • Who is responsible for reviewing and approving specific project quality system documents prior to use?
QC-20	Records updates	7.5.2	• Are document records regularly updated? • How are revisions and amendments controlled?
QC-21	Control of quality records	7.5.3	• How does the organization control the quality records?

(*Continued*)

Document No.	Document Title	Relevant Clause in 9001:2015	Typical Audit Questionnaire
			• Is there any procedure for the maintenance and control of quality records? • How are quality records archived and disposed of? • How does the organization control client supplied documents, material, software and proprietary items?
QC-22.1	Proposal (design, supervision) documents	8,2	• Does the organization have a proposal/ tender submission process? • How are the resource requirements worked out/calculated for consideration in the proposal to be submitted?
QC-22.2	Proposal review	8.2	• Is the proposal internally reviewed and analyzed prior to submission?
QC-22.3	Records of proposal changes	8.2	• How are the records of changes in the proposal documented and maintained?
QC-23.1	Planning of engineering design and quality plan	8.1/8.3/ 8.3.1/2/3	• How are the design quality requirements established? • How is the design quality management plan established? • How are specified codes, standards and HSE requirements planned and implemented? • How are sustainability requirements planned?

Document No.	Document Title	Relevant Clause in 9001:2015	Typical Audit Questionnaire
QC-23.2	Design development (Design-Bid-Build)	8.1/8.3/ 8.3.1/2/3	• Is there a standard procedure for the development of the design for the Design-Bid-Build type of project delivery system? • How are project specific quality requirements implemented?
QC-23.3	Design development (Design-Build)	8.1/8.3/ 8.3.1/2/3	• Is there a standard procedure for the development of the project design for the Design-Build type of project delivery system? • How are project specific quality requirements implemented? • Is the organization normally involved in the development of the concept design and documents for the Design-Build type of projects?
QC-23.4	Record of design development inputs	8.3.3	• How are the owner requirements collected and documented? • How is the data/ information collected to develop the design? • How are the codes, standards and regulatory requirements determined? • How is the record of design input maintained?
QC-23.5	Records of design development control	8.3.4	• How are design development activities monitored and controlled?

(Continued)

Document No.	Document Title	Relevant Clause in 9001:2015	Typical Audit Questionnaire
			• How is the design work progress monitored to ensure that the submission to the owner is as per schedule?
QC-23.6	Records of design development outputs	8.3.5	• How are design deliverables verified against design inputs and TOR?
QC-23.7	Records of design development changes	8.3.6	• How are the records for design development changes maintained?
QC-24.1	Selection and evaluation of subconsultant	8.4	• How is the specialist subconsultant selected? • What are the criteria for the evaluation of subconsultant/ subcontractor work?
QC-24.2	Control of subconsultant services	8.4	• How is the subconsultant's/ subcontractor's performance monitored?
QC-24.3	Communication with subconsultant	8.4.3	• What method is used to communicate with the subconsultant/ subcontractor?
QC-24.4	Evaluation and selection of equipment and provisions	8.4	• What is the procedure used to select the project material, system and equipment and how is their performance evaluated?
QC-25	Engineering design procedure	8.5	• Does the quality management system have a standard procedure for design development and how is the project design developed using the established procedure?

Document No.	Document Title	Relevant Clause in 9001:2015	Typical Audit Questionnaire
QC-25.1	Design review procedure	8.5	• Does the quality management system have a design review and analysis procedure? • Is the design reviewed at different stages? • How is the interdisciplinary coordination of design/drawings carried out?
QC-26.1	Construction supervision procedure	8.5	• Is the construction supervision manual/procedure developed considering QMS requirements? • How are the construction supervision activities/ responsibilities for a specific project established?
QC-26.2	Project management procedure	8.5	• Does the organization have a project management/ construction supervision procedure (manual)?
QC-26.3	Construction management procedure	8.5	• Does the organization have a construction management (if applicable) procedure (manual)?
QC-27	Submission of project documents	8.6	• How are design documents released and submitted to the owner/project manager for review and approval? • Is there any authorization matrix?

(Continued)

Document No.	Document Title	Relevant Clause in 9001:2015	Typical Audit Questionnaire
QC-28	Control of nonconforming work (design errors)	8.7	• How are the design errors/ omissions detected, documented and controlled?
QC-29.1	Project document review (management and control)	9.1	• Who is responsible for reviewing and approving project documents (design documents)? • Who is responsible for approving project (design) quality requirements? • Who is responsible for monitoring the effectiveness of the quality management system?
QC-29.2	Client/owner comments	9.1.2	• How are client/ owner comments addressed? • Who is responsible for taking action on client/owner comments? • Is there any agreed procedure?
QC-30	Internal quality audits	9.2	• Is an internal audit conducted at planned intervals, and are the results analyzed and actions taken?
QC-31	Management review	9.3	• How frequently does the management review the documents to ensure quality plans are implemented at all levels?
QC-32.1	Nonconformity and corrective action	10.2.2	• Who is responsible for taking action on nonconformities and taking corrective actions?

Document No.	Document Title	Relevant Clause in 9001:2015	Typical Audit Questionnaire
QC-32.2	Preventive action	10.2.2	• What action is taken to prevent the occurrence of errors/mistakes in the design and project documents? • How is the potential cause of nonconformance monitored and are preventive actions taken?
QC-33	Control of client complaints	10.2.2	• Is there any register maintained to record owner/client comments and the action taken?
QC-34	Continual improvement	10.3	• Is there any procedure for continual improvements?

Appendix C

Typical auditing questionnaire for quality manual documents – contractor

Document No.	Document Title	Relevant Clause in 9001:2015	Typical Audit Questionnaire
QC-0	Circulation list		
QC-00	Records of revision and ownership documents		
QC-01	Scope		
QC-02	Normative references		
QC-03	Terms and references		
QC-1.1	Understanding the organization and its context	4.1	• How were the strategic and business plans of organization established? • How was the mission and vision statement developed? • Is there any strategic policy register maintained?
QC-1.2	Monitoring and review of internal and external issues	4.1	• How does the organization monitor external influence on the organization's policy? • Are the organization's documents periodically monitored and reviewed? • How does the organization recognize and review internal issues including those of employees? • Does the organization maintain a list of internal and external issues?
QC-1.3	Review of QMS program	4.1	• Are there regular meetings conducted to discuss, review and analyze the organization's quality management program?

Document No.	Document Title	Relevant Clause in 9001:2015	Typical Audit Questionnaire
QC-2.1	Relevant requirements of stakeholders	4.2	• Does the organization identify and maintain stakeholder lists? • How are the customer/ project owner/ construction supervisor requirements communicated to the stakeholders? • How are the project owner's quality requirements developed, compiled and complied with?
QC-2.2	Monitoring and review of stakeholder's information	4.2	• How are the project owner's and other stakeholders' satisfactions taken into consideration? • Is there any customer satisfaction survey carried out periodically or at the end of each project? • How is the interest of the project owner protected? • How are the applicable requirements of regulatory and statutory authority monitored and protected?
QC-3	Scope of quality management system	4.3/4.4	• How is the quality requirement of the project established? • Is there an organizational quality management system covering the quality requirements of projects in general to meet ISO standards?
QC-4	Project quality management system	4.4	• How is the quality management system for each project documented? • How is the contractor's quality plan developed and documented? • Does the project quality management system differ from project to project?
QC-5	Management responsibilities	5.1	• Is the management regularly involved in maintaining the corporate quality management system?

(*Continued*)

Document No.	Document Title	Relevant Clause in 9001:2015	Typical Audit Questionnaire
QC -6	Customer focus	5.1.2	• Is the management involved in maintaining project quality and its commitment towards the project owner's satisfaction/compliance with contract documents and specifications? • How does the organization ensure that project quality complies with the project specifications and owner's requirements? • How does the organization measure customer/client/project owner satisfaction? • How are the risks related to customer quality requirements/satisfaction addressed? • How does the organization ensure that the facility/project complies with the regulatory authority's requirements? • What are the steps taken to enhance customer satisfaction? • How does the management strive to exceed customer satisfaction?
QC -7.1	Quality policy (organization)	5.2	• How does the management establish a quality policy that addresses the project requirements and satisfies the owner's goals and objectives? • Does the policy meet statutory and regulatory requirements, legislation and applicable laws? • Is the policy committed for continual improvement and reviewed frequently?

Document No.	Document Title	Relevant Clause in 9001:2015	Typical Audit Questionnaire
QC-7.2	Quality policy (project)	5.2.2	• How does the management establish a project quality policy that meets and satisfies the project specification requirements? • Does the management establish a project quality policy based on specific requirements of the project? • How is the project quality policy documented and communicated to related stakeholders? • Does the policy meet statutory and regulatory requirements, legislation and applicable laws?
QC-8	Organizational roles, responsibilities, and authorities (organization chart)	5.3	• Is there an organization chart, with names of the project team members/project staff that specifies the responsibilities and authorities of each member of the project team for general as well as the specific project?
QC-9	Preparation and control of project quality plan	6.0	• What quality requirements and other issues are considered while preparing contractor's quality plan and how are they implemented and controlled at the project site? • Does the contractor's quality plan include an organization chart specifying the responsibilities and authorities of each of the project team members? • Is the project quality plan developed considering the client's QMS? • Is there integration between the project quality plan and the company (corporate) quality manual?

(Continued)

Document No.	Document Title	Relevant Clause in 9001:2015	Typical Audit Questionnaire
QC-10.1	Project risk management (tendering)	6.1	• What are the major risk factors considered while preparing the tender for the project? • Is there any list of major risk factors that may affect the tender/quotation? • How is the risk related to conflict in the construction documents (tender documents) identified, analyzed and resolved? • How are risk management actions planned to address the risks to avoid, eliminate or reduce?
QC-10.2	Project risk management (construction)	6.1	• Is there any list of major risk factors that may affect the construction process? • Is the risk register maintained at the project site and updated whenever a risk is identified? • Are the identified risks reviewed qualitatively and quantified? • How are risk management actions planned to address the risks to avoid, eliminate or reduce?
QC-10.3	Project risk management (selection of subcontractor, supplier)	6.1	• What are the major risk factors considered while selecting the subcontractor and supplier for the project? • Is there any list of major risk factors that may affect the selection of the subcontractor and supplier? • How are risk management actions planned to address the risks to avoid, eliminate or reduce?
QC-11	Project quality objectives	6.2	• How are the project quality objectives established? • At what stage of the construction project is the project quality prepared?

Document No.	Document Title	Relevant Clause in 9001:2015	Typical Audit Questionnaire
QC-12.1	Change management (scope)	6.3	• Is there any provision in the corporate QMS about project specific quality? • How does the organization ensure that the project quality objectives are SMART? • Is there any process to submit the changes in the project scope of the contract (value engineering change proposal)? • How is the change of scope evaluated? • How are the changes implemented? • Is there any 'Variation Register' available at the project site?
QC-12.2	Change management (variation orders, site work instructions)	6.3	• Is there any process in the project process to submit the request for information (RFI) during the construction phase and changes in the contract? • Is there any process in the project process to submit the request for variation order during the construction phase and changes in the contract? • How are the site work instructions (SWI) addressed and implemented?
QC-13.1	Office resources (office staff, office equipment)	7.1	• How are the resources (office equipment) for the office as well as for the project site established? • How is it ensured that the office equipment is equipped with the latest technological standards and installed with approved software for use during the construction phase?

(Continued)

Document No.	Document Title	Relevant Clause in 9001:2015	Typical Audit Questionnaire
QC-13.2	Construction resources (Human resources, Equipment and Machinery)	7.1	• How are the project core staff for the construction process/phase established? • How are the resources (manpower) for the construction process/phases established? • How are the resources (equipment and machinery) for the construction process/phase established? • How are additional resources for the project, if required, obtained and managed?
QC-14.1	Infrastructure	7.1.3	• How does the organization maintain the infrastructure for the proper and smooth function of the works at the office as well as at the project site? • How does the organization determine the temporary site facility requirements?
QC-14.2	Work environment	7.1.4	• How does the organization determine that adequate facilities are available for safe and comfortable working at site? • How are the hygienic and HSE requirements for the project site considered and met? • Is the work environment congenial to work?
QC-15	Control of construction, material, measuring and test equipment	7.1.5	• How does the organization monitor that the construction equipment and machinery required as per the contract documents are available at site? • How does the organization monitor the proper utilization of construction equipment and machinery?

Document No.	Document Title	Relevant Clause in 9001:2015	Typical Audit Questionnaire
			• How does the organization ascertain that testing and measuring equipment have valid calibration? • How does the organization ensure that the installed software is original and genuine? • How is the utilization of construction material controlled to ensure that there is no wastage or surplus material ordered/purchased? • Is there any plan for preventive maintenance? • How is the calibration of equipment carried out? • Is the calibration certificate traceable with an approved national/international calibration/testing agency?
QC-16	Control of human resources	7.1.5	• How is the construction process controlled for optimum utilization of manpower?
QC-17	Organizational knowledge	7.1.6	• How does the organization determine that the organization has the necessary knowledge for execution of construction projects as per contract documents and project specifications to meet the project goals and objectives and owner satisfaction?
QC-18.1	Competence	7.2	• Does the organization possess the necessary competency to execute the construction projects as per the contract specification to achieve a qualitative, competitive and economical project? • Does the organization possess competency to complete the project as per the agreed-upon schedule and cost?

(Continued)

Document No.	Document Title	Relevant Clause in 9001:2015	Typical Audit Questionnaire
			• Does the organization have competent professionals with relevant certification to work as construction team members? • Does the organization have competent workforce to execute the works with qualitative workmanship?
QC-18.2	Training and development in quality system	7.2	• Does the organization conduct training for its staff/construction project team members in the execution of construction projects?
QC-18.3	Training in quality auditing	7.2	• Does the organization have a trained auditor/ lead auditor and provide training in auditing?
QC-18.4	Training in operational/ technical skills	7.2	• Does the organization conduct training to achieve skills and competency in the execution of construction projects? • Does the organization conduct training for its workforce to achieve skills and competency in their trade?
QC-18.5	Records of training, skills, experience and qualifications	7.2	• Does the organization maintain records of training conducted and identify and assess training needs?
QC-19	Communication internal and external	7.4	• How does the organization determine a communication method both internal and external that covers the project owner, construction supervisor, regulatory authorities, subcontractor and all the stakeholders? • Are there periodic progress and coordination meetings conducted to monitor the project progress?

Document No.	Document Title	Relevant Clause in 9001:2015	Typical Audit Questionnaire
QC-20.1	Control of documents for general application	7.5.2/3	• How are general documents controlled and archived? • Who is responsible for reviewing and approving quality system documents prior to use? • How are revisions and amendments controlled?
QC-20.2	Control of documents for specific projects	7.5.2/3	• How are project specific documents controlled and archived? • Who is responsible for reviewing and approving specific project quality system documents prior to use?
QC-21	Records updates	7.5.2	• Are document records regularly updated? • How are revisions and amendments controlled?
QC-22	Control of quality records	7.5.3	• How does the organization control the quality records? • Is there any procedure for the maintenance and control of quality records? • How are quality records archived and disposed of? • How the organization control client supplied documents, material, software and proprietary items? • How is the project record book prepared and maintained? • Is the project record book classified and has an index developed?
QC-23	Documents control (logs)	7.5.3.2	• How are the project control documents and logs maintained and controlled?
QC-24	Project planning and control	8.1	• What tools and techniques are used for project planning? • How is the contractor's construction schedule developed?

(Continued)

Document No.	Document Title	Relevant Clause in 9001:2015	Typical Audit Questionnaire
QC-25	Project specific requirements	8.2	• How is the project progress monitored and controlled? • Is the monitoring system traditional or digitized? • How are project specific requirements/contract requirements identified? • Who is responsible for reviewing construction documents and developing project execution requirements?
QC-26	Project specific quality control plan (contractor's quality control plan)	8.2	• How is the contractor's quality control plan (CQCP) developed? • Is there any procedure to develop the CQCP? • How are contract specific requirements integrated in the CQCP?
QC-27.1	Tender documents	8.2	• How does the organization get pre-qualified for bidding and tendering? • Is there any fee to collect the tender documents?
QC-27.2	Tender review	8.2	• Who is responsible for reviewing the tender documents? • How is the quotation prepared? • Who are the team members involved in the preparation of the tender? • Who approves the tender submission price? • How are the records of changes in the tender documented and maintained?
QC-27.3	Contract review	8.2.1	• What is the procedure to review the contract prior to the signing of the contract?
QC-28	Construction processes	8.2.2	• Does the organization have a construction process describing all the activities/process starting from the award/signing of the contract through to the issuance of the completion certificate?

Document No.	Document Title	Relevant Clause in 9001:2015	Typical Audit Questionnaire
QC-29	Variation review	8.2.3	• Are all the special processes 'qualified' as per the codes and standards? • How is the variation identified? • What is the procedure to submit a request for variation? • Is there any procedure to review and finalize the variation order?
QC-30.1	Engineering and shop drawings	8.3	• What is the procedure to prepare and submit shop drawings? • Does the organization prepare shop drawing in-house or from outside? • Who is responsible for reviewing and approve shop drawings for submittal?
QC-30.2	Records of engineering and shop drawing input	8.3.3	• What factors are considered when preparing shop drawings?
QC-31	Design developments for Design-Build projects	8.3	Please see Annexure A2.4
QC-32	Selection and evaluation of subcontractors	8.4	• Does the organization have a selection (bidding & tendering) process for subcontractor? • Does the organization maintain a list of subcontractors? • How are the nominated subcontractors selected? • What is the procedure for approval of subcontractors? • Is there a pre-qualification process to select vendors/suppliers?
QC-32.1	Selection and evaluation of suppliers	8.4.1	• Does the organization maintain a list of suppliers and manufacturers? • What is the process for approval of suppliers and manufacturers?

(Continued)

Document No.	Document Title	Relevant Clause in 9001:2015	Typical Audit Questionnaire
QC-33	Communication with subcontractors, material suppliers and vendors	8.4.3	• What method is used to communicate with the subcontractors, suppliers and vendors?
QC-34.1	Inspection of subcontracted work	8.4.2	• What is the procedure for inspection and approval of subcontracted works?
QC-34.2	Incoming material inspection and testing	8.4.3	• Is all the material inspected after receipt on site? • What procedure is followed to submit the material inspection report? • What inspection procedure is followed for off-site manufactured/assembled products? • Who is responsible for approving incoming material?
QC-35	Installation procedures	8.5	• Does the organization have an installation procedure? • Does the organization follow the method statement and manufacturer recommendations for installation of works?
QC-36	Product identification and traceability	8.5.2	• Does the material on site have identification labels and records for tracking?
QC-37	Identification of inspection and test status	8.5.2	• Is the inspected/tested and approved material labeled and signed?
QC-38	Control of owner supplied items	8.5.3	• What is the procedure to receive and store owner supplied items?
QC-39	Handling and storage	8.5.4	• How the organization ensure it keeps materials in safe conditions on site?
QC-40	Construction inspection, testing and commissioning	8.6	• Is the ITP (Inspection and Test Plan) approved by the client/construction supervisor? • What procedure is followed to identify and list major testing and commissioning activities?

Document No.	Document Title	Relevant Clause in 9001:2015	Typical Audit Questionnaire
QC-41	Project handover	8.6	• Does the organization maintain a log for the submission of testing and commissioning documents? • What is the procedure followed to handover the project?
QC-42.1	Control of nonconforming work	8.7	• How are nonconforming works controlled and corrective action taken?
QC-42.2	Control of nonconforming subcontractor	8.7	• How are subcontractors' nonconforming works controlled and corrective action taken?
QC-43.1	Project performance review	9.1.1	• Is there any checklist to review project performance and conformance to project quality requirements?
QC-43.2	Project quality assessment and measurement	9.1.2	• Does the organization conduct a customer satisfaction/assessment survey?
QC-44	Internal quality audits	9.2	• Is an internal audit conducted at planned intervals, results analyzed and actions taken?
QC-45	Management review	9.3	• How frequently the management review the documents to ensure that project quality plans are implemented at all the levels?
QC-46	New technology in construction	10.1	• How does the organization introduce new technology in construction, if there is such provision in the contract?
QC-47.1	Nonconformity and corrective action	10.2.2	• Who is responsible to take actions on nonconformities and take corrective action?
QC-47.2	Preventive action	10.2.2	• What action is taken to prevent occurrence of non-approval/rejection of executed works, material, shop drawings, core staff and subcontractors?

(*Continued*)

Document No.	Document Title	Relevant Clause in 9001:2015	Typical Audit Questionnaire
QC-48	Control of client complaints	10.2.2	• How are the potential causes of nonconformance monitored and preventive actions taken? • Is there any register maintained to record owner/client, construction supervisor comments and comments from regulatory authority and the action taken?
QC-49	Continual improvement	10.3	• Is there any procedure for continual improvements?
Annexure A2.4 (Additional documents for Design-Build Contracting)			
QC-10.4	Project risk management (tendering)	6.1	• What are the major risk factors considered while preparing the tender for the project? • Is there any list of major risk factors that may affect the tender/quotation? • How are risk management actions planned to address the risks to avoid, eliminate or reduce?
QC-10.5	Project risk management (design)	6.1	• What are the major risk factors considered while preparing the design proposal for the Design-Build type of project? • Is there any list of major risk factors that may affect the design proposal? • How are risk management actions planned to address the risks to avoid, eliminate or reduce?
QC-10.6	Project risk management (construction)	6.1	• What are major risks factors that may affect the Design-Build construction process? • Is a risk register maintained at the project site and updated whenever a risk is identified? • How are risk management actions planned to address the risks to avoid, eliminate or reduce?

Document No.	Document Title	Relevant Clause in 9001:2015	Typical Audit Questionnaire
QC-12.3	Change management (design changes)	6.3	• Is there any process to submit the variation/ additional fees claim for changes in the contract during the development of design if the concept design does not support proceeding to the schematic design development?
QC-13.3	Design resources (design team, design software)	7.1	• Does the organization have its own (in-house) design team or subcontractor? • How are the resources for the design development established? • How does the organization ensure that the office equipment is equipped with the latest technological standards and installed with approved software for design development?
QC-18.4	Competence	7.2	• Does the organization possess the necessary competency to design the Design-Build type of projects as per contract specifications? • Does the organization possess the competency to develop sustainable projects? • Does the organization have competent professionals with relevant certification to work as design team members?
QC-25.1	Planning of engineering design and quality plan	8.1/8.3	• How are the design quality requirements for the Design-Build type of project established?
QC-25.2	Design development	8.1/8.3	• Is there a standard procedure for the development of project design for the Design-Build type of project delivery system?

(Continued)

Document No.	Document Title	Relevant Clause in 9001:2015	Typical Audit Questionnaire
QC-25.3	Records of design development inputs	8.3.3	• How are project specific quality requirements implemented? • How are the owner requirements collected based on contracted concept design and specifications? • How is the data/ information collected to develop the design? • How are the codes, standards and regulatory requirements determined? • How is the record of design input maintained?
QC-41.1	Release of project design	8.6	• What is the procedure to release the designs and construction documents for simultaneous construction/execution works? • What is the procedure to review and verify documents for compliance with contract documents?
QC-43.3	Control of nonconforming work (design errors)	8.7	• How are design errors/ omissions detected and controlled?
QC-44.3	Client/Owner/ Project Manager/ Construction Manager Comments	9.1	• How are client/owner/ project manager/ construction manager comments addressed? • Who is responsible for taking action on client/ owner comments?

Appendix D

Typical auditing questionnaire for quality manual documents – manufacturer

Document No.	Document Title	Relevant Clause in 9001:2015	Typical Audit Questionnaire
QC-0	Circulation list		
QC-00	Records of revision and ownership documents		
QC-01	Scope		
QC-02	Normative references		
QC-03	Terms and references		
QC-1.1	Understanding the organization and its context	4.1	• How were the strategic and business plans of organization established? • How was the mission and vision statement developed? • Is there any strategic policy register maintained?
QC-1.2	Monitoring and review of internal and external issues	4.1	• How does the organization monitor external influence on the organization's policy? • How does the organization recognize and review internal issues including that of employees? • Does the organization maintain a list of internal and external issues?
QC-1.3	Review of QMS program	4.1	• Are there regular meetings conducted to discuss, review and analyze the organization's quality management program?

(*Continued*)

Document No.	Document Title	Relevant Clause in 9001:2015	Typical Audit Questionnaire
QC-2.1	Relevant requirements of stakeholders	4.2	• Does the organization identify and maintain stakeholder lists? • How are the customer requirements communicated to the stakeholders? • How are the customer's quality requirements developed, compiled and complied with?
QC-2.2	Monitoring and review of stakeholders' information	4.2	• How are the customer's, purchaser's and other stakeholders' satisfactions taken into consideration? • Is there any customer satisfaction survey carried out periodically or at the completion of supply or ordered product/material? • Is there any survey conducted to measure the satisfaction of the supplier/ service provider? • How are the applicable requirements of regulatory, statutory authority monitored and protected?
QC-3	Scope of quality management system	4.3/4.4	• How is the quality requirement of the product/material/system established? • Is there an organizational quality management system covering the quality requirement of products/materials/ systems used in the projects in line with ISO standards?
QC-4	Project quality management system	4.4	• How are the quality management systems for each product/system documented?

Document No.	Document Title	Relevant Clause in 9001:2015	Typical Audit Questionnaire
QC-5	Management responsibilities	5.1	• Is the contractual quality management system merged into the project quality management system? • Is the management regularly involved in maintaining the corporate quality management system? • Is the management involved in maintaining product quality and its commitment towards customer satisfaction/ compliance with contract documents and specifications?
QC-6	Customer focus	5.1.2	• How does the organization ensure that product quality complies with project specifications and requirements? • How does the organization measure customer satisfaction? • What are the steps taken to enhance customer satisfaction? • How are the risks related to customer quality requirements/satisfaction addressed? • How does the organization ensure that the product/system complies with the regulatory authority's requirements?
QC-7.1	Quality policy (organization)	5.2	• How does the management establish a quality policy that meets the required specifications and satisfies the customer needs?

(*Continued*)

Document No.	Document Title	Relevant Clause in 9001:2015	Typical Audit Questionnaire
QC-7.2	Quality policy (product supply)	5.2.2	• Does the policy meet statutory and regulatory requirements, legislation and applicable laws? • Is the policy committed to continual improvement and reviewed frequently? • How does the management establish a product/material/system quality policy that meets and satisfies the specification/customer requirements? • Does the management establish a product/material/system quality policy based on the specific requirements of the project and standards? • How is the product/material/system quality policy documented and communicated to related stakeholders? • Does the policy meet statutory and regulatory requirements, legislation and applicable laws? • How is the vendor evaluation carried out? What are the parameters and frequency of such evaluations?
QC-8	Organizational roles, responsibilities, and authorities (organization chart)	5.3	• Is there an organization chart, with names of production, quality management staff that specifies the responsibilities and authorities of each member?
QC-9	Preparation and control of quality plan	6.0	• What quality requirements and other issues are considered while preparing the quality plan and how they are implemented and controlled at the production and assembly line?

Document No.	Document Title	Relevant Clause in 9001:2015	Typical Audit Questionnaire
QC-10.1	Risk management (proposal/quotation submission)	6.1	• Is there any process sheet? • Is there any method statement? • What are the major risk factors considered while preparing the quotation/ proposal for the product/ material/system? • Is there any list of major risk factors that may affect the quotation/ proposal? • Are identified risks reviewed qualitatively and quantified and resolved? • How are risk management actions planned to address the risks to avoid, eliminate or reduce?
QC-10.2	Risk management (manufacturing process)	6.1	• Is there any list of major risk factors that may affect the manufacturing and assembling process? • Is a risk register maintained at the manufacturing/assembly plant and updated whenever a risk is identified? • How are risk management actions planned to address the risks to avoid, eliminate or reduce? • Is there any periodic risk review to assess the effectiveness of risk responses?
QC-10.3	Risk management (selection of vendor, supplier)	6.1	• What are the major risk factors considered while selecting the vendor/ supplier for the supply of parts/components? • Is there any list of major risk factors that may affect the selection of vendor/ supplier?

(Continued)

Document No.	Document Title	Relevant Clause in 9001:2015	Typical Audit Questionnaire
QC-11	Quality objectives	6.2	• How are risk management actions planned to address the risks to avoid, eliminate or reduce? • How are the product/material/system quality objectives established? • How does the organization ensure that the quality objectives are SMART?
QC-12.1	Change management	6.3	• How are the specification changes implements to produce the product/material/system?
QC-13.1	Office resources (office staff, office equipment)	7.1	• How are the resources (office equipment) for the office as well as for the plant established? • How is it ensured that the office equipment is equipped with latest technological standards and installed with approved software for design and production and assembly of the product/material/system?
QC-13.2	Plant resources (human resources, equipment and machinery)	7.1	• How are the resources (manpower) for the manufacturing and assembly plant established? • How are the resources (equipment and machinery) for the construction process/phase established?
QC-14.1	Infrastructure	7.1.3	• How does the organization maintain the infrastructure for the proper and smooth function of the works at the office as well as at the manufacturing and assembly plant?

Document No.	Document Title	Relevant Clause in 9001:2015	Typical Audit Questionnaire
QC-14.2	Work environment	7.1.4	• How does the organization determine that adequate facilities are available for safe and comfortable working at the plant? • How are the hygienic and HSE requirements for the plant considered and met? • Is the work environment congenial to work?
QC-15	Control of processes, material, measuring and test equipment	7.1.5	• How does the organization monitor that the plant equipment and machinery required are suitable for efficient production and assembly? • How does the organization monitor the proper utilization of equipment and machinery? • How does the organization ascertain that testing and measuring equipment are of the latest technology and have valid calibration? • Is there any plan for preventive maintenance? • How is the calibration of equipment carried out? • Is the calibration certificate traceable with an approved national/ international calibration/ testing agency?
QC-16	Control of human resources	7.1.6	• How is the manufacturing and assembly process controlled for optimum utilization of manpower?

(*Continued*)

Document No.	Document Title	Relevant Clause in 9001:2015	Typical Audit Questionnaire
QC-17	Organizational knowledge	7.1.6	• How does the organization determine that the organization has the necessary knowledge for manufacturing/ assembly of product/ material/system as per specific specifications to meet and satisfy the customers?
QC-18.1	Training and development in quality system	7.2	• Does the organization conduct training for its staff and plant members?
QC-18.2	Training in quality auditing	7.2	• Does the organization have a trained auditor/ lead auditor and provide training in auditing?
QC-18.3	Training in operational/ technical skills	7.2	• Does the organization conduct training to achieve skills and competency in manufacturing and assembly of the product/ material/system as per specific specifications? • Does the organization conduct training for its workforce to achieve skills and competency in their trade? • Does the organization develop training needs to improve skill of the people?
QC-18.4	Records of training, skills, experience and qualifications	7.2	• Does the organization maintain records of training conducted and identify and assess training needs?
QC-19	Communication internal and external	7.4	• How does the organization determine the communication method both internal and external? • Are there periodic meetings to appraise new developments?

Document No.	Document Title	Relevant Clause in 9001:2015	Typical Audit Questionnaire
QC-20.1	Control of documents for general application	7.5.2/3	• How are general documents controlled and archived? • Who is responsible for reviewing and approving quality system documents prior to use? • How are revisions and amendments controlled?
QC-20.2	Control of documents for specific product/ material/system	7.5.2/3	• How are documents related to a specific product/material/system controlled and archived? • Who is responsible for reviewing and approving specific product/material/ system quality system documents prior to use?
QC-21	Records updates	7.5.2	• Are document records regularly updated? • How are revisions and amendments controlled?
QC -22	Control of quality records	7.5.3	• How does the organization control the quality records?
QC-23	Process planning and control	8.1	• Does the organization have a process that conforms to the requisite specifications?
QC-24	Customer specific requirements	8.2	• How are the product/ material/system with specific requirements manufactured/ assembled?
QC-25	Customer specific quality control plan	8.2	• How does the organization develop a quality plan for manufacturing and assembly of a specific product/material/system
QC-26.1	Tender (quotation) documents	8.2	• How does the organization get pre-qualified/registered for bidding and tendering (submission of quotation)? • Is there any fee to collect the tender documents?

(*Continued*)

Document No.	Document Title	Relevant Clause in 9001:2015	Typical Audit Questionnaire
QC-26.2	Tender review	8.2	• Who is responsible for reviewing the tender (quotation) documents? • How is the quotation prepared? • Who approves the quotation/proposal? • Is there any authorization matrix?
QC-26.3	Contract review	8.2.1	• What is the procedure to review the contract/ purchase order prior to signing? • How are discrepancies resolved?
QC-27	Manufacturing processes	8.2.2	• Does the organization have a manufacturing process conforming to international standards?
QC-28	Engineering drawings	8.3	• What is the procedure to prepare engineering drawings for manufacturing and assembly of the product/ material/system? • How are the drawings reviewed against requirements? • How are interdisciplinary checks carried out?
QC-29	Selection and evaluation of vendors and suppliers	8.4	• Does the organization have a selection (bidding & tendering) process for vendors and suppliers? • Does the organization maintain a list of vendors and suppliers?
QC-30	Communication with material suppliers and vendors	8.4.3	• What method is used to communicate with the vendors and suppliers?
QC-31.1	Inspection of outsourced work	8.4.2	• What procedure is followed for inspection and approval of outsourced items?
QC-31.2	Incoming material inspection and testing	8.4.3	• Is all the material inspected after being received at the plant?

Document No.	Document Title	Relevant Clause in 9001:2015	Typical Audit Questionnaire
QC-32	Fabrication/ assembling procedures	8.5	• Does the organization have a standard procedure for fabrication/assembly?
QC-33	Product identification and traceability	8.5.2	• Does the product/ material/system have identification labels and records for tracking
QC-34	Identification of inspection and test status	8.5.2	• Is the inspected/tested and approved product/ material/system labeled and signed?
QC-35	Handling and storage	8.5.4	• Is there separate storage facility for finished goods and raw materials?
QC-36	Product inspection and testing	8.6	• What is the procedure for product/material/system inspection and testing? • Does the organization use ansampling method for inspection and testing? • How is inspection monitored? • What is the system of product being released upon completion of inspection?
QC-37	Product packing and dispatch	8.6	• What type of packing material is used? • Is the product/material/ system directly dispatched from the assembly line or from a dispatch store?
QC-38.1	Control of nonconforming work	8.7	• How are nonconforming works controlled and corrective action taken?
QC-38.2	Control of nonconforming work	8.7	• How are the vendor's and supplier's nonconforming works controlled and corrective action taken?
QC-39.1	Production performance review	9.1.1	• Is there any checklist to review product/material/ system performance and conformance to the specification requirements?

(*Continued*)

Document No.	Document Title	Relevant Clause in 9001:2015	Typical Audit Questionnaire
QC-39.2	Product quality assessment and measurement	9.1.2	• Who is responsible for reviewing and approving the product/material/system? • Does the organization conduct a customer satisfaction/assessment survey?
QC-40	Internal quality audits	9.2	• Is an internal audit conducted at planned intervals, results analyzed and actions taken?
QC-41	Management review	9.3	• How frequently does the management review the documents to ensure that product/material/system quality is maintained and implemented?
QC-42	New technology in manufacturing	10.1	• Does the organization have a R&D department to introduce new technology and new products?
QC-43.1	Nonconformity and corrective action	10.2.2	• Who is responsible for taking action on nonconformities and taking corrective actions?
QC-43.2	Preventive action	10.3	• What action is taken to prevent the occurrence of rejection of a manufactured/assembled product/material/system?
QC-44	Control of client/customer complaints	10.2.2	• Is there any register maintained to record client/customer complaints?
QC-45	Continual improvement	10.3	• Is there any procedure for continual improvements?

3 Auditing standards for construction projects

Mustafa Shraim

3.1 Introduction

Given the size of investments and the increasingly demanding schedule for construction projects, companies are competing based on cost (budget) and completing projects on time (schedule). As a result, a structured quality management system (QMS), such as ISO 9001-based, to standardize processes, reduce waste and improve internal processes is needed. Additionally, this system has provisions in place to verify the performance over time through audits as well as customer satisfaction data.

For an organization to be compliant with or certified to the ISO 9001 standard, it must conduct internal audits at planned intervals. The aim of these audits is twofold: first is to ensure that the QMS conforms to the organizational requirements as well as all applicable ISO 9001 requirements. The second is to ensure that the QMS is effectively implemented throughout the organization. This means that the system is producing the outputs sought or called for in the documentation. For example, the ISO 9001 standard requires that the organization evaluate its suppliers (external providers). The organization may have additional requirements – such as different criteria for different types of suppliers. Subcontractors, for example, may be qualified and evaluated differently than service providers. In any case, the qualification and evaluation criteria must be documented appropriately in the QMS (i.e. in the QMS manual or an operating procedure). For implementing the system effectively, the QMS must show, for example, supplier evaluation reports or supplier audit reports.

3.1.1 Management system standards – auditing requirements

The internal audit section (9.2) in the ISO 9001:2015 standard starts with requiring any organization seeking compliance with the standard to conduct internal audits "at planned intervals." In other words, the organization must plan its audits and do so on a regular basis. However, the exact frequency of performing these audits is not specified and left for the individual organization to decide. Based on the authors' experience, most companies tend to audit their entire QMS every year with fewer opting to do so on either a higher or lower frequency. This doesn't mean that the frequency could not change over time. An organization may start with an annual frequency and switch to a semiannual basis after a few years. They may also audit certain areas more frequently than others. For example, it is not uncommon to audit areas such as "control of nonconforming output" or "corrective action" at a higher frequency than "purchasing," for example.

As indicated above, the ISO 9001:9015 standard further explains that the objective of such audits is to provide information on whether the QMS conforms to requirements. These requirements could originate from:

- The ISO 9001 standard itself
- Other requirements not mentioned in the standard but essential for operations
- Applicable regulatory and statutory requirements.

Conformance to requirements is an indication that the QMS is comprised of documented processes to address them adequately. For example, handling incidents of nonconformance generated internally may be addressed in a form of general procedure or a combination of documented information. Such a procedure or other documented information must cover all aspects for handling a nonconforming product (or service) including identification, containment and disposition. Furthermore, a procedure that includes provisions to handle requirements must be communicated to all affected personnel. The communication may take the form of official training on how to handle nonconforming output.

Having provisions such as procedures in place does not mean that things get done or get done correctly. How do we know that affected employees are following established processes? In the nonconforming output example, how can we know that the responsible associate is identifying nonconforming product or service? As indicated in the last part of Section 9.2.1 of the ISO 9001 standard, the internal audits are also there to ensure that the provisions established through procedures, instructions, etc., are implemented effectively and maintained. This can be verified through records, observing work being done and interviews with employees. For nonconforming output, records include reports of disposition and status tags attached to nonconforming product or output. In a manufacturing environment, an area for nonconforming or suspect products is often created to place the suspect material until disposition. In construction projects, not using the correct materials by a contractor to complete a task on the project schedule can be viewed as a case of nonconformance. Dealing with such a nonconformance may take the form of informing the client and making the corrections. It may also require a corrective action if it is a repeated problem.

3.2 ISO standards for auditing

While some form of internal auditing is included as a requirement in each contractual management system standard, no guidelines within these standards are given on how to set up and implement audit programs. That is, standards like ISO 9001,

ISO 14001 and ISO 13485 require that the organization perform internal audits on a regular basis with slight differences. However, the sections stipulating such requirements do not elaborate on how to set up such programs. Thus, guidelines on auditing management systems were needed.

The International Organization for Standardization first published guidelines on auditing management systems in 2002. These guidelines were only applicable to quality and environmental management systems. As the number of management system standards grew over the years including, but not limited to, ISO 13485 (medical devices) and ISO 22000 (food safety), broader guidelines for auditing were needed. Most of these standards have a common structure and similar base requirements. As a result, ISO 19011 guidelines were updated in 2011 and then in 2018 to be applicable to all management system standards.

3.2.1 ISO 19011:2018 – Guidelines for Auditing Management Systems

The latest revision of ISO 19011 published in the summer of 2018 provides guidelines to plan and conduct audits for any management system. Though the title of ISO 19011 is *Guidelines for Auditing Management Systems,* it is not a contractual standard. This means that these are just guidelines and not requirements for any type of certification. However, most auditor training programs, from internal to third party, use this standard as the basis of their body of knowledge.

The ISO 19011 standard can be used by all sizes and types of organization and for any types of audit. However, it does concentrate on internal auditing and second party audits. That is, organizations can use this standard to set up their audit programs for their organizations.

ISO 19011 is based on several clauses starting with the following three informational ones: scope of the standard including applicability issues, normative references and definitions of key terms used in the body of the standard. The rest of the clauses (4 through 7) include the actual guidelines for auditing management systems:

- Principles of Auditing
- Managing Audit Programs
- Conducting Audits
- Auditor Competence and Evaluation

Principles of auditing are considered the pillars for making audits effective and reliable. When adhered to, these principles ensure that similar conclusions can be reached by a number of auditors working independently of each other and auditing the same process or system. At a glance, these principles seem logical and should be easy to adhere to. However, and in practice, audit programs may turn into a paperwork exercise if auditees do not cooperate for one reason or another. Therefore, when auditors receive their training, examples and scenarios of how such principles could be compromised should be worked on.

The first of the auditing principles is *integrity*. The Oxford dictionary defines integrity as "The quality of being honest and having strong moral principles" and "The condition of being unified or sound in construction." For auditors, this means that they should do their work ethically – following the code of ethics presented by the granting entity. For the American Society of Quality, which grants certification for Quality Auditors, certified auditors must adhere to the code of ethics, besides being the right thing to do in the first place.

Auditors should also report on findings truthfully and accurately without omitting significant findings or exaggerating others. This should be done with due diligence and the ability to make decisions in all types of auditing situations.

Confidentiality is another principle that many organizations are increasingly sensitive to. The Oxford dictionary defines confidentiality as "The state of keeping or being kept secret or private." It is common for many organizations to require signing a confidentiality agreement before accessibility to information is granted. Auditors are exposed to all types of company information which must be protected. This includes propriety, security-related or personnel information, among others. If auditors need information to examine later at an off-site location, they usually require permission to do so, in addition to the confidentiality agreement.

One of the fundamental principles of auditing is *independence*. That is, the auditor must be independent of the function or process being audited. The reason for this is obvious; when the rule of independence is violated, impartiality and objectivity of results may be compromised. For example, someone from procurement or the purchasing department should not audit the process of qualification and evaluation of external providers. The question is *how independent should an auditor be?* This question comes up often particularly when there are many functions within a department. In the procurement example, what if the auditor does not directly work with external providers (suppliers or contractors)? Does that mean he or she is independent? The answer to this question depends on the level of involvement and who you ask. However, a good rule of thumb is not to audit functions within the same department or under the same manager. It is much harder to do so in smaller organizations where one employee could be responsible for multiple functions. In such cases, management should make every effort to ensure objectivity, including subcontracting audits to an external source if necessary.

Another principle of importance is being *fact-based with objective evidence.* Audit findings should not be based on opinions and the facts stated must verifiable. This means that auditors should reference the samples taken with enough information that, if needed, someone can return to it later and find such evidence. Additionally, auditors should take enough samples within the period of interest, including current ones, to ensure reliable conclusions are reached.

Since contractual management standards, such as ISO 9001, now include risk-based thinking, ISO 19011 also includes reference to risks and opportunities under its principles. It states that "the risk-based approach should substantively influence the planning, conducting, and reporting of audits in order to ensure that audits are focused on matters that are significant to the client."

ISO 19011 also addresses *management of audit programs*. According to the standard, an audit program is defined as "arrangements for a set of one or more audits planned for a specific time frame and directed towards a specific purpose." In practice, audit programs are established and maintained for ongoing management system audits. One audit program may be established for one or more management systems such as environmental and quality. In any case, the extent of an audit program depends on the size and complexity of the organization. For example, an organization with multiple sites and outsourcing vendors or contractors requires more details and resources than a simple manufacturing process. Auditing a multi-site construction project with vendors, consultants and contractors is an example of higher complexity. For this, the audit program includes auditing both internal processes as well as external providers (e.g. contractors). Priorities should be given to processes and entities that pose higher risks for achieving organizational and project objectives.

Managing audit programs involve establishing audit objectives, evaluating risks, identifying responsibilities for those involved, selecting auditors, managing program records, and evaluating and improving the auditing process as well as the overall audit program. The first step in establishing an audit program is establishing its aims. This will help in determining required resources including auditors. The following questions can be asked to help determine the overall objectives:

- What are the needs and expectations of clients/customers?
- What are the needs and expectations of other interested parties including employees (e.g., internal auditors), external providers (e.g., contractors), among others?
- What are the competencies needed for those managing the audit program?
- What are the current issues and risks with the processes being audited?
- What are the results of previous audits conducted on each of the processes?

The objectives for an audit program will vary from organization to organization. For a construction/engineering organization, the objectives could include the following:

- Identifying current shortcomings of different processes (nonconformance) regarding meeting quality management system requirements as well as contract requirements
- Identifying opportunities for improvement
- Identifying capabilities of established processes and external providers. In other words, are the processes capable of delivering to customer requirements consistently? In construction projects cost and time represented by budget and schedule, respectively, are key attributes for successful project completion
- Complying with legal and statutory requirements
- Aligning project objectives with overall organizational objectives

The program should incorporate and monitor these objectives over time to assess the role of auditing in achieving them. After all, the audit program is a part of process performance evaluation, so countermeasures can be implemented, and improvements can be made.

Another important aspect of establishing an audit program is the selection of auditors. In addition to including an appropriate number of auditors to cover the scope, it is essential to consider factors such as competence and knowledge in the processes being audited and availability. In most cases, internal QMS auditors have other functions within the organization limiting their availability. Some organizations supplement their internal audit team with subcontracted external auditors.

Generally, organizations certified to ISO 9001 and/or other management system standards have a procedure for their QMS audit program. Such a procedure covers all aspects of *conducting audits*, including:

- Auditor Selection
- Audit Planning
- Audit Scheduling
- Conducting Audits
- Audit Reporting
- Audit Follow-ups

Table 3.1 is an example of an internal audit procedure.

The process flow for managing an audit program is displayed in Figure 3.1 where it is illustrated using the Plan-Do-Check-Act (PDCA) cycle. Clauses 5 and 6 from the standard are incorporated into this flow diagram.

Table 3.1 Typical procedure for QMS audits

Purpose & Scope
This procedure describes the process of planning, conducting and reporting internal quality audits.
Responsibilities
The Program Manager is responsible for ensuring that internal audits are conducted according to schedule and for reporting to management on the status of the quality management system as a result of internal audits. Qualified auditors will conduct audits and report on findings.
Procedure
1   Auditor selection:
      Auditor must be selected by the Program Manager according to the following criteria:
        a Training: each auditor must have had responsible training in auditing. This training can be provided internally or externally.
        b Independence: an auditor cannot audit his/her immediate areas of work. Auditors can also be subcontracted from an outside source.
        c Knowledge: auditors must be scheduled based on their knowledge and experience in the function being audited.

---

2   Audit planning:
    Audits must be planned so that the entire quality system is audited each year. Items that indicated problems in prior audits or other nonconformities should be included in current audits. The Program Manager should inform auditors of items to be included as part of their checklists or provide them with applicable checklists.
3   Audit scheduling:
    The Program Manager prepare audit schedule quarterly. Each audit schedule must at least include:
    a Area to be audited
    b Audit dates
    c Auditors name
    d Special instruction, if any.
4   Audit findings:
    Auditors perform audits through questioning, observing and verifying records. The following findings must be recorded:
    a Area for Improvement: this occurs when a system shows some weakness but not necessarily nonconformance to establish requirement. These weaknesses may lead to nonconformance if not treated.
    b Minor Nonconformance: this occurs when there is only a lapse on the system given that that requirement shows conformance most of the time.
    c Major Nonconformance: this occurs when there is a breakdown in the system. It is also possible that few minors within the same area at one time could constitute a major.
5   Audit reporting:
    a The auditee must acknowledge the findings by signing the specified space on the Internal Audit Report. If there is a conflict, the Program Manager should resolve it.
    b A copy of the audit sheet should be given to the auditee after getting the signature.
    c The Program Manager issues formal corrective action requests to the auditee for any nonconformance immediately by using the Nonconformance Database.
    d The areas of improvements should be left for the auditee to decide on when to start such a process of improvement. Records for such improvements must be kept with the auditee for future reference.
    e The Program Manager provides management with reports on status of the quality management system regarding number of nonconformances, frequencies, trends, etc., and corrective action taken.
    f Nonconformance and areas for improvement uncovered may be included as items in subsequent audits.

Records   Audit schedules
              Audit reports

Before conducting audits, it is imperative for the audit program manager to consider planning factors such as auditee availability, audit feasibility and resources, among others. Table 3.2 can be used as a checklist for this purpose. Once a checklist is completed, an action list could be prepared for items not addressed or completed.

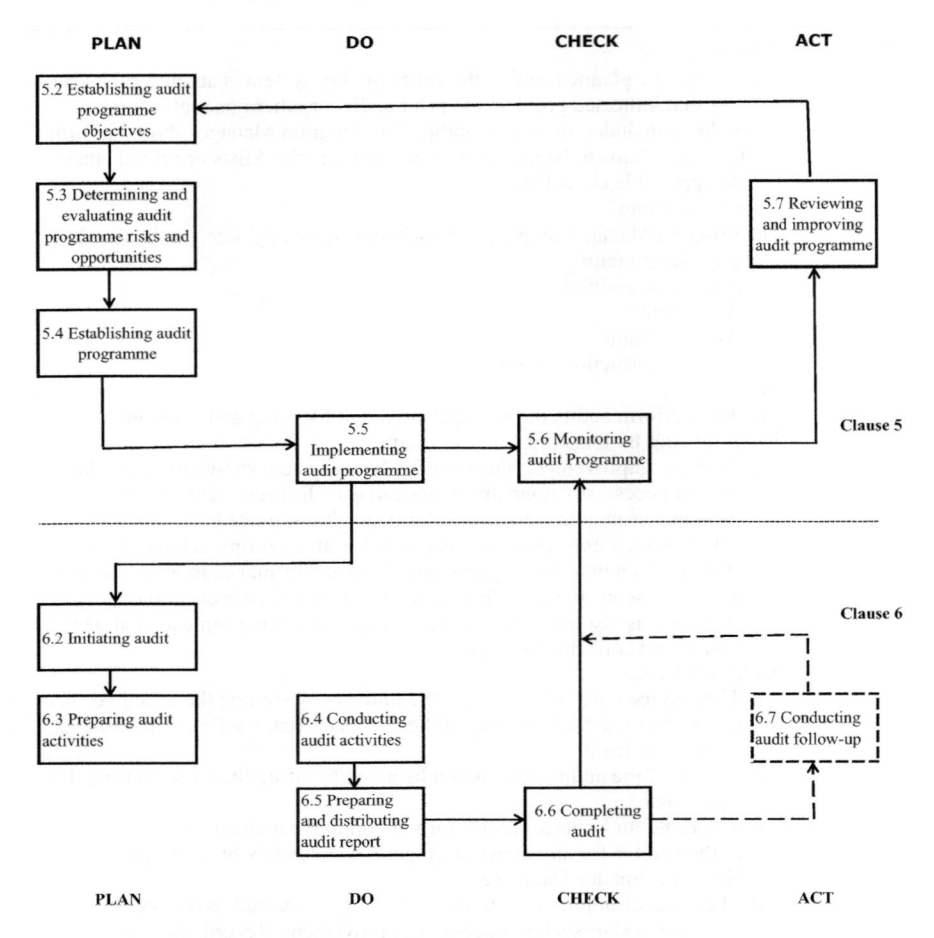

Figure 3.1 Process flow for managing audit programs

Source: ASQ/ANSI/ISO 19011:2018 *Guidelines for Auditing Management Systems* (2018), ASQ Quality Press, Milwaukee, WI, USA. Reprinted with permission from ASQ

Table 3.2 Checklist for audit preparation

Serial Number	Preparation Item	Y/N/NA
1	Auditee contacted with details?	
2	Auditee confirmed audit time(s) and dates?	
3	Guides can be available if needed?	
4	Auditee confirmed opening and closing meeting times and dates?	
5	Auditee communicated back confidentiality requirements?	

Serial Number	Preparation Item	Y/N/NA
6	Auditee communicated back safety and health requirements?	
7	Auditor independent of the process being audited?	
8	Auditor is knowledgeable of process being audited?	
9	Criteria for audits (procedures, contracts and standards) are available for review	
10	Adequate time is allowed?	
11	Other resources (e.g. computer access) are available to perform the audit?	
12	Risk audit poses on audit process, if any, are considered?	
13	Sampling process of objective evidence is determined?	
14	Working language of the auditee is manageable by the auditor?	

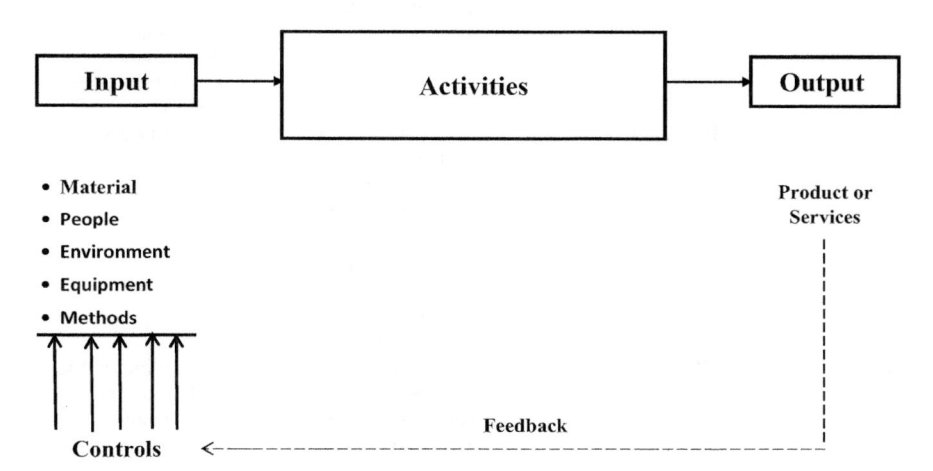

Input → Activities → Output

- Material
- People
- Environment
- Equipment
- Methods

Controls

Product or Services

Feedback

Figure 3.2 Process approach in auditing

Audits should be conducted using the process approach and not just a predictable checklist looking for the same items every time. The process approach demands that we look at the process as a whole with all of its component and not in "silos." As shown in Figure 3.2 (process-based audits), auditors verify the control mechanisms for all components of the process. Taking the activities of a construction project as an example, the following may be considered:

- Material/information: this includes all the raw materials needed where the control mechanism is the identification of materials and supplier control
- Equipment where the control mechanism is preventive maintenance processing equipment and calibration activities for the measuring equipment such as those used for surveying

- Employees where the control is training and perhaps certification for certain skills
- Methods where the control is document control manifested in the revisions of procedures, instructions and drawings
- Environment where the control could be adherence to safety requirements

Table 3.3 Auditor evaluation methods

Serial Number	Evaluation method	Objectives	Examples
1	Review of records	To verify background of the auditor	Analysis of records of education, training, employment, professionals
2	Feedback	To provide information about how the performance of the auditor is perceived	Surveys, questionnaires, personal references, testimonials, complaints, performances, performance evaluation, peer review
3	Interview	To evaluate desired professional behavior and communication skills, to verify information and test knowledge and to acquire additional information	Personal interviews
4	Observation	To evaluate desired professional behavior and the ability to apply knowledge and skills	Role playing, witness audits, on-the-job performance
5	Testing	To evaluate desired behavior and knowledge and skills and their application	Oral and written exams, psychometric testing
6	Post-audit review	To provide information on the auditor performance during the audit activities, identify strengths and opportunities for improvement	Review of the audit report, interviews with the audit team leader, the audit team and, if appropriate, feedback from the auditee

Source: ASQ/ANSI/ISO 9011:2018, *Guidelines for Auditing Management Systems*, (2018), ASQ Quality Press, Milwaukee, WI, USA. Reprinted with permission from ASQ

When conducting audits, records of controls are verified by the auditor, in addition to interviews and observation of the work being performed.

The last of the guidelines addressed by the ISO 19011 standard is *the competence and evaluation of auditors*. These guidelines provide details on factors that should be considered in selecting an audit team. To do so, the standard suggests a variety of methods for evaluating potential auditors and is summarized in Table 3.3. Generally, the program manager may use the following steps in selecting auditors:

1 Identify competence needs: this includes both auditing skills and knowledge about processes being audited. So competence would include personal behavior attributes as well as skills and knowledge. Many organizations select potential internal auditors based on interest and personal behavior, among other factors. For skills and knowledge, appropriate training is provided.
2 Prepare evaluation criteria to make the selection. The criteria could be selecting from different departments (distribution), years of experience, skills and knowledge, and personal behavior, among others.
3 Us the proper evaluation methods: this should include at least two from Table 3.3 as recommended by the standard.
4 Conduct the evaluation, keeping in mind that some may need additional training before being selected.
5 Maintain current auditors' list and provide continual opportunities to for skill improvement and professional development.

One of the new aspects of auditing in the ISO 19011:2018 standard is the inclusion of provisions for auditing from a remote location. In addition to traditional on-site methods, remote audit methods through means such as video conferencing include:

- Interviews
- Observing work being completed
- Conducting document reviews with and without auditee involvement
- Analyzing data presented remotely by the auditee

4 Auditing fundamentals

Shirine L. Mafi

4.1 Introduction

In today's global competitive environment, and with an increase in customer demand for higher performance requirements, organizations are facing many challenges. They are finding that their survival in the competitive market is increasingly in doubt. To achieve competitive advantage, effective quality improvement is critical for an organization's growth. This can be achieved through the development of long-range strategies for quality. The assessment of existing quality systems which are being implemented provides the factual information to developing long-term organizational strategies for quality.

The development and implementation of any system needs adequate assessment. Audits are used to determine the effectiveness of the system and to aim at exposing any pain points in order to improve them. According to the ASQ Guidelines for Auditing Management Systems (2018), an audit can be conducted against a range of audit criteria, separately or in combination, including but not limited to:

- Requirements defined in one or more management systems standards
- Policies and requirements specified by relevant interested parties
- Statutory and regulatory requirements
- Management system plans relating to the provision of specific outputs of a management system (e.g. quality plan, project plan)

Figure 4.1 illustrates the same.

This chapter provides guidance for all sizes and types of organizations and audits of varying scopes and scales, including those completed by large audit teams and those by small teams or single auditors, whether in large or small organizations, and whether for service or manufacturing entities.

A quality audit is defined as a systematic and independent examination to determine whether quality activities and their outcomes comply with the plan and whether the plan is achieving its goals. Quality audits in the construction industry are used in two broad categories. The first is auditing the quality of workmanship such as code compliance. The second category pertains to assessment of the

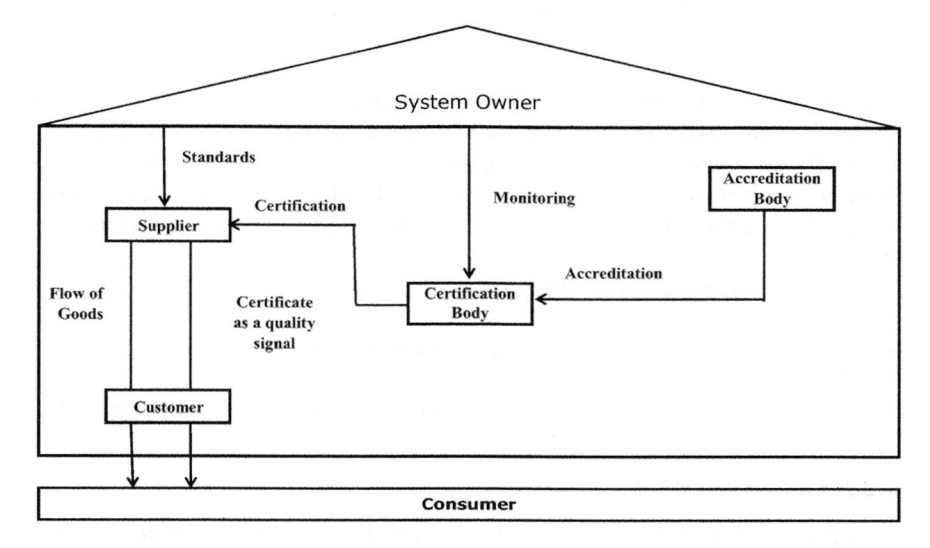

Figure 4.1 Basic structure of certification

Source: Albersmeier, F., Shulzae, H., Jahn, G,. and Spiller, A. (2009), The reliability of Third Party certification in the food chain, Checklist to Risk-oriented Auditing, *Food Control* 20(10): 927–935. Reprinted with permission from Elsevier

quality management system (QMS). This is necessary to validate the system and its processes and discover any gaps that may exist.

Many quality standards stipulate that a quality system needs to be audited by third parties who are independent of the organization providing the system. An exception to this is when companies use internal auditors sometimes selected from within the organization itself to find the inadequacies prior to external auditors' assessment.

An auditor is a person or a team of people who have been trained in quality auditing and/or have experience in audit. Auditors can be from the same company (internal audit), from other independent organizations (i.e. American Society for Quality – ASQ), government agencies (Nuclear Regulatory Agency – NRA) or even auditors sent by clients who are interested in learning more about the organization's quality system and whether or not it is effective.

4.2 Categories of auditing

This element of assessment is related to the evaluation of present quality management system related activities. Assessment of present quality activities can be evaluated with two perspectives:

1   Assessment that focuses on customer satisfaction results but includes an evaluation of the present quality system

2. Assessment that focuses on evaluation of the present quality system with little emphasis on customer satisfaction results

In either case the assessment can be performed by the organization itself or by an external body. The assessment performed by the organization is known as "self-assessment," whereas an assessment performed by an external body is referred as a quality audit. A quality audit is the formal or methodical examining, reviewing and investigating of the existing system to determine whether agreed-upon requirements are being met.

Audits are mainly classified as:

- First Party – Audit one's own organization (internal audit)
- Second Party – Customer audits the supplier
- Third Party – Audits performed by an independent audit organization

4.2.1 First party auditing

First party audits are usually conducted by employees of the company who are experts on assessment. They usually are selected from other parts of the company to minimize conflict of interest. Experts from outside may be called in to conduct audits. Most frequently these are volunteer members of the local quality association who come from everywhere for 1 to 3 days to conduct audits. Those auditors have some type of quality certification to assess the system of the organization. At the conclusion of the audit, the lead auditor will present the audit findings to the leadership of the organization including the staff in charge of quality compliance.

4.2.2 Second party auditing

Second party audits are usually external audits performed by a party outside the organization. These audits are usually conducted when a prospective client is interested in assessing the quality system of the supplier prior to a purchase agreement. These audits could range from being process-specific to an overall system audit. Clients or customers could waive such audits (or limit them) if the organization is already certified to an ISO 9001 or similar management system standard.

4.2.3 Third party auditing

Third party audits are conducted by agencies or certification and/or accreditation bodies. These could be government or regulatory agencies such as the Food and Drug Administration (FDA) in the United States. They could also be auditing services through independent entities specialized in quality system certification. A key feature of a certification system is that inspections are carried out by independent bodies (third party certification) in accordance with standards laid down by external organizations.

There are many certification bodies worldwide that provide certification services for the ISO 9001 or other management system standards. If the audit is successful, then the certification body such as BSI, TUV, NQA, among others, awards a certificate to be renewed every three years through a recertification audit. Between the three-year cycles, surveillance audits are conducted annually to ensure the QMS is operating effectively. The surveillance audits are not as comprehensive as the certification or re-certification audits. If the organization being audited has multiple locations, then surveillance may be conducted on some every year so that all are audited within a certification cycle.

Sometimes firms opt for a pre-audit before the audit. Due to the high cost of third party certification, often a pre-audit is done to ensure that the company is well prepared and equipped before the major third party certification audit.

4.3 Auditing in construction projects

In the construction sector today, quality workmanship provides a competitive advantage in terms of customer satisfaction and loyalty. Quality-conscious organizations provide a safe working environment for their crews and a safe living environment for future users of their construction projects. Although construction site accidents are not as common as in developed countries, even one accident is one too many (Mafi, Huber and Shraim, 2016). The loss of human life, property, credibility and resources are often the result of poor quality control. One tragic example of human loss and suffering was experienced when the Raza Plaza collapsed in 2013 in Dakha, Bangladesh, killing 1134 people and maiming many more (*New York Times*, May 22, 2013). This accident was due to a structural failure. The crew used substandard materials, and the city officials blatantly disregarded the building codes and approved the permit for construction and occupancy.

Quality in the construction industry has a lengthy history comparable to manufacturing, but with its own unique characteristics (Arditi and Gunaydin, 1997):

- Almost all construction projects are different from each other and constructed in varying conditions
- The life cycle of a construction project is longer than the life cycle of most manufactured products
- There is no uniform standard used to evaluate overall construction quality as in manufacturing; therefore, evaluations of quality of the construction projects tend to be more subjective
- Various stakeholders in construction projects, from owner to designer, to general contractor and subcontractor, and to materials supplier, have different ideas of what quality workmanship is

Because of these complexities, the construction industry has not been able to implement quality systems as effectively as the manufacturing sector. Defect and accident prevention, however, is desirable for both sectors. The rule of quality cost is called 1–10–100, meaning it only costs one unit to detect quality issue in the design phase. It will cost 10 times more if the defect is identified during the installation stage and 100 times more if the defect is discovered after the product is installed or purchased by the customer (Sowards, 2013).

Another problem is that the costs of poor construction are not really known because defects are often not recorded (Mafi, Huber and Shraim, 2016). In order to reduce defects, construction managers need to implement a system to better manage their processes. One easy way to do this is to adopt a quality framework, such as the International Organization for Standardization (ISO) framework for planning and design, load computations, operations and maintenance.

4.3.1 Purpose

Loss of life and property makes compliance to standards of quality critical. Therefore, the purpose of auditing is to ensure the system is working as it is supposed to be working. If not, the fail points need to be discovered and be addressed. The Rana Plaza collapse could have been totally prevented had there been an audit process in place. Since audit is only looking at a portion (sample) of the system and not the entire system, it is possible that certain deficiencies will be not uncovered. So whenever deficiencies are uncovered, similar areas of concern should also be investigated using a risk-based analysis. See Chapter 3 for risk in auditing.

4.4 Auditing process

Auditors must understand that auditing a management system is auditing an organization's processes and their interactions in relation to one or more management system standards. Consistent and predictable results are achieved more effectively and efficiently when activities are understood and managed as interrelated processes that function as a coherent system. Figure 4.2 shows the relationship between system, process and product/service from the point of view of the auditing process.

Audits are intended to identify any gaps that may exist between what the organization says it is doing to what is actually being done in meeting clients' expectations. So it is a fact-finding mission. Audits can be formal or informal; however the intent is to verify the "management by fact." The expected outcomes of any audit are independent, accurate, verifiable and traceable facts and observations.

There is a lead auditor who puts together a team of auditors to conduct an audit. The lead auditor is in charge of setting the audit date and, based on input from the auditee, the scope of audit and the areas to be audited is selected. Prior to getting to the audit site, the lead auditor needs to provide the client with the names and qualifications of the auditors including current required licenses or certificates. In any audit, there is an opening meeting, the audit itself and the closing meeting. Table 4.1 gives an example of an one-day audit site visit and the areas to be audited as agreed upon by both parties.

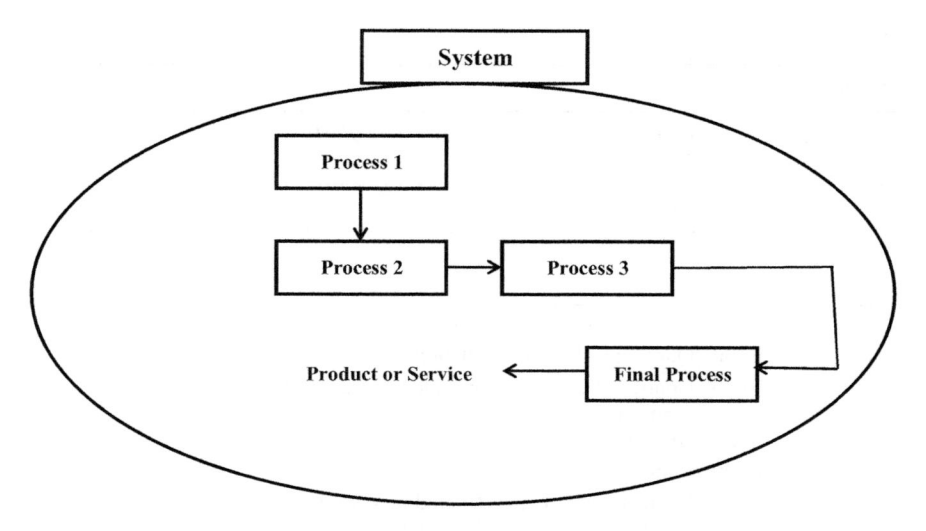

Figure 4.2 Relationship between system, process and output as defined by the audit process

Table 4.1 Audit schedule

AUDIT SCHEDULE			
Audit Date(s): (1 Day)		*Organization Name:*	*Standard: ISO 9001:2015*
		Auditor Names:	
			Auditor's Name
Facility Location	8:30	**Opening meeting:** • Review any changes in the organization or to the procedure and/or manual since the last audit • Review scope • Review number of employees • Review how the sequence and interaction of processes are identified • Any nonconformance cited during the last audit	
	8:45	**Management process** • Management review • Internal audits • Customer communication/satisfaction • Continuous improvement • Corrective actions • Risk management • Identification & expectations of interested parties	

(*Continued*)

Table 4.1 (Continued)

AUDIT SCHEDULE			
Audit Date(s): (1 Day)	Organization Name:	Standard: ISO 9001:2015	
	Auditor Names:		
			Auditor's Name
	9:30	Customer satisfaction Process	
	10:00	Quoting/order review/planning Process	
	10:30	Control of externally provided processes, products and services	
	11:30	Auditor time	
	12:00	Lunch	
	1:00	Value-stream process (including equipment maintenance and calibration)	
	3:15	Workforce competence/training	
	4:15	Auditor time	
	4:45	Closing meeting	
	5:00	Depart	

4.5 Audit tools

There are a variety of tools used in auditing, including interviews, review of documents and observation. Applying a variety and combination of different audit tools and methods can optimize the efficiency and effectiveness of the audit process and its outcome.

Interviews – when interviewing, auditors should compare answers to what is being done and/or what is being documented. The auditor should try to put the interviewee at ease by first introducing themself, explaining the purpose of the visit and avoiding being perceived as "I know everything and I am here to get you."

In interviewing, the auditor should use proper questioning techniques. Examples include *what, where, when, why, who, how* or *explain to me*, rather than *did you or didn't you?* Also it is critical that the auditor verifies what is being said to ensure the observation is accurately documented and reported later in the meeting. Once the audit is completed, the common courtesy of thanking the auditee for their time is highly recommended. It is also good practice to praise what auditors find as excellent practice to share with the management. This encourages employees to strive for excellence.

Verifying objective evidence – the auditor should be asking for samples of records including the most recent ones in order to compare these with what the system says. For instance if the quality system calls for a six-month interval for a lab equipment calibration, the auditee should be able to provide a record that shows the calibration dates. That is proving that the equipment is being calibrated according to its schedule, and there is a date stamp, and a signature attesting to this protocol. Another critical point is the auditor needs to make meticulous notes on what equipment was audited (write the serial number) and where it was audited. This helps with the traceability of the audit process in case they have to go back and reexamine the same equipment.

Observing physical evidence – the auditor should also pay attention to the physical surroundings including machinery parameters, inventory tags and how an associate is doing the job. The auditor should also include in this assessment whether the safety precautions as stated by the quality manual are being observed. For instance, if the associate is working with hazardous material, are they dressed according to the guidelines?

Some audit tools are as follows:

- Standard audit checklist of specific documents, policies, procedures, instructions, log sheets, etc.
- Client generated checklist (usually geared to their specific needs)
- Flow charts for process analysis as well as forward and backward tracing
- Audit sampling plans
- Data gathering with qualitative as well as quantitative unbiased evidence

Once the audit of each area is concluded, the auditor has to record the findings.

4.6 Audit findings

Based on the result of the audited area, an auditor will make one of the following three decisions:

1. OK – Everything is implemented according to the specified arrangement
2. Opportunity/Area for Improvement – No problems at this point but may develop into a problem if not addressed
3. Nonconformance – violating requirements. This category has two options: Minor and Major. A minor nonconformance is a lapse in the system while a major nonconformance is a breakdown in the system and a much more serious problem.

Additionally, auditors may note "good practices" that should be recognized and may be applied in similar areas.

4.7 Risks in auditing

When a statistical sample is developed, the level of sampling risk that the auditor is willing to take is an important consideration. This is often referred as the sampling risk or the probability of committing a Type 1 error. For instance a sampling risk of 5% corresponds to an acceptable confidence of 95% meaning that the auditor is willing to accept the risk that 5 out of 100 (or 1 in 20) of the samples examined will not reflect the actual values that would be seen if the entire population was examined. Therefore there is always risk inherent in the sampling process.

When statistical sampling is used, auditors should appropriately document the work performed including a description of the population that was intended to be sampled, the sampling criteria used for evaluation, the statistical parameters and methods that were utilized, the number of samples evaluated and the results obtained. Auditors need to convey that the results of the audit are based on judgment-based samples and the management should be cautious in terms of generalizing the findings. For instance if the auditor does not find a noncompliance in the surveyed sample, that does not mean that there was no compliance in the population.

4.8 Training for auditing

Certified auditors undergo an approved auditing training. Although not necessary, some auditors may have experience in the particular type of work being examined or assessed. Beside their technical skills, auditors should have good social skills. They should be a good communicator including having excellent listening abilities and patience, and being confident, objective, tactful and trustworthy. Additionally, good auditors should have the necessary knowledge, sound judgment and tenacity, keep an eye on the time allowed and have excellent planning skills.

Since the organization is allowing access to their system, the auditors have to respect the confidentiality of information shared with them and protect the auditee property entrusted to them. Auditors often may be asking for sensitive issues that may not be easy for the auditee to discuss with an auditor. Furthermore it is important for the auditing team to stay within the scope of the audit as agreed prior to the start of the audit. A sufficient number of samples should be assessed to ensure a proper conclusion based on evidence. The auditors are the guests of the auditee, therefore they have to observe all the rules regarding safety (watching safety videos prior to embarking on the audit and wearing required masks and clothing) or restricted access. Once the audit is completed and the report is being presented to the organization, the results of the audit should be traceable to the requirements under which the audit is being conducted. The auditors must at all times be independent of the function being audited. If this rule is violated, it may cause bias or, worse, suppression of results. Some undesirable characteristics of auditors are being argumentative, opinionated, easily influenced, impulsive, devious, inflexible, prescriptive and a poor planner and communicator. Table 4.2 lists the certification for auditors.

Tables 4.3 through 4.6 demonstrate different examples of actual audit reports.

Table 4.2 List of certification for quality audits

Serial Number	Certification	Description
1	ASQ Certified Quality Auditor (CQA)	The Certified Quality Auditor is a professional who understands the standards and principles of auditing and the auditing techniques of examining, questioning, evaluating and reporting to determine a quality system's adequacy and deficiencies. The Certified Quality Auditor analyses all elements of quality system and judges its degree of adherence to the criteria of industrial management and quality evaluation and control systems.
2	Certified HACCP Auditor (CHA)	The Certified HACCP Auditor (CHA) is a professional who understands the standards and principles of auditing of the hazard analysis and critical control points (HACCP) based (or process-safety) system. A HACCP auditor uses various tools and techniques to examine, evaluate and report on the system's adequacy and deficiencies. The HACCP auditor analyzes all elements of the system and reports on how well it adheres to the criteria for management and control of process safety.
3	IRCA Lead Auditor	The International Register of Certified Auditors (IRCA) is the international auditor certification body. IRCA/CQI (Chartered Quality Institute) conduct certification program for Auditor/Lead Auditor. The Lead Auditor should have full knowledge and skills to conduct a full audit of an organization's Quality Management System (QMS) to ISO 9001:2015. In order for any professional to be Lead Auditor he/she should posses certain years of experience (as per IRCA/CQI rules for certification) in conducting the audit.

Table 4.3 An example of an internal audit schedule

Internal Audit Schedule

Period	2017				
Location	ID	Process		Auditor	Target Date
	147	VSP-4 Construction Inspection and Testing			9/8/2017
	148	QSP-1 Document & Record Control			9/8/2017
	145	QSP-6 Maintenance & Calibration			6/12/2017
	144	VSP-4 Construction Inspection and Testing			6/12/2017

(*Continued*)

Table 4.3 (Continued)

Internal Audit Schedule

Period	2017				
Location	ID	Process		Auditor	Target Date
Columbus					
	154	VSP-1 Design Services			12/8/2017
	149	QSP-6 Maintenance & Calibration			12/8/2017
	150	QSP-2 Management Review			12/8/2017
	152	QSP-3 Purchasing & Supplier Control			12/8/2017

Table 4.4 An example of audit findings

Audit Result

Period	2017				
Decision	Location	ID	Process	Auditor	Completion Date
Area for Improvement		156	QSP-10 Client Feedback Management		12/15/2017
		161	QSP-8 Control of Nonconformance		12/8/2017
Nonconformance		162	QSP-9 Corrective and Preventive Action		12/15/2017
Ok		150	QSP-2 Management Review		12/8/2017
		151	QSP-5 Feasibility Review		12/15/2017
		152	QSP-3 Purchasing & Supplier Control		12/8/2017
		153	VSP-5 Information Technology		12/8/2017
		154	VSP-1 Design Services		12/8/2017

Table 4.5 An example of a nonconformance report

Internal Audits Nonconformance Report

Period	2017					
Location	ID	Process	Auditor	Summary	Completion Date	Corrective Action Requested
	162	QSP-8 Corrective and Preventive Action		**Nonconformance** Corrective Action # 34 was a complaint from INDOT. It is marked as Corrective Action	12/15/2017	√

Internal Audits Nonconformance Report

Period	2017						
Location	ID	Process	Auditor	Summary		Completion Date	Corrective Action Requested
				in the database and countermeasures have been completed. However root cause analysis was not completed. This was an office complaint and per procedure (QSP-9) root cause should have been documented. **Area for Improvement** Add 'nonconformance" to the drop down menu in the QMS Action Tracker database.			

Table 4.6 Audit report example (area for improvement)

Internal Audits
Areas for Improvement Report

Period	2017					
Location	ID	Process	Auditor	Summary	Completion Date	Corrective Action Requested
	161	QSP-4 Control of Nonconformance		For NC # 34 countermeasures were entered and on involved retraining for crews. Make sure there is evidence to show that each countermeasure was completed. For this training, there wasn't evidence readily available with the nonconformance information.	12/8/2017	☐

(Continued)

132 Shirine L. Mafi

Table 4.6 (Continued)

Internal Audits Areas for Improvement Report						
Period 2017						
Location	ID	Process	Auditor	Summary	Completion Date	Corrective Action Requested
	154	QSP-10 Client Feedback Management		Consider whether sending of cards with the last invoice of the project might solicit stronger percentage of return. Currently sending out with every invoice throughout the project.	12/8/2017	☐

Audit exercises

1. Management review minutes are the documented output of the management review. When the CEO was asked for the minutes, she indicated that she did not have them documented in one report yet. Is this a nonconformance? If so minor or major? Why or why not?
2. When the Purchasing Manager was asked about supplier evaluation, she indicated that this is done but no records are kept. Is this a nonconformance? If so, is this a minor or major finding? Why or why not?
3. A copy of a form used in Maintenance was not very clear. It had grease all over it. Is this a nonconformance? If so, is this a minor or major finding? Why or why not?
4. The Maintenance Manager ensured that all internal QMS audits conducted on the Maintenance Department were conducted by either himself or his assistant. Is this a nonconformance? If so, is this a minor or major finding? Why or why not?
5. When the Auditor asked the Maintenance Manager about the preventive maintenance record, he was able to produce 4 out of the 5 that the auditor asked for. The Maintenance Manager contended that his score was 80% which means "No Problem." The auditor was to give her response. Is this a nonconformance? If so, is this a minor or major finding? Why or why not?
6. When verifying records of product performance, you found one characteristic (dimension) to be above specs on 2 out of 5 records. The quality technician told you that the customer agreed to accept these products and that they were shipped "as is." What would your reaction or follow-up be as an auditor?

5 Auditor/auditing team selection for construction projects

5.1 Construction project auditing stakeholders

A stakeholder is anyone who has involvement, interest or impact on the construction project processes in a positive or negative way. Stakeholders play a vital role in the determination, formulation and successful implementation of project processes. Stakeholders can mainly be classified as:

- Direct stakeholders
- Indirect stakeholders
- Positive stakeholders
- Negative stakeholders
- Legitimacy and power

Construction projects have the direct involvement of the following three stakeholders:

1. Owner
2. Designer
3. Contractor

However there are many other stakeholders who have significant influence/impact on the outcome of a construction project. It is important to identify stakeholders who have interest and significant influence in the project. The stakeholders include members from within the organization and people, agencies and authorities outside the organization. A stakeholder register/log is developed using different types of classification models for stakeholder analysis. The stakeholder register/log is maintained and updated throughout the life cycle of the project. Figure 5.1 illustrates stakeholders having involvement or interest in the construction project.

Each of these stakeholders has their own quality management system and also quality auditing system.

Traditional construction projects involve the following three parties:

1. Owner/client
2. Designer/consultant
3. Contractor

134 *Abdul Razzak Rumane*

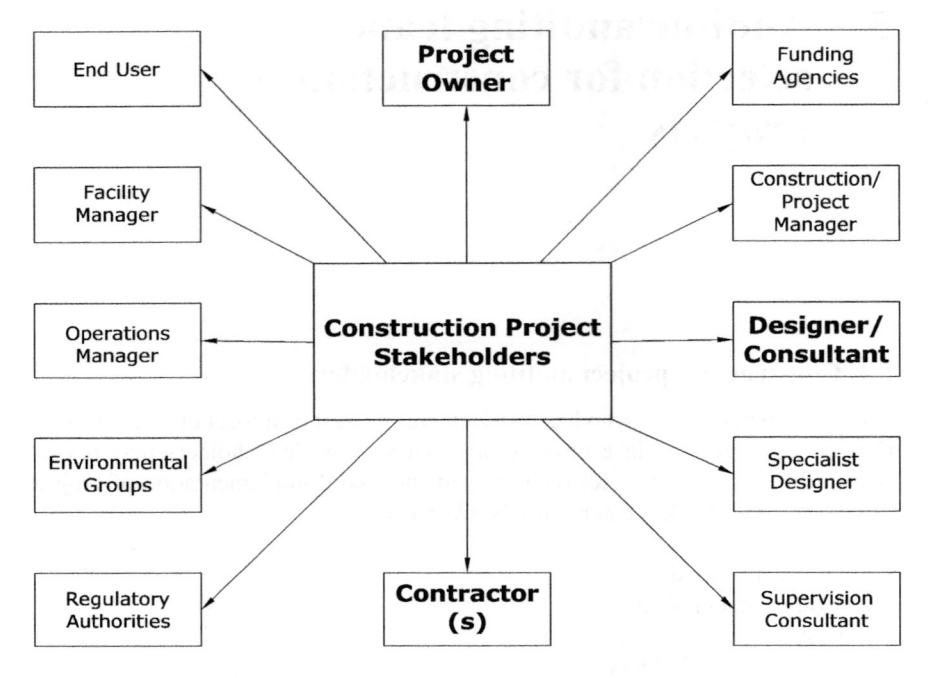

Figure 5.1 Construction project stakeholders

Source: Abdul Razzak Rumane (2016). *Handbook of Construction Management: Scope, Schedule, and Cost Control*. Reprinted with permission from Taylor & Francis Group

Tables 2.2–2.4 in Chapter 2 list example contents of a quality management system for these three traditional parties.

The quality auditing function involves three groups who may interrelate in a number of ways. These are:

1 Client (the organization desiring the audit)
2 Auditor (the person/organization who plans and carries out the audit)
3 Auditee (the organization that is to be audited)

Figure 5.2 illustrates the auditing structure of an audit group.

1 Client team

    •   The team member(s) depends whether it is an external or internal audit

2 Auditor team

The number of auditor team members depends on the following:

    a   Size of the organization
    b   Nature of business of the organization

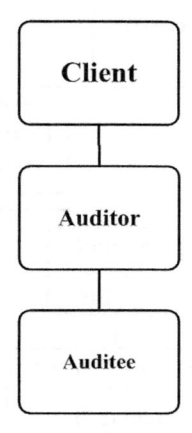

Figure 5.2 Audit structure

    c   Complexity of audit
    d   Audit objective
    e   Levels of details required from the audit
    f   Type and size of project
    g   Types of audit

        i   Product
        ii   Process
        iii System
        iv Compliance
        v   Adequacy

    h   Time schedule to complete the audit
    i   Audit program
    j   Regulatory requirements
    k   Statutory requirement

Audits are classified as:

1   Internal

    i   First party

2   External

    i   Second party
    ii   Third party

Figure 5.3 illustrates the classification of audits and Figure 5.4 illustrates an example of an organizational structure for an internal audit team.

Internal audits (also known as first party audits) are performed by company staff/employees of the company that are trained to conduct audits, with enough

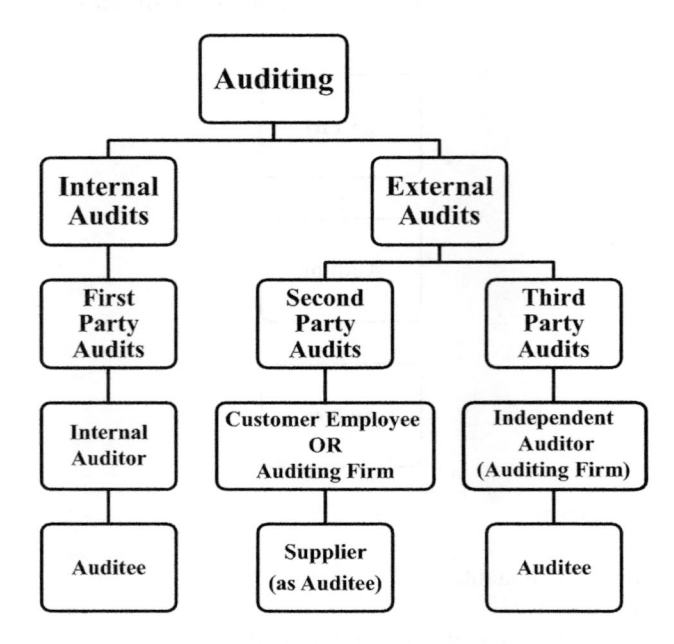

Figure 5.3 Classification of audits

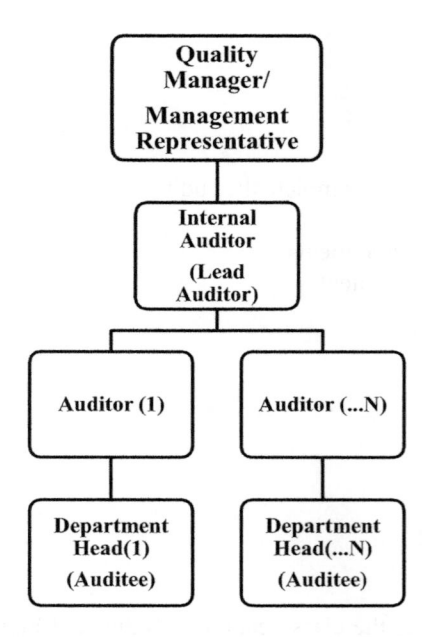

Figure 5.4 Organization structure of a first party (internal auditing) audit team

Auditor selection for construction projects 137

knowledge about company activities but no relationship with or interest in the works to be audited. Normally an internal auditor performs the role of lead auditor and the auditees are department heads.

External audits can be either second party audits or third party audits. Figure 5.5 illustrates an example of an organizational structure for a second party audit team.

Every audit team has a lead auditor or audit team leader. In order to conduct audits of big organizations/major projects, there may be a lead auditor as well as audit team leader(s). Normally the following are members of a third party audit:

i   Lead auditor
ii Audit team leader
iii Auditors
iv Subject matter experts
v   Auditor-in-training

Figure 5.6 illustrates an example of a third party audit team.

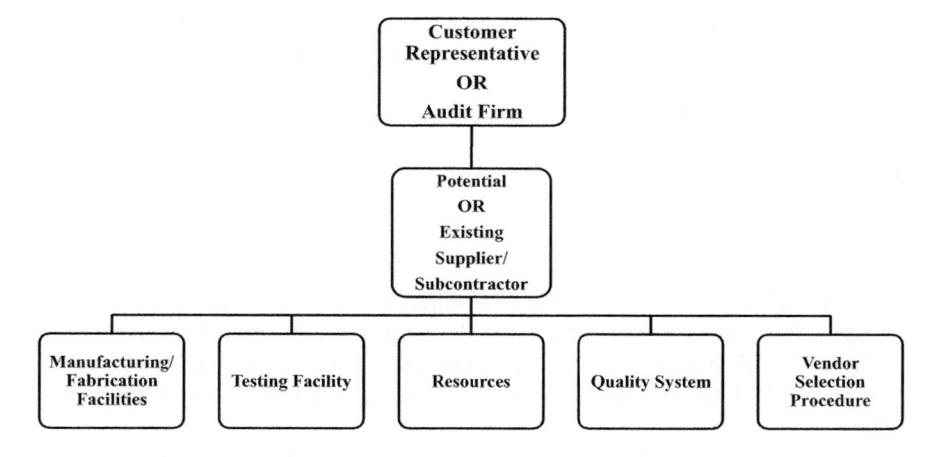

Figure 5.5 Organization structure of a second party audit team

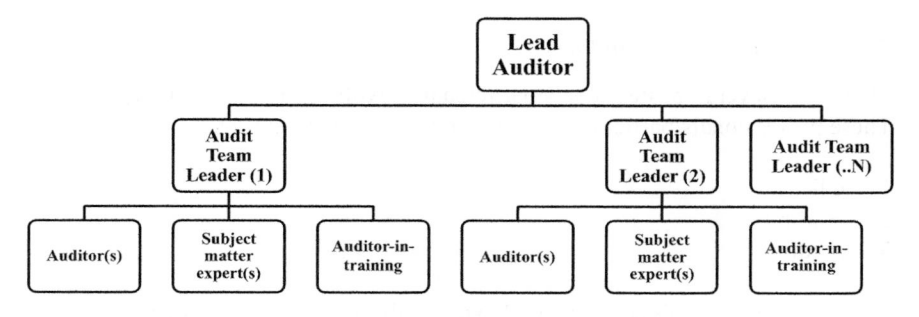

Figure 5.6 Organization structure of a third party audit team

138 Abdul Razzak Rumane

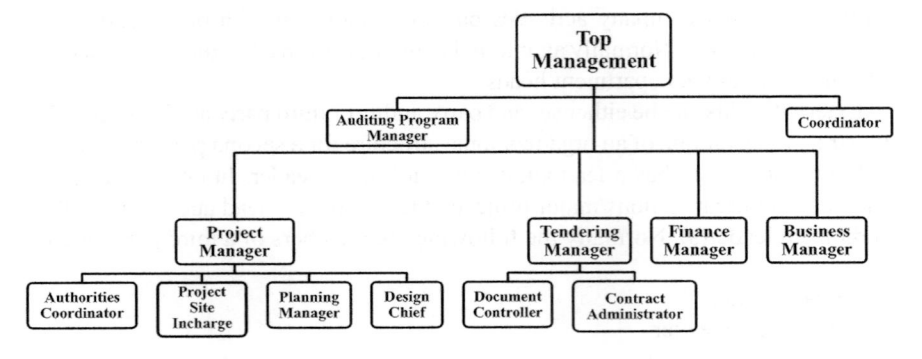

Figure 5.7 Organization structure of a project owner auditee team

The number of personnel needed in the auditor team depends on the amount of audit work to be covered, the experience and qualifications of the team members to conduct the audit and the availability of audit resources.

The members of the auditee team depend on the size of organization to be audited, the type of business of the organization and the type of audits to be performed.

Construction projects involve three parties. These are:

1   Owner/Client
2   Designer (Consultant)
3   Contractor

Figure 5.7 illustrates an example of a project owner auditee team.
Figure 5.8 illustrates an example of a designer (consultant) auditee team.
Figure 5.9 illustrates an example of a contractor (head office) auditee team.
Figure 5.10 illustrates an example of a contractor (construction site) auditee team.

The number of personnel from each trade depends on the magnitude of the audit.

5.1.1 Roles and responsibilities

There are several professionals/participants involved in the auditing process. These professionals/participants belong to the following groups:

1   Client
2   Auditor
3   Auditee

These professionals/participants have to fully coordinate and cooperate with each other for a successful and effective audit to achieve the audit objectives. Each of these professionals/participants has different roles and responsibilities

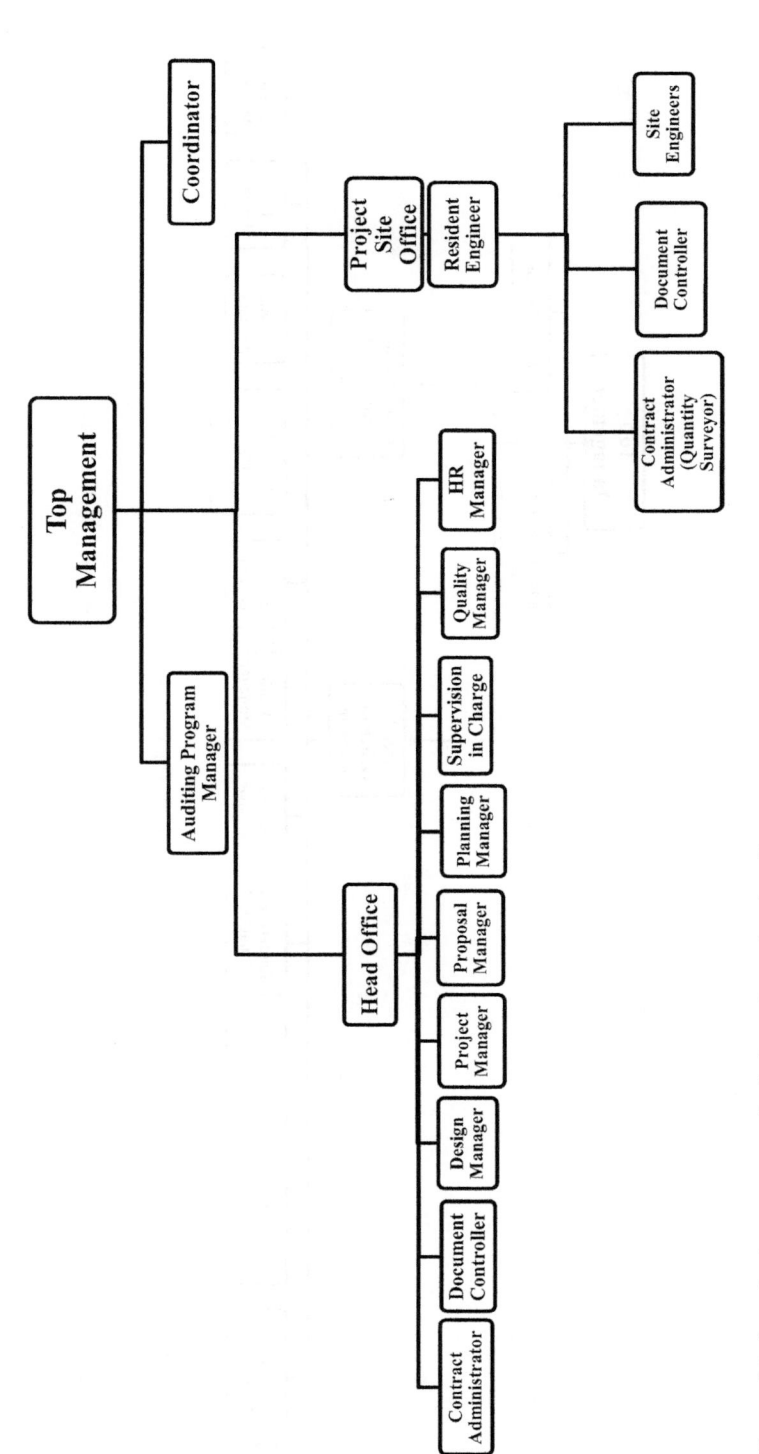

Figure 5.8 Organization structure of a designer (consultant) auditee team

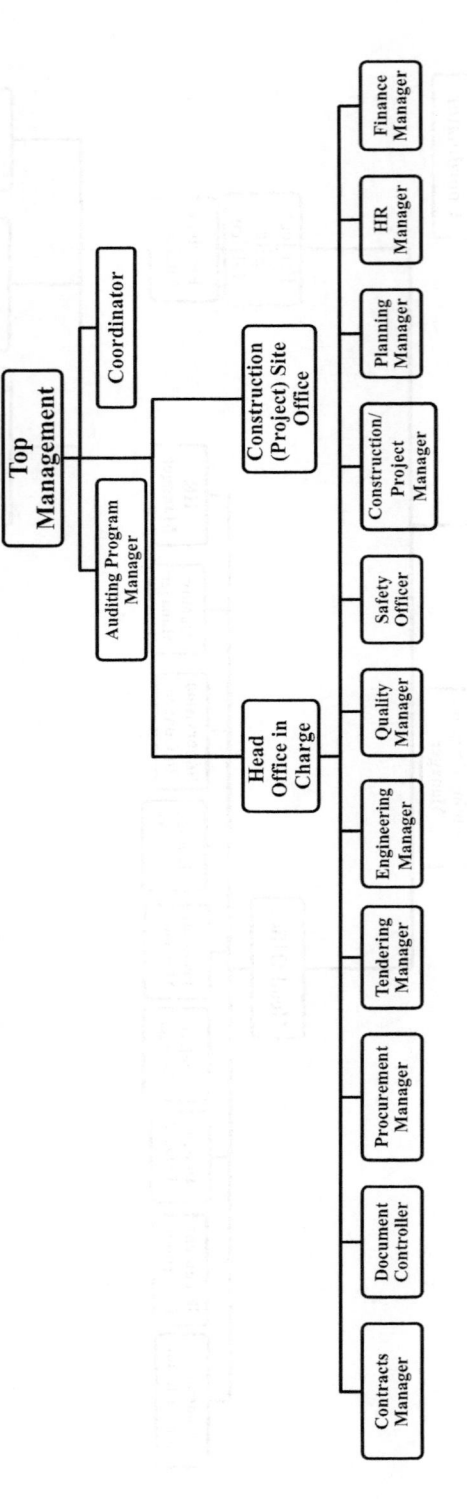

Figure 5.9 Organization structure of a contractor (head office) auditee team

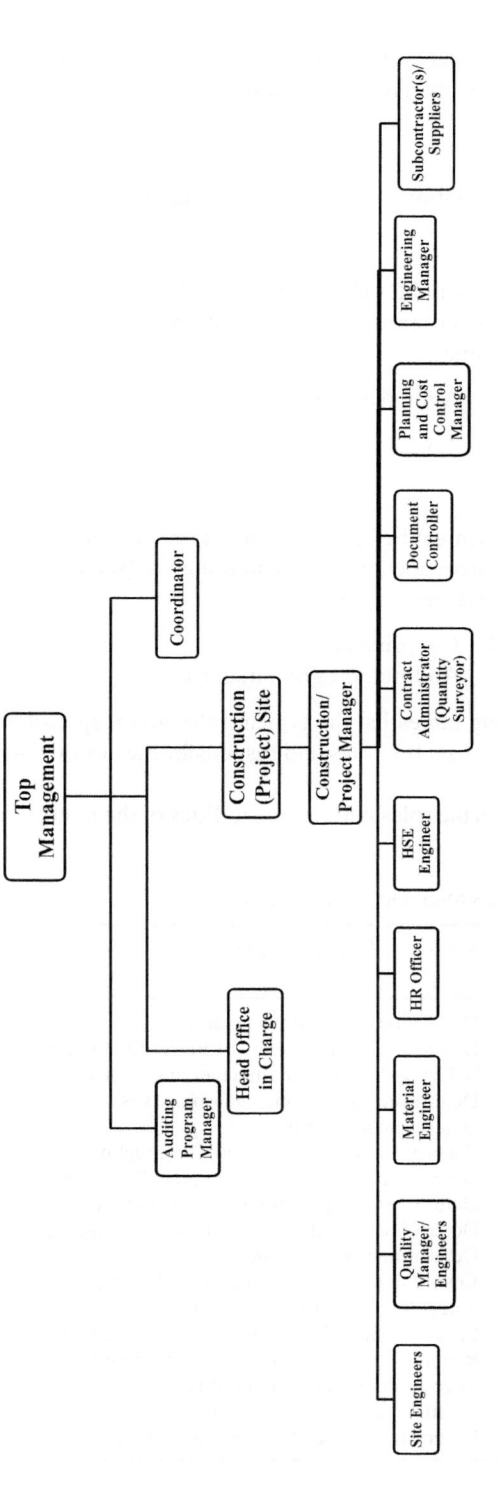

Figure 5.10 Organization structure of a contractor (construction site) auditee team

that depend on which functional group they are affiliated to. The following are the main participants involved in performing major project audits:

1   Client

    1.1   Organization or person desiring audit

2   Auditor

    2.1   Internal auditor (first party audits)
    2.2   Customer representative/external auditor (second party audits)
    2.3   Lead auditor

        2.3.1   Audit team leader

            2.3.1.1   Auditors

3   Auditee

    3.1   Department heads/management review body (first party audits)
    3.2   Supplier/subcontractor representative (second party audits)
    3.3   Management representative

        3.3.1   Coordinator
        3.3.2   Auditing program manager

The following tables list the typical roles and responsibilities of audit participants. Table 5.1 lists the roles and responsibilities of the client in performing the audit.

Table 5.2 lists the roles and responsibilities of the internal auditor in performing the audit.

Table 5.1 Client's roles and responsibilities

Serial Number	Roles and Responsibilities
1	Determine the need for an audit
2	Determine the frequency of the audits to be performed
3	Select audit program manager and coordinator to oversee audit activities
4	Determine audit scope and objectives
5	Determine audit criteria
6	Specify what standards, tools and techniques to be followed to do audit
7	Determine the auditing location (office or project site or both)
8	Determine the departments to be audited (extent of audit)
9	Determine the activities, processes, products, systems to be audited
10	Determine auditing timeline
11	Ensure that audit resources are adequate
12	Select the auditor organization
13	Specify which follow-up actions the auditee should take
14	Receive and review the reports prepared by the auditor
15	Specify distribution of audit reports
16	To attend audit process meetings
17	Determine the need for follow-up actions

Table 5.2 Internal auditor's roles and responsibilities

Serial Number	Roles and Responsibilities
1	Prepare audit plan and manage all the phases of the audit
2	Determine priorities and establish the most cost effective means of achieving audit objectives
3	Identify organization's objectives of the audit
4	Identify audit areas
5	Review the scope and objectives of audit
6	Liaison with management, quality manager, audit team members and relevant departments regarding areas to be audited
7	Select audit team members and allocate the audit activities
8	Ensure that necessary resources are available
9	Assign audit tasks to team members
10	Prepare the scope and audit objectives for individual audit member assignment
11	Brief the auditors by defining the audit requirements
12	Monitor audit schedule
13	Establish communication system between auditors and relevant departments
14	Conduct opening and closing meetings
15	Ensure uniformity in the performance of different auditors
16	Motivate team members to accomplish team goals
17	Prepare audit checklists and necessary forms
18	Assist auditors in preparing audit reports
19	Manage and control conflicts during auditing process
20	Report major obstacle encountered during auditing process
21	Report major nonconformities immediately
22	Determine if any follow-up action is required
23	Assist audit team members to prepare reports
24	Report and present audit reports
25	Identify areas for improvement

Table 5.3 lists the roles and responsibilities of the lead auditor in performing the audit.

Table 5.4 lists the roles and responsibilities of the auditor in performing the audit.

Table 5.5 lists the roles and responsibilities of the auditee in performing the audit.

Table 5.6 lists roles and responsibilities of the audit program manager in performing the audit.

5.1.2 Competencies and expertise

There are three categories of audits. These are:

1. First party audits
2. Second party audits
3. Third party audits

Table 5.3 Lead auditor's roles and responsibilities

Serial Number	Roles and Responsibilities
1	Manage the audit
2	Review the scope and objectives of audit
3	Liaison with client, audit program manager, team leaders, auditors and auditee team members
4	Select audit team members and allocate the audit activities
5	Ensure that necessary resources are available
6	Assign audit tasks to team members
7	Prepare the scope and audit objectives for individual audit member assignment
8	Prepare audit plan and manage all the phases of the audit
9	Brief the auditors by defining the audit requirements
10	Monitor audit schedule
11	Orient audit team with auditee
12	Establish communication system between auditors and auditee
13	Conduct opening and closing meetings
14	Ensure uniformity in the performance of different auditors
15	Motivate team members to accomplish team goals
16	Prepare audit checklists and necessary forms
17	Assist auditors in preparing audit reports
18	Manage and control conflicts during auditing process
19	Report major obstacle encountered during auditing process
20	Report major nonconformities immediately
21	Determine if any follow-up action is required
22	Assist audit team members to prepare reports
23	Report and present audit reports

Table 5.4 Auditor's roles and responsibilities

Serial Number	Roles and Responsibilities
1	Liaison with lead auditor, team leader (as the case may be) and auditee
2	Understand the scope and objectives of the audit
3	Prepare for the audit
4	Participate in planning of audit
5	Perform assigned audit tasks
6	Evaluate quality system and collect evidence to verify conformance or nonconformance of the quality system
7	Apply auditing techniques as specified
8	Find out if quality objectives are being achieved
9	Determine whether quality policy is being applied
10	Participate in opening and closing meetings
11	Keep to the audit program and schedule
12	Attend team meetings to discuss audit progress
13	Maintain confidentiality
14	Report findings to lead auditor, team leader as the case may be
15	Ensure that collected data and information is accurate and factual
16	Document the results of audit observations and findings

Serial Number	Roles and Responsibilities
17	Prepare records and reports of the audit
18	Safeguard audit documents, records and reports
19	Provide any information requiring follow-up actions
20	Report conflict of interest
21	Comply with audit requirements
22	Keep to the audit program and schedule
23	Be ethical and observe code of conduct

Table 5.5 Auditee's roles and responsibilities

Serial Number	Roles and Responsibilities
1	Liaise with lead auditor, audit team leader (as the case may be) for audit work
2	Explain the scope, objectives of the audit to the concerned department heads/personnel
3	Cooperate with the auditors
4	Assign personnel to assist auditors
5	Ensure that all the responsible personnel related to audit extend full cooperation to auditors for success of audit
6	Provide logistic support, working area needed to do the audit
7	Provide necessary resources that is needed by audit team to do the audit
8	Ensure availability of relevant documents needed to do the audit
9	Allow auditors to examine/review all the documents and records to do the audit
10	Attend opening and closing meetings
11	Avoid conflicts and misunderstandings
12	Correct and prevent if any problem identified during audit
13	Receive audit reports and findings
14	Take corrective actions on audit findings

Table 5.6 Audit program manager's roles and responsibilities

Serial Number	Roles and Responsibilities
1	Plan and schedule audit programs
2	Establish audit program goals and objectives
3	Ensure selection of audit teams to do the audit
4	Assign auditors to scheduled audit
5	Communicate audit program to relevant audit teams
6	Establish procedures, criteria for an effective and efficient audit program
7	Establish process for evaluation of auditors and their performance in implementation of audit plan
8	Establish communication system between the client, auditors and auditee team members
9	Monitor audit schedule for completion within agreed-upon time
10	Ensure availability of all resources to the audit team

(*Continued*)

Table 5.6 (Continued)

Serial Number	Roles and Responsibilities
11	Coordinate with client about auditing practices
12	Communicate audit progress with the client
13	Ensure ethical behavior of audit teams
14	Ensure the control of audit reports and records
15	Maintain audit reports and records
16	Review and approval of audit reports and records
17	Coordinate with the client to take necessary action on audit findings

Table 5.7 Internal auditor's competencies and expertise

Serial Number	Description
1	Establishing, implementing and maintaining quality management system (QMS) of the organization in compliance with ISO Standards
2	Managing, auditing and monitoring compliance to the quality management system of the organization
3	Ability to understand the organization's mission, strategy and objectives
4	Audit planning and scheduling
5	Identify, analyze and manage strategic risk of the organization
6	Performing audits as per ISO Standards at all the business areas and locations of the organization
7	Ability to understand the audit strategy
8	Ability to identify and solve the problems
9	Technical capability to understand and apply principles, procedures, requirements and regulation related to specialized expertise to do the audit
10	Ability to evaluate strength and weakness in the internal control of the organization
11	Capable of gathering information and references to evaluate objective evidence
12	Ability to evaluate the processes for compliance with quality system requirements
13	Professional judgment
14	Analysis of audit processes and results
15	Ability to withstand and resolve conflicts
16	Ethical conducts: integrity, honesty, confidentiality,
17	Interpersonal skills
18	Perform audit without fear and favor
19	Proper communication skills both oral and written
20	Interview techniques and effective communication
21	Ability to perceive situations in a realistic way
22	Preparation and presentation of reports
23	Review and input actions to improve the organization's overall business performance
24	Ensure reporting system is working properly
25	Professional development
26	Competency certificate in quality auditing profession
27	Knowledge of systematic auditing process
28	Knowledge of tools and techniques used for auditing
29	Knowledge of statutory requirements for audits

Each of these audits is performed by auditors who have different functional roles and responsibilities. The competencies and expertise of the auditor depends on the audit work and their involvement to do the audits. It also depends on the job criteria such as knowledge, experience, skills, attitude and aptitude (KESAA) specified by the organization (client). The following designations are normally used to relate auditors with these categories:

1 Internal auditor – first party audits
2 Auditor (company employees or outside auditor) – second party audits
3 Lead auditor/audit team leader/auditor – third party audits

Table 5.7 illustrates the typical competencies and expertise of an internal auditor.
Table 5.8 illustrates the typical competencies and expertise of a lead auditor.
Table 5.9 illustrates the typical competencies and expertise of an auditor.
Table 5.10 illustrates the typical competencies and expertise of a subject matter expert auditor.

Table 5.8 Lead auditor's competencies and expertise

Serial Number	Description
1	Manage team of auditors ensuring that the audits comply with the applicable standards, regulations and QMS to achieve the objectives
2	Ability to identify and solve the problems
3	Strong analytical and problem solving capability
4	Audit planning and scheduling
5	Ability to explain and illustrate the main concept of QMS to do audit
6	Performing audits as per ISO Standards at all the business areas and locations of the organization
7	Ability to understand the audit strategy
8	Sequence of auditing
9	Technical capability to understand and apply principles, procedures, requirements and regulation related to specialized expertise to do the audit
10	Organizational understanding
11	Ability to evaluate strength and weakness in the internal control of the organization being audited
12	Capable of gathering information and references to evaluate objective evidence
13	Professional judgment
14	Ability to withstand and resolve conflicts
15	Ethical conducts: integrity, honesty, confidentiality,
16	Interpersonal skills
17	Perform audit without fear and favor
18	Proper communication skills both oral and written
19	Interview techniques and effective communication
20	Ability to perceive situations in a realistic way
21	Preparation and presentation of reports

(*Continued*)

Table 5.8 (Continued)

Serial Number	Description
22	Review and input actions to improve the organization's overall business performance
23	Ensure reporting system is working properly
24	Professional development
25	Experience and knowledge of performing internal audits
26	Specific process and products knowledge of the audits to be performed
27	Competency certificate in quality auditing profession
28	Knowledge of systematic auditing process
29	Knowledge of tools and techniques used for auditing
30	Ability to build strong relationship within the organization
31	Knowledge of statutory requirements for audits

Table 5.9 Auditor's competencies and expertise

Serial Number	Description
1	Technical capability to understand and apply principles, procedures, requirements and regulation related to the audit objectives
2	Performing audits as per ISO Standards at the assigned business areas and locations of the organization
3	Ability to identify and solve the problems
4	Strong analytical and problem solving capability
5	Ability to understand the audit strategy
6	Organizational understanding
7	Capable of gathering information and references to evaluate objective evidence
8	Professional judgment
9	Ability to withstand and resolve conflicts
10	Ethical conducts: integrity, honesty, confidentiality,
11	Interpersonal skills
12	Perform audit without fear and favor
13	Proper communication skills both oral and written
14	Interview techniques and effective communication
15	Ability to perceive situations in a realistic way
16	Preparation and presentation of reports
17	Experience and knowledge of performing audits
18	Specific process and products knowledge of the audits to be performed
19	Competency certificate in quality auditing profession
20	Knowledge of systematic auditing process
21	Knowledge of tools and techniques used for auditing
22	Ability to build strong relationship within the organization
23	Knowledge of statutory requirements for audits

Table 5.10 Subject matter expert auditor's competencies and expertise

Serial Number	Description
1	Technical capability to understand and apply principles, procedures, requirements and regulation related to specialized expertise to do the audit
2	Performing audits as per ISO Standards at the assigned business areas and locations of the organization
3	Knowledge of the auditing criteria relevant to specific process, system, product that has been assigned for audit
4	Knowledge of audit tools and techniques applicable to the assigned audit areas
5	Ability to identify and solve the problems
6	Strong analytical and problem solving capability
7	Ability to understand the audit strategy
8	Organizational understanding
9	Capable of gathering information and references to evaluate objective evidence
10	Professional judgment
11	Ability to withstand and resolve conflicts
12	Ethical conducts: integrity, honesty, confidentiality,
13	Interpersonal skills
14	Perform audit without fear and favor
15	Proper communication skills both oral and written
16	Interview techniques and effective communication
17	Ability to perceive situations in a realistic way
18	Preparation and presentation of reports
19	Experience and knowledge of performing audits
20	Specific process and products knowledge of the audits to be performed
21	Competency certificate in quality auditing profession
22	Knowledge of systematic auditing process
23	Knowledge of statutory requirements for audits

5.2 Selection of the auditor for project life cycle phases

The construction project life cycle of a major construction project has seven phases. These are:

1 Conceptual Design
2 Schematic Design
3 Design Development
4 Construction (Contract) Documents
5 Bidding and Tendering
6 Construction
7 Testing, Commissioning and Handover

Each phase is further subdivided into the Work Breakdown Structure (WBS) principle to reach a level of complexity where each element/activity can be treated as a single unit that can be conveniently managed. Division of these phases improves the control and planning of the construction project at every stage before a new phase starts. Each phase is composed of activities, elements which have a functional relationship to achieve a common objective for useful purpose.

The Project Management Body of Knowledge (PMBOK® Guide) published by the Project Management Institute describes the application of project management processes during the life cycle of projects to enhance the chances of success over a wide range of projects. The PMBOK® Guide-Fifth Edition identifies and describes five project management process groups required for the successful completion of any project. These are:

1. Initiating Process Group
2. Planning Process Group
3. Executing Process Group
4. Monitoring and Controlling Process Group
5. Closing Process Group

Figure 5.11 illustrates an overview of project management process groups. These process groups are independent of application areas or industry focus.

The methodology applied in this book is based on seven phases. These phases are treated as a project itself with all the five process groups operating as they do for the overall project. Based on the principles of project management processes defined in the PMBOK® Guide and Construction Extension to the PMBOK® Guide Third Edition there are 13 knowledge areas that are categorized into management processes to manage and control various processes and activities to be performed in construction project management. These are as follows:

1. Integration Management
2. Stakeholder Management
3. Scope Management
4. Schedule Management
5. Cost Management
6. Quality Management

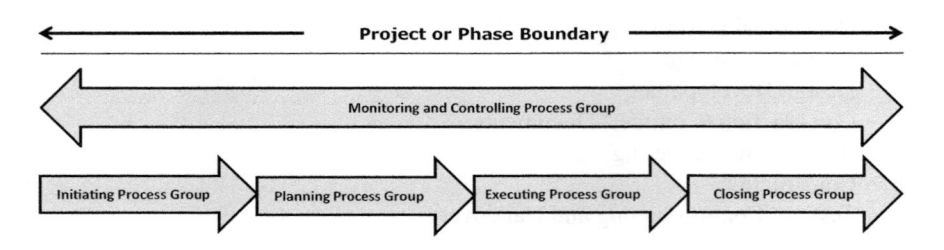

PROJECT MANAGEMENT PROCESS GROUPS

Figure 5.11 Overview of project management process groups

7   Resource Management
8   Communication Management
9   Risk Management
10 Contract Management
11 Health, Safety and Environment Management (HSE)
12 Financial Management
13 Claim Management

Each of these knowledge areas consists of management processes that are applied during the management of the project to enhance the success of the project. These management processes can be divided into construction related technical activity/activities which can further be subdivided into element(s) to enable the efficiency and effectiveness of construction management to be improved. These management processes can be performed in any order as long as the required inputs are available.

While developing the construction projects, each of these activities has to be reviewed, verified, evaluated and assessed during and/or after execution/implementation to ensure compliance to project quality requirements.

The assessment (audit) of these activities can be performed either by the organization itself (internal auditor) or by an external agency (audit firm). The auditing at different stages/phases of the project has to be performed by the auditor with knowledge and expertise in those activities that are executed/implemented/performed at that particular stage/phase.

The audits during project life cycle phases are mainly systematic examination, assessment of organizational compliance to the quality management system, procedural follow-up of construction management processes, the organization's specific factors, project specific requirements and operational management to successfully achieve project goals and objectives, improve performance and enhance competitive advantage.

The following sections discuss the selection of the auditor(s) to carry out audits for these stages/phases.

5.2.1 Auditor for the bidding and tendering process

In construction projects, the involvement of outside companies/parties such as consultants, designers, project managers and construction managers starts at an early stage of the project development process. The owner/client has to decide which work is to be procured, analyzed, designed and constructed by others. Every organization has its procurement system and strategy to procure services, contracts and products from others. Figure 5.12 illustrates the stages at which the outside agency (consultant, designer, project manager, construction manager, contractor) is selected as per the procurement (bidding and tendering) strategy for the particular type of project delivery system. At each of these stages the procurement management process (bidding and tendering process) takes place. Figure 5.13 illustrates the bidding and tendering process to select the contractor (outside agency/firm).

152   Abdul Razzak Rumane

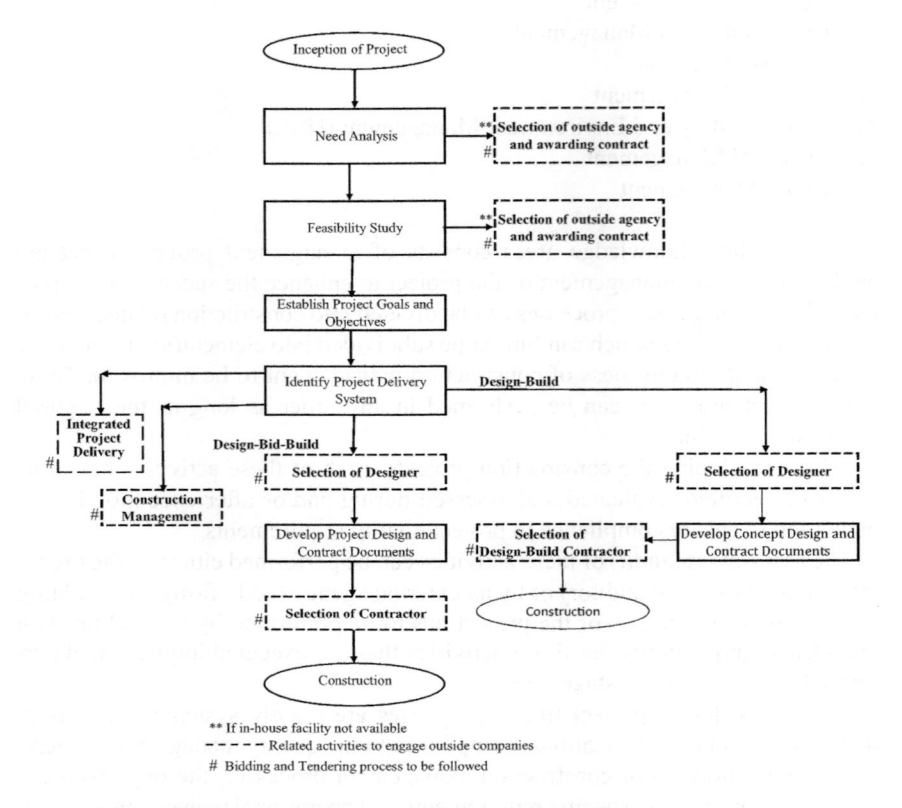

Figure 5.12 Bidding and tendering stages for construction projects
Source: Abdul Razzak Rumane. (2013). *Quality Tools for Managing Construction Projects.* Reprinted with permission of Taylor & Francis Group

The following are common procurement methods for the selection of project teams:

1   Low Bid
    - Selection is based solely on the price
2   Best Value
    a   Total Cost
        - Selection is based on total construction cost and other factors
    b   Fees
        - Selection is based on weighted combination of fees and qualification
3   Qualification Based Selection
    - Selection is based solely on qualification

Figure 5.13 Bidding and tendering process

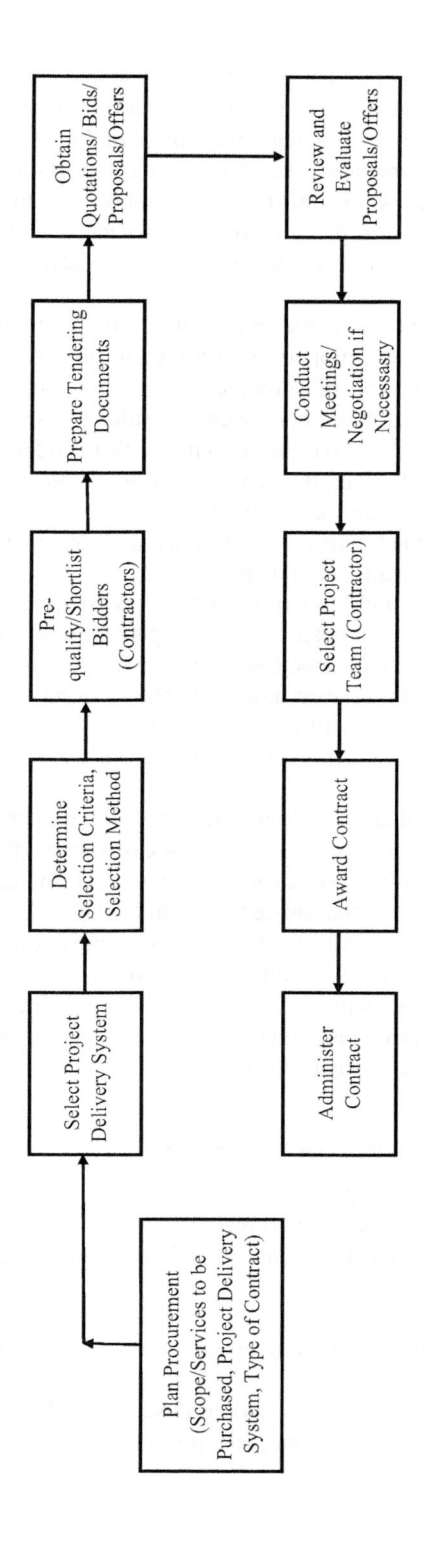

Source: Abdul Razzak Rumane (2016). *Handbook of Construction Management: Scope, Schedule, and Cost Control*. Reprinted with permission of Taylor & Francis Group

Complex and major construction projects have many challenges such as delays, changes, disputes, accidents on site, etc. and therefore need efficient management of the project from the beginning to the end of construction of the facility/project to meet the intended use and owner's expectations. The owner/client may not have the necessary staff/resources in-house to manage the planning, design and construction of the construction project to achieve the desired results. Construction projects are executed based on a predetermined set of goals and objectives.

Business need assessment and a feasibility study are essential to ensure the owner's business case has been properly considered to prepare an accurate and comprehensive client brief (Terms of Reference) to achieve a qualitative and competitive project. It is essential that a competent consultant is selected to prepare Terms of Reference (TOR) that give the designer a clear understanding for the development of the project. Further, the owner has to select the designer (consultant) to develop the project design for construction.

The audit of the consultant selection procedure and designer selection procedure is a process of verification/evaluation of the procurement process to the extent to which the bidding and tendering process has complied with organizational strategy and policy. It is to ensure that the organizational strategy and process for the selection method and selection criteria is correctly followed to select the outside agency to perform relevant functions. The auditing of the bidding and tendering process is carried out with a view to improving construction management processes and as a measure for improving the success of the construction project.

The auditor for the consultant selection procedure audit can be either an internal auditor or external auditor (audit firm). In both cases the auditor should have the requisite experience, qualification, expertise, knowledge and understanding of all the major activities/elements that relate to the study stage.

Similarly the auditor for the designer selection procedure audit can be either an internal auditor or external auditor (audit firm). In both cases the auditor should have the requisite experience, qualification, expertise, knowledge and understanding of all the major activities that are related to the development of the design for all phases of the construction project life cycle.

5.2.1.1 Auditor for the consultant selection process

The inception of a construction project starts with the identification of the business case and its needs. The need of the project is linked to the available financial resources to develop the facility. The owner's needs are quite simple and are based on the following:

- To have the best facility for the money i.e. to have maximum profit or services at a reasonable cost
- On time completion i.e. to meet the owner's/user's schedule
- Completion within budget i.e. to meet the investment plan for the facility.

The owner's need must be well defined indicating the minimum requirements of quality and performance, required completion date and an approved main budget. Sometimes, the project budget is fixed and therefore the quality of the building system, materials and finishes of the project need to be balanced with the budget. A business case typically addresses the business need for the project and the value the project brings to the business (project value proposition). A value proposition is a promise of value to be delivered by the project. The following questions address the value proposition:

a   How the project solves the current problems or improves the current situation
b   What specific benefits the project will deliver
c   Why the project is the ideal solution for the problem

Generally the project owner engages a specialist consultant involved in carrying out a need analysis/assessment.

Once the need analysis is carried, out a need statement is prepared and a feasibility study is conducted to assist the project owner/decision makers in making the decision that will be in the best interests of the owner.

A feasibility study is defined as an evaluation or analysis of the potential impact of the identified need of the proposed project. A feasibility study is performed to define more clearly the viability and form of the project that will produce the best or most profitable results. The feasibility study assists the decision makers (investors/owners/clients) in determining whether or not to implement the project. The feasibility study may be conducted in-house if the capability exists. However, the services of specialists involved in the preparation of economic and financial studies are usually commissioned by the owner/client to perform such a study. Since the feasibility study stage is a very crucial stage, in which all kinds of professionals and specialists are required to bring many kinds of knowledge and experience into a broad-ranging evaluation of feasibility, it is necessary to engage a firm which has expertise in the related fields. The feasibility study establishes the broad objectives for the project and so exerts an influence throughout the subsequent stages. The successful completion of the feasibility study marks the first of several transition milestones and is therefore most important to determine whether or not to implement a particular project or program. The feasibility study decides the possible design approaches that can be pursued to meet the need.

After completion of the feasibility study and its approval by the owner, project goals and objectives are prepared taking into consideration the final recommendations/outcome of the feasibility study. Clear goals and objectives provide the project team with appropriate boundaries to make decisions about the project and ensure the project/facility will satisfy the owner/end user's requirements fulfilling the owner's needs. Establishing properly defined goals and objectives is the most fundamental element of project planning. Therefore the project goals and objectives must be:

- Specific (Is the goal specific?)
- Measurable (Is the goal measurable?)

- Agreed upon/achievable (Is the goal achievable?)
- Realistic (Is the goal realist or result oriented?)
- Time (cost) limited (Does the goal have a time element?)

Based on the project goals and objectives, the owner selects a project delivery system that mainly depends on the project size, complexity of the project, innovation, uncertainty, urgency and the degree of involvement of the owner.

An audit during the study stage is mainly to examine the processes used by the consultant or owner's in-house team members to prepare the business case for the project and in defining the goals and objectives of the project.

The process for selecting an auditor firm mainly consists of the following stages:

1. Screening of audit firms by performing shortlisting (prequalification) process
2. Request for proposal/quotation
3. Proposals/quotation from the bidders
4. Evaluation of bids

5.2.1.1.1 EVALUATION OF BIDS

While selecting the auditor for a consultant selection procedure audit, the project owner has to review and evaluate the bid documents for compliance with the following requirements:

1. Procedural requirements

    - Bidding and tendering process
    - Organizational strategy
    - Organizational policy
    - Selection method
    - Selection criteria
    - Regulatory requirement

2. Full compliance to the tender requirements that includes:

    - Related questionnaires in the tender documents are completed
    - Requisite documents are attached
    - Bid bond

3. Licensing

    - The auditor/firm has the regulatory approval/license/certificate to perform auditing

4. Independence

    - The auditor/firm is independent for the purposes for which it is hired

5. Experience

Auditor selection for construction projects 157

The auditor/firm has:

- Requisite experience to do the audit of the major elements of the study stage of the construction project, as illustrated in Table 5.11.
- Experience in performing audits of a similar type and nature of projects
- Ability to understand organizational strategy towards the development of the new project
- Knowledge of regulatory requirements
- Knowledge of environmental protection agency requirements
- Knowledge of health and safety requirements
- Ability to assess project need
- Knowledge of various indices to set up a new project
- Knowledge of market conditions
- Experience in strategic and analytical analysis
- Knowledge of analytical approach and background
- Ability to collect large number of important and necessary data
- Subject matter experts in the auditing team
- Ability to review and analyze market information
- Ability to identify and analyze project risks
- Ability to properly define goals and objectives of the construction project

Table 5.11 Major elements of the study stage

Serial Number	Elements	Description	Related Task
1	Problem statement/ need identification	Project needs, goals and objectives	Strategic objectives, policies and priorities
2	Need assessment	Identification of needs Prioritization of needs Leveling of needs Deciding what needs to be addressed	Ensure that owner's business case has been properly considered
3	Need analysis	Perform project need analysis/ study to outline the scope of issues to be considered in the planning phase	Perform need analysis
4	Need statement	Develop project need statement	Develop need statement
5	Feasibility study	Technical studies, economics assessment, financial assessment, scheduling, market demand, risk, environmental and social assessment	Perform feasibility study and statement

(*Continued*)

Table 5.11 (Continued)

Serial Number	Elements	Description	Related Task
6	Establish project goals and objectives	Scope, time, cost, quality	Project initiation documents developed on SMART concept
7	Identify alternatives	Identify alternatives based on a predetermined set of performance measures	Select conceptual alternatives
8	Preliminary schedule	Estimate the duration for completion of project/facility	Establish project schedule
9	Preliminary financial implications	Preliminary budget estimates of total project cost (life cycle cost) on the basis of any known research and development requirements. This will help arrange the finances (Funding agency)	Determine project budget
10	Preliminary resources	Estimate resources	Confirm availability of resources, manpower, material, equipment
11	Project risk	Identify project risk, constraints	Establish risk response, mitigation plan
12	Authorities clearance	Identify issues, sustainability, impacts and potential approvals (environmental, authorities, permits) required for subsequent design and authority approval processes	Establish requirements for statutory approvals and other regulatory authorities
13	Select preferred alternative	Assess technological and economical feasibility and compare to the preferred option/alternative to prepare business case	Discuss relative merits of various alternative schemes and evaluate the performance measures to meet owner's needs/requirements. Consider social, economical, environmental impact, safety, reliability and functional capability

Serial Number	Elements	Description	Related Task
14	Identify project delivery system	Establish how the participants, owner, designer (A/E), and contractor will be involved to construct the project/facility. (Design/Build/Bid, Design/Build, Guaranteed Maximum Price, CM type, PM type, BOT, Turnkey, etc.)	Select suitable project delivery system as per strategic decision and suitability of appropriate system
15	Identify type of contracting system	Select contract pricing system such as firm fixed price or lump sum, unit price, cost reimbursement (cost plus), reimbursement, target price, time and material, guaranteed maximum price	Select contracting/pricing most appropriate for the benefit of owner
16	Identify project team	Select designer (A/E) firm if Design/Bid/Build type of contract system is selected. Select other team members based on project delivery system requirements	Select project team considering the selected type of project delivery system and procurement process of the organization
17	Project launch	Project charter	Prepare Terms of Reference (TOR)

5.2.1.2 Auditor for the designer selection process

Generally the owner selects the designer/consultant on the basis of their qualifications (qualifications based system) and prefers to use one they have used before and with whom they have had satisfactory results. Figure 5.14 illustrates the logic flow diagram for the selection of a designer (A/E) and Table 5.12 illustrates the selection criteria for a designer (A/E).

An audit for the designer selection procedure is mainly to evaluate whether the organizational strategy and process has been adhered to when selecting the designer.

The selection process to select the auditor mainly consists of the following stages:

1   Screening of audit firms by performing shortlisting (prequalification) process
2   Request for proposal/quotation
3   Proposals/quotation from the bidders
4   Evaluation of bids

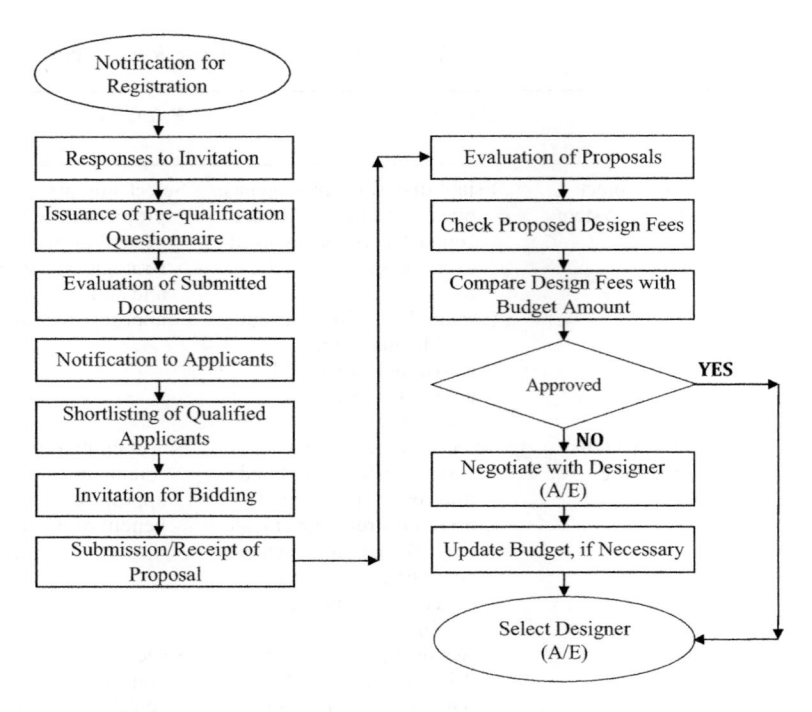

Figure 5.14 Logic flow diagram for the selection of a designer (A/E)

Table 5.12 Designer's (A/E) selection criteria

Serial Number	Evaluation Criteria	Weightage	Notes
1	**General Information**		
	a Company information		
2	**Business**	10%	
	a LEED or similar certification	5%	
	b ISO certification	5%	
3	**Financial**	20%	
	a Turnover	5%	
	b Financial standing	5%	
	c Insurance and bonding limit	10%	
4	**Experience**	30%	
	a Design experience	10%	
	b Similar type of projects	10%	
	c Current projects	10%	
5	**Design capability**	10%	
	a Design approach	5%	
	b Design capacity	5%	
6	**Resources**	20%	
	a Design team qualification	10%	
	b Design team composition	5%	
	c Professional certification	5%	
7	**Design quality**	5%	
8	**Safety consideration in Design**	5%	

Note: The weightage mentioned in the table is indicative only. The % can be determined as per the owner's strategy.

5.2.1.2.1 EVALUATION OF BIDS

While selecting the auditor for the designer selection procedure audit, the project owner has to review and evaluate the bid documents for compliance with the following requirements:

1. Procedural requirements

    - Bidding and tendering process
    - Organizational strategy
    - Organizational policy
    - Selection method
    - Selection criteria
    - Regulatory requirement

2. Full compliance to the tender requirement that includes:

    - Related questionnaires in the tender documents are completed
    - Requisite documents are attached
    - Bid bond

3. Licensing

    - The auditor/firm has the regulatory approval/license/certificate to perform auditing

4. Independence

    - The auditor/firm is independent for the purposes for which they are hired

5. Experience

The auditor/firm has:

- Ability to understand Terms of Reference (TOR), Scope of Work and compliance by the designer
- Requisite knowledge about the qualification requirements of the designer to make a selection on a quality based system. Table 5.13 lists the criteria for the selection of a construction project designer on a quality based system.
- Experience in performing audits of similar type and nature of projects
- Knowledge of different types of contracts documents used to hire the designer
- Ability to understand organizational strategy towards development of project design and drawings
- Knowledge of regulatory requirement
- Knowledge of regulatory approval system
- Knowledge of environmental protection agency requirements
- Knowledge of health and safety requirements
- Experience in strategic and analytical analysis
- Knowledge of analytical approach and background
- Knowledge of phases of life cycle of construction project
- Knowledge about project life cycle costing

Table 5.13 Consultant's qualification for feasibility study

Serial Number	Description
1	Experience in conducting feasibility study
2	Experience in conducting feasibility study in similar type and nature of projects
3	Fair and neutral with no prior opinion about what decision should be made
4	Experience in strategic and analytical analysis
5	Knowledge of analytical approach and background
6	Ability to collect large number of important and necessary data via work sessions, interviews, surveys and other methods.
7	Market knowledge
8	Ability to review and analysis of market information
9	Knowledge of market trend in similar type of projects/facility
10	Multidisciplinary experienced team having proven record in following fields: a Financial analyst b Engineering/technical expertise c Policy experts d Project scheduling
11	Experience in review of demographic and economic data

Source: Abdul Razzak Rumane (2013). *Quality Tools for Managing Construction Projects.* Reprinted with permission of Taylor & Francis Group

- Knowledge about sustainable project development processes
- Knowledge of construction project design process
- Ability to review design drawings
- Ability to review specification documents
- Ability to review contract documents
- Ability to analyze project risks
- Knowledge of project management process groups, knowledge areas and major construction activities in each process group
- Project planning and scheduling
- Knowledge of project quality requirements
- Ability to review and understand designer's quality management plan
- Knowledge about resource deployment during design period
- Knowledge about project constraints
- Functional as well as subject matter experts in the auditing team

Once the audit firm for the auditing of the project design is contracted, it is important to select specific personnel (auditors) who will do the actual audit work of the project. The following sections discuss specific criteria for the selection of a subject matter expert in the audit team.

5.2.2 Auditor for the conceptual design phase

Conceptual design is the first phase of the construction project life cycle. This phase/stage is concerned with the preparation of the concept design. Conceptual design is often viewed as most critical to achieving outstanding project performance. Most important decisions about planning, organization and type of contract take place during this phase.

The conceptual design phase commences once the need is recognized. In this phase the idea is conceived and given initial assessment. During the conceptual phase, project goals and objectives are established, a preliminary project plan is prepared, alternatives are analyzed and the preferred alternative is selected, the environment is examined, forecasts are prepared, and the cost and time objectives of the project are performed. Figure 5.15 illustrates the logic flow diagram for the conceptual design phase.

Figure 5.16 illustrates the major activities relating to the concept design phase based on the project management process groups methodology.

The most significant impacts in the quality of the project begin during the conceptual phase. This is the time when specifications, statement of work, contractual agreements and initial design are developed. Initial planning has the greatest impact on a project because it requires the commitment of processes, resources, schedules and budgets. A small error that is allowed to stay in the plan is magnified several times through subsequent documents that are second or third in the hierarchy.

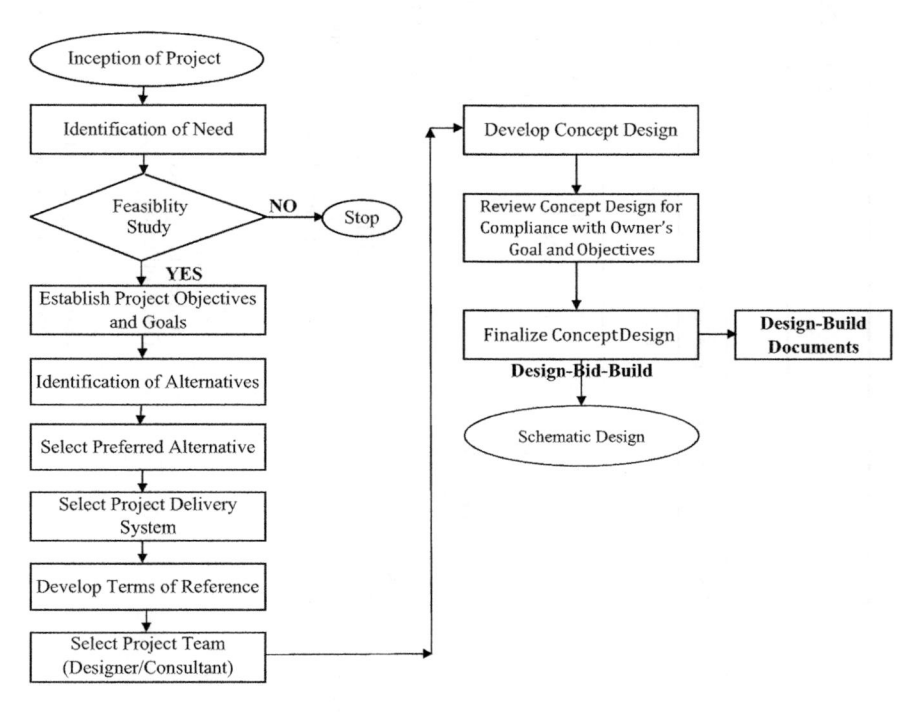

Figure 5.15 Logic flow process for the conceptual design phase

Conceptual Design Phase

Management Processes	Project Management Process Groups				
	Initiating Process	Planning Process	Execution Process	Monitoring & Controlling Process	Closing Process
Integration Management	Problem Statement/Need Statement Feasibility Study Project Goals and Objectives Project Terms of Reference	Preliminary Project Management Plan	Development of Alternatives Development of Concept Design Implement Changes Concept Drawings, Reports, Models Design Performance Report	Monitor Concept Design Progress Manage Owner Need, Changes Review of Concept Design Design Progress Design Status	Concept Design Deliverables
Stakeholder Management	Project Delivery System Project Team Members Regulatory Aothorities	Responsibility Matrix Establish Stakeholders Requirements		Evaluation and Approval of Alternative Aesthetics, Constructability Sustainability, Economy, Environmental Compatibility Authorities' Approval Compliance to Owner's Need	
Scope Management		Identification of Alternatives Concept Design Scope Data Collection Owner's Requirements Design Deliverables			

Schedule Management	Preliminary Schedule		Approval of Project Schedule
Cost Management	Conceptual Estimate		Control Project Cost
Quality Management	Quality Codes and Standards, Regulatory Requirements	Design Compliance to Codes, Standards and Regulatory Requirements	Conformance to Technical and Functional Capability, Energy Efficiency,
Resouce Management	Assign Project Design Team	Manage Team Members from Different Disciplines	Performance of Team Members	Assign New Phase/ Project
	Roles and Responsibilities of Team Members		
Communication Management	Design Progress Information	Liaison and Coordination with All Parties	Design Status Information
		Coordination Meetings	
Risk Management	Management of Design Risk		Control Design Risk
Contract Management	Project Delivery System	Design to Comply Contract Type/Pricing	Administer Project Delivery System Requirements	Demobilize Team Members
	Contracting System		
	Selection of Consultant for Feasibility Study		Check for Contracting System
	Selection of Design Team (Consultant)		

Figure 5.16 Major activities relating to the conceptual design process

Note: These activities may not be strictly sequential, however the breakdown allows implementation of project management function more effective and easily manageable at different stages of the project phase.

Conceptual Design Phase ↑

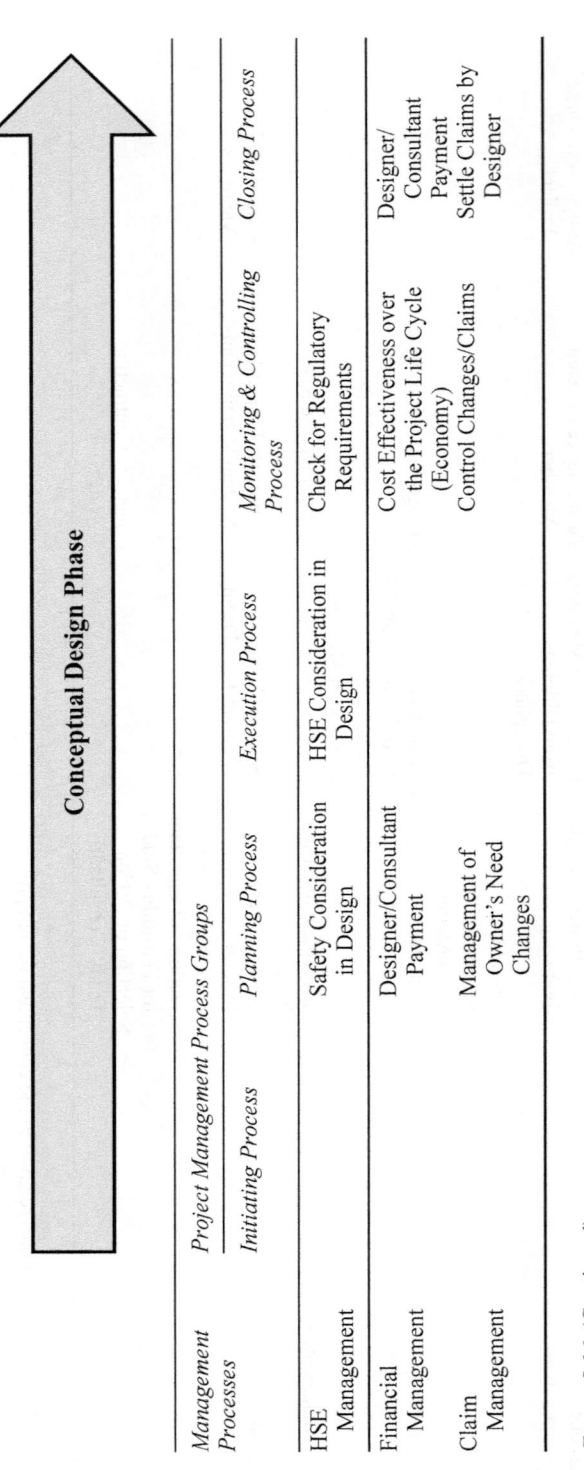

Management Processes	Project Management Process Groups				
	Initiating Process	*Planning Process*	*Execution Process*	*Monitoring & Controlling Process*	*Closing Process*
HSE Management		Safety Consideration in Design	HSE Consideration in Design	Check for Regulatory Requirements	
Financial Management		Designer/Consultant Payment		Cost Effectiveness over the Project Life Cycle (Economy)	Designer/Consultant Payment
Claim Management		Management of Owner's Need Changes		Control Changes/Claims	Settle Claims by Designer

Figure 5.16 (Continued)

5.2.2.1 Auditor for the study stage

Once the audit firm for the auditing of the study stage is contracted, it is important to select specific personnel (auditors) who will do the actual audit work of the project. The auditor should have the necessary qualifications, experience and track record of performing audits on the study stage of the project. The auditor should have relevant competence and expertise according to the auditor category as discussed in Section 5.1.2. In addition to the competence and expertise discussed in Section 5.1.2 the client (project owner or organization/authority desiring the audit) has to consider industry specific experience to ascertain the auditor's competence and expertise to audit the specific work. Table 5.14 lists the major selection criteria that need to be considered while selecting the auditor for the audit of the study stage of a construction project.

5.2.2.2 Auditor for the concept design

The audit team to perform concept design audits should include functional as well as subject matter experts. The team members should be familiar with concept

Table 5.14 Auditor selection criteria for the study stage audit

Serial Number	Competencies and Experience
1	Knowledge of activities to be performed during study stage as listed in Figure 5.15
2	Experience in need assessment
3	Review of business case
4	Review of feasibility study
5	Establishing goals and objectives of construction project
6	Preparation of Terms of Reference (TOR)
7	Identification of possible alternatives
8	Project schedule
9	Project budget
10	Knowledge about availability of resources
11	Knowledge of market conditions
12	Knowledge of various indices to develop new project
13	Capability to identify and measure areas of improvement to achieve project goals and objectives
14	Understanding of project life cycle cost analysis
15	Knowledge of regulatory requirements and approval criteria
16	Knowledge of environmental protection agency requirements
17	Knowledge of project safety regulations
18	Familiar with different types of project delivery system and method selection
19	Knowledge of different types of contracts used in construction projects
20	General understanding of the project
21	Identification and analysis of project risks
22	Understanding about project constraints and boundaries
23	Understanding of organizational strategy to develop new project
24	Familiarity with ISO audit procedures and requirements
25	Experience on similar types of projects

design practices applicable to construction projects. The auditor should have the necessary qualifications, experience and track record of performing audits on the concept design phase of the project. To perform audits for the concept design phase, the auditor should be a subject matter expert with experience in the relevant activities of the project phase/stage. The auditor should have diverse experience of concept design phase auditing that mainly involves the evaluation of a wide range of activities in the following fields:

1 Technical
2 Contractual
3 Time and cost
4 Quality requirements to achieve project goals and objectives
5 Risk identification and management

The auditor should have relevant competence and expertise according to the auditor category as discussed in Section 5.1.2. In addition to the competence and expertise discussed in Section 5.1.2 the client (project owner or organization/authority desiring the audit) has to consider industry specific experience to ascertain the auditor's competence and expertise to audit the specific work. Table 5.15 lists the major selection criteria that need to be considered while selecting the auditor for the audit of the concept design of a construction project.

Table 5.15 Auditor selection criteria for the concept design phase audit

Serial Number		Competencies and Experience
1		Knowledge of activities to be performed during conceptual design phase as listed in Figure 5.16
2		Aware of regulatory requirements
3		Knowledge of different codes and standards
4		Understanding of project goals and objectives
5		Familiar with usage of project
6		Knowledge of HSE requirements
7		Technical
	7.1	Familiar with alternatives
	7.2	Review of site conditions and related data
	7.3	Data collection methods
	7.4	Understanding of concept design
	7.5	Review and understanding of drawings
	7.6	Familiar with LEED requirements
	7.7	Technical report
	7.8	Sketches
	7.9	Models
8		Contractual
	8.1	Understanding of TOR (Project Charter)
	8.2	Bill of quantities
	8.3	Change management
	8.4	Specifications

Serial Number	Competencies and Experience
8.5	Different types of contracting systems
8.6	Stakeholder management
9	Knowledge of schedule levels
10	Knowledge of cost estimation methods
11	Understanding of project quality
11.1	Design quality management plan
11.2	Owner's quality requirements
12	Familiar with resource estimation and availability of resources
13	Identification of project risks and recommendations to mitigate and improvements
14	Familiar with ISO audit procedures and requirements
15	Experience on similar types and size of projects

5.2.3 Auditor for the schematic design phase

Schematic design is mainly a refinement of the elements in the conceptual design phase. Schematic design is also known as **preliminary design**. During this phase, design intent documents which quantify functional performance expectations and parameters for each system are commissioned. It is traditionally labeled as 30% design. Schematic design adequately describes information about all proposed project elements in sufficient detail to obtain the regulatory approvals, necessary permits and authorization. At this phase, the project is planned to the level where sufficient details are available for initial cost and schedule. This phase also include the initial preparation of all documents necessary to build the facility/construction project. The primary goal of this phase is to develop a clearly defined design based on the client's requirements. Figure 5.17 illustrates the logic flow process for the schematic design phase.

Figure 5.18 illustrates the major activities relating to the schematic design phase based on the project management process groups methodology.

The audit team to perform schematic design audits should include functional as well as subject matter experts. The team members should be familiar with schematic design practices applicable to construction projects. The auditor should have the necessary qualifications, experience and track record of performing audits on the schematic design of the project. To perform audits for the schematic design phase, the auditor should be a subject matter expert with experience in the relevant activities of the project phase/stage. The auditor should have diverse experience of schematic design phase auditing that mainly involves the evaluation of a wide range of activities in the following fields:

1 Technical
2 Contractual
3 Time and cost
4 Project quality requirements to achieve project goals and objectives
5 Risks identification and management
6 Design deliverables
7 Value engineering

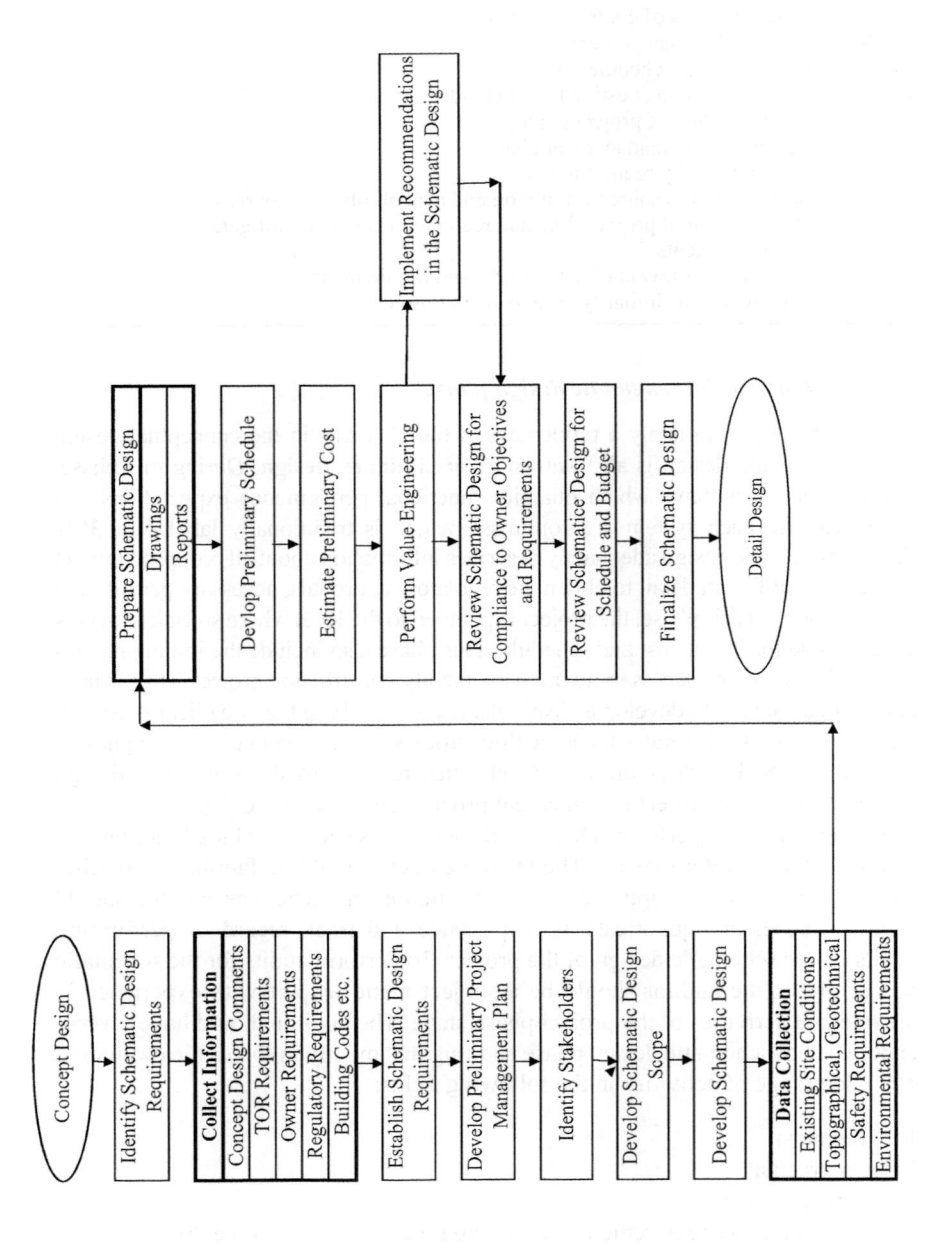

Figure 5.17 Logic flow process for the schematic design phase

Schematic Design Phase

Management Processes	Project Management Process Groups				
	Initiating Process	*Planning Process*	*Execution Process*	*Monitoring & Controlling Process*	*Closing Process*
Integration Management	Concept Design Deliverables Concept Design Comments TOR Requirements Authorities' Requirements	Preliminary Project Management Plan	Develop General Layout of Project (Site Plans) Architectural Plans Structural Scheme Plans Electromechanical Services Landscape and Infrastructure Develop Schematic Design Owner's Need Considerations	Building Code Requirements Concept Design Comments Existing Conditions Design Calculations Compliance to Regulatory/Athorities' Requirements Constructability	Schematic Design Deliverables
Stakeholder Management	Identify Project Team Identify Design Team	Stakeholders Requirements		Stakeholders' Requirements	

Figure 5.18 Major activities relating to the schematic design process

Note: These activities may not be strictly sequential, however the breakdown allows the implementation of project management function to be more effective and easily manageable at different stages of the project phase.

Schematic Design Phase

| Management Processes | Project Management Process Groups |||||
	Initiating Process	Planning Process	Execution Process	Monitoring & Controlling Process	Closing Process
Scope Management		Site Conditions Energy Conservation Requirements Technical and Functional Capability Outline Specifications System Schematics Value Engineering Design Deliverables		Authorities' Approval Compliance to Owner's Needs Stakeholders' Approval	
Schedule Management		Project Schedule (CPM/Bar Chart)	Preliminary Schedule	Project Schedule	
Cost Management		Cost of Activities Cost of Resources Preliminary Estimate	Preliminary Cost	Project Cost Estimate	
Quality Management		Design Criteria Codes and Standards Authorities Requirements	Design Coordination with All Disciplines	Compliance to Codes, Standards, and Authorities	

Resource Management	Assign Project Design Team	Manage Team Members of Difference Discipline	Performance of Team Members	Assign New Phase/ Project
	Estimate Resources			
Communication Management	Design Progress Information	Liaison and Coordination with All Parties Coordination Meetings	Design Status Information	
Risk Management	Management of Design Risk		Control Design Risk	
Contract Management	Contract Terms and Conditions	Preliminary Contract Documents	Check for Contracting System	
HSE Management	Safety Considerations in Design	Environmental Requirements	Life Safety in Design Requirements	
Financial Management	Designer/Consultant Payment			Designer/Consultant Payment
Claim Management	Design Change Payments		Control Changes	Settle Claim by Designer

Figure 5.18 (Continued)

The auditor should have the relevant competence and expertise according to the auditor category as discussed in Section 5.1.2. In addition to the competence and expertise discussed in Section 5.1.2 the client (project owner or organization/authority desiring the audit) has to consider industry specific experience to ascertain the auditor's competence and expertise to audit the specific work. Table 5.16 lists the major selection criteria that need to be considered while

Table 5.16 Auditor selection criteria for the schematic design phase audit

Serial Number		Competencies and Experience
1		Knowledge of activities to be performed during schematic design phase as listed in Figure 5.18
2		Aware of regulatory requirements and approvals
3		Knowledge of different codes and standards
4		Understanding of project goals and objectives
5		Familiar with usage of project
6		Knowledge of HSE requirements
7		Technical
	7.1	Review of concept design comments
	7.2	Knowledge about site investigations and data collection and analysis methods
	7.3	Understanding of schematic design
	7.4	Knowledge of quality tools used for design mistake proofing
	7.5	Ability of reviewing and understanding of drawings
	7.6	Familiar with LEED requirements
	7.7	Familiar with energy conservation requirements
	7.8	Familiar with review of technical reports
	7.9	Understanding of sketches/perspective
	7.10	Understanding of graphic presentation
8		Contractual
	8.1	Understanding of TOR (Project Charter)
	8.2	Bill of quantities
	8.3	Change management
	8.4	Specifications
	8.5	Different types of contracts
	8.6	Stakeholder management
	8.7	Design deliverables
9		Knowledge of schedule levels
10		Knowledge of cost estimation methods
11		Understanding of project quality
	11.1	Design quality management plan
	11.2	Owner's quality requirements
12		Familiar with resource estimation and availability of resources
13		Identification of project risks and recommendations to mitigate and improvements
14		Aware of value engineering process
15		Familiar with ISO audit procedures and requirements
16		Experience on similar types and size of projects

selecting the auditor for the audit of the schematic design phase of a construction project.

5.2.4 Auditor for the design development phase

Design development is the third phase of the construction project life cycle. It follows the preliminary design phase and takes into consideration the configuration and the allocated baseline derived during the preliminary phase. The design development phase is also known as **detail design/detailed engineering**. The client approved schematic (preliminary) design is the base for the preparation of design development or development of detail design. All the comments and suggestions on the schematic design from the client and regulatory bodies are reviewed and resolved to ensure that changes will not detract from meeting the project design goals/objectives. Detail design involves the process of successively breaking down, analyzing and designing the structure and its components so that it complies with the recognized codes and standards of safety and performance while rendering the design in the form of drawings and specifications that will tell the contractors exactly how to build the facility to meet the owner's need. Figure 5.19 illustrates the logic flow process for the design development phase.

Figure 5.20 illustrates the major activities relating to the design development phase based on the project management process groups methodology.

The audit team to perform detail design/design development phase audits should include functional as well as subject matter experts. The team members should be familiar with schematic design practices applicable to construction projects. The auditor should have the necessary qualifications, experience and track record of performing audits on the detail design of the project. To perform audits for the design development phase, the auditor should be a subject matter expert with experience in the relevant activities of the project phase/stage. The auditor should have diverse experience of design development phase auditing that mainly involves the evaluation of a wide range of activities in the following fields:

1 Technical
2 Contractual
3 Time and cost
4 Project quality requirements to achieve project goals and objectives
5 Risks identification and management
6 Design deliverables

The auditor should have relevant competence and expertise according to the auditor category as discussed in Section 5.1.2. In addition to the competence and expertise discussed in Section 5.1.2 the client (project owner or organization/authority desiring the audit) has to consider industry specific experience to ascertain the auditor's competence and expertise to audit the specific work. Table 5.17

176 Abdul Razzak Rumane

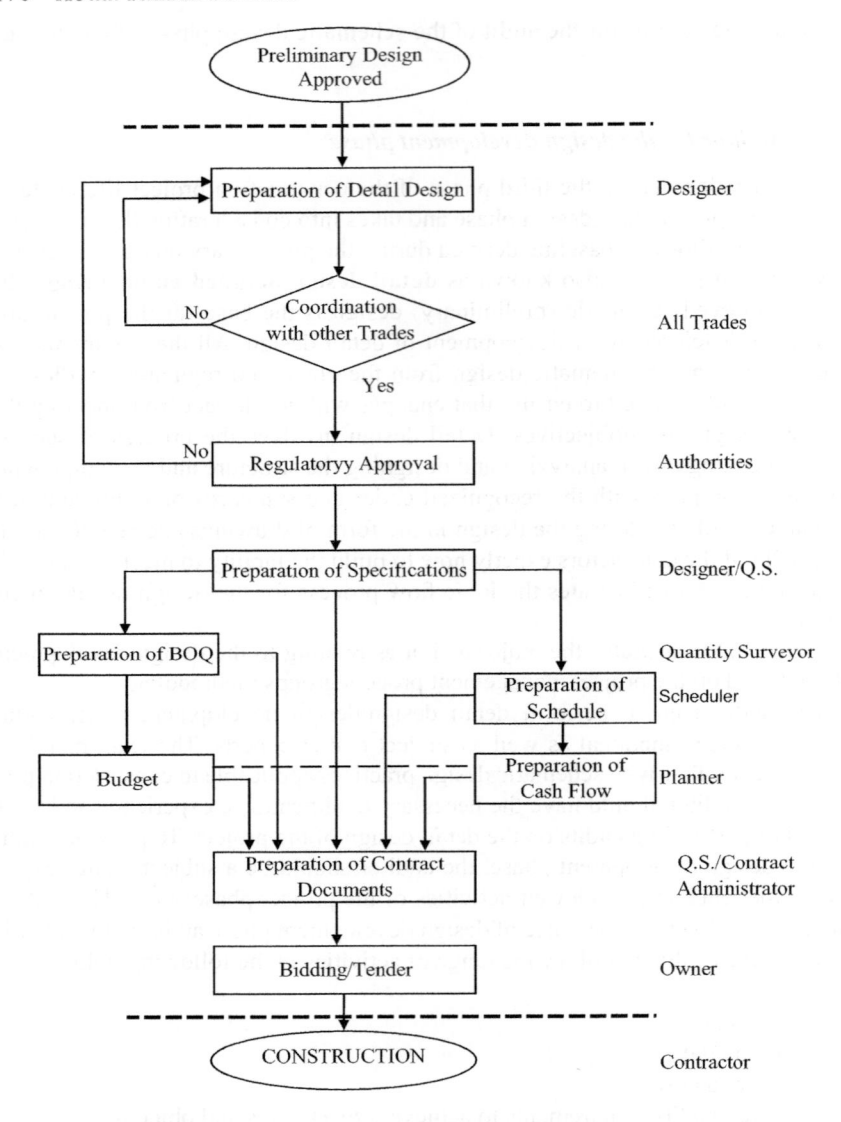

```
                    ┌─────────────────────┐
                   (  Preliminary Design   )
                   (      Approved         )
                    └──────────┬──────────┘
        ─ ─ ─ ─ ─ ─ ─ ─ ─ ─ ─ ─│─ ─ ─ ─ ─ ─ ─ ─ ─ ─ ─ ─
                               ▼
              ┌─────────────────────────────┐
      ┌──────▶│ Preparation of Detail Design │       Designer
      │       └──────────────┬──────────────┘
      │                      ▼
      │              ╱─────────────╲
      │    No       ╱ Coordination  ╲
      └────────────◀  with other Trades ▶           All Trades
                   ╲               ╱
                    ╲─────────────╱
                           │ Yes
                           ▼
           No     ┌───────────────────┐
      ┌──────────▶│ Regulatoryy Approval│                Authorities
      │          └──────────┬────────┘
      │                     ▼
      │          ┌─────────────────────────┐
      ├──────────│ Preparation of Specifications │──┐   Designer/Q.S.
      │          └─────────────────────────┘     │
      │                                          │
┌──────────────┐                        ┌──────────────┐
│Preparation of BOQ│                    │Preparation of│  Quantity Surveyor
└──────┬───────┘                        │   Schedule   │  Scheduler
       │                                └──────┬───────┘
       │                                       │
┌──────────────┐                        ┌──────────────┐
│    Budget    │                        │Preparation of│   Planner
└──────┬───────┘                        │  Cash Flow   │
       │                                └──────┬───────┘
       │        ┌───────────────────────┐     │
       └───────▶│ Preparation of Contract │◀──┘        Q.S./Contract
                │      Documents          │             Administrator
                └────────────┬────────────┘
                             ▼
                  ┌────────────────────┐
                  │   Bidding/Tender   │                Owner
                  └──────────┬─────────┘
        ─ ─ ─ ─ ─ ─ ─ ─ ─ ─ ─│─ ─ ─ ─ ─ ─ ─ ─ ─ ─ ─ ─
                             ▼
                    (   CONSTRUCTION   )                Contractor
```

Figure 5.19 Logic flow process for design development phase

lists the major selection criteria that need to be considered while selecting the auditor for the audit of the design development phase of a construction project.

5.2.5 Auditor for the construction documents phase

The construction documents phase is the fourth phase of the construction project life cycle. During this phase, the drawings and specifications prepared during the

Design Development Phase ↑

Management Processes	Process Management Groups				
	Initiating Process	Planning Process	Execution Process	Monitoring & Controlling Process	Closing Process
Integration Management	Schematic Design Deliverables Comments on Schematic Design TOR Requirements Authorities' Requirements	Project Management Plan	Data Collection/Site Investigations Detail Design Drawings Bill of Quantities Model	Design Calculations Interdisciplinary Coordination Compliance to TOR	Detail Design Drawings
Stakeholder Management	Identify Design Team	Identify Stakeholders Stakeholders' Requirements Stakeholders Matrix		Design Progress	
Scope Management		Owner's Needs, Project Goals and Objectives Design Development Design Documents Design Deliverables Bill of Quantity	Project Specifications	Authorities' Approval Compliance to Owner's Need Stakeholders' Approval	

Figure 5.20 Major activities relating to the design development phase

Note: These activities may not be strictly sequential, however the breakdown allows the implementation of the project management function to be more effective and easily manageable at different stages of the project phase.

Design Development Phase

Process Management Groups

Management Processes	Initiating Process	Planning Process	Execution Process	Monitoring & Controlling Process	Closing Process
Schedule Management		Activity Duration Precedence Diagram Construction Schedule		Project Schedule	
Cost Management		Price Analysis Bill of Quantities Resources Detail Estimate		Project Budget	
Quality Management		Codes and Standards Regulatory Requirements Design Crieteria Well-defined Specifications Plan Design Quality	Design Coordination with All Disciplines Assure Design Quality	Design Compliance to Owner's Goals and Objectives Coordination with all Disciplines Control Design Quality	
Resource Management		Estimate Project Resources	Manage Team Members from All Discipline	Performance of TeamProject Members	

Communication Management	Communication Matrix	Liaison with All Disciplines Coordination Meetings	Design Status Information
Risk Management	Identification of Risk during Bidding, Construction, Testing and Commissioning		Design Risk Control
Contract Management	Bidding and Tendering Documents		Check for Contracting System	Contract Documents
HSE Management	Safety in Design Environmental Compatibility	Safety Requirements Environmental Requirements	HSE Compliance in Design
Financial Management	Designer/Consultant Payment		Payment to Designer/Consultant
Claim Management	Design Change Payment		Control Changes	Settle Designer's Claim

Figure 5.20 (Continued)

Table 5.17 Auditor selection criteria for the design development phase audit

Serial Number		Competencies and Experience
1		Knowledge of activities to be performed during design development phase as listed in Figure 5.20
2		Aware of regulatory requirements and approvals
3		Knowledge of different codes and standards
4		Understanding of project goals and objectives
5		Familiar with usage of project
6		Knowledge of HSE requirements
7		Technical
	7.1	Understanding of schematic design review comments
	7.2	Knowledge about site investigations and data collection and analysis methods
	7.3	Understanding of detail design
	7.4	Knowledge of quality tools used for design mistake proofing
	7.5	Ability to review and understanding of drawings
	7.6	Familiar with quality tools for design coordination
	7.7	Familiar to resolve design conflict between different trades
	7.8	Knowledge of detail design review methods
	7.9	Familiar with review of technical reports and design calculation
	7.10	Familiar with analysis of design drawing quality
	7.11	Familiar with method statements in construction processes
	7.12	Safety in design
8		Contractual
	8.1	Understanding of TOR (Project Charter)
	8.2	Bill of quantities
	8.3	Change management
	8.4	Specifications
	8.5	Different types of contracts
	8.6	Stakeholder management
	8.7	Design deliverables
9		Knowledge of project schedule
10		Knowledge of definitive cost estimation
11		Understanding of project quality
	11.1	Design quality management plan
	11.2	Owner's quality requirements
12		Familiar with resource estimation and availability of resources
13		Identification of project risks and recommendations to mitigate and improvements
14		Bidding and tendering documents
15		New technologies in construction industry
16		Familiar with ISO audit procedures and requirements
17		Experience on similar types and size of projects

design development phase are further developed into working drawings. All the drawings, specifications, documents and other related elements necessary for the construction of the project are assembled and subsequently released for bidding and tendering. Figure 5.21 illustrates the logic flow process for the construction documents phase.

Figure 5.22 illustrates the major activities relating to the construction documents phase based on the project management process groups methodology.

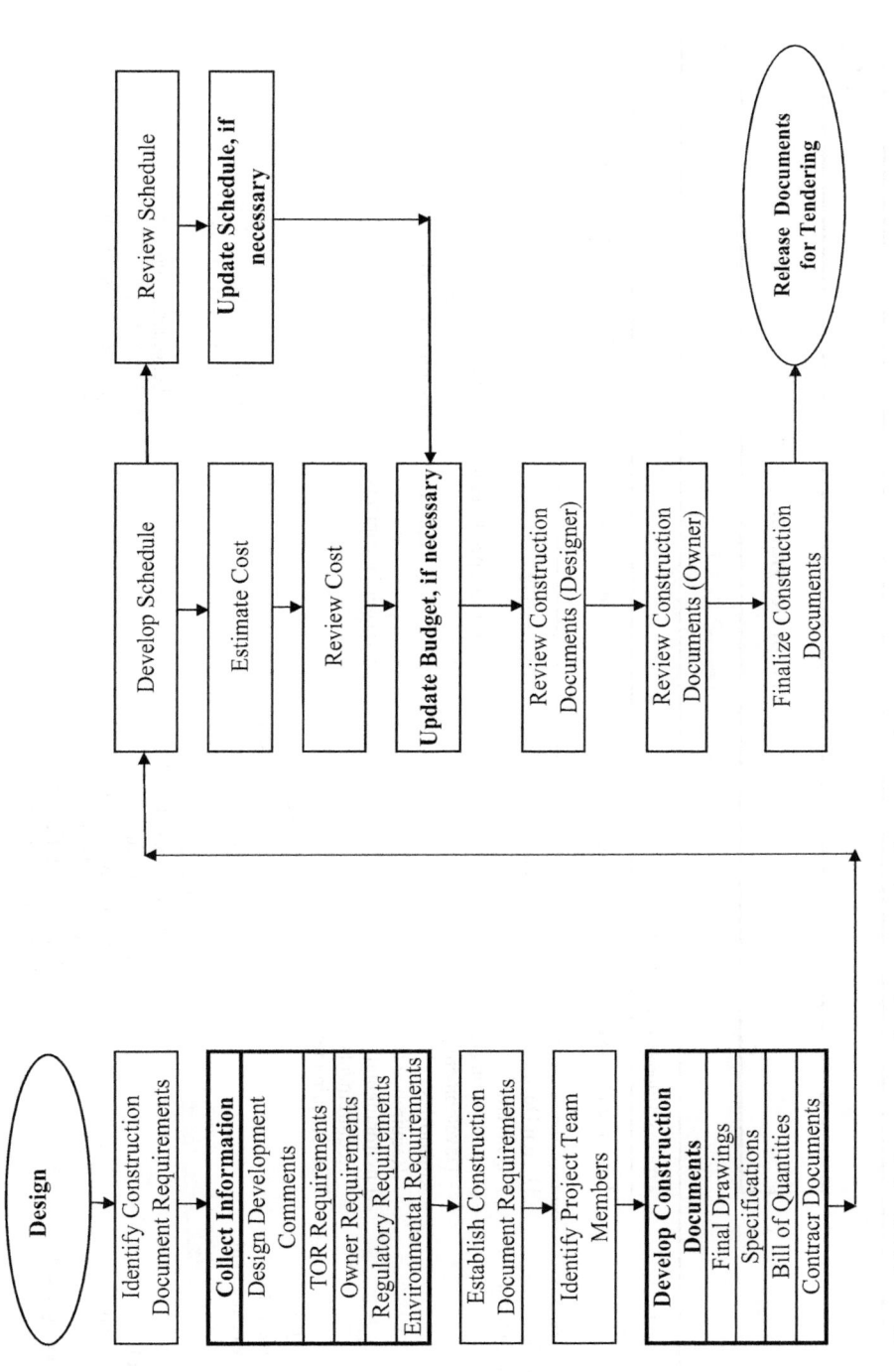

Figure 5.21 Logic flow process for the construction documents phase

Construction Document Phase ↑

Management Processes	Process Management Groups				
	Initiating Process	*Planning Process*	*Execution Process*	*Monitoring & Controlling Process*	*Closing Process*
Integration Management	Design Development Deliverables Comments on Design Development Authorities' Requirements Environmental Requirements TOR Requirements	Project Management Plan	Working Drawings Project Specifications Bill of Quantities Tender Documents	Design Calculations Interdisciplinary Coordination Compliance to TOR Design Progress Authorities Approval Compliance to Owner's Need Stakeholders' Approval Project Schedule Construction Document Schedule Project Budget	Working Drawings
Stakeholder Management		Project Stakeholders Stakeholders' Requirements			
Scope Management		Construction Documents Construction Documents Deliverables Tender Documents			
Schedule Management		Project Schedule Construction Document Schedule Deliverables			
Cost Management		Bill of Quantities Price Analysis Definitive Estimate			

Quality Management	Codes and Standards	Design Coordination with All Disciplines	Design Compliance to Owner's Goals and Objectives
	Regulatory Requirements	Assure Design Quality	Coordination with all Disciplines
	Design Quality		Control Design Quality
	Well-defined Specifications		
	Documents Quality		
Resource Management	Supervision Team Requirements	Manage Team Members from All Disciplines	Performance of Team/ Project Members	Assign New Project
	Contractor's Core Team		
	Contractor's Manpower		
Communication Management	Communication Matrix	Liaison with All Disciplines Coordination Meetings	Design Status Information	
Risk Management	Identification of Risk during Bidding, Construction, Testing and Commissioning		Design Risk Control	
Contract Management	Bidding and Tendering Documents		Check for Contracting System	Contract Documents
				Tender Documents
HSE Management	Safety in Design	Safety Requirements	HSE Compliance in Design	
	Environmental Compatibility	Environmental Requirements		
Financial Management	Designer/Consultant Payment		Payment to Designer/ Consultant	
Claim Management	Design Change Payment		Control Changes	Settle Designer's Claim

Figure 5.22 Major activities relating to the construction documents phase

Note: These activities may not be strictly sequential, however the breakdown allows the implementation of the project management function to be more effective and easily manageable at different stages of the project phase.

The audit team to perform construction documents phase audits should include functional as well as subject matter experts. The team members should be familiar with the relevant construction documents to be prepared in construction projects for the bidding and tendering of the construction project. The auditor should have the necessary qualifications, experience and track record of performing audits on the construction documents phase of the project. To perform audits for the construction documents phase, the auditor should be a subject matter expert with experience in the relevant activities of the project phase/stage. The auditor should have diverse experience of construction documents phase auditing that mainly involves the evaluation of a wide range of activities in the following fields:

1 Technical
2 Contractual
3 Time and cost
4 Project quality requirements
5 Risks identification and management
6 Availability of resources
7 Construction documents

The auditor should have the relevant competence and expertise according to the auditor category as discussed in Section 5.1.2. In addition to the competence and expertise discussed in Section 5.1.2 the client (project owner or organization/ authority desiring the audit) has to consider industry specific experience to ascertain the auditor's competence and expertise to audit the specific work. Table 5.18 lists the major selection criteria that need to be considered while selecting the auditor for the audit of the construction documents phase of a construction project.

Table 5.18 Auditor selection criteria for the construction documents phase audit

Serial Number		Competencies and Experience
1		Knowledge of activities to be performed during construction documents phase as listed in Figure 5.22
2		Aware of regulatory requirements and approvals
3		Understanding of TOR requirements
4		Knowledge about the construction documents to be submitted for bidding and tendering
5		Understanding of detail drawings and their compliance with TOR requirements
6		Technical
	6.1	Knowledge of detail design review methods
	6.2	Design quality
7		Contractual
	7.1	Bill of quantities
	7.2	Specifications
	7.3	Different types of contracts
	7.4	Stakeholder management

Serial Number	Competencies and Experience
8	Knowledge of project schedule
9	Knowledge of definitive cost estimation
10	Familiar with project quality requirements
11	Familiar with resource estimation and availability of resources
12	Identification of project risks
13	Project report
14	Familiar with ISO audit procedures and requirements
15	Experience on similar types and size of projects

5.2.6 Auditor for the bidding and tendering for the construction contractor

During this phase, tender documents are released for bidding and a contract is awarded to the successful bidder. Figure 5.23 illustrates the logic flow process for the bidding and tendering phase.

Figure 5.24 illustrates the major activities relating to the bidding and tendering phase based on the project management process groups methodology.

The audit team to perform bidding and tendering phase audits should include functional as well as subject matter experts. The team members should be familiar with the relevant bidding and tendering procedures to manage the bidding and tendering phase of the construction project. The auditor should have the necessary qualifications, experience and track record of performing audits on the bidding and tendering of the project. To perform audits for the bidding and tendering phase, the auditor should be a subject matter expert with experience in the relevant activities of the project phase/stage. The auditor should have diverse experience of bidding and tendering phase auditing that mainly involves the evaluation of a wide range of activities in the following fields:

1 Technical
2 Contractual
3 Time and cost
4 Project quality requirements
5 Risks identification and management
6 Availability of resources
7 Tendering documents

The auditor should have the relevant competence and expertise according to the auditor category as discussed in Section 5.1.2. In addition to the competence and expertise discussed in Section 5.1.2 the client (project owner or organization/authority desiring the audit) has to consider industry specific experience to ascertain the auditor's competence and expertise to audit the specific work. Table 5.19 lists the major selection criteria that need to be considered while selecting the auditor for the audit of the bidding and tendering phase of a construction project.

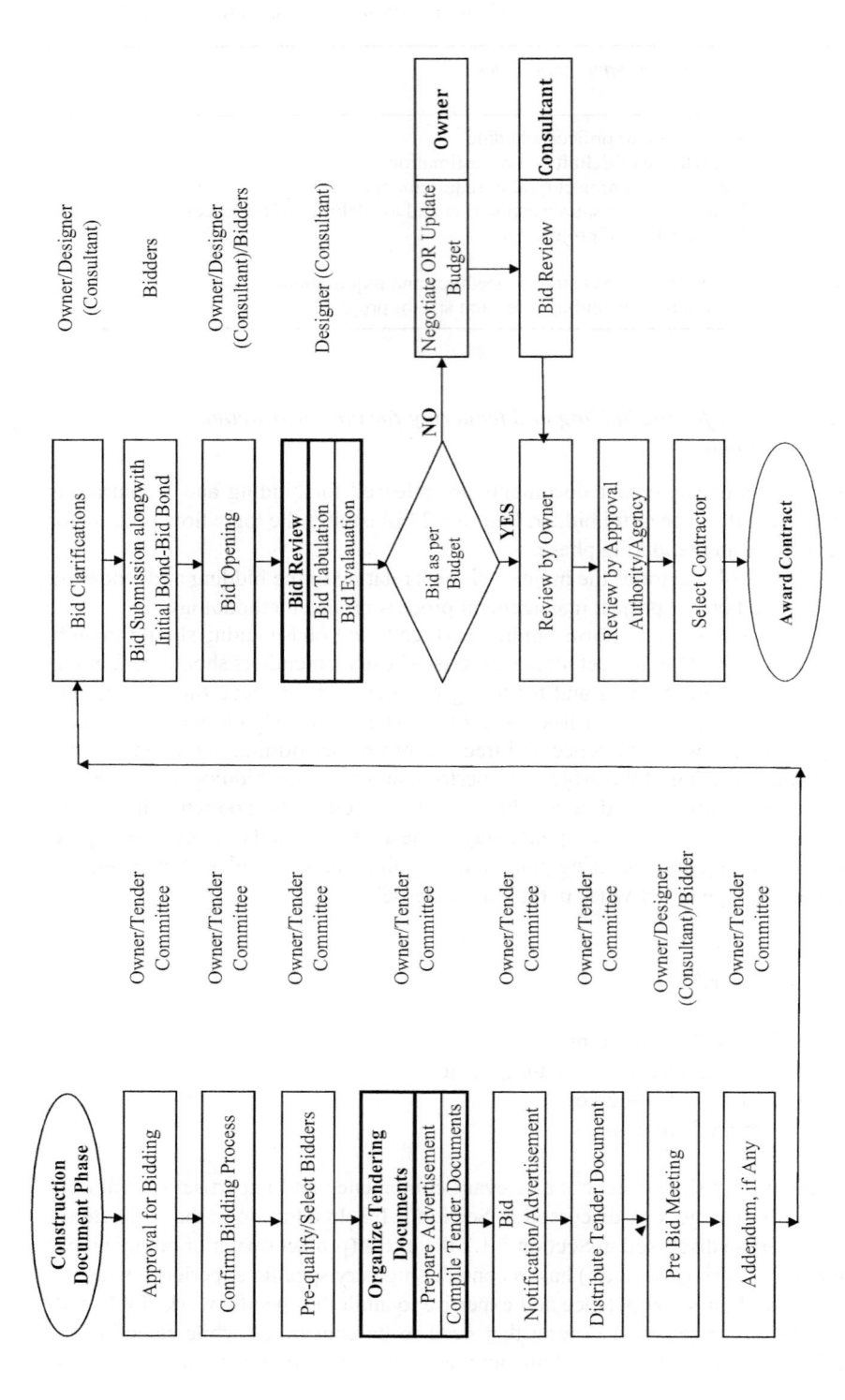

Figure 5.23 Logic flow process for the bidding and tendering phase

Bidding and Tendering Phase

Management Processes	Process Management Groups				
	Initiating Process	Planning Process	Execution Process	Monitoring & Controlling Process	Closing Process
Integration Management	Construction Documents TOR Requirements Owner's Requirements Authorities' Requirements	Organize Tender Documents	Tendering Documents		Award Contract
Stakeholder Management	Identify Bidders				
Scope Management		Bidder Selection Procedure Bid Review Procedure	Contractor Selection Bid Review	Addendum	
Schedule Management		Bid Period Bid Review Duration		Monitor Bid Duration Monitor Review Duration	

Figure 5.24 Major activities relating to the bidding and tendering phase

Note: These activities may not be strictly sequential, however the breakdown allows the implementation of the project management function to be more effective and easily manageable at different stages of the project phase.

Bidding and Tendering Phase ↑

Management Processes	Process Management Groups				
	Initiating Process	Planning Process	Execution Process	Monitoring & Controlling Process	Closing Process
Cost Management		Estimate Bid Price		Control Bid Value	
Quality Management					
Resource Management					
Communication Management		Advertize Tender	Conduct Meetings		
Risk Management			Manage Risk	Control Risk	
Contract Management		Select Bidders	Prepare Construction Contract		Signed Contract
HSE Management					
Financial Management				Update Project Finances	
Claim Management					

Figure 5.24 (Continued)

Table 5.19 Auditor selection criteria for the bidding and tendering phase audit

Serial Number		Competencies and Experience
1		Aware of logic flow of bidding and tendering procedure as per Figure 5.23
2		Knowledge of activities to be performed during bidding and tendering phase as listed in Figure 5.24
3		Aware of regulatory requirements and approvals
4		Understanding of TOR requirements
5		Knowledge about the tender documents to be submitted for bidding and tendering
6		Understanding of construction documents and drawings and their compliance with TOR requirements
7		Technical
	7.1	Knowledge of design review methods
8		Contractual
	8.1	Bill of quantities
	8.2	Specifications
	8.3	Different types of contracts
	8.4	Managing tender process
	8.5	Bid analysis and review
	8.6	Stakeholder management
9		Knowledge of project schedule
10		Knowledge of project estimation
11		Familiar with resource estimation and availability of resources
12		Identification of risks in tendering and selection of contractor
13		Bidder (contractor) selection procedure
14		Tender distribution and receiving procedures
15		Familiar with ISO audit procedures and requirements
16		Experience on similar types and size of projects

5.2.7 *Auditor for the construction phase*

Construction is the translation of the owner's goals and objectives, by the contractor, to build the facility as stipulated in the contract documents, plans and specifications within budget and on schedule. Construction is the sixth phase of the construction project life cycle and is an important phase in construction projects. A majority of the total project budget and schedule is expended during construction. Similar to costs, the time required to construct the project is much higher than the time required for the preceding phases. Construction usually requires a large number of workforce and variety of activities. Construction activities involve erection, installation or construction of any part of the project. Construction activities are actually carried out by the contractor's own workforce or by subcontractors. Construction therefore requires more detailed attention to its planning, organization, monitoring and control of the project schedule, budget, quality, safety and environment concerns. Figure 5.25 illustrates the logic flow process for the construction phase.

Figure 5.26 illustrates the major activities relating to the construction phase based on the project management process groups methodology.

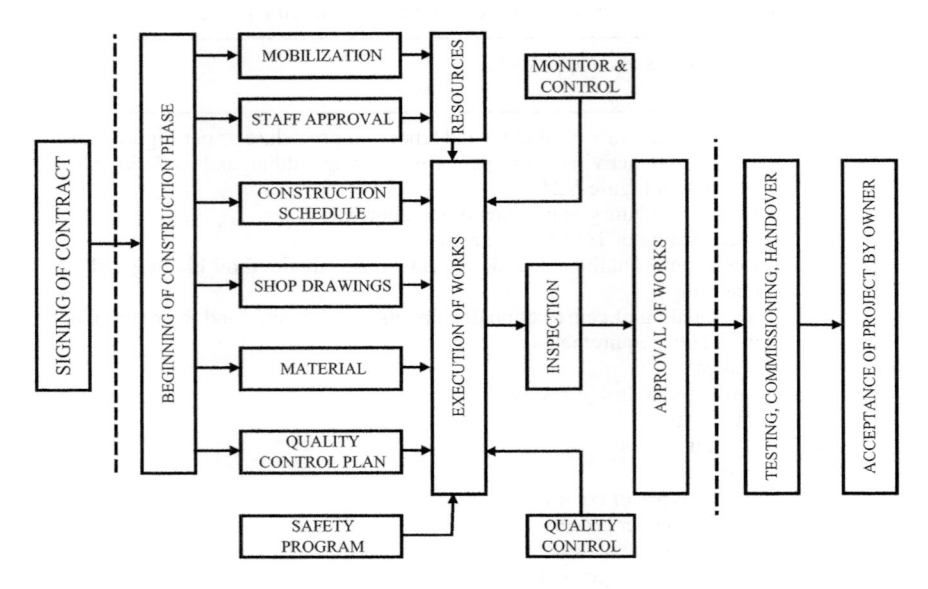

Figure 5.25 Logic flow process for the construction phase

The audit team to perform construction phase audits should include functional as well as subject matter experts. The team members should be familiar with the relevant construction process to manage the construction phase of the construction project. The auditor should have the necessary qualifications, experience and track record of performing audits on the construction phase of the project. To perform audits for the construction phase, the auditor should be a subject matter expert with experience in the relevant activities of the project phase/stage. The auditor should have diverse experience of construction phase auditing that mainly involves the evaluation of a wide range of activities in the following fields:

1. Technical
2. Contractual
3. Planning and scheduling
4. Budget
5. Quality management plans
6. Resource management
7. Risks management
8. Stakeholder management
9. Communication and documentation
10. Monitoring and control
11. HSE requirements

The auditor should have the relevant competence and expertise according to the auditor category as discussed in Section 5.1.2. In addition to the competence

Construction Phase ↑

Project Management Process Groups

Management Processes	Initiating Process	Planning Process	Execution Process	Monitoring & Controlling Process	Closing Process
Integration Management	Contract Documents Tender Documents Notice to Proceed	Construction Management Plan	Mobilization Submittals Execution Process Construction Work	Compliance to Contract Documents Change Management	Executed Project
Stakeholder Management	Owner's Representative Supervision Team Contrator's Core Staff Subcontractor Authorities	Responsibility Matrix Stakeholder Requirements Reports Meetings		Project Status/ Performance Report Payments Variation Orders Conflict Resolution	
Scope Management		Scope Change Management Preventive and Corrective Actions	Design Changes	Authorities' Approval Stakeholders' Approval Scope Change Control	

Figure 5.26 Major activities relating to the construction phase

Note: These activities may not be strictly sequential, however the breakdown allows the implementation of the project management function to be more effective and easily manageable at different stages of the project phase.

Construction Phase ↑

Management Processes	Project Management Process Groups				
	Initiating Process	*Planning Process*	*Execution Process*	*Monitoring & Controlling Process*	*Closing Process*
Schedule Management		Contractor's Construction Schedule		Alternate Material Site Work Instruction Variation Orders Preventive and Corrective Actions Plan Updates Schedule Monitoring Schedule Control Work Progress Monitoring Submittals Monitoring	
Cost Management		Contracted Value of Project Construction Budget		Cost Control Cash Flow Progress Payment Variation Orders	
Quality Management		Contractor's Quality Control Plan	Quality Assurance	Quality Control	

Resource Management	Resource Management Plan	Shop Drawings Builders Drawings Composite Drawings	
	Project Manpower	Material Approvals	Quality Auditing
	Subcontractor	Method Statement	Material Inspection
	Material and Equipment	Training of Project Team Members	Work Inspection/Testing Rework
		Manage Project Team	Regulatory Compliance
			Performance of Team Members
			Dispute Resolution
			Demobilization of Workforce
Communication Management	Kick off Meeting		
	Communication Plan	Site Administration Matrix	Performance of Workforce Meetings
	Submittals		
	Documentation		
	Correspondance		
Risk Management	Risk Management Plan		Control Risk
	Construction Risks Register		Risk Audit
Contract Management	Contract Management	Contract Documents	Inspection	Finalise Work Performed
	Plan Purchase of Material/Equipment	Selection of Subcontractor(s)	Check List	Finalise Material/Equipment Supplier's Contract
		Material, Systems and Equipment		
HSE Management	Safety Management Plan	Site Safety	Accident Prevention Measures	
	Waste Management Plan	Temporary Fire Fighting	Loss Prevention during Construction	

Figure 5.26 (Continued)

Construction Phase →

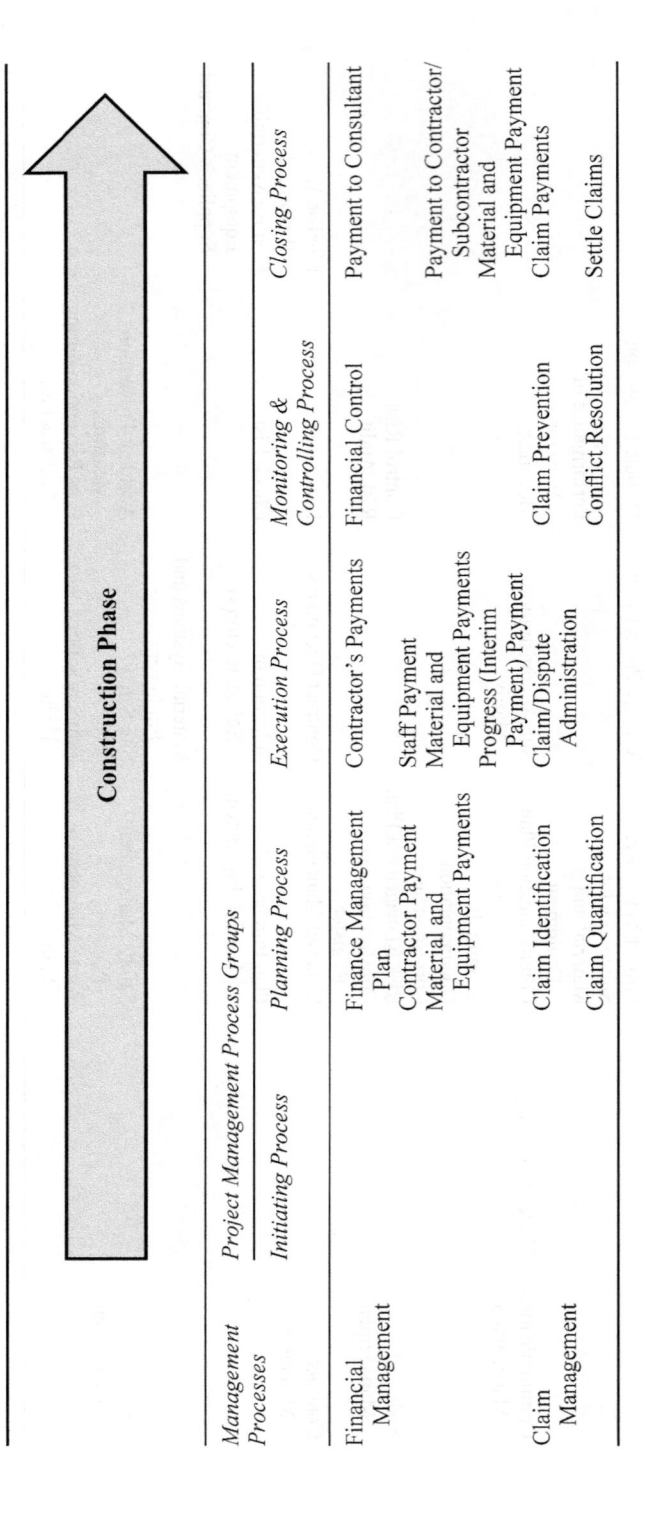

| Management Processes | Project Management Process Groups |||||
	Initiating Process	Planning Process	Execution Process	Monitoring & Controlling Process	Closing Process
Financial Management		Finance Management Plan Contractor Payment Material and Equipment Payments	Contractor's Payments Staff Payment Material and Equipment Payments Progress (Interim Payment) Payment	Financial Control	Payment to Consultant Payment to Contractor/Subcontractor Material and Equipment Payment
Claim Management		Claim Identification Claim Quantification	Claim/Dispute Administration	Claim Prevention Conflict Resolution	Claim Payments Settle Claims

Figure 5.26 (Continued)

Table 5.20 Auditor selection criteria for the construction phase audit

Serial Number		Competencies and Experience
1		Knowledge of activities to be performed during construction phase as listed in Figure 5.26
2		Aware of regulatory requirements and approvals
3		Understanding of contract requirements
4		Understanding of construction management processes
5		Understanding of construction documents and drawings
6		Technical
	6.1	Understanding of shop drawings, composited drawings, coordination drawings
	6.2	Familiar with construction method statements
	6.3	Material specifications, codes and standards
	6.4	Evaluation of substitute, alternative material
7		Contractual
	7.1	Bill of quantities
	7.2	Specifications
	7.3	Contract documents
	7.4	Logs and documentations
	7.5	Progress payments
	7.6	Change management,
	7.7	Stakeholder management
	7.8	Communication and administration matrix
8		Understanding of construction management plans
9		Knowledge of project schedule control
10		Knowledge of project cost control
11		Knowledge of quality management plan
10		Familiar with resource management and availability of resources
11		Identification of risks and management of risk
12		Contract management
13		Knowledge of HSE requirements
14		Claim management
15		Familiar with ISO audit procedures and requirements
16		Experience on similar types and size of projects

and expertise discussed in Section 5.1.2 the client (project owner or organization/authority desiring the audit) has to consider industry specific experience to ascertain the auditor's competence and expertise to audit the specific work. Table 5.20 lists the major selection criteria that need to be considered while selecting the auditor for the audit of the construction phase of a construction project.

5.2.8 *Auditor for the testing, commissioning and handover phase*

Testing, commissioning and handover is the last phase of the construction project life cycle. This phase involves the testing of electro-mechanical systems, commissioning of the project, obtaining authorities' approval, training of user's personnel and handing over of technical manuals, documents and as-built drawings to the owner/owner's representative. During this period the project is transferred handed over to the owner/end user for their use and a substantial completion certificate is issued to the contractor. Figure 5.27 illustrates the testing, commissioning and handover process.

Figure 5.28 illustrates the major activities relating to the testing, commissioning and handover phase based on the project management process groups methodology.

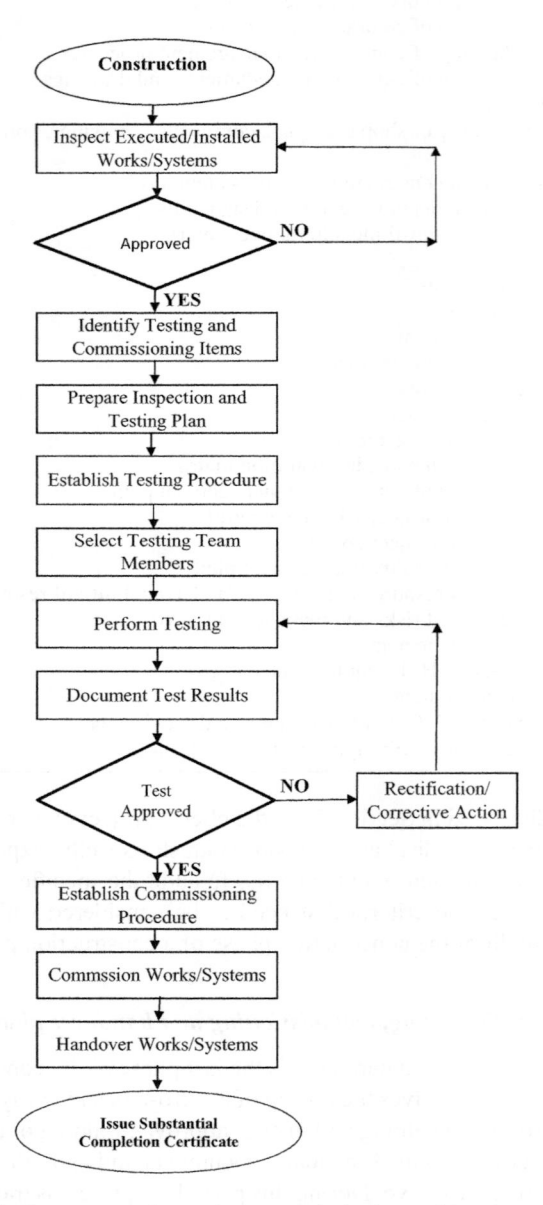

Figure 5.27 Logic flow process for testing, commissioning and handover phase

Testing, Commissioning and Handover Phase

Management Processes	Project Management Process Groups				
	Initiating Process	Planning Process	Execution Process	Monitoring & Controlling Process	Closing Process
Integration Management	Executed Project/ Facility Testing and Commissioning Program Contract Documents	Testing, Commissioning and Handover Plan	Testing and Commissioning Authorities' Approval Punch List/Snag List As-Built Drawings Manuals Spare Parts Move In Plan	Punch List/Snag List Testing and Commissioning Requirements	Handover of Project/ Facility
Stakeholder Management	Identify Stakeholders	Stakeholders' Requirements			
Scope Management		Contract Documents		Authorities' Approval Stakeholders' Approval	Project Acceptance/ Takeover
Schedule Management		Testing Schedule Commissioning Schedule			Lessons Learned
Cost Management					
Quality Management				Project Quality	

Figure 5.28 Major activities relating to testing, commissioning and handover phase

Note: These activities may not be strictly sequential, however the breakdown allows the implementation of the project management function to be more effective and easily manageable at different stages of the project phase.

Testing, Commissioning and Handover Phase

Management Processes	Project Management Process Groups				
	Initiating Process	Planning Process	Execution Process	Monitoring & Controlling Process	Closing Process
Resource Management		Demobilization Plan			Assign New Project/ Termination
Communication Management					
Risk Management		Test Results		Control Risk	
Contract Management		Plan Start-up Risk Prepare Contract Closeout Documents	Finalize Closeout Documents	Check Documents for Compliance to Contract Requirements	Close Contracts
HSE Management					
Financial Management		Financial Administration and Records	Payment to All Contractors and Subcontractors		Payment to All Contractors and Subcontractors Payment towards all Purchases
Claim Management		Claim Resolution		Check for Claims	Settlement of Claims

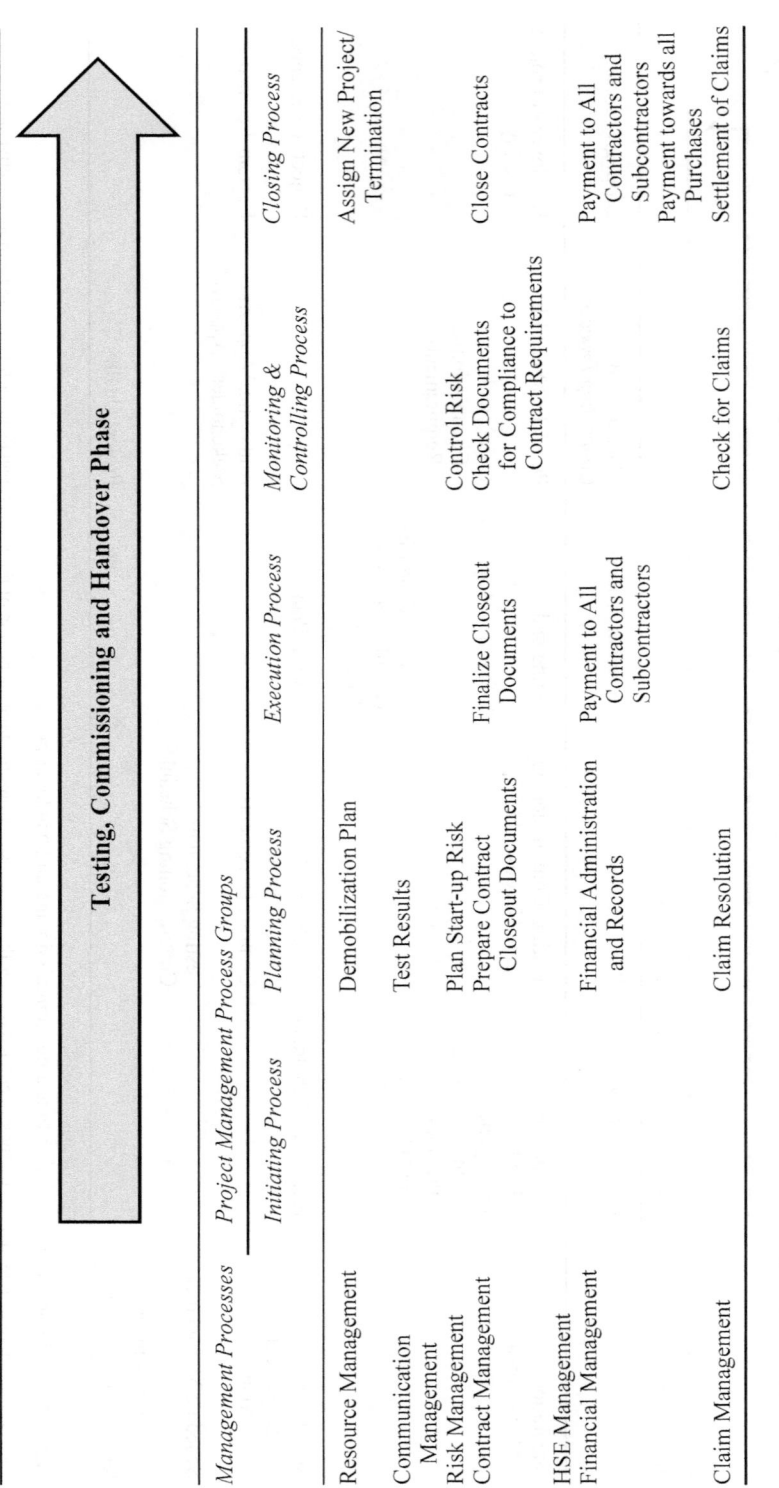

Figure 5.28 (Continued)

The audit team to perform audits of the testing, commissioning and handover phase should include functional as well as subject matter experts. The team members should be familiar with relevant testing and commissioning processes of all the systems and products installed in the project. The auditor should have the necessary qualifications, experience and track record of performing audits on the testing, commissioning and handover phase of the project. To perform audits for construction phase, the auditor should be a subject matter expert with experience in the relevant activities of the project phase/stage. The auditor should have diverse experience as an auditor, having performed audits of this phase that mainly involve evaluation of a wide range of activities and processes, in the following fields:

1 Technical
2 Contractual
3 Quality management plans

Table 5.21 Auditor selection criteria for the testing, commissioning and handover phase audit

Serial Number	Competencies and Experience
1	Knowledge of activities to be performed during testing, commissioning, and handover phase as listed in Figure 5.28
2	Aware of regulatory requirements and approvals
3	Understanding of contract requirements
4	Understanding of project goals and objectives
5	Understanding of project usage
6	Understanding of project quality
7	Technical
7.1	Testing procedures
7.2	Commissioning procedures
7.3	Punch list
7.4	Understanding of as-built drawings
7.5	Understanding of technical manuals
7.6	Spare parts
8	Contractual
8.1	Contract closeout documents
8.2	Handover and takeover procedure
8.3	Testing schedule
8.4	Commissioning schedule
8.5	Settlement of claims
8.6	Settlement of payments
8.7	Stakeholder management
8.8	Warrantees, guarantees
8.9	Substantial completion
8.10	Demobilization
9	Identification of startup risks and management of risk
10	Knowledge of HSE requirements
11	Familiar with ISO audit procedures and requirements
12	Experience on similar types and size of projects

4 Resource management
5 Risks management
6 Stakeholder management
7 Communication management
8 HSE requirements
9 Regulatory requirements

The auditor should have the relevant competence and expertise according to the auditor category as discussed in Section 5.1.2. In addition to the competence and expertise discussed in Section 5.1.2 the client (project owner or organization/ authority desiring the audit) has to consider industry specific experience to ascertain the auditor's competence and expertise to audit the specific work. Table 5.21 lists the major selection criteria that need to be considered while selecting the auditor for the audit of the testing, commissioning and handover phase of a construction project.

6 Auditing processes for project life cycle phases

6.1 Development of auditing processes for the construction project life cycle

Quality audit is a systematic, planned, independent and documented process to verify or evaluate and report on the degree of compliance to the agreed-upon quality criteria, or the specification or contract requirements of the product, services or project. Figure 6.1 illustrates the PDCA (Plan-Do-Check-Act) cycle for the auditing process, Figure 6.2 illustrates the typical quality auditing process for construction projects and Table 6.1 lists the typical contents of an audit plan.

Construction is the translation of the owner's goals and objectives, by the contractor, to build the facility as stipulated in the contract documents, plans and specifications on schedule and within budget .

A project is a plan or program performed by people with assigned resources to achieve an objective within a finite duration. The duration of a project is finite, projects are not ongoing efforts and the project ceases when its declared objectives have been attained.

Construction projects are custom oriented and custom designed, having specific requirements (defined scope) set by the customer/owner to be completed within a finite duration and assigned budget.

Construction projects comprise a cross-section of many different participants. These participants are both influenced by and depend on each other in addition to "other players" involved in the construction process. Therefore efforts are required to ensure the completion of a construction project within the agreed-upon schedule and within the approved budget to satisfy the owner's/client's/end user's intended need (defined scope). This phenomenon is also known as triple constraints. Figure 6.3 illustrates this.

6.1.1 Auditing performance criteria

Quality auditing in construction projects is a systematic and planned examination, evaluation, assessment and review of project management processes, project activities, project deliverables, systems, products and project status/compliance with respect to the required project performance by an internal or external auditor

- IMPLEMENT CORRECTIVE ACTION
- UPDATE QUALITY SYSTEM
- ARCHIVE RECORDS

- ESTABLISH OBJECTIVES/CRITERIA
- DETERMINE CATEGORY OF AUDIT
- IDENTIFY TEAM MEMBERS (AUDITOR)
- IDENTIFY TEAM MEMBERS (AUDITEE)
- IDENTIFY RESOURCES
- PLAN AUDIT
- METHODOLOGY

4. ACT | 1. PLAN
3. CHECK | 2. DO

- IDENTIFY NEED FOR CORRECTIVE ACTION
- IDENTIFY OPPORTUNITIES FOR IMPROVEMENT

- PERFORM AUDIT
- REVIEW DOCUMENTS, RECORDS
- DOCUMENT AUDIT FINDINGS
- PREPARE AUDIT REPORT
- NON COMPLIANCE REPORT

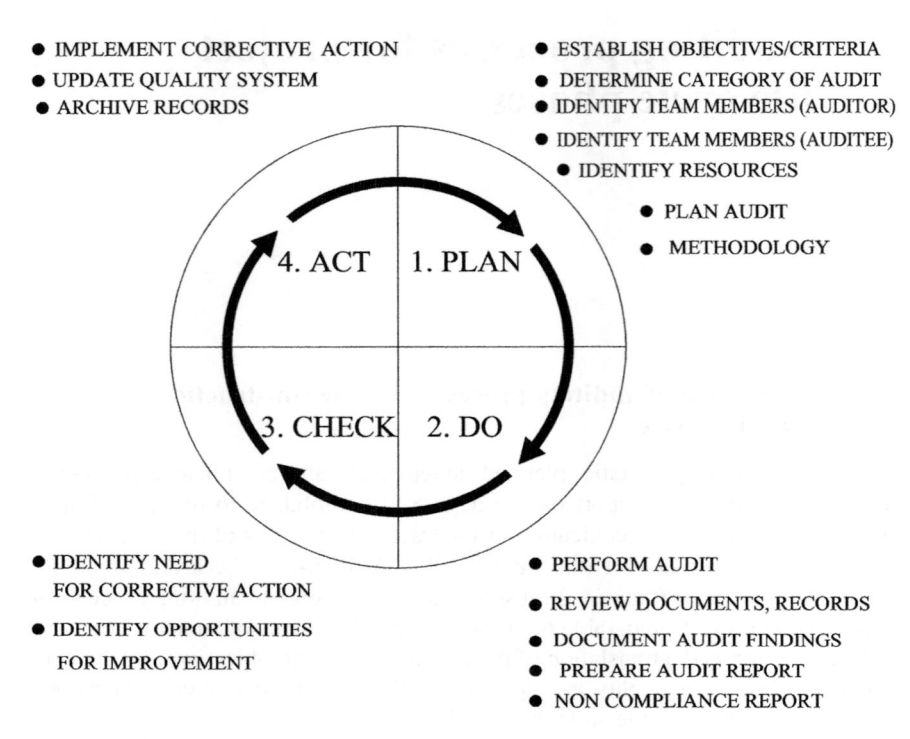

Figure 6.1 PDCA cycle for the auditing process

for providing the information to those (clients) in need of assurance in respect to their defined scope to achieve the goals and objectives of the project.

The quality audit process helps the completion of construction projects by achieving construction project quality (Figure 1.3) requirements as well as balancing triple constraints (Figure 6.3).

In order to plan, manage and control construction projects, the project life cycle is divided into a number of phases depending on construction processes, technologies, processes and complexity. The process audit process can be structured and performed in accordance with the individual phases.

Audit performance is the fieldwork carried out by the auditor at the audit location. It mainly consists of the following activities:

- Opening meeting
- Communication with relevant parties that will be involved in the audit
- Understanding the system and process to be followed
- Understanding of audit purpose and objectives
- Confirmation of audit plan
- Understanding of audit methodology
- Review of documents and records

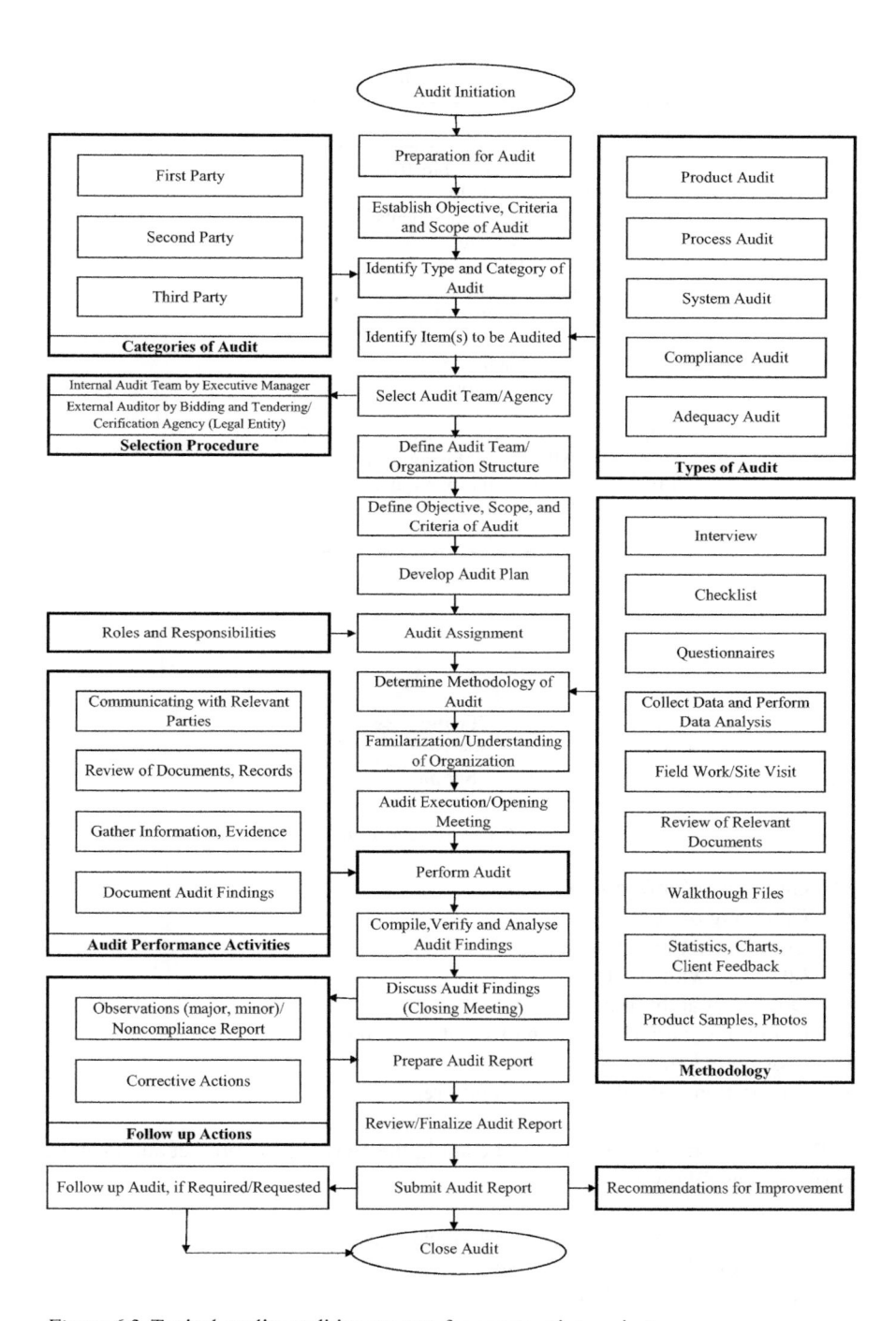

Figure 6.2 Typical quality auditing process for construction projects

Table 6.1 Contents of an audit plan

Section	Topic
1	Purpose and objective
2	Scope of audit
3	Applicable documents
4	Audit schedule
	4.1 Date, duration
	4.2 Location
5	Audit methodology, tools
6	Logic arrangements
7	Team members
	7.1 Auditor
	7.2 Auditee
8	Roles and responsibilities, assignment
9	Allocation of resources
10	Confidentiality requirements
11	Distribution of report
12	Follow-up action

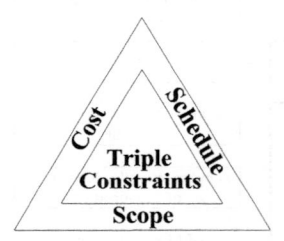

Figure 6.3 Triple constraints

- Gathering of information; data
- Gathering of evidence
- Communication with team members
- Interview with auditee
- Document audit findings
- Compiling, verifying and analysis of audit finding
- Closing meeting
- Preparation and submission of audit report.

Depending on the objectives of the audit, the construction project audit can be performed as follows:

- During the execution of each individual phase
- After completion of each phase
- After completion of the whole project

The audit for construction project life cycle phases can be performed by a first party, third party and, for certain activities, by a second party.

There are several types of project delivery systems in a construction project. Table 6.2 illustrates the most common project delivery systems followed in construction projects.

The methodology applied in this book covers quality auditing processes mainly in the following project delivery systems:

1. Design-Bid-Build
2. Design-Build
3. Project Manager
4. Construction Management (Agency Construction Management)

Table 6.2 Categories of project delivery systems

Serial Number	Category	Classification	Sub-classification
1	Traditional System (Separated & Cooperative)	Design-Bid-Build Variant of Traditional System	Design-Bid-Build Sequential Method Accelerated Method
2	Integrated System	Design-Build Design-Build	Design-Build Joint Venture (Architect and Contractor)
		Variant of Design-Build System	Package Deal
		Variant of Design-Build System	Turnkey Method (EPC) (Engineering, Procurement, Construction)
		Variant of Design-Build System (Turnkey)	Build-Operate-Transfer (BOT) Build-Own-Operate-Transfer (BOOT) Build-Transfer-Operate (BTO) Design-Build-Operate-Maintain (DBOM)
		Variant of Design-Build System (Funding Option)	Lease-Develop-Operate (LDO) Wraparound (Public-Private Partnership)
		Variant of Design-Build System	Build-Own-Operate (BOO) Buy-Build-Operate (BBO)
3	Management Oriented System	Management Contracting Construction Management	Project Manager (Program Management) Agency Construction Manager Construction Manager-At-Risk
4	Integrated Project Delivery System	Integrated Form of Contract	

Source: Abdul Razzak Rumane. (2013). *Quality Tools for Managing Construction Projects*. Reprinted with permission of Taylor & Francis Group

Figures 6.4a–d illustrate the contractual relation between different parties in these types of project delivery system.

The following sections discuss the audit process for each phase of the construction project life cycle in these types of project delivery system.

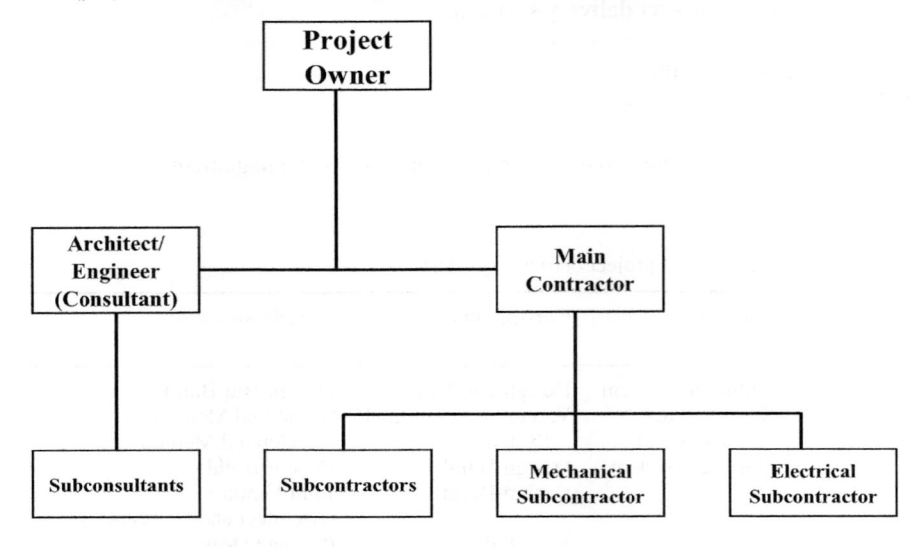

Figure 6.4a Design-Bid-Build (traditional contracting system) contractual relationship

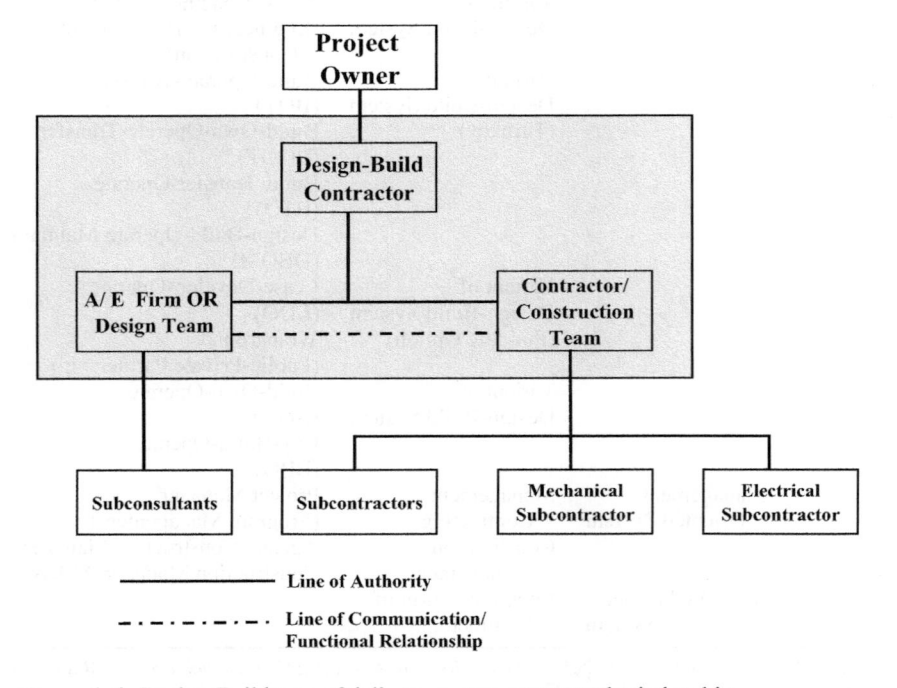

Figure 6.4b Design-Build type of delivery system contractual relationship

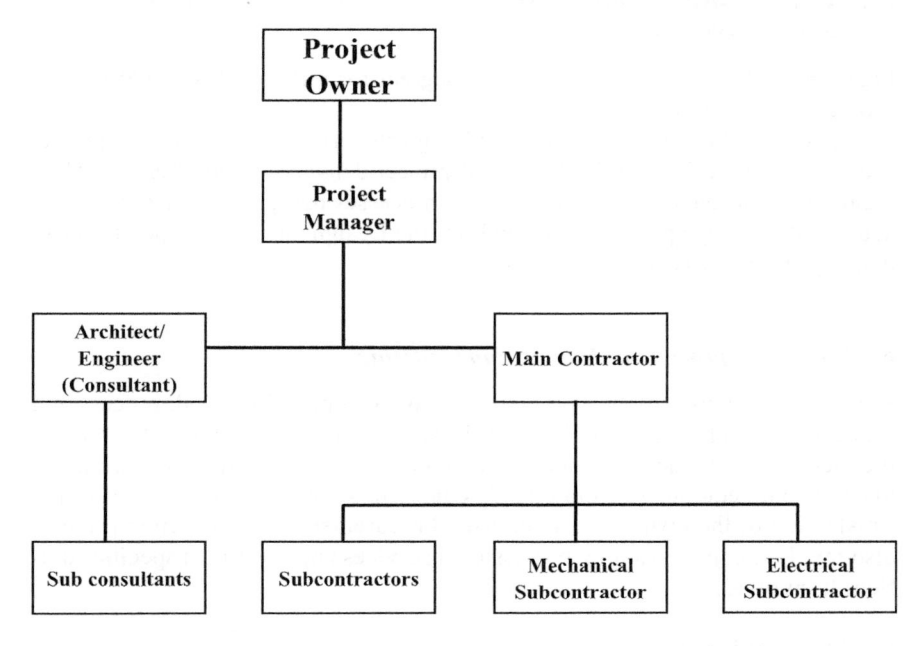

Figure 6.4c Project Manager type of delivery system contractual relationship

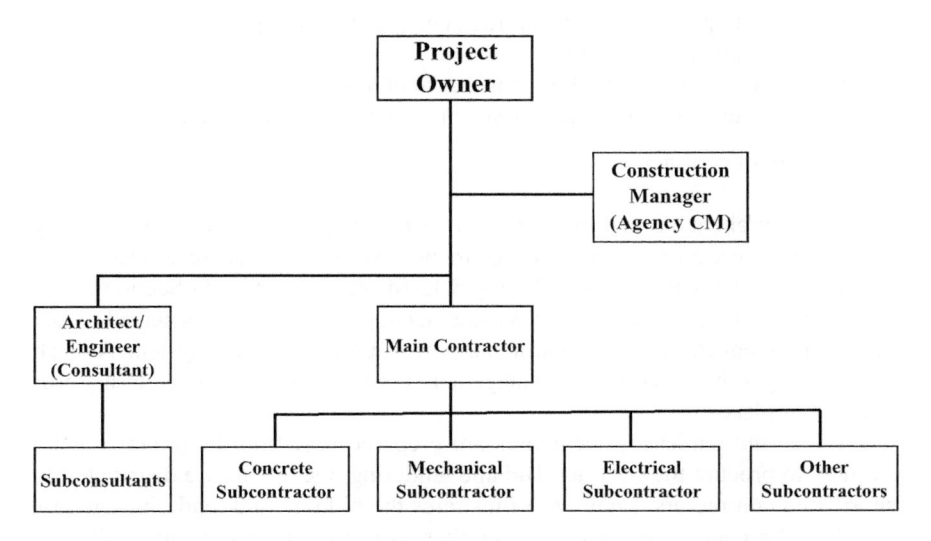

Figure 6.4d Agency construction management contractual relationship

6.2 Auditing process for the Design-Bid-Build type of project delivery system

Figure 6.5a shows a typical logic flow diagram for a Design-Bid-Build type of project delivery system.

Figure 6.5b illustrates the quality auditing categories in a construction project life cycle of a Design-Bid-Build type of project delivery system. Figure 6.5b is mainly to understand the quality auditing processes using these categories. The audit methodology applied in this book for the Design-Bid-Build type of project delivery system is as per Figure 6.5a.

6.2.1 Auditing process for bidding and tendering

Procurement in construction projects is an organizational method, process and procedure to obtain the required services, systems and products. It includes the bidding and tendering process to acquire all the related services, products/ materials and equipment from outside contractors/consultant/companies to the satisfaction of the owner/client/end user. Procurement in construction projects also involves commissioning professional services and creating a specific solution. It involves:

- identification of:
    - What services are available in-house
    - What services are to be procured from outside agencies/organizations
    - How to procure (direct contract, competitive bidding)
    - How much to procure
    - How to select a supplier/consultant/contractor
    - How to arrive at an appropriate price, terms and conditions
- Signing of contract

The owner/client has to decide which work is to be procured and constructed by others. Every organization has its procurement system to procure services, contracts and products from others. Figure 5.12 (discussed earlier in Section 5.2.1) illustrates the stages at which the outside agency (contractor) is selected as per the procurement strategy for a particular type of project delivery system. At each of these stages the procurement management process (bidding and tendering process) takes place.

Bidding and tendering documents are prepared based on the organizational strategy to procure the contract. Bid and tendering documents are distributed to shortlisted contractors. Figure 6.6 illustrates the bidding and tendering process to select the contractor/consultant/supplier as per the procurement method in accordance with the contractor selection strategy for the selection of specific services.

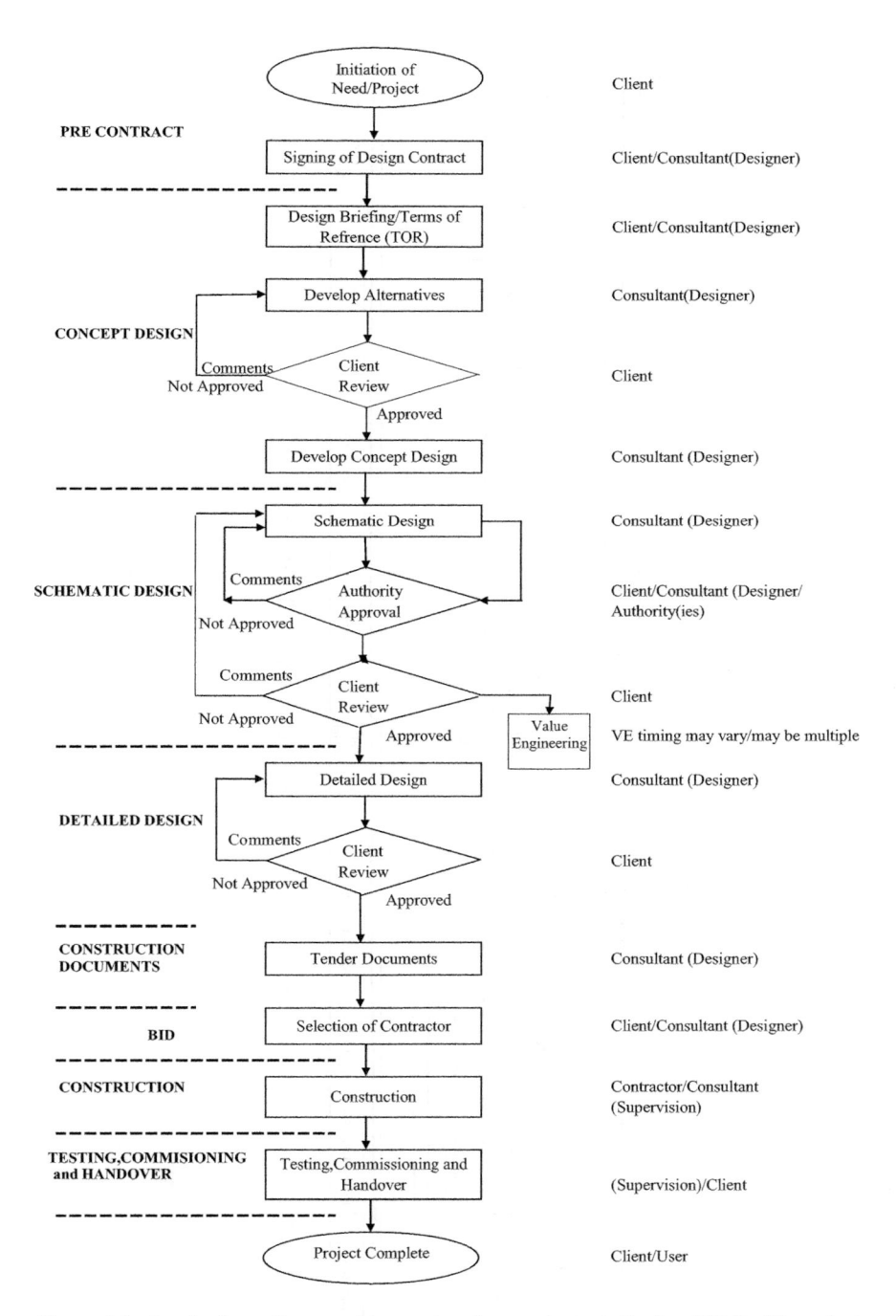

Figure 6.5a Logic flow diagram for construction projects – Design-Bid-Build project delivery system

Source: Abdul Razzak Rumane (2010), *Quality Management in Construction Projects.* Reprinted with permission from Taylor & Francis Group

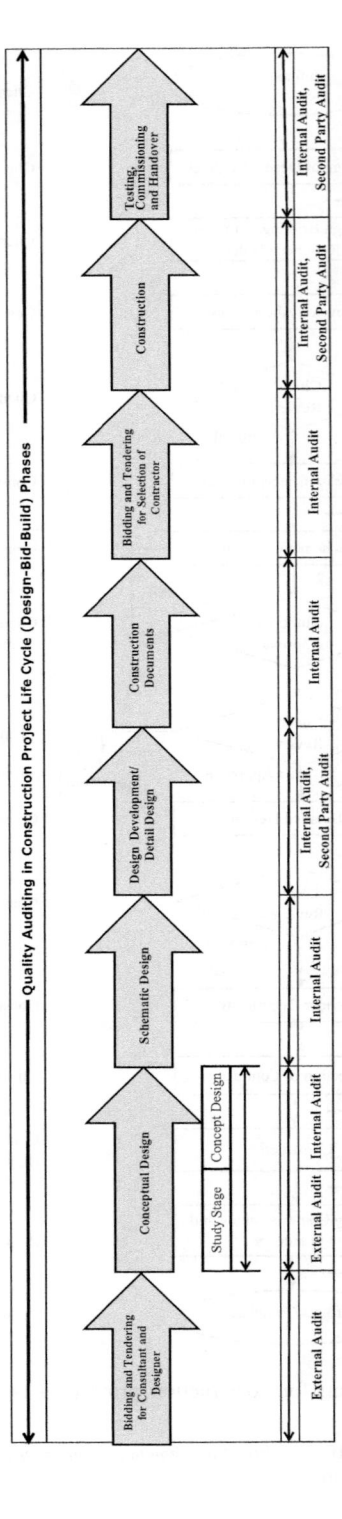

Figure 6.5b Quality auditing categories in the construction project life cycle (Design-Bid-Build) phases

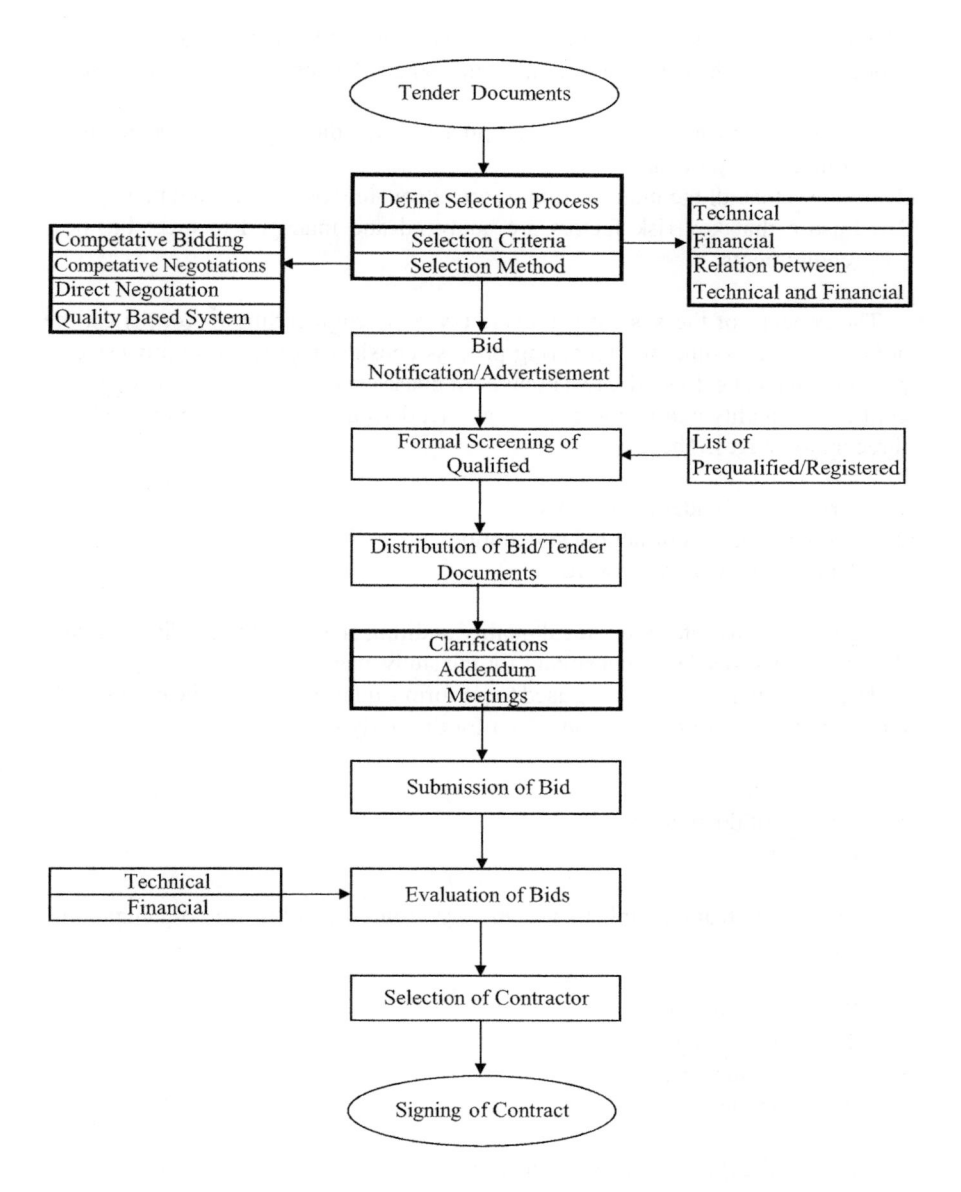

Figure 6.6 Bidding and tendering (procurement) process

6.2.1.1 Auditing process for the selection of the consultant for the study stage

The main objectives of performing a quality audit for the bidding and tendering process to select the consultant for the study stage of a construction project are to:

1. Evaluate and assess process compliance with the organizational strategic policies and procedure
2. Assess that all the major activities are adequately performed and managed
3. Ensure that key risk factors are analyzed and managed to meet business requirements

The concept of the system life cycle (systems engineering approach) can be applied to the bidding and tendering process considering the entire process as a project that can be divided into a number of phases/stages. In order to manage and control the quality audits, the bidding and tendering process can be divided into three major stages. These are:

1. First stage: Tender documents
2. Second stage: Contract bid solicitation
3. Third stage: Contract award

Figure 6.7 illustrates the typical quality auditing stages of the bidding and tendering process to select a consultant for the study stage.

The following audit tools are used to perform a quality audit for the bidding and tendering process to select a consultant for the study stage:

- Checklists
- Review of documents
- Interviews

The auditee team members are mainly from the following organizational groups:

- Business manager
- Tendering manager
- Financial manager
- Project manager

Based on the typical stages as shown in Figure 6.7, the audit methodology can also be divided into three stages.

Table 6.3 illustrates the audit methodology to assess/evaluate the process compliance at stage I, Table 6.4 illustrates the audit methodology to assess/evaluate the process compliance at stage II and Table 6.5 illustrates the audit methodology to assess/evaluate the process compliance at stage III.

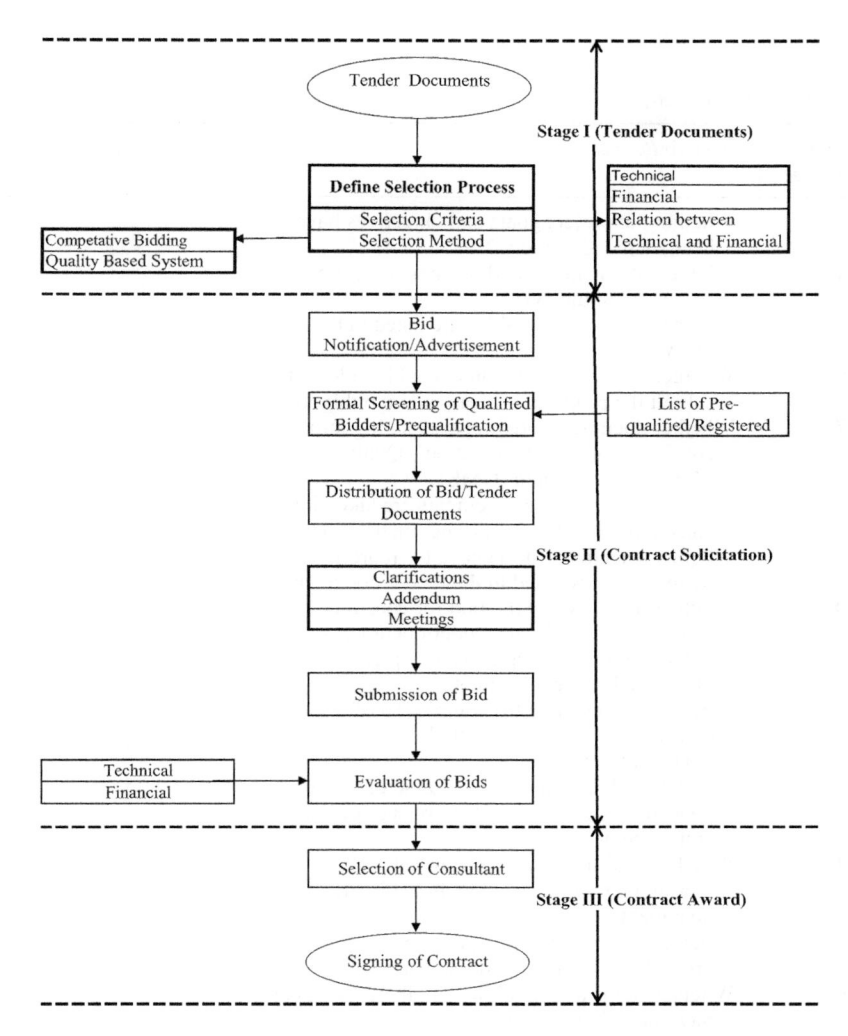

Figure 6.7 Quality auditing stages of the bidding and tendering (procurement) process for the selection of the consultant

Table 6.3 Audit methodology for the assessment of the bidding and tender process for the selection of a consultant – Stage I (tender documents)

Serial Number	Item To Be Assessed	Yes	No	Comments
A. Checklist				
1	Whether the documents specify conditions/ qualifications for participation in the tender			
2	Whether the statement of work (SOW) is properly defined			

(*Continued*)

Table 6.3 (Continued)

Serial Number	Item To Be Assessed	Yes	No	Comments
3	Whether the project goals and objectives have been included in the SOW			
4	Whether all the needs and business case are addressed in the SOW			
5	Whether all the deliverables are listed in the SOW			
6	Whether all the information to enable bidders to submit the proposal is clear and unambiguous			
7	Whether documents are prepared taking into consideration the organization's Quality Management System (QMS)			
8	Whether regulatory requirements are taken into consideration while preparing the documents			
9	Whether selection criteria have been properly defined and specified in the tender documents			
10	Whether the selection method for consultant is clearly specified in the tender documents			
11	Whether tender opening and closing dates are mentioned in the documents			
12	Whether the schedule for completion of consultancy services is included in the documents			
13	Whether budget approval is obtained for consultancy services prior to release of bid announcement			
14	Whether qualifications of personnel for consultancy services are included in the documents			
15	Whether project quality requirements are properly defined in the documents			
16	Whether rights and liabilities of all the parties are mentioned in the documents			
17	Whether change/variation clauses are included in the documents			
18	Whether cancellation/termination clauses are included in the documents			
19	Whether bond and insurance clauses are included in the documents			
20	Whether the tender documents are as per international standard documents used by the construction industry			
21	Whether evaluation and assessment criteria are well defined			
22	Whether review and analysis of documents are included			
23	Whether agreement to be signed for consultancy services is included in the documents			
24	Whether all the records are properly maintained			
25	Whether the request to provide the bidder's contact details is included in the documents			

B. Interview Questionnaires

Serial Number	Description	Response	Comments
1	Whether any outside agency was involved in preparation of tender documents		
2	What guidelines were adopted to prepare the tender documents		
3	Whether any specific format/template was followed		
4	Whether all the requirements of tender documents were coordinated and agreed by relevant stakeholders		
5	Whether risks of disputes and associated financial costs were considered while preparing the SOW		
6	Whether problems and challenges were identified while preparing tender documents		
7	Who was responsible for authorizing bidding requirements		

Table 6.4 Audit methodology for the assessment of the bidding and tender process for the selection of a consultant – Stage II (contract bid solicitation)

Serial Number	Item To Be Assessed	Yes	No	Comments
A. Checklist				
1	Whether tender documents are properly organized			
2	Whether there is a list of prequalified/registered consultants to participate in the tender			
3	Whether the tender is announced in leading news media as per the organization's policy			
4	Whether the tender is notified to all the prequalified consultants			
5	Whether all the prequalified/registered bidders participated in the tender			
6	Whether the attendance sheet for the meeting is signed by all the attendees			
7	Whether minutes of the meeting are circulated to all the attendees/participating bidders			
8	Whether sufficient time is provided to prepare and submit the proposal/quotation			
9	Whether the tender submission date was extended and reason for extension			

(*Continued*)

Table 6.4 (Continued)

Serial Number	Item To Be Assessed	Yes	No	Comments
A. Checklist				
10	Whether an addendum (if any) is issued/notified to all bidders			
11	Whether there are any requests for extension of submission date			
12	Whether the clarification meeting is attended by all participating bidders			
13	Whether any changes are made to the originally announced tender documents			
14	Whether received bid envelopes are placed in safe custody			
15	Whether technical and financial envelopes are submitted at same time			
16	Whether there are any delayed submissions			
17	Whether tenders are opened as per the announced date and time			
18	Whether tenders are opened in presence of all the bidders			
19	Whether received tenders are acknowledged			
20	Whether received tenders are clearly identified and recorded			
21	Whether applicable fees/bid bond are submitted by all the bidders who participated in the tender			
22	Whether tenders are fairly evaluated as per the announced policy and procedure			
23	Whether comparison of bids is tabulated			
24	Whether evaluation criteria, weightage and methodology are as specified			
25	Whether the bid cost is higher than the budgeted cost			
26	Whether any submitted tender is found to be incomplete			
27	Whether there are any late submissions			
28	Whether all the scheduled pages are signed by the bidder			
29	Whether risk factors are considered while evaluating the bids			
30	Whether reasons for non-acceptance of tender are conveyed to the unsuccessful bidders			

B. Interview Questionnaires

Serial Number	Description	Response	Comments
1	How tenders were submitted; • Hand delivery • By post • Courier service		
2	Whether tender was available for electronic distribution		
3	How was the response to participate in tender		
4	Whether tender boxes were placed at secured places and monitored through electronic surveillance system		
5	How much time was allowed to open the bids after the notified tendering closing time		
6	Whether all the tender envelopes/packages were sealed with marking of tenders and contain original tender documents and required number of sets as per tender conditions		

Table 6.5 Audit methodology for the assessment of the bidding and tender process for the selection of a consultant – Stage III (contract award)

Serial Number	Item To Be Assessed	Yes	No	Comments
A. Checklist				
1	Whether the selection of the consultant satisfies all the conditions to be a successful bidder			
2	Whether the selected bidder (consultant) is a legal entity			
3	Whether the selection is made by the selection committee			
4	Whether the selection is approved by the relevant applicable authority, if mandated by the organization's policy			
5	Whether the selected consultant is capable of carrying out the specified work			
6	Whether risk factors are considered prior to the signing of the contract			
7	Whether the standard contract format is used for signing the contract			

(*Continued*)

Table 6.5 (Continued)

Serial Number	Item To Be Assessed	Yes	No	Comments
A. Checklist				
8	Whether the contract between the owner and designer (A/E) has the following key elements: i Project definition ii Project schedule iii Scope of work iv Design deliverables v Owner responsibilities vi Fees for the services vii Variation order viii Penalty for delay ix Liability toward design errors x Insurance xi Arbitration/dispute resolutions xii Taxes xiii Appointment of subconsultant xiv Selection of team members xv Duties (responsibilities) xvi Compliance to authority's requirements xvii Suspension of contract xviii Termination xix Glossary			
9	Whether contract terms and conditions are clearly written and unambiguous			
10	Whether an addendum and minutes of meetings are included in the contract documents			
11	Whether the contract period/schedule is properly described			
12	Whether staff qualifications are properly defined			
13	Whether a performance bond is submitted by the consultant			
14	Whether variation/change management clauses are included			
15	Whether dispute resolution and conflict resolution clauses are properly defined			
16	Whether contract documents are reviewed prior to signing			
17	Whether tender documents and contract are properly archived			
18	Whether a letter of award is sent to the selected consultant			
19	Whether final results are published and announced			

B. Interview Questionnaires

Serial Number	Description	Response	Comments
1	On what basis the bidder (consultant) was selected: low bid or quality based system		
2	Was there any tie between two or more bidders		
3	Was there any dispute or objection raised by any of the bidders before the award of the contract to the successful bidder		
4	Whether the final price was negotiated or as per the bid received		

6.2.1.2 Auditing process for the selection of the designer

Normally, design professionals (A/E) are hired on the basis of **qualifications**. The qualification based selection can be considered as meeting one of the 14 points of Deming's principles of transformation which states "End the practice of awarding business on the basis of price alone." The basis of selection is solely based on demonstrated competence, professional qualification and experience for the type of services required. In quality based selection, the contract price is negotiated after the selection of the best qualified firm.

Figure 5.12 (discussed in Section 5.2.1) illustrate the stages at which a designer (A/E) is selected as per the procurement strategy for a particular type of project delivery system. The scope of work for the designer (A/E) varies according to the type of project delivery system. The scope of work for the designer of a Design-Bid-Build type of project delivery system is different to that of a designer for a Design-Build type of project delivery system. The tender document differs as the works to be performed are different in each case. The designer for a Design-Bid-Build type of project delivery is responsible for:

1. Development of concept design
2. Development of preliminary design
3. Development of detail design
4. Development of construction documents
5. Coordinating with project owner for bidding and tendering

The designer's responsibilities in a Design-Build type of project delivery system consist of:

1. Development of concept design
2. Development of construction documents
3. Coordinating with project owner for bidding and tendering

Figure 6.8 illustrates the typical quality auditing process for bidding and tendering for the selection of a designer (A/E).

The main objectives of performing a quality audit for the bidding and tendering process to select the designer (A/E) for a design construction project are to:

1. Evaluate and assess process compliance with the organizational strategic policies and procedure
2. Assess conformance to QMS and that processes meet established quality system requirements
3. Verify that all the major activities are adequately performed and managed
4. Ensure that key risk factors are analyzed and managed to meet business requirements
5. Check proper functioning and implementation of the project management system

The following audit tools are used to perform a quality audit for the bidding and tendering process to select a designer:

- Checklists
- Review of documents
- Interviews
- Questionnaires
- Analysis of audit documentation

The auditee team members are mainly from the following organizational groups:

- Business manager
- Tendering manager
- Financial manager
- Project manager

In order to manage and control the bidding and tendering process for the selection of a designer, the process can be divided into four major stages. These are:

1. First stage: Shortlisting/prequalification of designers
2. Second stage: Proposal documents
3. Third stage: Contract bid solicitation
4. Fourth stage: Contract award

Figure 6.9 illustrates the bidding and tendering process stages for the selection of a designer (A/E). The four stages discussed in Figure 6.9 are developed taking into considerations the basics discussed earlier as illustrated in Figure 5.14 (Section 5.2.1.2). Table 6.6 lists the prequalification questionnaires (PQQ) to register a designer (A/E).

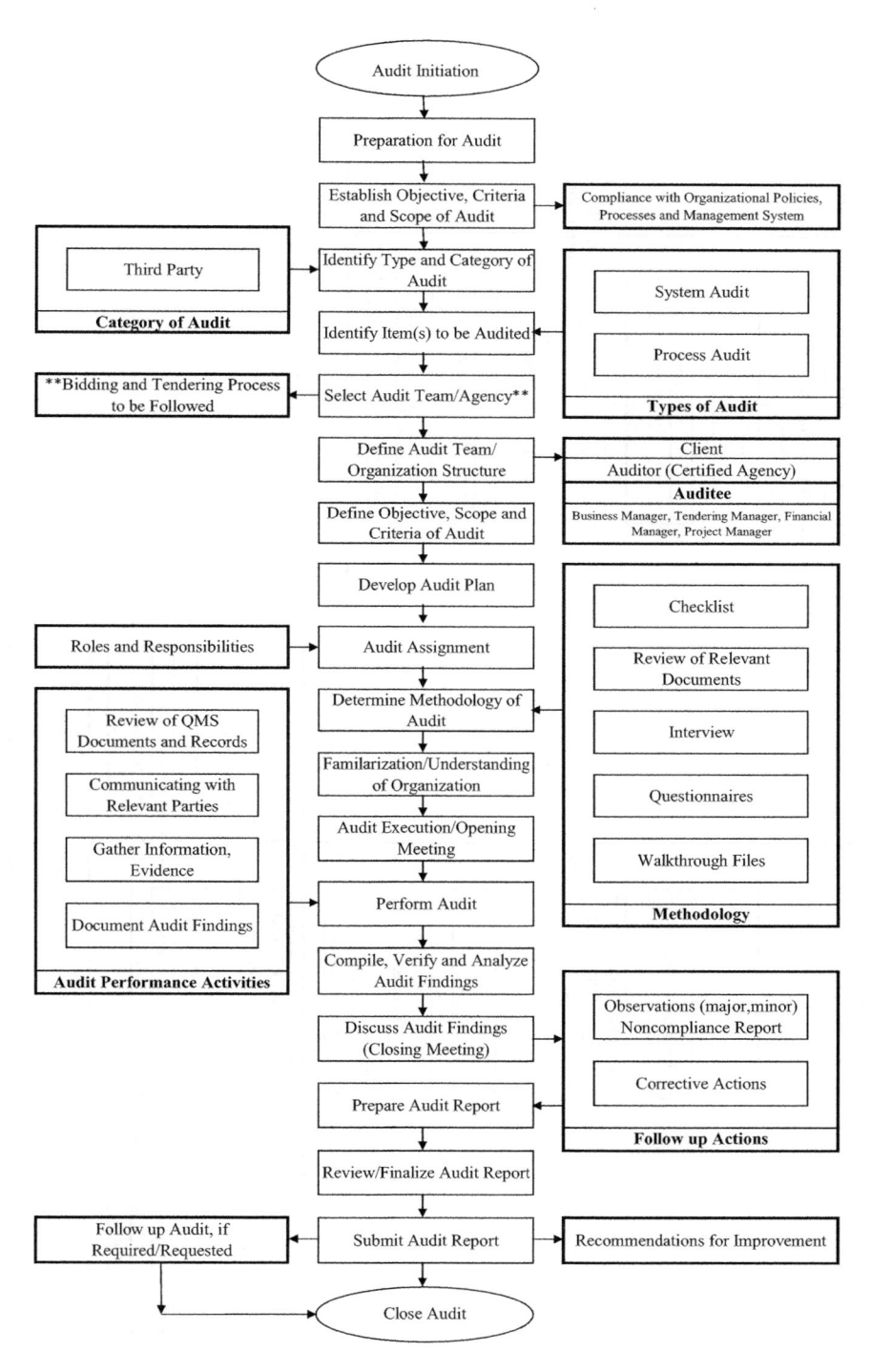

Figure 6.8 Typical quality auditing process for the bidding and tendering procedure for the selection of the designer (A/E)

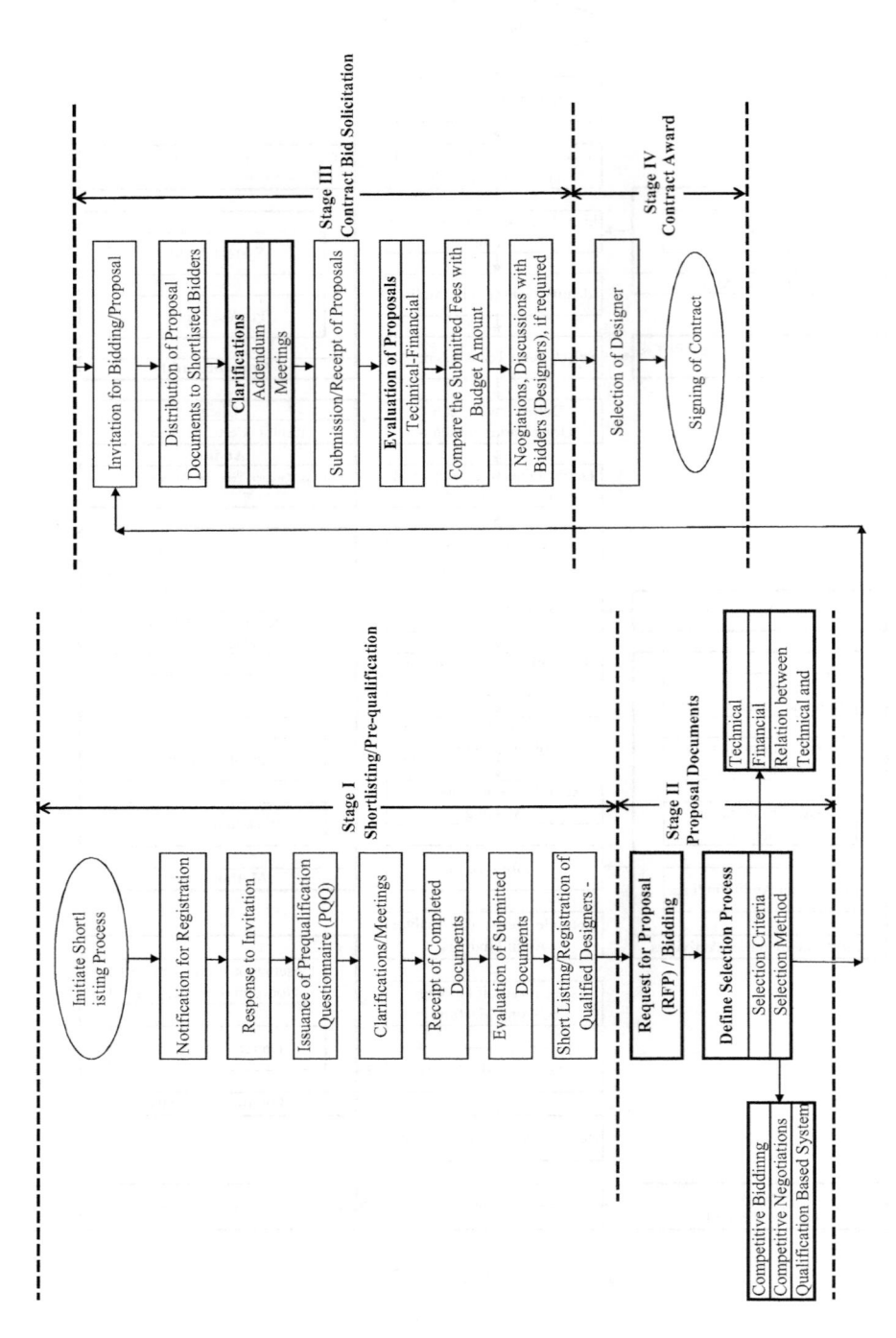

Figure 6.9 Bidding and tendering (procurement) process stages for the selection of the designer (A/E)

Table 6.6 Prequalification questionnaires (PQQ) for the registration of a designer (A/E)

Serial Number	Question	Answer
1	Name of the organization and address	
2	Organization's registration and license number	
3	ISO certification	
4	LEED or similar certification	
5	Total experience (years) in designing following type of projects 5.1 Residential 5.2 Commercial (mixed use) 5.3 Institutional (governmental) 5.4 Industrial 5.5 Infrastructure 5.6 Design-Build (specify type)	
6	Size of project (maximum amount single project) 6.1 Residential 6.2 Commercial (mixed use) 6.3 Institutional (governmental) 6.4 Industrial 6.5 Infrastructure 6.6 Design-Build (specify type)	
7	List successfully completed projects 7.1 Residential 7.2 Commercial (mixed use) 7.3 Institutional (governmental) 7.4 Industrial 7.5 Infrastructure 7.6 Design-Build	
8	List similar type (type to be mentioned) of projects completed 8.1 Project name and contracted amount 8.2 Project name and contracted amount 8.3 Project name and contracted amount 8.4 Project name and contracted amount 8.5 Project name and contracted amount	
9	Total experience in green building design	
10	Joint venture with any international organization	
11	Resources 11.1 Management 11.2 Engineering 11.3 Technical 11.4 Design equipment 11.5 Latest software	
12	Design production capacity	
13	Design standards	
14	Present work load	
15	Experience in value engineering (list projects)	
16	Financial capability (turnover for last 5 years)	
17	Financial audited report for last 3 years	
18	Insurance and bonding capacity	
19	Organization details 19.1 Responsibility matrix 19.2 CVs of design team members	
20	Design review system (quality management during design)	
21	Experience in preparation of contract documents	
22	Knowledge about regulatory procedures and requirements	

(*Continued*)

224   Abdul Razzak Rumane

Table 6.6 (Continued)

Serial Number	Question	Answer
23	Experience in training of owner's personnel	
24	List of professional awards	
25	Litigation (dispute, claims) on earlier projects	

Based on the typical stages as per Figure 6.9, the audit methodologies can also be performed in four stages.

Table 6.7 illustrates the audit methodology to assess/evaluate the process compliance at stage I, Table 6.8 illustrates the audit methodology to assess/evaluate the process compliance at stage II, Table 6.9 illustrates the audit methodology to assess/evaluate the process compliance at stage III and Table 6.10 illustrates the audit methodology to assess/evaluate the process compliance at stage IV.

6.2.1.3 Auditing process for the designer's proposal submission process

Figure 6.10 illustrates a typical submittal procedure by the designer

Figure 6.11 illustrates a typical quality auditing process for the designer's proposal submission process.

The main objectives of performing a quality audit of the designer's proposal submission process are to:

1. Evaluate and assess process compliance with the organizational strategic policies and procedure
2. Assess conformance to QMS and that processes meet established quality system requirements
3. Review the processes used to prepare the proposal
4. Verify that the proposal preparation and submission comply with the organization's proposal submittal process
5. Validate the response to the contents of the Request for Proposal (RFP) is complete in all respect. Table 6.11 illustrates the contents of a RFP for the designer.
6. Check that the availability of resources for the development of the design is considered
7. Review the process used to prepare the proposal
8. Validate the schedule for completion of the project design
9. Validate the cost of the proposal
10. Review that the key risk factors are analyzed and managed
11. Check compliance with regulatory requirements

Table 6.7 Audit methodology for the assessment of the bidding and tender process for the selection of a designer (A/E) – Stage I (shortlisting/registration of designers)

Serial Number	Item To Be Assessed	Yes	No	Comments
A. Checklist				
1	Whether the notification for registration was announced in all technical newsletters and leading news media as per the organization's policy			
2	Whether there are registered designers to participate in the tender			
3	Whether prequalification questionnaires (PQQ) are issued to all the intending bidders			
4	Whether there are any requests for bid clarification			
5	Whether the meeting for bid clarification is attended by all the intending bidders			
6	Whether the attendance sheet for the meeting is signed by all the attendees			
7	Whether minutes of the meeting are circulated to all the attendees/participating bidders			
8	Whether sufficient time is provided to prepare and submit completed PQQ			
9	Whether there are any requests for extension to submission date			
10	Whether all the intending bidders submitted the response to PQQ			
11	Whether received bid envelopes are placed in safe custody			
12	Whether there are any delayed submissions			
13	Whether received responses are acknowledged			
14	Whether received responses are clearly identified and recorded			
15	Whether responses are fairly evaluated as per the organization's selection policy (please refer to Table 6.6 as a guideline for the registration of the designer)			
16	Whether evaluation criteria, weightage and methodology are as specified (please refer to Table 5.12 as guidelines for selection)			
17	Whether any of the submitted responses are found to be incomplete			
18	Whether there are any late submissions			
19	Whether risk factors are considered while evaluating the response and registration of designers			
20	Whether reasons for non-registration are conveyed to unsuccessful participants			
21	Whether shortlisted designers are allotted a code and registration number			

(*Continued*)

Table 6.7 (Continued)

B. Interview Questionnaires

Serial Number	Description	Response	Comments
1	How responses were submitted: • Hand delivery • By post • Courier service		
2	Whether PQQ was available for electronic distribution		
3	How was the response for registration/shortlisting		

Table 6.8 Audit methodology for the assessment of the bidding and tender process for the selection of a designer (A/E) – Stage II (proposal documents)

Serial Number	Item To Be Assessed	Yes	No	Comments
A. Checklist				
1	Whether the Request for Proposal (RFP) includes all the relevant information required to select the designer (A/E)			
2	Whether the RFP properly defines the project goals, objectives and designer's Scope of Work (SOW)			
3	Whether all the deliverables are listed in the RFP			
4	Whether all the information to enable bidders/designers to submit the proposal is clear and unambiguous			
5	Whether documents are prepared taking into consideration organization's Quality Management System (QMS)			
6	Whether regulatory/authority requirements are taken into consideration while preparing the RFP			
7	Whether the selection criteria and selection method for designer selection are clearly specified in the organization's QMS			
8	Whether budget approval is obtained for design services prior to the release of the RFP announcement			
9	Whether the RFP opening and closing dates are mentioned in the announcement			
10	Whether the schedule for completion of design services is included in the RFP			
11	Whether qualifications of the personnel for design services is requested in the RFP			

Serial Number	Item To Be Assessed	Yes	No	Comments

A. Checklist

12	Whether project quality requirements are properly defined in the documents			
13	Whether rights and liabilities of all the parties are mentioned in the documents			
14	Whether change/variation clauses are included in the RFP			
15	Whether cancellation/termination clauses are included in the documents			
16	Whether bond and insurance clauses are included in the documents			
17	Whether the RFP documents are as per international standard documents used by the construction industry			
18	Whether evaluation and assessment criteria are well defined			
19	Whether review and analysis of RFP are included			
20	Whether agreement to be signed for design services is included in the documents			
21	Whether all the records are properly maintained			
22	Whether the request to provide the designer's contact details is included in the documents			

B. Interview Questionnaires

Serial Number	Description	Response	Comments
1	Whether any outside agency was involved in preparation of tender documents		
2	What guidelines were adopted to prepare the tender documents		
3	Whether any specific format/template was followed		
4	Whether all the requirements of RFP documents were coordinated and agreed by relevant stakeholders		
5	Whether risks of disputes and associated financial costs were considered while preparing the RFP		
6	Whether problems and challenges were identified while preparing RFP		
7	Who was responsible for authorizing bidding requirements		

Table 6.9 Audit methodology for the assessment of the bidding and tender process for the selection of a designer (A/E) – Stage III (contract bid solicitation)

Serial Number	Item To Be Assessed	Yes	No	Comments
A. Checklist				
1	Whether RFP documents are properly organized			
2	Whether the RFP is distributed to all the prequalified/shortlisted/registered designers (A/E) to participate in the tender			
3	Whether the tender is notified to all the prequalified consultants			
4	Whether all the prequalified/shortlisted/registered bidders participated in the tender			
5	Whether the attendance sheet for the meeting was signed by all the attendees			
6	Whether minutes of the meeting were circulated to all the attendees/participating bidders/designers			
7	Whether sufficient time is provided to prepare and submit the proposal/quotation			
8	Whether the tender submission date is extended from the date originally announced			
9	Whether an addendum (if any) is issued/notified to all the participating bidders			
10	Whether there are any requests for extension of the submission date			
11	Whether the clarification meeting is attended by all participating bidders			
12	Whether any changes are made to the original RFP			
13	Whether received bid envelopes are placed in safe custody			
14	Whether technical and financial envelopes are submitted at the same time			
15	Whether there are any delayed submissions			
16	Whether proposals are opened as per the announced date and time			
17	Whether proposals are opened in the presence of all the bidders			
18	Whether received tenders are acknowledged			
19	Whether received tenders are clearly identified and recorded			
20	Whether applicable fees/bid bond are submitted by all the bidders who participated in the tender			
21	Whether tenders are fairly evaluated as per the announced policy and procedure			
22	Whether the comparison of bids is tabulated			
23	Whether evaluation criteria, weightage and methodology are as specified			
24	Whether the bid cost is higher than the budgeted cost			
25	Whether any submitted tender is found to be incomplete			
26	Whether there are any late submissions			

Serial Number	Item To Be Assessed	Yes	No	Comments
A. Checklist					
27	Whether all the scheduled pages are signed by the bidder			
28	Whether risk factors are considered while evaluating the bids			
29	Whether reasons for non-acceptance of the tender was conveyed to the unsuccessful bidders			

B. Interview Questionnaires

Serial Number	Description	Response	Comments
1	How the notification was announced		
2	How tenders were submitted: • Hand delivery • By post • Courier service		
3	Whether the tender was available for electronic distribution		
4	How was the response to participate in the tender		
5	Whether tender boxes were placed in secured places and monitored through an electronic surveillance system		
6	How much time was allowed to open the bids after the notified tender closing time		
7	Whether the tender submission date was extended and what was the reason for extension		
8	Whether all the tender envelopes/packages were sealed with the marking of the tenders and contain the original tender documents and required number of sets as per the tender conditions		

Table 6.10 Audit methodology for the assessment of the bidding and tender process for the selection of a designer (A/E) – Stage IV (contract award)

Serial Number	Item To Be Assessed	Yes	No	Comments
A. Checklist					
1	Whether the selection of the designer (A/E) satisfies all the conditions to be a successful bidder			
2	Whether the selected bidder (designer) is a legal entity			
3	Whether the selection was made by the selection committee			

(*Continued*)

Table 6.10 (Continued)

Serial Number	Item To Be Assessed	Yes	No	Comments
A. Checklist				
4	Whether the selection of the designer is approved by the relevant applicable authority, if mandated by the organization's policy			
5	Whether the selected designer is capable of carrying out the specified work and has the necessary resources			
6	Whether risk factors are considered prior to the signing of the contract			
7	Whether the standard contract format is used for signing the contract			
8	Whether the contract terms and conditions are clearly written and unambiguous			
9	Whether an addendum and minutes of the meeting are included in the contract documents			
10	Whether clarifications to queries are included as part of the contract documents			
11	Whether the contract period/schedule is properly described			
12	Whether staff qualifications are properly defined			
13	Whether a performance bond is submitted by the consultant			
14	Whether variation/change management clauses are included in the contract agreement			
15	Whether dispute resolution and conflict resolution clauses are properly defined			
16	Whether contract documents are reviewed prior to signing			
17	Whether tender (proposal) documents and contract are properly archived			
18	Whether a letter of award is sent			
19	Whether the final results are published and announced			

Serial Number	Description	Response	Comments
B. Interview Questionnaires			
1	On what basis the bidder (designer) was selected: low bid, competitive bid or quality based system		
2	Was there any tie between two or more bidders		
3	Was there any dispute or objection raised by any of the bidders before the award of the contract to the successful bidder		
4	Whether the final price was negotiated or as per the bid (proposal) received		

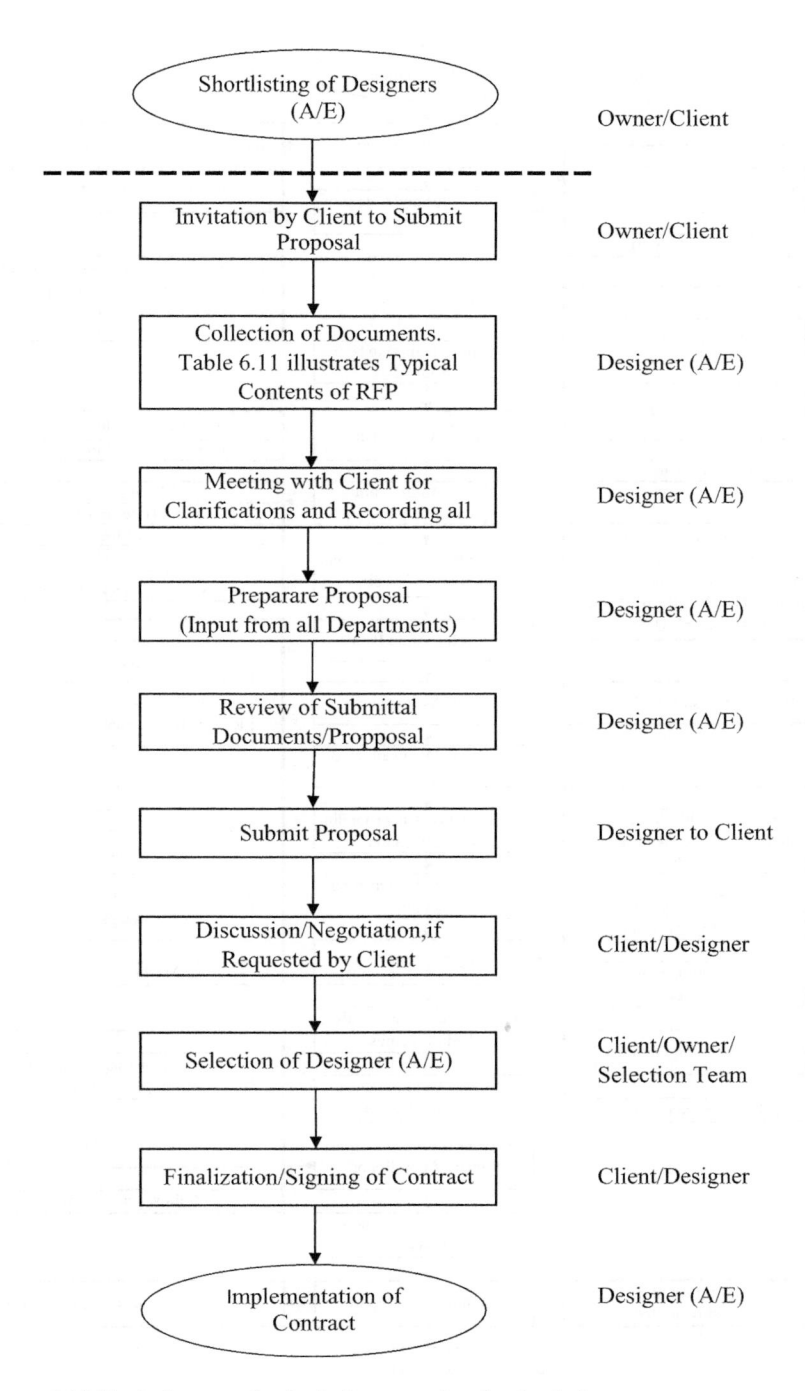

```
Shortlisting of Designers          Owner/Client
         (A/E)
- - - - - - - - - - - - - - - - - - - - - - - - -
             ↓
Invitation by Client to Submit     Owner/Client
          Proposal
             ↓
   Collection of Documents.
Table 6.11 illustrates Typical     Designer (A/E)
      Contents of RFP
             ↓
   Meeting with Client for         Designer (A/E)
Clarifications and Recording all
             ↓
     Preparare Proposal            Designer (A/E)
 (Input from all Departments)
             ↓
   Review of Submittal             Designer (A/E)
  Documents/Propposal
             ↓
      Submit Proposal              Designer to Client
             ↓
 Discussion/Negotiation, if        Client/Designer
   Requested by Client
             ↓
                                   Client/Owner/
  Selection of Designer (A/E)      Selection Team
             ↓
 Finalization/Signing of Contract  Client/Designer
             ↓
     Implementation of             Designer (A/E)
         Contract
```

Figure 6.10 Typical proposal submission procedure by the designer

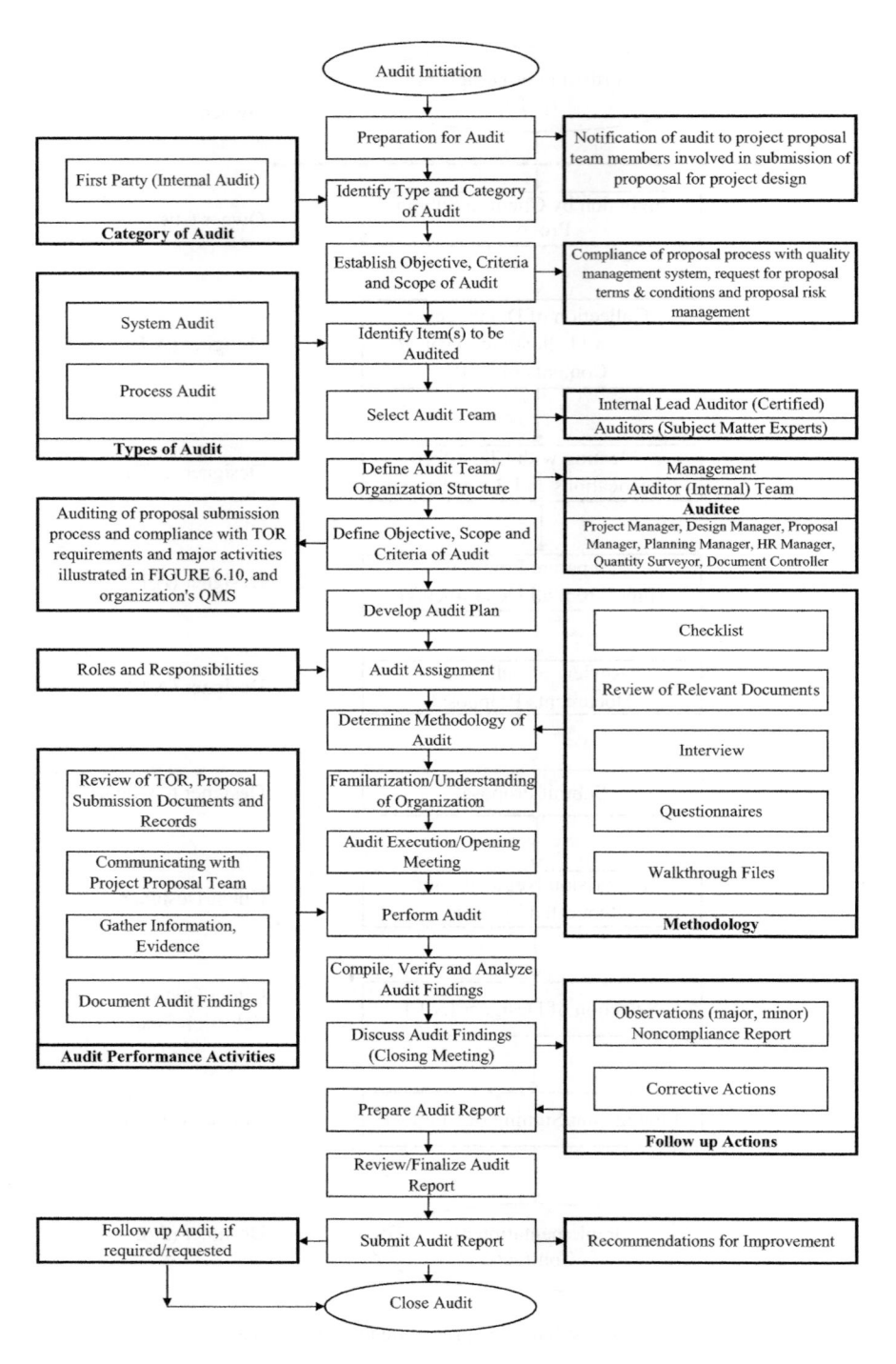

Figure 6.11 Typical quality auditing process for the designer's proposal submission process

Table 6.11 Contents of a Request for Proposal (RFP) for the designer (A/E)

Serial Number	Content
Project details (Project objectives)	
1	Introduction
2	Project description
3	Project delivery system
4	Designer's/consultant's scope of work
5	Preliminary project schedule
6	Preliminary cost of project
7	Type of project delivery system
Sample questions (information for evaluation)	
1	Consultant name
2	Address
3	Quality management system certification
4	Organization details
5	Type of firm such as partnership or limited company
6	Is the firm listed in stock exchange
7	List of awards, if any
8	Design production capacity
9	Current workload
10	Insurance and bonding
11	Experience and expertise
12	Project control system
13	Design submission procedure
14	Design review system
15	Design management plan
16	Design methodology
17	Submission of alternate concept
18	Quality management during design phase
19	Design firm's organization chart a) Responsibility matrix b) CVs of design team members
20	Designer's experience with green building standards or highly sustainable projects
21	Conducting value engineering
22	Authorities' approval
23	Data collection during design phase
24	Design responsibility/professional indemnity
25	Designer's relationship during construction
26	Preparation of tender documents/contract documents
27	Review of tender documents
28	Evaluation process and criteria
29	Any pending litigation
30	Price schedule

Source: Abdul Razzak Rumane (2013). *Quality Tools for Managing Construction Projects.* Reprinted with permission of Taylor & Francis Group

The following audit tools are used to perform a quality audit of the designer's proposal submission process:

- Analytical tools
- Checklists
- Review of documents
- Interviews
- Questionnaires

The auditee team members are mainly from the following organizational groups:

- Project Manager
- Design Manager
- Proposal Manager
- Quality Manager
- Contract Administrator/Quantity Surveyor
- HR Manager
- Document Controller
- Financial Manager

Table 6.12 illustrates the audit methodology for the assessment of the designer's proposal submission process and Figure 6.12 illustrates a logic flow diagram for the quality auditing of the designer's proposal submission process.

Table 6.12 Audit methodology for the assessment of the designer's bid (proposal) submission process

Serial Number	Item To Be Assessed	Yes	No	Comments
A. Checklist				
1	Whether the company is already registered as a designer to participate in the tender			
2	Whether there are any fees to collect tender (RFP) documents			
3	Whether the scope of work is clearly defined in the RFP			
4	Whether the TOR (Terms of Reference) clearly defines all the scope			
5	Whether there is any request for clarification raised			
6	Whether the response to the RFP is complete			
7	Whether sufficient time is provided to prepare and submit the completed RFP			

Serial Number	Item To Be Assessed	Yes	No	Comments
8	Whether there are any requests for extensions to the submission date			
9	Whether there are any delayed submissions			
10	Whether the RFP is prepared taking into consideration/referring to the organization's quality management system			
11	Whether the design quality requirements are considered while preparing the proposal			
12	Whether the proposal was reviewed prior to submission			
13	Whether the proposal was approved by the relevant committee/authorized person			
14	Whether the risk factors were considered and analyzed			

B. *Interview Questionnaires*

Serial Number	Description	Response	Comments
1	Whether the notification for the proposal submission was announced in all technical newsletters and leading news media or it was only for registered bidders (designers)		
2	How the proposal was submitted: • Hand delivery • By post • Courier service		
3	Whether the RFP was available for electronic distribution		
4	What tools and techniques were used to price the proposal		
5	Whether the organization's design management plan was considered to evaluate the price schedule		
6	Whether the availability of qualified resources (technical) was considered while preparing the price schedule		
7	Whether the design schedule was properly planned by taking into account the availability of qualified manpower and their production output		
8	Whether the regulatory approval period was properly considered		
9	What risk factors were considered while evaluating the price schedule		

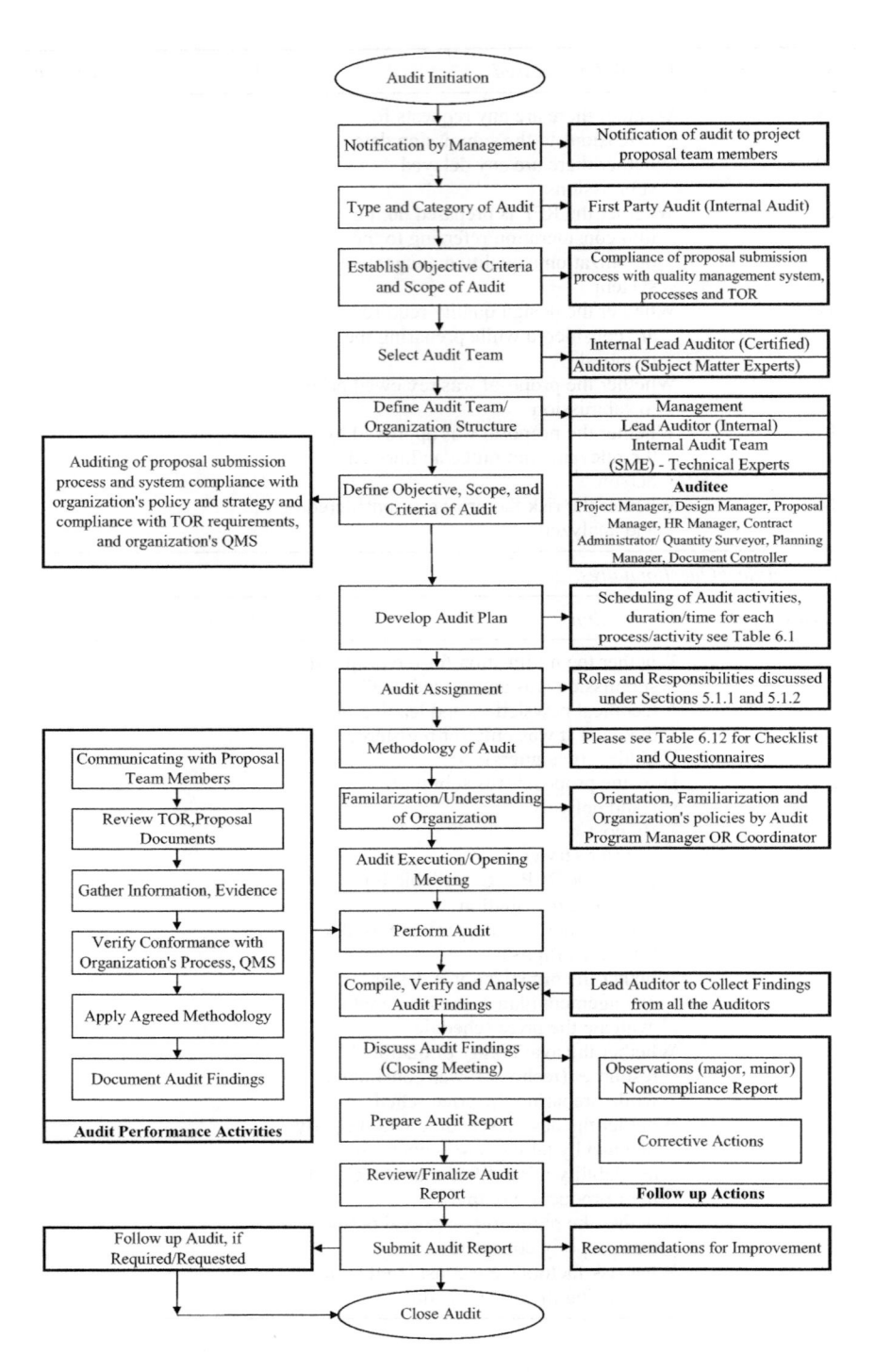

Figure 6.12 Logic flow diagram for the quality auditing process for the designer's proposal submission process

6.2.2 Auditing process for design phases

Most construction projects are custom-oriented, having a specific need and a customized design. It is always the owner's desire that their project should be unique and better. There are innumerable processes that make up the construction process. A systems engineering approach to construction projects helps to understand the entire process of project management and to manage and control its activities at different levels of various phases to ensure the timely completion of the project with economical use of resources to make the construction project the most qualitative, competitive and economical. In order to process the construction project in an effective and efficient manner and to improve control and planning, construction projects are divided into various phases. The life cycle of a major construction project has seven phases. Each phase can be further subdivided on the Work Breakdown Structure (WBS) principle to reach a level of complexity where each element/activity can be treated as a single unit that can be conveniently managed. As construction projects are unique and non-repetitive in nature, they need specified attention to maintain quality. Each project has to be designed and built to serve a specific need.

Construction project life cycle phases have construction related technical activities that need to be assessed, verified, reviewed and evaluated to achieve a successful project. The quality audit process for each of these phases depends on the objectives of the audits. The audit process of each phase should cover all the areas/activities to achieve the audit objectives.

The quality audit objectives of construction projects vary in line with clients' requirements and to what extent the details are to be assessed and verified. Some of the objectives of an audit of the design phases are as follows:

- Project goals and objectives have been properly defined
- Terms of Reference (TOR) is properly established
- Design management activities are adequately performed and managed
- Project management processes relating to design are properly followed
- Quality management procedures are fully complied
- Design deliverables are identified and produced
- Construction project quality elements and constraints are considered while developing the design and construction documents
- Risk factors are considered

The following sections discuss the auditing processes for design phases.

6.2.2.1 Auditing process for the conceptual design phase

The conceptual design phase commences once a need for the project is recognized. The conceptual design phase basically has two stages:

1   Study stage:

The study stage activities include:

- Identification of a need by the owner, and establishment of the main goals

- Feasibility study which is based on the owner's objectives
- Identification of the project delivery system
- Identification of the project team by selecting other members and allocation of responsibilities
- Identification of alternatives
- Selection of preferred alternative
- Time schedule
- Financial implications based on an estimation of the life cycle cost of the favorable alternative
- Resources
- Project risk
- Authorities' approval
- Terms of Reference (TOR)

2 Concept design:

- Development of concept design

The most significant impacts in the quality of the project begin during the conceptual phase. This is the time when specifications, statement of work, contractual agreements and initial design are developed. Initial planning has the greatest impact on a project because it requires the commitment of processes, resources schedules and budgets. A small error that is allowed to stay in the plan is magnified several times through subsequent documents that are second or third in the hierarchy.

6.2.2.1.1 AUDITING PROCESS FOR THE STUDY STAGE

Figure 6.13 illustrates a typical quality auditing process for the study stage of conceptual design.

The main objectives of performing a quality audit for the study stage of the conceptual design phase are to:

1. Evaluate and assess process compliance with the organizational strategic policies and procedure
2. Assess conformance to QMS and that processes meet established quality system requirements
3. Review the process used to prepare business case and justification for the project
4. Examine that key risk factors are analyzed and managed to establish business case
5. Validate expected schedule and cost of the project

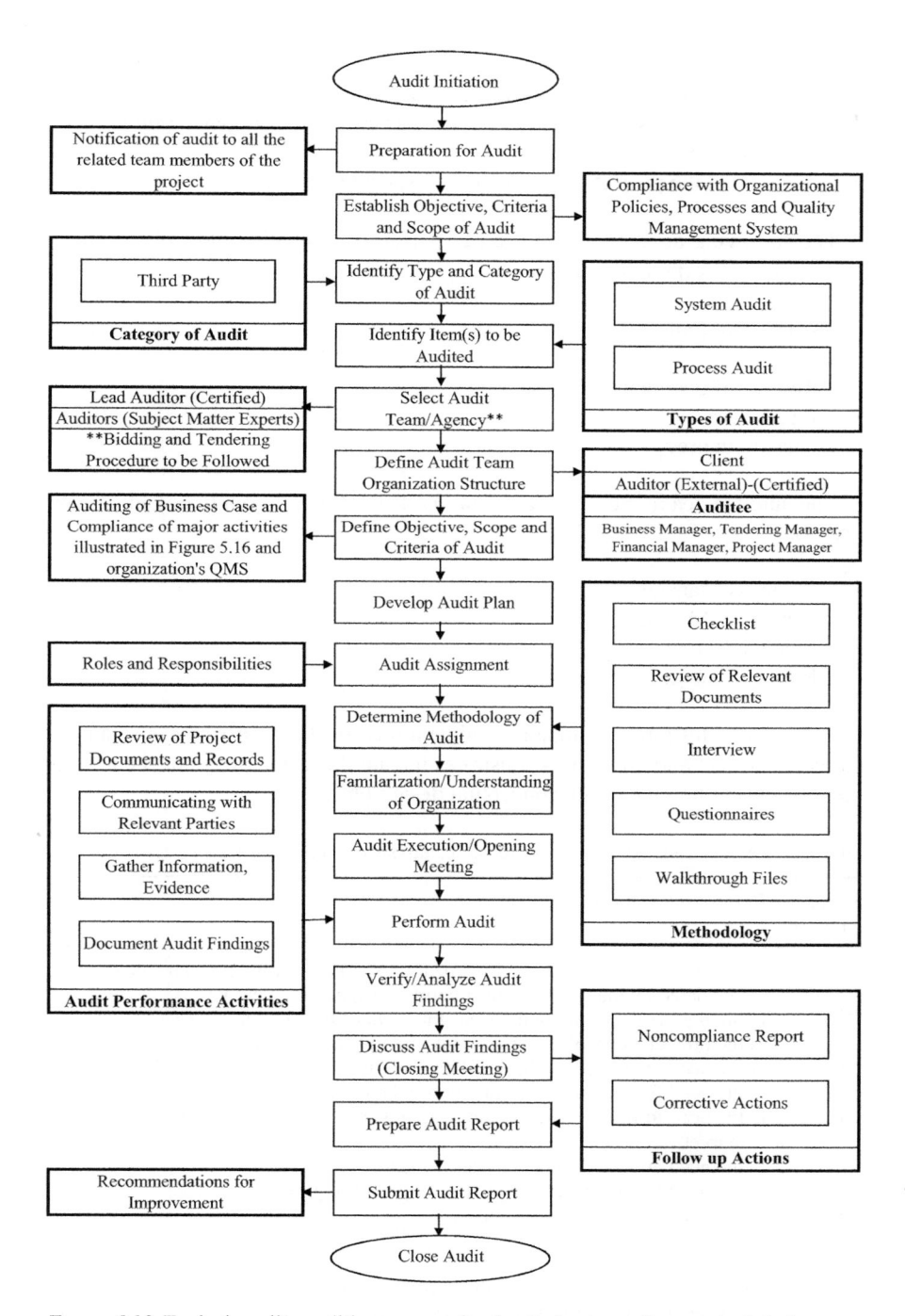

Figure 6.13 Typical quality auditing process for the study stage of conceptual design

6 Assess that all the major activities adequately performed and managed
7 Assess proper functioning of project management system
8 Identify if there is any weakness in the current state of organization to manage and execution of the project
9 Assess that the project will deliver the expected benefits

The following audit tools are used to perform the quality audit of the study stage of the conceptual design phase:

- Analytical tools
- Checklists
- Review of documents
- Interviews
- Questionnaires

The auditee team members are mainly from the following organizational groups:

- Business Manager
- Tendering Manager
- Financial Manager
- Project Manager

Table 6.13 illustrates the major considerations for the need analysis of a construction project and Table 6.14 illustrates the audit methodology for the assessment of the study stage of the conceptual design phase.

Table 6.13 Major considerations for the need analysis of a construction project

Serial Number	Points To Be Considered
1	Is the project in line with the organization's strategy/strategic plan and mandated by management in support of a specific objective?
2	Is the project a part of the mission statement of the organization?
3	Is the project a part of the vision statement of the organization?
4	Is the need mandated by the regulatory body?
5	Is the need for meeting government regulations?
6	Is the need to fulfill the deficiency/gap of such a type of project(s) in the market?
7	Is the need created to meet market demand?
8	Is the need to meet research and development requirements?
9	Is the need for technical advances?
10	Is the need generated to construct a facility/project which is innovative in nature?
11	Are the need aims to improve the existing facility?
12	Is the need a part of mandatory investment?
13	Is the need to develop infrastructure?

Serial Number	Points To Be Considered
14	Will the need serve the community and fulfill social responsibilities?
15	Is the need created to resolve specific problems?
16	Will the need have an effect on the environment?
17	Does the need have any time frame to implement?
18	Does the need have financial constraints?
19	Does the need have major risks?
20	Is the need within the capability of the owner/client, either alone or in cooperation with other organizations?
21	Can the need be managed and implemented?
22	Is the need realistic and genuine?
23	Is the need measurable?
24	Will the need be beneficial?
25	Does the need comply with environmental protection agency requirements?
26	Does the need comply with the government's health and safety regulations?

Source: Abdul Razzak Rumane (2013). *Quality Tools for Managing Construction Projects.* Reprinted with permission of Taylor & Francis Group

Table 6.14 Audit methodology for the assessment of the study stage of the conceptual phase

Serial Number	Item To Be Assessed	Yes	No	Comments
A. Checklist				
1	Whether the project need is established taking into consideration the SMART (Specific, Measurable, Attainable, Realistic, Time bound) concept			
2	Whether the project is in line with the organization's strategic policy/strategic plans			
3	Whether the project is mandated by the management in support of a specific objective			
4	Whether the business need has been clearly defined			
5	Whether risk factors are considered prior to establishing the project need			
6	Whether the project conforms with government regulations			
7	Whether assessment/analysis of need is carried out as per the guidelines listed in Table 6.13			
8	Whether a feasibility study is performed in the following areas; a Technical b Economical c Financial d Market demand e Time scale			

(*Continued*)

Table 6.14 (Continued)

Serial Number	Item To Be Assessed	Yes	No	Comments
A. Checklist				

	f Resource availability
	    g Risk analysis
	    h Environmental
	    i Ecological
	    j Sustainability
	    k Political
	    l Legal
	    m Social			

9	Whether the business case is clearly defined after considering the following points as a minimum:
	    a Project need
	    b Stakeholders
	    c Project benefits
	    d Financial benefits
	    e Estimated cost
	    f Estimated time
	    g Justification			

10	Whether life cycle costing of the project has been carried out and considered			

11	Whether project goals and objectives are established on SMART concepts considering the following information:
	    a Project scope and project deliverables
	    b Preliminary project schedule
	    c Preliminary project budget
	    d Specific quality criteria the deliverables must meet
	    e Type of contract to be employed
	    f Design requirements
	    g Regulatory requirements
	    h Potential project risks
	    i Environmental considerations
	    j Logistic requirements			

12	Whether preferred alternatives are selected after analyzing and evaluating each of the identified alternatives			
13	Whether preferred alternatives are approved by the owner/end user			
14	Whether high level requirements (scope, schedule, cost, and quality) for the project are established			
15	Whether the environmental authority's clearance certificate is obtained			
16	Whether the project delivery system is selected as per the strategic decision and stability of the appropriate system			

Serial Number	Item To Be Assessed	Yes	No	Comments
A. Checklist				
17	Whether the selected contracting/pricing is beneficial to the owner			
18	Whether authorities' approval is obtained			
19	Whether problems/challenges are identified and resolved			
20	Whether the project team is selected to suit the adopted project delivery system			
21	Whether the project team is selected as per the organization's bidding and tendering process			
22	Whether the Terms of Reference (TOR) cover all the requirements to fulfill the owner's project objectives and goals			
23	Whether the project boundaries and basic parameters are clearly defined			
24	Whether the project is approved by all the concerned stakeholders			

B. Interview Questionnaires

Serial Number	Description	Response	Comments
1	Why the project was initiated		
2	Was the project created to meet market demand		
3	Whether the project was mandated by the regulatory authority or as part of the vision statement of the organization		
4	Was there any need for such project		
5	Whether the project duration is in line with the organization's expected requirements		
6	Whether the project cost was within the reach of the organization		
7	Whether there was delay in getting authorities' approval		
8	Was there was any difficulty in approving the project by the concerned authority/team/owner		

6.2.2.1.2 AUDITING PROCESS FOR THE CONCEPT DESIGN

Figure 6.14 illustrates a typical quality auditing process for the concept design of the conceptual design phase.

The main objectives of performing a quality audit for the concept design of the conceptual design phase are to:

1 Evaluate and assess process compliance with the organizational strategic policies and procedure

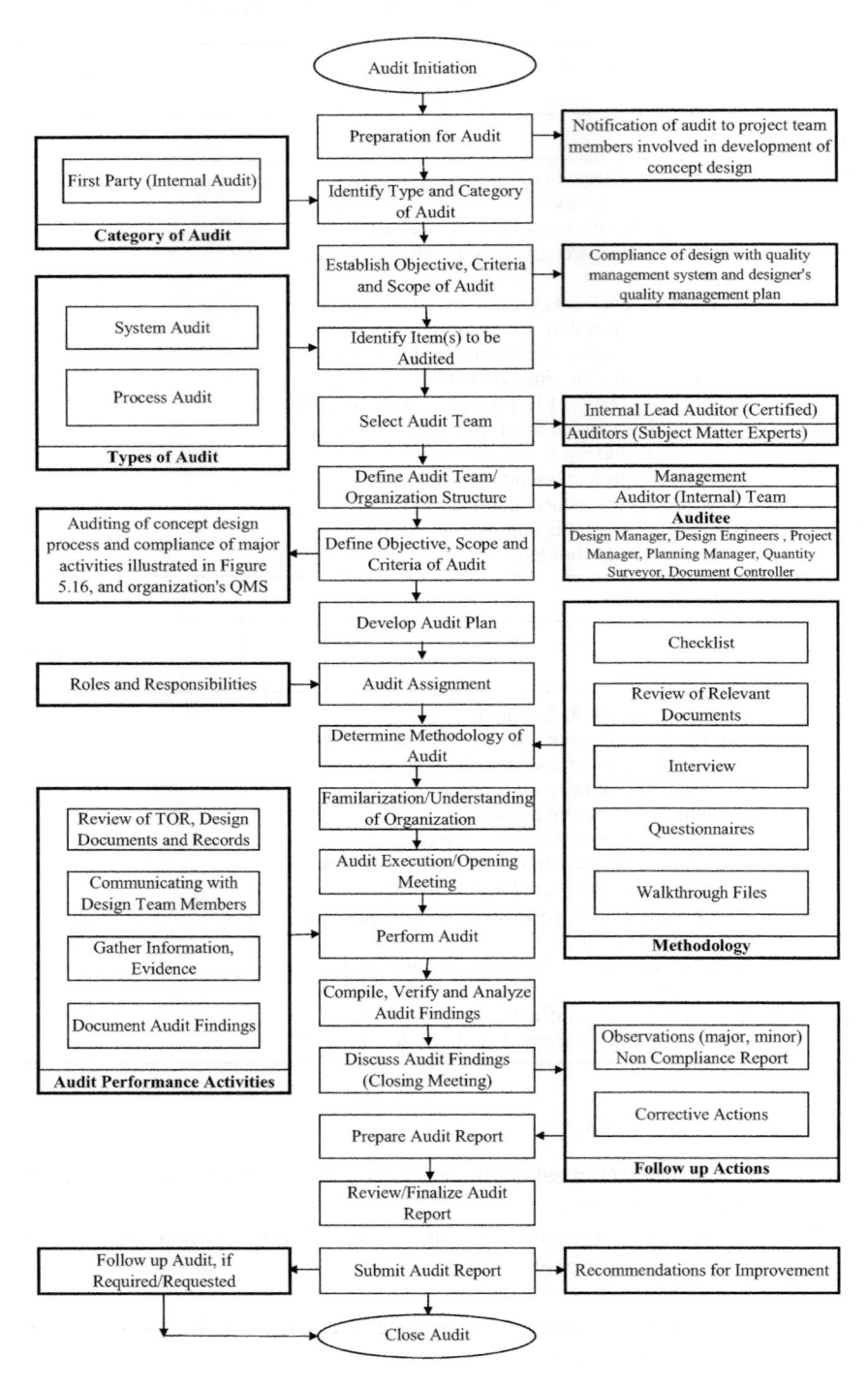

Figure 6.14 Typical quality auditing process for concept design

Auditing processes for project life cycle phases 245

2 Assess conformance to QMS and that processes meet established quality system requirements
3 Review the process used to prepare the concept design
4 Verify that the concept design complies with project goals and objectives and fully meets the owner's requirements
5 Check for compliance with TOR requirements. Table 6.15 lists the typical contents of TOR for building construction projects.
6 Review the process used to prepare the design report and model
7 Validate the estimated schedule and cost of the project meeting the budget
8 Validate compliance with quality standards and codes
9 Check for compliance with regulatory requirements
10 Check for sustainability considerations in the development of the design
11 Check that the conservation of energy requirements is considered
12 Check that facility management requirements are considered
13 Review the key risk factors to be analyzed and managed to develop the concept design
14 Assess that all major activities are adequately performed and managed
15 Assess the proper functioning of the project management system

Table 6.15 Typical contents of Terms of Reference (TOR) documents

Serial Number	Topics
1	Project Objectives 1.1 Background 1.2 Project Information 1.3 General Requirements 1.4 Special Considerations

2	Project Requirements
	2.1 Scope of Work
	2.2 Work Program
	    2.2.1 Study Phase
	    2.2.2 Design Phase
	    2.2.3 Tender Stage
	    2.2.4 Construction Phase
	2.3 Reports and Presentations
	2.4 Schedule of Requirements
	2.5 Drawings
	2.6 Energy Conservation Considerations
	2.7 Cost Estimates
	2.8 Time Program
	2.9 Interior Finishes
	2.10 Aesthetics
	2.11 Mechanical
	2.12 HVAC
	2.13 Lighting
	2.14 Engineering Systems

3	Opportunities and Constraints

(*Continued*)

Table 6.15 (Continued)

Serial Number	Topics
	3.1 Site Location
	3.2 Site Conditions
	3.3 Land Size and Access
	3.2 Climate
	3.3 Time
	3.4 Budget
4	Performance Target
	4.1 Financial Performance
	4.1.1 Performance Bond
	4.1.2 Insurance
	4.1.3 Delay Penalty
	4.2 Energy Performance Target
	4.2.1 Energy Conservation
	4.3 Work Program Schedule
5	Environmental Considerations
6	Design Approach
	6.1 Procurement Strategy
	6.2 Design Parameters
	6.2.1 Architectural Design
	6.2.2 Structural Design
	6.2.3 Mechanical Design
	6.2.4 HVAC Design
	6.2.5 Electrical Design
	6.2.6 Information and Communication Technology
	6.2.7 Conveying System
	6.2.8 Landscape
	6.2.9 External Works
	6.2.10 Parking
	6.3 Sustainable Architecture
	6.4 Engineering Systems
	6.5 Value Engineering Study
	6.6 Design Review by Client
	6.7 Selection of Products/Systems
7	Specifications and Contract Documents
8	Project Control Guidelines
9	Submittals
	9.1 Reports
	9.2 Drawings
	9.3 Specifications
	9.4 Models
	9.5 Sample Boards
	9.6 Mock up
10	Presentation
11	Project Team Members
	11.1 Number of Project Personnel
	11.2 Staff Qualification
	11.3 Selection of Specialists
12	Visits

Source: Abdul Razzak Rumane (2013). *Quality Tools for Managing Construction Projects*. Reprinted with permission of Taylor & Francis Group

The following audit tools are used to perform a quality audit of the concept design:

- Analytical tools
- Checklists
- Review of documents
- Interviews
- Questionnaires

The auditee team members are mainly from the following organizational groups:

- Project Manager
- Design Manager
- Design Engineers (All Trades)
- Planning Manager
- Contract Administrator/Quantity Surveyor
- Document Controller

Table 6.16 illustrates the audit methodology for the assessment of the concept design of the conceptual design phase and Figure 6.15 illustrates an example house of quality for the development of concept design.

Table 6.16 Audit methodology for assessment of the concept design

Serial Number	Item To Be Assessed	Yes	No	Comments
A. Checklist				
1	Whether the scope of work for development of the concept design is established based on TOR requirements			
2	Whether an organization chart with responsibilities to develop the concept design is established			
3	Whether the concept design supports the owner's project goals and objectives			
4	Whether the owner's requirements are fully considered			
5	Whether the concept design meets all the elements specified in the TOR			
6	Whether the design complies with organizational strategic policies			
7	Whether the design is based on specified quality standards and codes			
8	Whether the design meets all the performance requirements			

(*Continued*)

Table 6.16 (Continued)

Serial Number	Item To Be Assessed	Yes	No	Comments
A. Checklist					
9	Whether the data and other information are gathered prior to start of the design			
10	Whether the owner's preferred requirements are collected to prepare the design and take care of the requirements in the design			
11	Whether regulatory/statutory requirements are considered			
12	Whether the design team has specific qualifications and experience to design such a project			
13	Whether the project schedule is achievable in practice			
14	Whether the project cost is properly estimated			
15	Whether the concept design is prepared taking into consideration project quality requirements			
16	Whether the design is coordinated for the requirements of all trades			
17	Whether the designer has considered the availability of resources during the entire project life cycle			
18	Whether design risks have been identified and analyzed, and responses planned for mitigation			
19	Whether health and safety requirements are considered in the design			
20	Whether environmental constraints are considered			
21	Whether the design meets LEED requirements			
22	Whether cost-effectiveness over the entire project life cycle is considered			
23	Whether reports are prepared to meet TOR requirements			
24	Whether energy conservation is considered			
25	Whether sustainability is considered in the design			
26	Whether all reasonable alternative options/systems are considered for the design			
27	Whether all reasonable alternative options are considered for project economy			
28	Whether constructability has been considered			
29	Whether the designer used the quality function deployment (house of quality) technique to develop concept design (Figure 6.15 is an example house of quality)			
30	Whether the model meets the design objectives			
31	Whether the design supports proceeding to the next design stage			
32	Whether drawings were reviewed and checked for quality compliance prior to submission			
33	Whether reports and documents were reviewed prior to submission			
34	Whether the reports are complete and include adequate information about the project			

Serial Number	Item To Be Assessed	Yes	No	Comments

A. Checklist

35	Whether the report is properly formatted and there is a table of contents for each report			
36	Whether the designer has considered and followed quality management process activities related to concept design as listed in Figure 5.16			
37	Whether the design drawings, reports and other documents are properly labeled and archived			

B. Interview Questionnaires

Serial Number	Description	Response	Comments
1	Whether there was any delay in collecting the data		
2	Whether there was any delay in collection of the owner's preferred requirements		
3	What method was adopted to collect the owner's requirements and compliance with all the functional, parking, landscape areas		
4	Whether the concept design was prepared within the agreed-upon schedule		
5	Whether a work progress schedule was prepared to monitor progress		
6	Whether a time sheet was maintained to record the time spent by each member of staff engaged in the preparation of the concept design		
7	Was there any delay in getting authorities' approval		
8	Whether risk factors were considered while developing the concept design		
9	Whether the concept design was submitted in time for the owner's review		
10	Whether a number of sets were prepared as per the TOR and submitted		
11	Whether the owner issued a concept design acknowledgement letter		
12	Whether the concept design has any major comments from the owner		
13	Whether the concept design was not approved by the owner		
14	Whether the reasons (causes) for comments, non-approval, if any, were analyzed and corrective/preventive action taken		

6.2.2.2 Auditing process for the schematic design phase

Schematic/preliminary design is the basic responsibility of the architect (designer/consultant or A&E). The central activity of schematic/preliminary design is the architect's design concept of the owner's objective which can help in making detailed engineering and design for the required project/facility. Schematic/

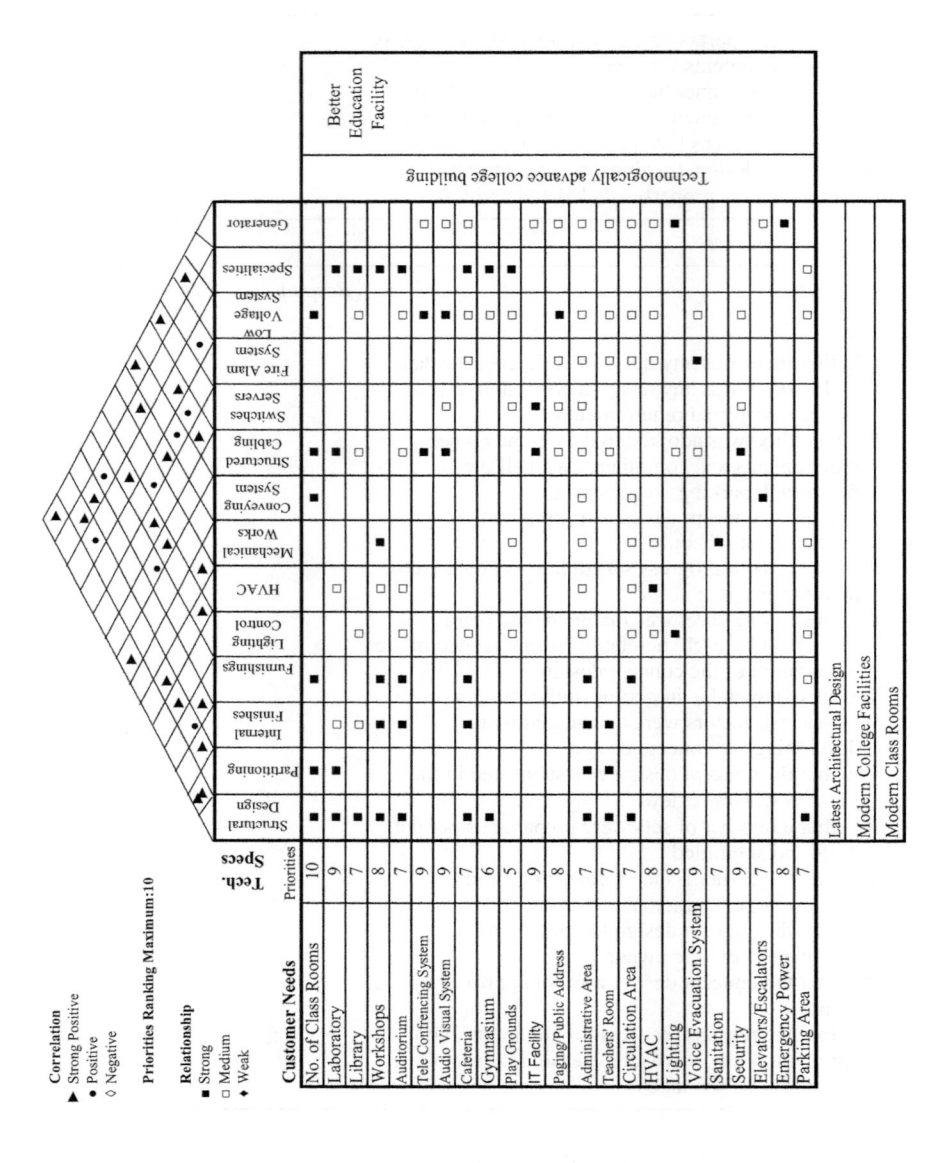

Figure 6.15 House of Quality for the development of the concept design

Source: Abdul Razzak Rumane (2013). *Quality Tools for Managing Construction Projects*. Reprinted with permission of Taylor & Francis Group

Auditing processes for project life cycle phases 251

preliminary design is a subjective process transforming ideas and information into plans, drawings and specifications of the facility to be built. Component/equipment configurations, material specifications and functional performance are decided during this phase.

At the preliminary design stage, the scope must define deliverables, that is, what will be furnished. It should include a schedule of dates for delivering drawings, specifications, calculations and other information, forecasts, estimates, contracts, materials and construction.

Figure 6.16 illustrates a typical quality auditing process for the schematic design phase.

The main objectives of performing a quality audit of the schematic design phase are to:

1. Evaluate and assess process compliance with the organizational strategic policies and procedure
2. Assess conformance to QMS and that processes meet established quality system requirements
3. Review the process used to prepare the schematic design
4. Check that the schematic design complies with the project goals and objectives
5. Verify that design deliverables are as per TOR requirements. Table 6.17 is an example list of schematic design deliverables.
6. Validate expected schedule and cost of the project meeting the project boundaries and defied parameters
7. Validate the compliance with quality standards and codes
8. Assess compliance with the design quality management system
9. Check for compliance with regulatory requirements
10. Check for sustainability considerations in the development of the design
11. Check that the conservation of energy requirements is considered
12. Check that facility management requirements are considered
13. Review that the key risk factors are analyzed and managed to develop the schematic design
14. Review preliminary specifications and contract documents
15. Assess that all the major activities are adequately performed and managed as listed in Figure 5.18
16. Assess the proper functioning of the project management system
17. Check the value engineering study is performed.

The following audit tools are used to perform a quality audit of the schematic phase:

- Analytical tools
- Checklists
- Review of documents
- Interviews
- Questionnaires

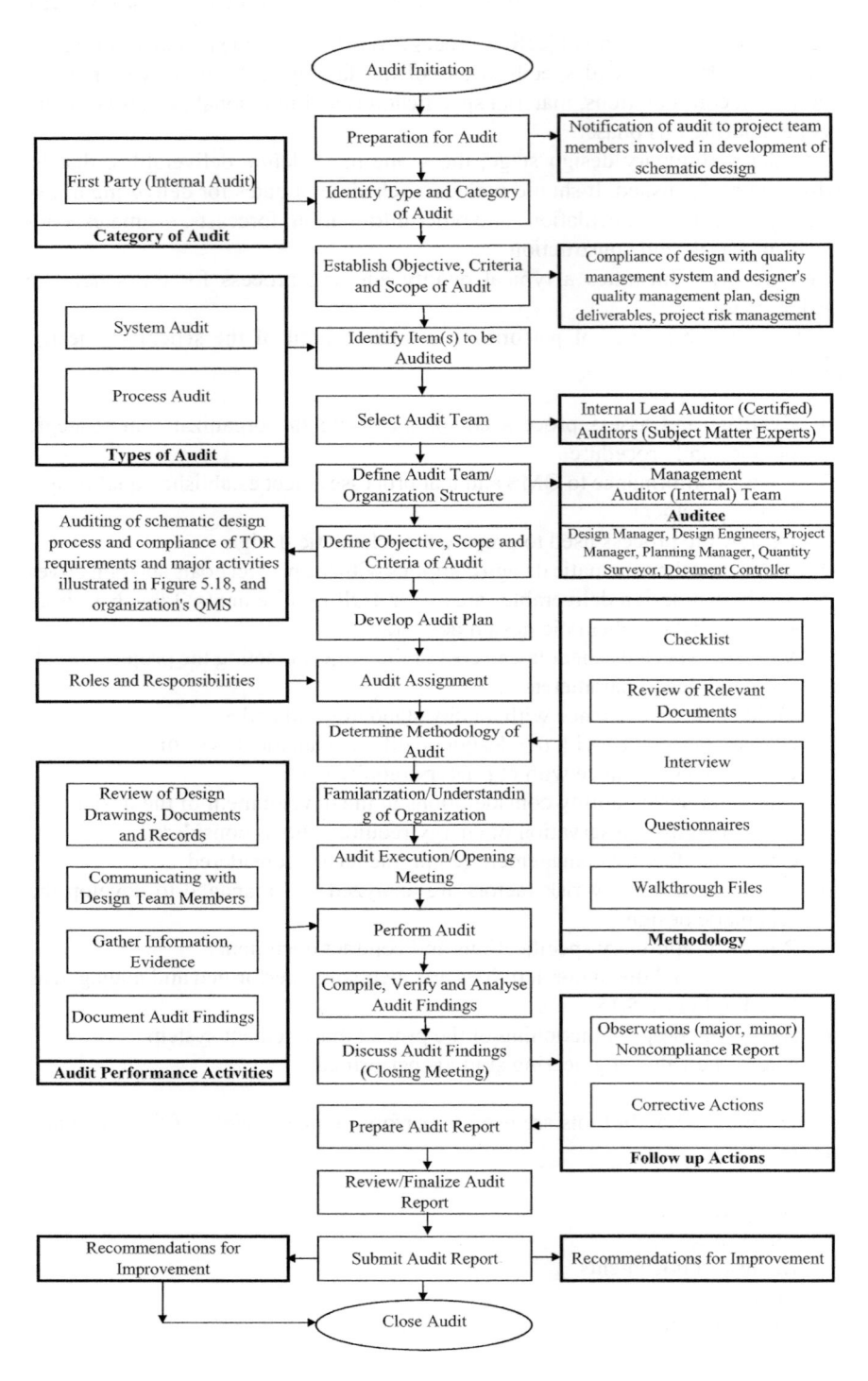

Figure 6.16 Typical quality auditing process for the schematic design phase

Table 6.17 Schematic design deliverables

Serial Number	Deliverables
1	General
1.1	a Preliminary/outline specifications
	b Zoning
	c Permits and regulatory approvals
	d Energy code requirements
	e Construction methodology narration
	f Descriptive report of environmental, health and safety requirements
	g Estimate construction period (preliminary schedule)
	h Estimated cost
	i Project risk assessment
	j Value engineering suggestions and resolutions
	k Life safety requirements
	l Sketches/perspective
	i Interior
	ii Exterior
	m Graphic presentation
2	Preliminary design drawings
2.1	Architectural
2.1.1	Overall site plans
2.1.2	Floor plans
2.1.3	Roof plans
2.1.4	Elevations
2.2	Structural
2.3	Elevator
2.4	Plumbing and fire suppression
2.5	HVAC
2.6	Electrical
2.7	Landscape
2.8	External
3	Narrative report
4	Model

Source: Abdul Razzak Rumane (2013). *Quality Tools for Managing Construction Projects*. Reprinted with permission of Taylor & Francis Group

The auditee team members are mainly from the following organizational groups:

- Project Manager
- Design Manager
- Design Engineers (All Trades)
- Planning Manager
- Contract Administrator/Quantity Surveyor
- Document Controller

Table 6.18 illustrates the audit methodology for the assessment of the schematic design and Figure 6.17 illustrates a logic flow diagram for the quality auditing of the schematic design.

Table 6.18 Audit methodology for the assessment of the schematic design

Serial Number	Item To Be Assessed	Yes	No	Comments
A. Checklist				
1	Whether the scope of work for development of the schematic design is established based on TOR requirements			
2	Whether an organization chart with responsibilities to develop the schematic design is established			
3	Whether the schematic design supports the owner's project goals and objectives			
4	Whether the design includes all the functional, external works, parking and landscape areas			
5	Whether the design complies with the organizational strategic policies			
6	Whether the owner's requirements are fully considered			
7	Whether comments on concept design are taken into account while preparing the schematic design			
8	Whether the schematic design meets all the elements specified in TOR			
9	Whether regulatory/statutory requirements are considered			
10	Whether the design is based on specified quality standards and codes			
11	Whether the design meets all the performance requirements			
12	Whether the designer considered project boundaries			
13	Whether constructability has been considered			
14	Whether technical and functional capabilities have been considered			
15	Whether sustainability has been considered in the design			
16	Whether the design meets LEED requirements			
17	Whether energy conservation is considered			
18	Whether the designer has considered the facility management requirements			
19	Whether project objectives in respect of time, cost and quality have been considered while developing the design			
20	Whether valid assumptions and constraints have beenconsidered			
21	Whether the project schedule is achievable in practice			
22	Whether the project cost is properly estimated			

Serial Number	Item To Be Assessed	Yes	No	Comments
A. Checklist				
23	Whether the schematic design prepared taking into consideration the project quality requirements			
24	Whether the designer has considered the availability of resources during the entire project life cycle			
25	Whether the design team has specific qualifications to design such projects			
26	Whether design risks have been identified and analyzed, and responses planned for mitigation			
27	Whether design was fully coordinated with all the trades			
28	Whether health and safety requirements in the design are considered			
29	Whether the design confirms with fire and egress requirements			
30	Whether environmental constraints and requirements have been considered			
31	Whether the specification was prepared taking into consideration international construction contract documents			
32	Whether a value engineering study was performed and recommendations taken care of			
33	Whether cost-effectiveness over the entire project life cycle is considered			
34	Whether the designer has considered project boundaries			
35	Whether the drawings, specifications and documents have been prepared to meet TOR requirements			
36	Whether the design deliverables conform as listed in Table 6.17			
37	Whether there is any approved design review procedure that was followed to review and quality check			
39	Whether all the drawings are numbered			
40	Whether the reports are complete and include adequate information about the project			
41	Whether the report is properly formatted and there is a table of contents for each report			
42	Whether drawings were reviewed prior to submission			
43	Whether the designer has considered and followed management process activities related to schematic design as listed in Figure 5.18			

(Continued)

Table 6.18 (Continued)

Serial Number	Item To Be Assessed	Yes	No	Comments
A. Checklist				
44	Whether the design complies with designer's quality management plan			
45	Whether the scope of the design work was properly established			
46	Whether design criteria were properly established			
47	Whether the design supports proceeding to the next design development stage			
48	Whether the design drawings, reports and other documents are properly labeled and archived			

B. Interview questionnaires			
Serial Number	Description	Response	Comments
1	Whether there was any delay in collecting the data		
2	Whether there was any delay in site investigation		
3	Was there any delay in getting authorities' approval		
4	Whether the schematic design was prepared within the agreed-upon schedule		
5	Whether a work progress schedule was prepared to monitor progress		
6	Whether a time sheet was maintained to record the time spent by each member of staff engaged for the preparation of the schematic design		
7	Whether coordination meetings were held among all the team members		
8	Whether there is a quality organization with responsibilities matrix		
9	Whether the schematic design was submitted in time for the owner's review		
10	Whether the numbers of sets were prepared as per TOR and submitted		
11	Whether the owner issued a schematic design acknowledgement letter		
12	Whether the schematic design has any major comments from the owner		
13	Whether the schematic design was not approved by the owner		
14	Whether the reasons (causes) for comments, non-approval, if any, were analyzed and corrective/preventive action taken		

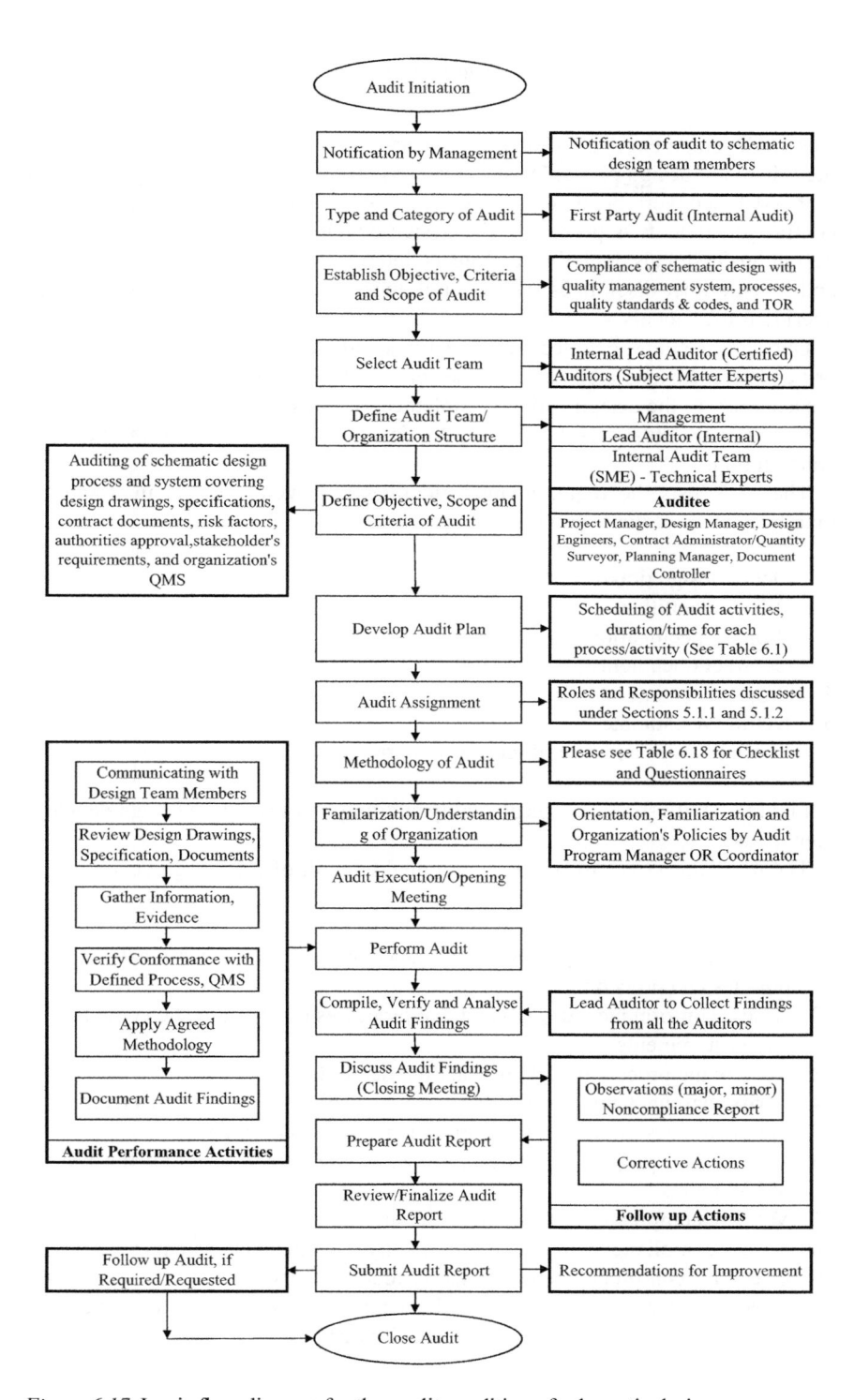

Figure 6.17 Logic flow diagram for the quality auditing of schematic design

6.2.2.3 Auditing process for the design development/detail design phase

Design development/detail design is enhancement of the work carried out during the schematic design phase. During this phase a comprehensive design of works with a detailed work breakdown structure of design, drawings, specifications and contract documents is prepared. The design development phase is the realm of design professionals, including architects, interior designers, landscape architects and several other disciplines such as civil, mechanical, electrical and other engineering professionals as needed.

During this phase, detailed plans, sections and elevations are drawn to scales, principle dimensions are noted and design calculations are checked to conform the accuracy of the design and its compliance to codes and standards.

Figure 6.18 illustrates the stages to develop a detail design for building construction.

The TOR lists the requirements guidelines to develop the detail design. It mainly consists of the following:

1. Detail design drawings
2. Bill of quantities
3. Project specifications
4. Contract documents
5. Project schedule
6. Definitive estimate
7. Format, scales and size of reports and drawings

Figure 6.19 illustrates a typical quality auditing process for the design development phase.

The main objectives of performing a quality audit of the design development phase are to:

1. Evaluate and assess process compliance with the organizational strategic policies and procedure
2. Assess conformance to QMS and that processes meet established quality system requirements
3. Review the process used to prepare the detail design
4. Verify that the detail design complies with project goals and objectives
5. Verify that design deliverables are as per TOR requirements.
6. Validate the project schedule and cost of the project meeting the project requirements
7. Validate compliance with quality standards and codes
8. Assess compliance with the design quality management system
9. Check for compliance with regulatory requirements
10. Check for sustainability considerations in the development of the design
11. Check that conservation of energy requirements is considered

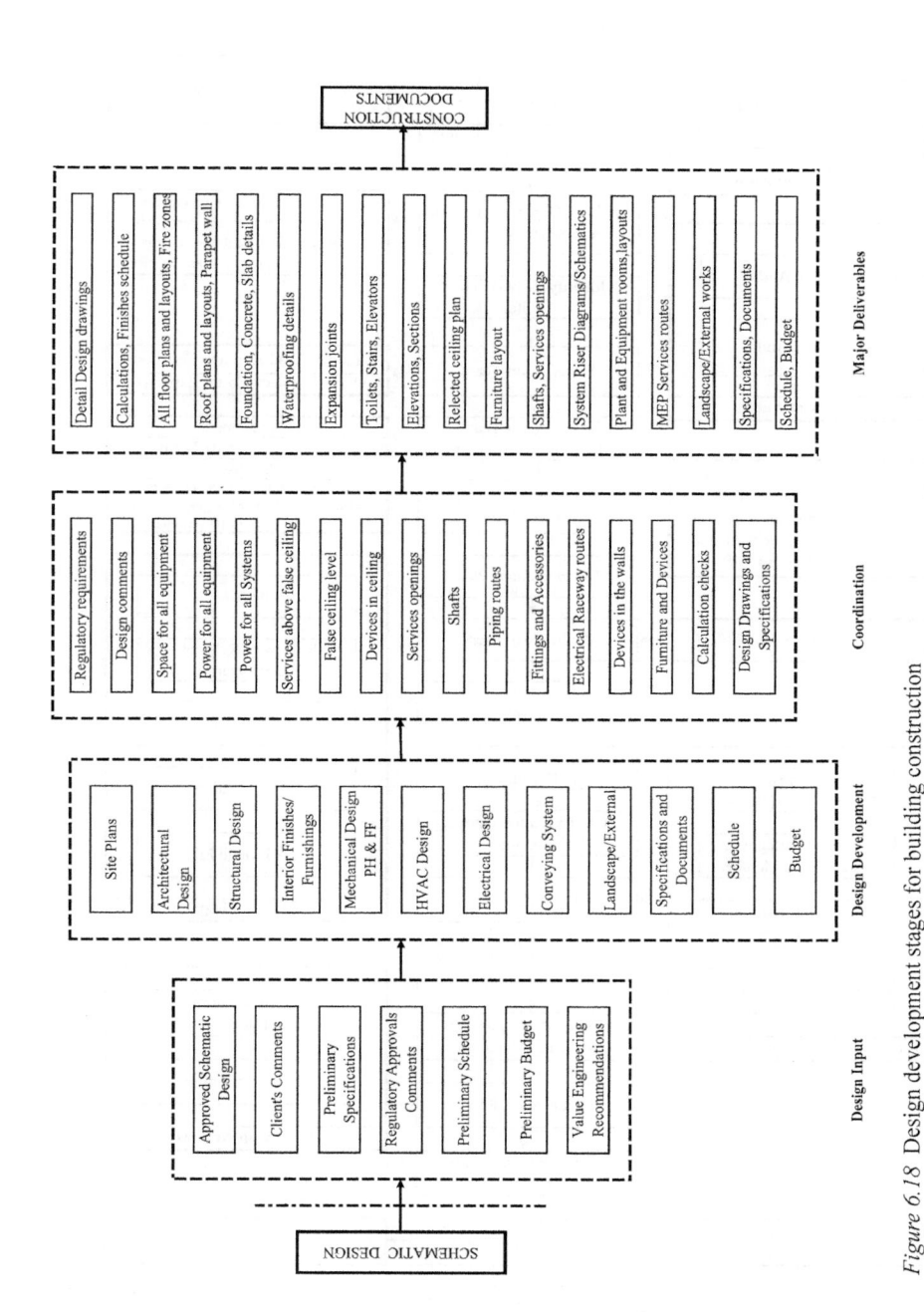

Figure 6.18 Design development stages for building construction

Source: Abdul Razzak Rumane (2013). *Quality Tools for Managing Construction Projects*. Reprinted with permission of Taylor & Francis Group

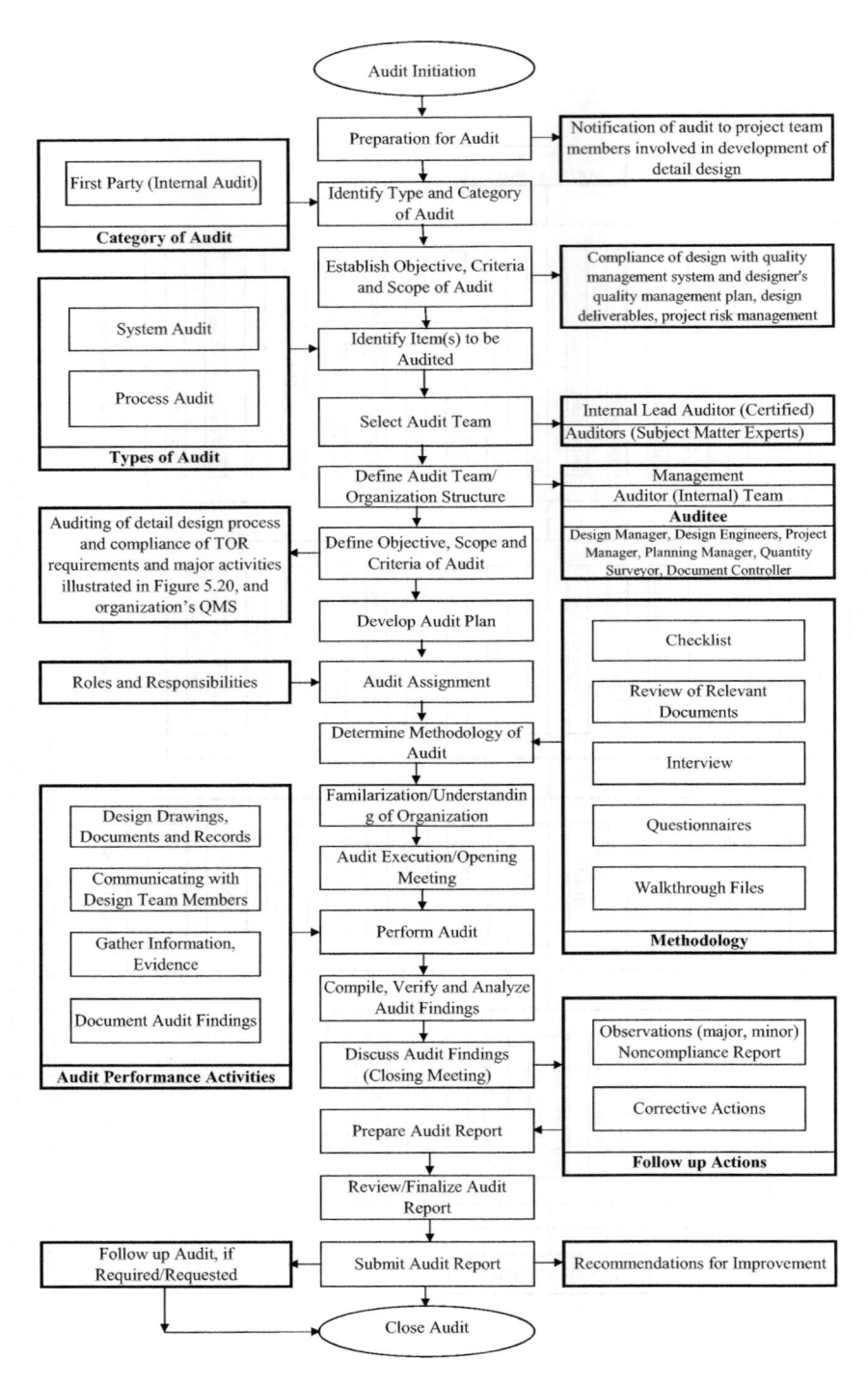

Figure 6.19 Typical quality auditing process for the design development phase

12 Check that facility management requirements are considered
13 Review that the key risk factors are analyzed and managed during the design development phase for the execution of the project/facility
14 Assess compliance to safety consideration in the design
15 Review design drawings, specifications and contract documents
16 Assess that all the major activities are adequately performed and managed as listed in Figure 5.20
17 Assess the proper functioning of the project management system

The following audit tools are used to perform a quality audit of the design development phase:

- Analytical tools
- Checklists
- Review of documents
- Interviews
- Questionnaires

The auditee team members are mainly from the following organizational groups:

- Project Manager
- Design Manager
- Design Engineers (All Trades)
- Planning Manager
- Contract Administrator/Quantity Surveyor
- Document Controller
- AutoCAD Technicians

Figure 6.20 illustrates a framework for the quality auditing of the detail design phase, Figure 6.21 illustrates the design review procedure, Table 6.19 illustrates mistake proofing to eliminate design errors, Table 6.20 illustrates the audit methodology for the assessment of the design development phase and Figure 6.22 illustrates a logic flow diagram for the quality auditing of the detail design phase.

6.2.2.4 Auditing process for the construction documents phase

The construction documents phase provides a complete set of working drawings of all the disciplines, site plans, technical specifications, Bill of Quantities (BOQ), schedule, (except the standards specifications, documents for insertions normally added during bidding and tendering phase) and related graphic and written information to bid for the project. It is necessary that utmost care is taken to develop and assemble all the documents and ensure their accuracy and correctness to meet the owner" objectives.

262 Abdul Razzak Rumane

Auditing Task	Lead Auditor (Internal)								
	Auditors (Subject Matter Experts –Technical Experts)								
	Design and Drawings	Specifications and Documents	Schedule	Cost	Quality	Risk	Safety	Other Areas (General)	
Auditee (Project Team Members)	Project Manager	Project Manager	Project Manager	Project Manager	Quality Manager	Project Manager	Project Manager	Related Team Members	
	Design Manager	Contract Administrator	Planning Manager	Costing Manager	Design Manager	Design Manager	Design Manager		
	Design Engineers	Design Manager		Quantity Surveyor	Project Manager	Planning Manager			
	Technicians	Document Controller		Cost Estimator		Costing Manager			
Perform Audit	Audit Execution/Opening Meeting								
	Review, Verify Documents & Records for Conformance								
	Checklist, Questionnaire, Interview								
	Document Audit Findings								
	Closing Meeting								
	Audit Report								

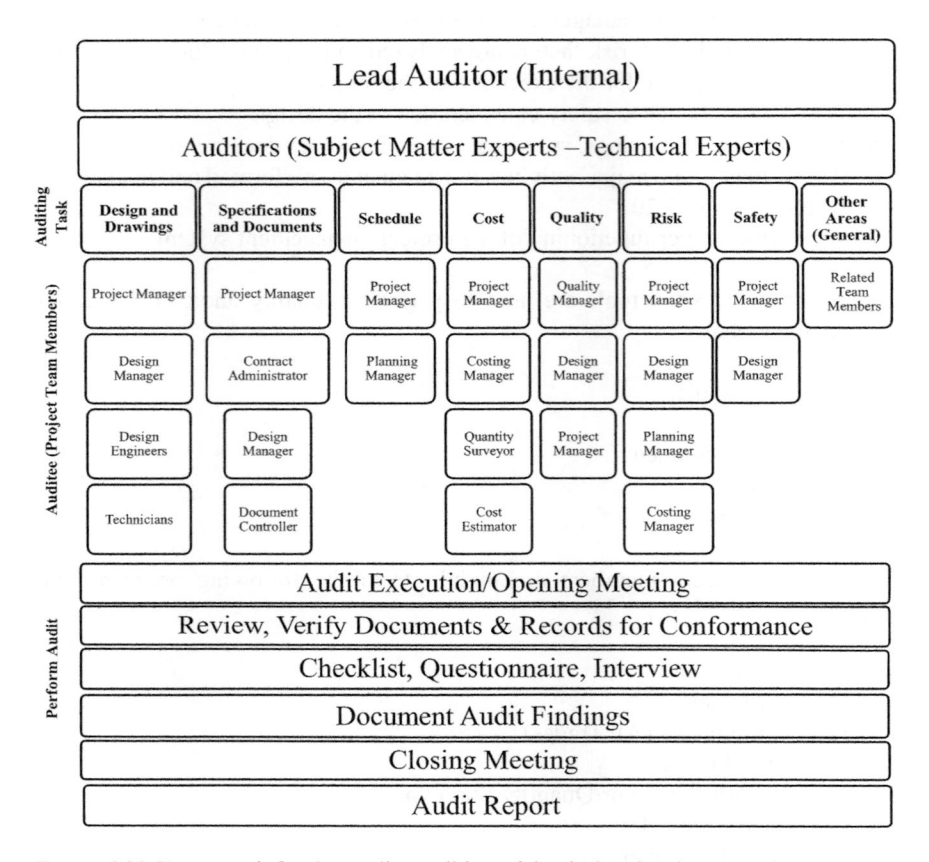

Figure 6.20 Framework for the quality auditing of the design development phase

In order to identify the requirements to assemble contract documents, the designer has to gather the comments on the submitted detail design by the owner/project manager, collect TOR requirements and regulatory requirements, identify the owner requirements and all other related information to ensure nothing is missed.

The following is the list of construction documents listed in the Terms of Reference (TOR) that are to be developed by the designer (A/E):

1. Final drawings (working drawings) to be prepared to the required scales, format with necessary logo, client name, location map, north orientation, project name, designer name, drawing title, drawing number, contract reference number, date of drawing, revision number, drawing scale, duly signed by the designer.
2. Bill of Quantities and schedule of rates
3. Contract documents

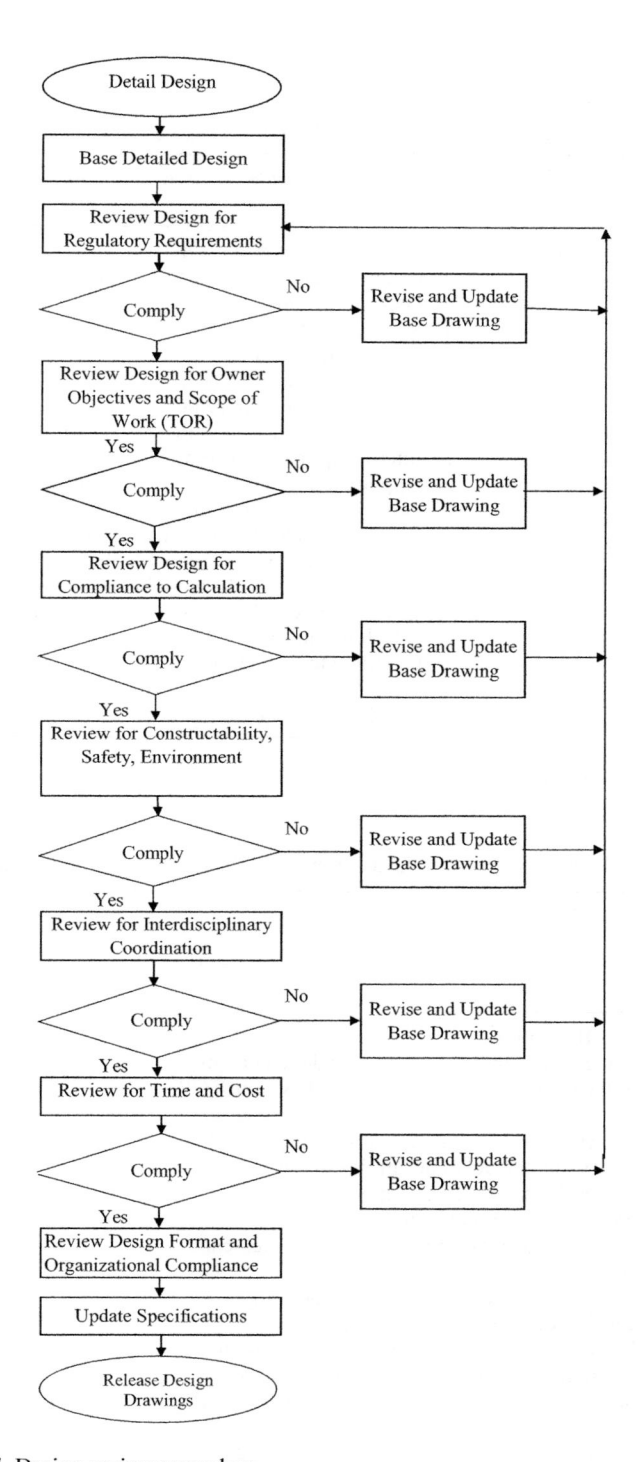

Figure 6.21 Design review procedure

Source: Abdul Razzak Rumane (2013). *Quality Tools for Managing Construction Projects.* Reprinted with permission of Taylor & Francis Group

Table 6.19 Mistake proofing for eliminating design errors

Serial Number	Items	Points To Be Considered To Avoid Mistakes
1	Information	1 Terms of Reference (TOR) 2 Client's preferred requirements matrix 3 Data collection 4 Regulatory requirements 5 Codes and standards 6 Historical data 7 Organizational requirements
2	Mismanagement	1 Compare production with actual requirements 2 Interdisciplinary coordination 3 Application of different codes and standards 4 Drawing size of different trades/specialist consultants
3	Omission	1 Review and check design with TOR 2 Review and check design with client requirements 3 Review and check design with regulatory requirements 4 Review and check design with codes and standards 5 Check for all required documents
4	Selection	1 Qualified team members 2 Available material 3 Installation methods

Source: Abdul Razzak Rumane (2013). *Quality Tools for Managing Construction Projects.* Reprinted with permission of Taylor & Francis Group

Table 6.20 Audit methodology for the assessment of the design development phase

Serial Number	Item To Be Assessed	Yes	No	Comments
A. Checklist				
A-1 Design and Drawings				
1	Whether the scope of work for the development of the detail design is established based on TOR requirements			
2	Whether an organization chart with responsibilities to develop the detail design is established			
3	Whether the detail design supports the owner's project goals and objectives			
4	Whether the detail design meets all the elements specified in the TOR			
5	Whether the owner's requirements are fully considered			
6	Whether the design includes all the functional, external works, parking and landscape areas			
7	Whether the design complies with the organizational strategic policies			
8	Whether comments on schematic design are considered while preparing the detail design			

Serial Number	Item To Be Assessed	Yes	No	Comments

A. Checklist

9	Whether regulatory/statutory requirements are considered			
10	Whether the design is based on specified quality standards and codes			
11	Whether the design meets all the performance requirements			
12	Whether constructability has been considered			
13	Whether technical and functional capability is considered			
14	Whether sustainability is considered in developing the design			
15	Whether the design meets LEED requirements			
16	Whether energy conservation is considered			
17	Whether the designer has considered the facility management requirements			
18	Whether the designer has taken into consideration project boundaries and assumptions			
19	Whether value engineering study recommendations are considered while developing the design			
20	Whether project objectives in respect of time, cost and quality are considered while developing the design			
21	Whether the design supports proceeding to the next design development stage			

A-2 Specifications and Documents

1	Whether the specification is prepared taking into consideration international construction contract documents			
2	Whether the specifications and documents are prepared to meet TOR requirements			
3	Whether the reports are complete and include adequate information about the project			
4	Whether the report is properly formatted and there is a table of contents for each report			
5	Whether the division numbers in the specifications matches with the BOQ			
6	Whether the numbers of sets comply with TOR requirements			

A-3 Schedule

1	Whether the project schedule is achievable in practice			
2	Whether all the activities are considered while preparing the schedule			
3	Whether activity durations for all the activities are considered while preparing the schedule			
4	Whether the relationship and dependency between activities are identified			

(Continued)

Table 6.20 (Continued)

Serial Number	Item To Be Assessed	Yes	No	Comments

A. Checklist

5	Whether the schedule developed considering specific internationally recommended schedule level practices			
6	Whether the detail design is prepared within the agreed-upon schedule			
7	Whether a work progress schedule was prepared to monitor progress			
8	Whether a time sheet was maintained to record the time spent by each member of staff engaged for the preparation of concept design			

A-4 Cost

1	Whether the project cost is properly estimated			
2	Whether cost-effectiveness over the entire project life cycle is considered			
3	Whether the cost is developed considering specific internationally recommended cost estimation level practices for construction projects			

A-5 Quality

1	Whether the design is fully coordinated for conflict between different trades			
2	Whether the designer has considered the availability of resources during the entire project life cycle			
3	Whether the design complies with the designer's quality management plan			
4	Whether designs were prepared using authenticated and approved software			
5	Whether all the drawings are numbered			
6	Whether drawings are reviewed prior to submission			
7	Whether the design was reviewed using mistake proofing tool to eliminate/mitigate design errors (Table 6.19 is an example mistake proofing quality tool)			
8	Whether the design matches with the property limits			
9	Whether legends match with layout			
10	Whether design drawings are properly numbered			
11	Whether design drawings have owner logo, designer logo as per standard format			
12	Whether design calculation sheets are included in the set of documents			
13	Whether the project name and contract reference are shown on the drawing			
14	Whether BOQ is verified with design drawings and specifications			
15	Whether the detail design was reviewed as per design review procedure in Figure 6.22			

Serial Number	Item To Be Assessed	Yes	No	Comments

A. Checklist

A-6 Risk
1. Whether design risks have been identified and analyzed, and responses planned for mitigation
2. Whether valid assumptions and constraints are considered
3. Whether there is any risk register maintained

A-7 Safety
1. Whether health and safety requirements in the design are considered
2. Whether environmental constraints and requirements are considered
3. Whether the design confirms with fire and egress requirements
4. Whether environmental compatibility is considered while developing the design

A-8 General
1. Whether the scope of design work was properly established
2. Whether design criteria were properly established
3. Whether design deliverables are established based on TOR
4. Whether all the related data and information are collected to develop the design
5. Whether site investigations are done prior to the development of the detail design
6. Whether design team members are qualified and experienced in developing the design of such a project
7. Whether there is an organization chart for the design team
8. Whether a responsibilities matrix is developed
9. Whether regulatory approval is obtained and comments, if any, incorporated and all review comments responded to
10. Whether the designer has considered and followed the management process activities related to design development as listed in Figure 5.20
11. Whether design is approved by all the concerned stakeholders

B. Interview Questionnaires

Serial Number	Description	Response	Comments
1	Whether there was any delay in collecting the data		
2	Whether there was any delay in site investigation		

(*Continued*)

268 Abdul Razzak Rumane

Table 6.20 (Continued)

B. Interview Questionnaires

Serial Number	Description	Response	Comments
3	Was there any delay in getting authorities' approval		
4	Whether the comments on schematic design were reviewed and taken into consideration		
5	Was there any approved design submission plan		
6	Whether the BOQ was fully coordinated with the design drawings and specifications		
7	Whether the availability of resources during the construction phase was explored and considered		
8	Whether coordination meetings were held among all the team members		
9	Whether there is a quality organization with responsibilities matrix		
10	Which international contract documents were followed to prepare specification		
11	Whether there is any approved design review procedure		
12	Whether the detail design was submitted in time for the owner's review		
13	Whether the numbers of sets were prepared as per TOR and submitted		
14	Whether the owner issued a detail design acknowledgement letter		
15	Whether the detail design has any major comments from the owner		
16	Whether the detail design was not approved by the owner		
17	Whether the reasons (causes) for comments, non-approval, if any, were analyzed and corrective/preventive action taken		
18	Whether any subconsultant was involved, if yes, what was the selection and design review criteria		
19	Whether the design drawings, reports and other documents are properly labeled and archived		

4   Technical specifications
5   Project schedule
6   Cost estimates
7   Summary report

Figure 6.23 illustrates a typical quality auditing process for the construction documents phase.

The main objectives of performing a quality audit of the construction documents phase are to:

1 Evaluate and assess process compliance with the organizational strategic policies and procedure

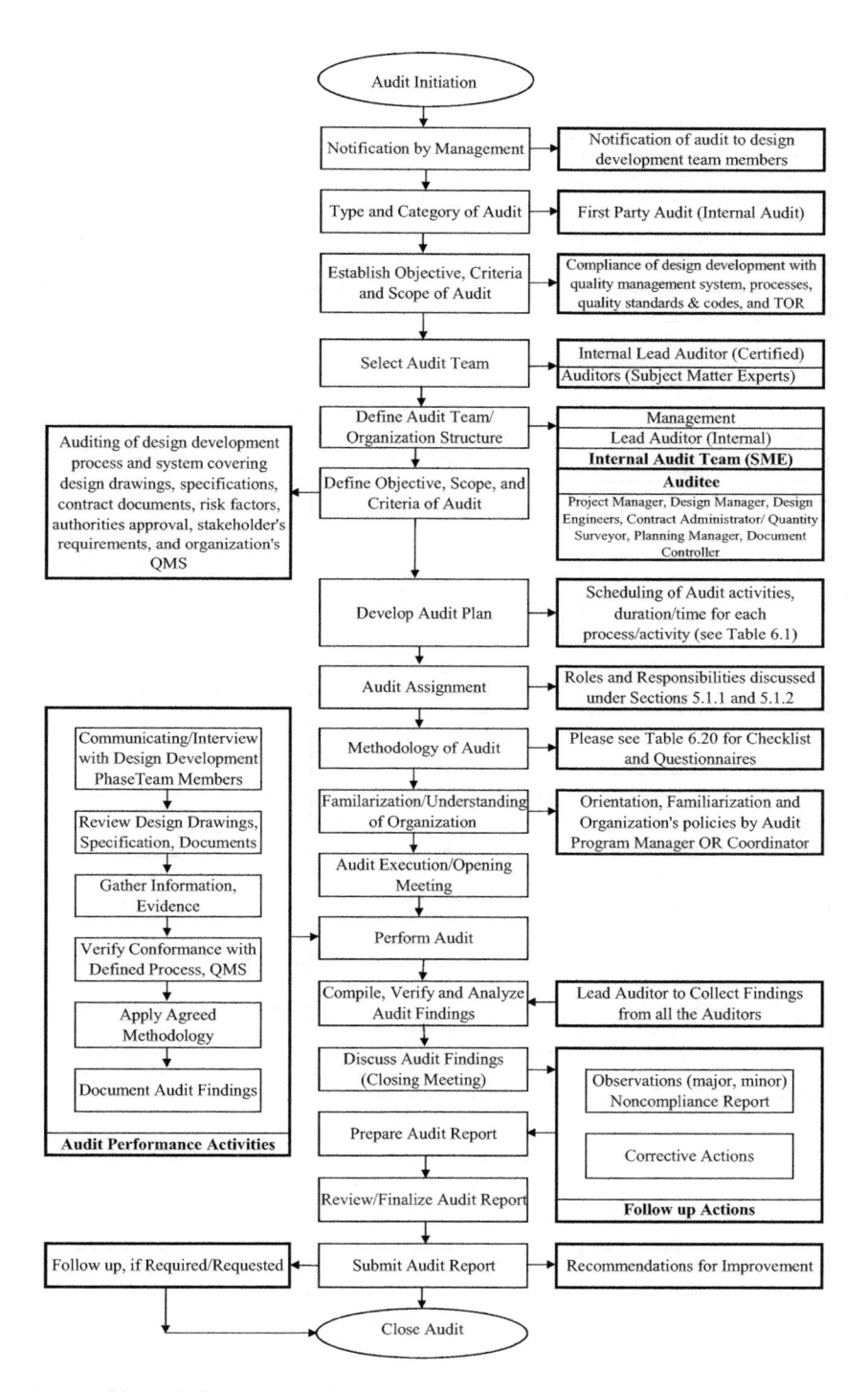

Figure 6.22 Logic flow diagram for the quality auditing of the design development phase

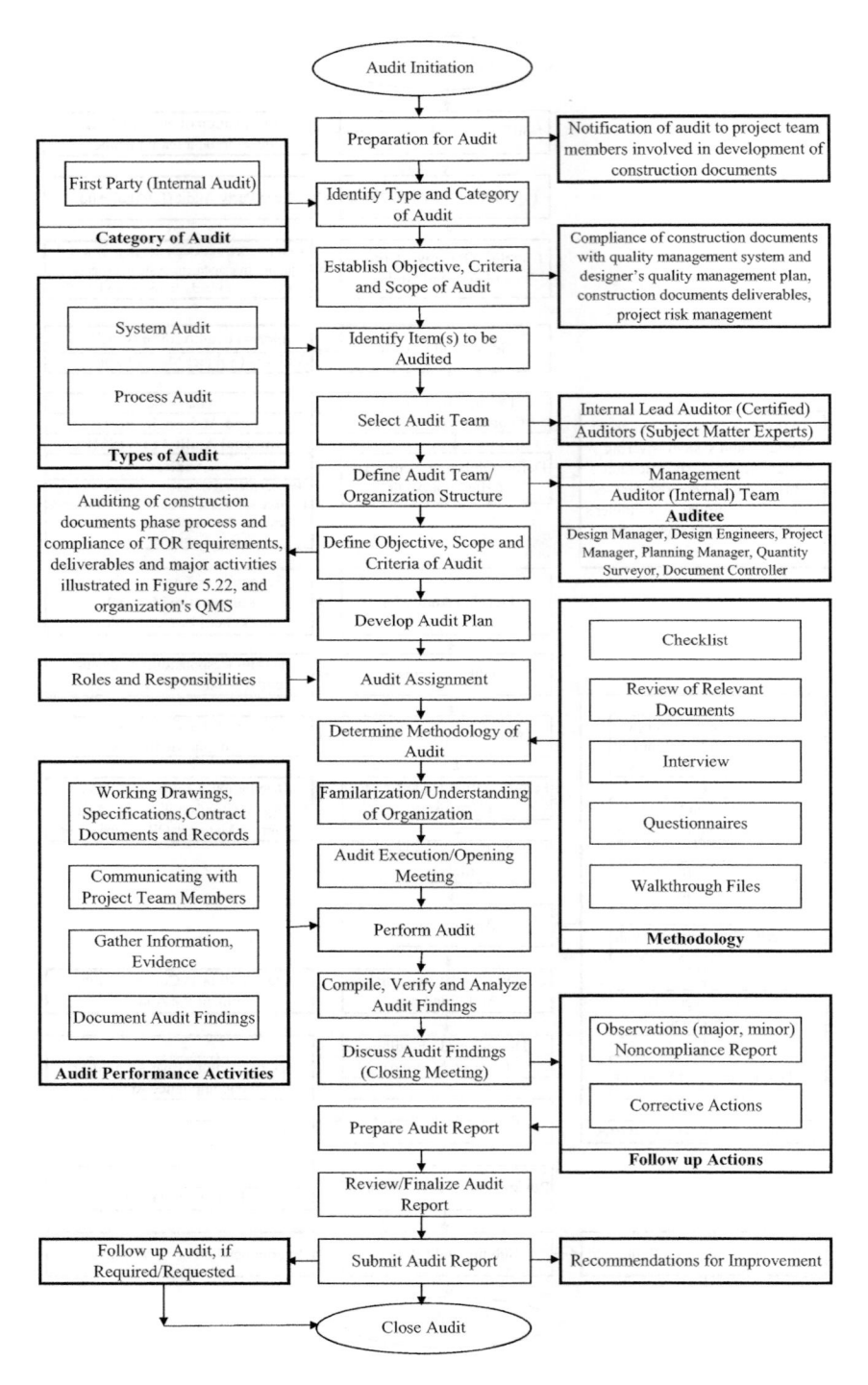

Figure 6.23 Typical quality auditing process for the construction documents phase

Auditing processes for project life cycle phases 271

2. Assesss conformance to QMS and that processes meet established quality system requirements
3. Review the process used to prepare construction documents
4. Verify that the construction documents comply with the requirements for the bidding and tendering process
5. Verify that construction documents deliverables are as per TOR requirements and as listed in Table 6.21
6. Validate the project schedule meeting the project requirements
7. Validate the cost of the project meeting the project requirements
8. Validate that working drawings and documents comply with quality standards and codes
9. Assess the compliance with working drawings with the design quality management system
10. Check for compliance with regulatory requirements
11. Check for sustainability considerations in the development of the design
12. Check that conservation of energy requirements is considered
13. Check that facility management requirements are considered

Table 6.21 Construction documents deliverables

Serial Number	Deliverables
1	Document I
1.1	Tendering procedure
	i Invitation to tender
	ii Instructions to bidders
	iii Forms for tender and appendix
	iv List of equipment and machinery
	v List of contractor's staff
	vi Contractor's certificate of work statement
	vii List of subcontractor(s) or specialist(s)
	viii Initial bond
	ix Final bond
	x Forms of agreement
2	Document II
2.1	Conditions of contract
	II-1 General conditions
	II-2 Particular conditions
	II-3 Public tender laws
3	Document III
	III-1 General specifications
	III-2 Particular specifications
	III-3 Drawings
	III-4 Schedule of rates and bill of quantities
	III-5 Analysis of prices
	III-6 Addenda
	III-7 Tender requirements (if any) and any other instructions issued by the owner

14 Review that the key risk factors are analyzed and managed while preparing construction documents
15 Assess compliance to safety considerations in the working drawings
16 Review working drawings, specifications and contract documents
17 Assess that all the major activities are adequately performed and managed as listed in Figure 5.22
18 Assess proper functioning of the project management system

The following audit tools are used to perform a quality audit of the design development phase:

- Analytical tools
- Checklists
- Review of documents
- Interviews
- Questionnaires

The auditee team members are mainly from the following organizational groups:

- Project Manager
- Design Manager
- Design Engineers (All Trades)
- Planning Manager
- Contract Administrator/Quantity Surveyor
- Document Controller

Table 6.22 illustrates the audit methodology for the assessment of the construction documents phase and Figure 6.24 illustrates a logic flow diagram for the quality auditing of the construction documents phase.

6.2.3 Auditing process for the bidding and tendering phase

In many countries, it is a legal requirement that government-funded projects employ the competitive bidding method. This requirement gives the opportunity for all qualified contractors to participate in the tender, and normally the contract is awarded to the lowest bidder. Private-funded projects have more flexibility in evaluating the tender proposal. Private owners may adopt the competitive bidding system, or the owner may select a specific contractor and negotiate the contract terms. Negotiated contract systems have flexibility of pricing arrangement as well as the selection of the contractor based on their expertise or the owner's past experience with the contractor successfully completing one of their projects. The negotiated contract systems are based on the following forms of payment:

1 Cost plus contracts: a type of contract in which the contractor agrees to do the work for the cost of time and material, plus an agreed-upon amount of profit. The following are the different types of cost plus contracts:

Table 6.22 Audit methodology for assessment of construction documents

Serial Number	Item To Be Assessed	Yes	No	Comments
A. Checklist				
1	Whether the scope of work for the development of construction documents is established based on TOR requirements			
2	Whether an organization chart with responsibilities to develop construction documents is established			
3	Whether the construction documents support the owner's project goals and objectives			
4	Whether the construction documents comply with organizational strategic policies			
5	Whether the construction documents meet all the elements specified in the TOR			
6	Whether comments on detail design are taken into account while preparing the construction documents			
7	Whether regulatory/statutory requirements are considered			
8	Whether the design has functional and technical compatibility			
9	Whether the design is constructible			
10	Whether the working drawings meet all the performance requirements			
11	Whether the working drawings meet the project objectives in respect of time, cost and quality			
12	Whether project boundaries are considered in the working drawings			
13	Whether the owner's preferred requirements are included in the working drawings			
14	Whether the design includes all the functional, external works, parking and landscape areas			
15	Whether the project schedule is achievable in practice			
16	Whether the project cost is properly estimated			
17	Whether the BOQ is coordinated with the specifications and contract documents			
18	Whether the construction documents phase complies with the designer's quality management plan			
19	Whether the working drawings are based on specified quality standards and codes			
20	Whether working drawings are coordinated with all the trades			

(*Continued*)

Table 6.22 (Continued)

Serial Number	Item To Be Assessed	Yes	No	Comments
A. Checklist				
21	Whether the following information is provided in each of the trade drawings: • Working drawings produced at different scales and format • Plans • Sections • Elevations • Schedule • Drawing index			
22	Whether all the drawings are numbered.			
23	Whether the following information is included on all the drawings: • Client name • Client logo • Location map • North orientation • Project name • Drawing title • Drawing number • Date of drawing • Revision number • Drawing scale • Contract reference number • Signature block • Signed by the designer for check and approval			
24	Whether the designer has considered the availability of resources during the construction and testing and commissioning phase			
25	Whether the recommended material meets the owner's objectives			
26	Whether the construction documents conform with the project delivery system			
27	Whether the construction documents conform to the type of contract/pricing with the adopted methodology			
28	Whether the following documents are included along with the working drawings: • Existing site conditions/site plan • Site surveys • Design calculations • Studies and reports			
29	Whether the specification is prepared taking into consideration international construction specifications and division numbers			
30	Whether construction documents are prepared taking into consideration international construction contract documents			

Serial Number	Item To Be Assessed	Yes	No	Comments
A. Checklist				
31	Whether risks has been identified and analyzed, and responses planned for mitigation of risk while preparing the working drawings			
32	Whether health and safety requirements in the working drawings are considered			
33	Whether the design confirms with fire and egress requirements			
34	Whether environmental constraints are considered			
35	Whether the working drawings, specifications and documents are prepared to meet the TOR requirements			
36	Whether construction documents deliverables conform as listed in Table 6.19			
37	Whether the reports are complete and include adequate information about the project			
38	Whether the report is properly formatted and there is a table of contents for each report			
39	Whether there is any approved review procedure			
40	Whether working drawings are reviewed prior to submission			
41	Whether the management process activities related to prepare construction documents as listed in Figure 5.22 are followed			
42	Whether the working drawings support construction of the project/facility			
43	Whether the documents include a penalty for delay in completion of the project			
44	Whetehr the documents include a penalty for failing to provide the required resources (manpower) for the project			
45	Whether the documents include a penalty for delay and violation of safety during the execution of project			
46	Whether the construction documents are properly labeled and archived			

Serial Number	Description	Response	Comments
B. Interview Questionnaires			
1	What procedure was followed to coordinate between working drawings and specifications		
2	How the availability of resources during the construction phase were ascertained		

(*Continued*)

Table 6.22 (Continued)

B. Interview Questionnaires

Serial Number	Description	Response	Comments
3	Was there any delay in getting authorities' approval		
4	Whether construction documents were prepared within the agreed-upon schedule Whether a work progress schedule was prepared to monitor the progress		
5	Whether a time sheet was maintained to record the time spent by each staff member engaged in the preparation of construction documents		
6	Whether construction documents were submitted in time for the owner's review		
7	What method was used to prepare the project schedule		
8	What methodology was used to estimate the definitive cost		
9	Whether risk factors are considered while developing the construction documents		
10	Whether the numbers of sets were prepared as per the TOR and submitted		
11	Whether the owner issued a construction documents acknowledgement letter		
12	Whether the construction documents have any major comments from the owner		
13	Whether the construction documents were not approved by the owner		
14	Whether the reasons (causes) for comments, non-approval, if any, were analyzed and corrective/preventive action taken		

- a Cost plus percentage fee contract
- b Cost plus fixed fee contract
- c Cost plus incentive fee contract

2. Reimbursement contracts: a type of contract in which the contractor agrees to do the work for the cost per schedule of rates, or bill of quantities, or bill of material.
3. Fixed price contracts: with this type of contract, the contractor agrees to work with a fixed price (it is also called *lump sum*) for the specified and contracted work. Any extra work is executed only upon receipt of instruction from the owner. Fixed price contracts are generally inappropriate for work involving major uncertainties, such as work involving new technologies.
4. Target cost contracts: this is based on the concept of a top-down approach, which provides a fixed price for an agreed range of out-turn costs around the

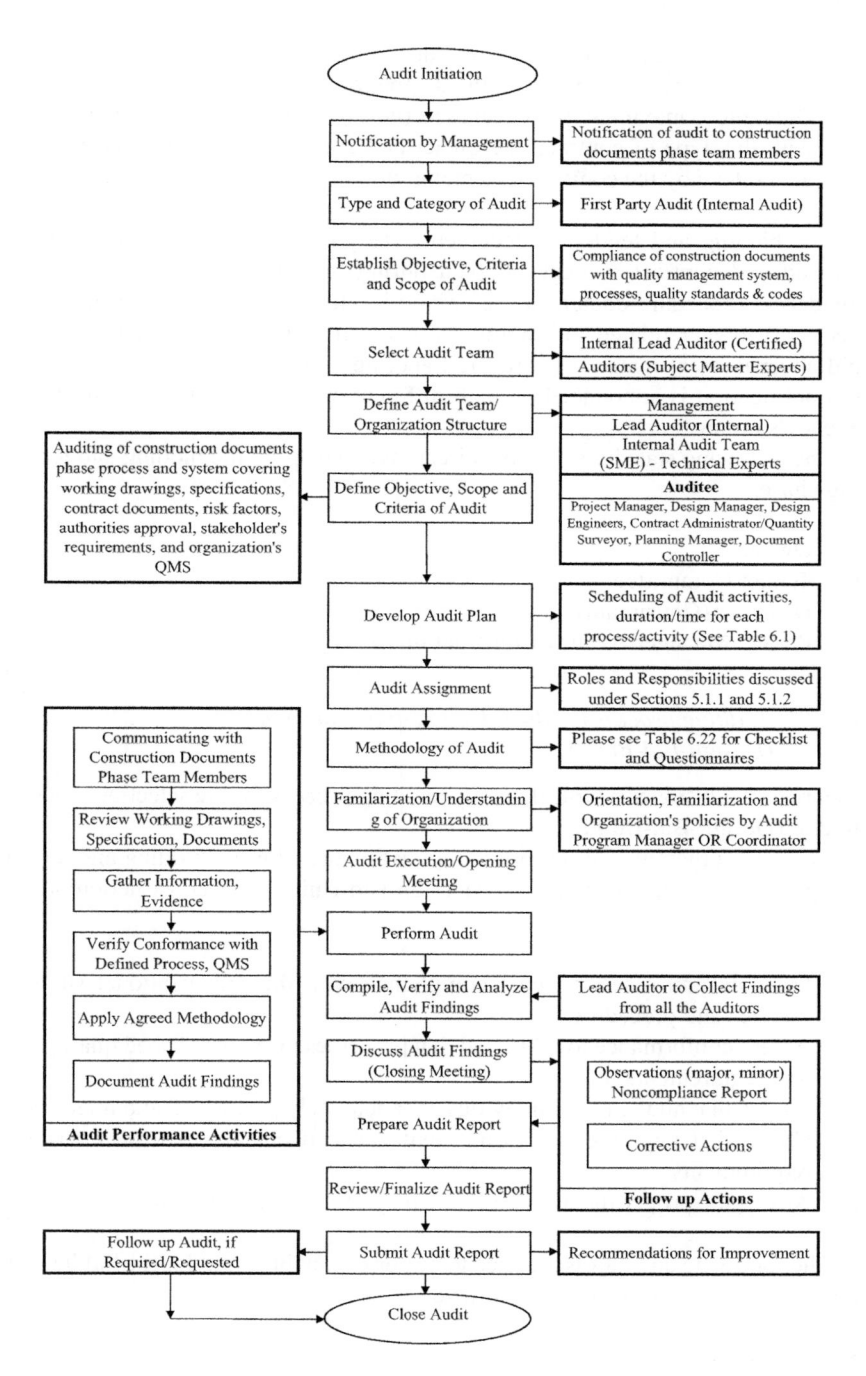

Figure 6.24 Logic flow diagram for the quality auditing of the construction documents phase

target. In this type of contract, overrun or underspend are shared by the owner and the contractor at predetermined agreed-upon percentages.
5   Guaranteed maximum price contracts (cost plus guaranteed maximum price): with this type of contract, the owner and contractor agree to a project cost guaranteed by the contractor as maximum.

The bidding and tendering documents are prepared as per the procurement method and contract strategy adopted during the early stages of the project. Tendering procedure documents submitted by the designer are updated and the necessary owner related information is inserted in the tender documents. The bid advertisement material is prepared and, upon approval from the owner, the bid notification is announced through different media as per the organization's/agency's policy.

The following stakeholders have direct involvement in the bidding and tendering phase:

- Owner
- Tender Committee
- Designer (Consultant)
- Project Manager/Agency Construction Manager (as applicable)

6.2.3.1 Auditing process for the selection of a contractor (Design-Bid-Build)

Figure 6.25 illustrates a typical quality auditing process for the selection of a contractor (Design-Bid-Build).

The main objectives of performing a quality audit for the bidding and tendering process to select the contractor (Design-Bid-Build) for a construction project are to:

1   Evaluate and assess process compliance with the organizational strategic policies and procedure
2   Assess conformance to QMS and that processes meet established quality system requirements
3   Assess that all the major activities are adequately performed and managed
4   Check that key risk factors are analyzed and managed to meet business requirements
5   Assess the proper functioning of the project management system

The following audit tools are used to perform a quality audit for the bidding and tendering process to select the contractor:

- Checklists
- Review of documents
- Interviews

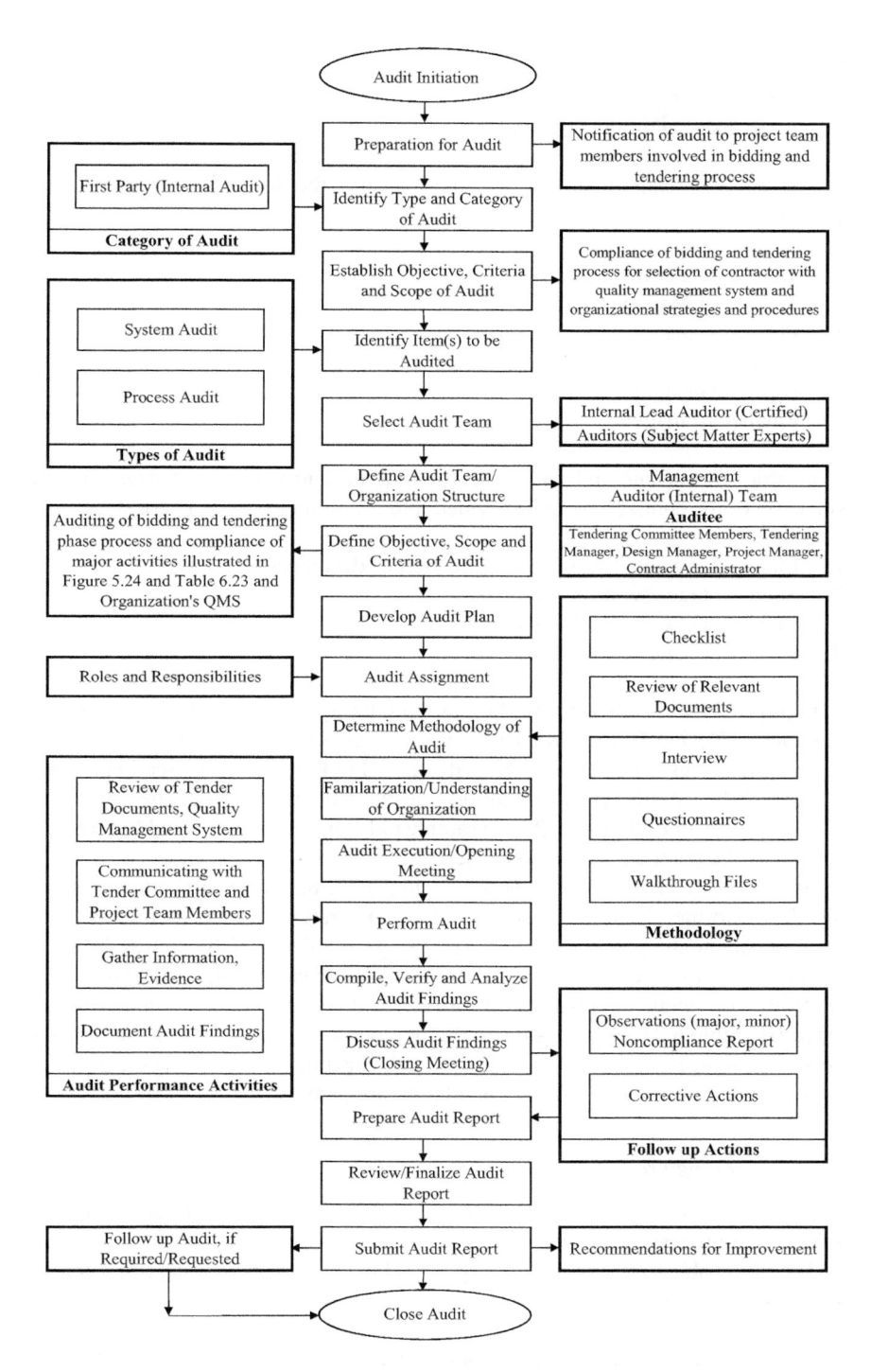

Figure 6.25 Typical Quality Auditing Process for the Bidding and Tendering Procedure for the Selection of the Contractor (Design-Bid-Build)

280 Abdul Razzak Rumane

- Questionnaires
- Analysis of audit documentation

The auditee team members are mainly from the following organizational groups:

- Tendering Committee
- Tendering Manager
- Project Manager
- Contract Administrator
- Design Manager
- Finance Manager

In order to manage and control it, the bidding and tendering process for the selection of the contractor can be divided into four major stages. These stages are:

1. First stage: Shortlisting/prequalification of designers
2. Second stage: Tender documents
3. Third stage: Contract bid solicitation
4. Fourth stage: Contract award

Figure 6.26 illustrates the bidding and tendering process stages to select the contractor (Design-Bid-Build). The four stages are developed taking into consideration the basics discussed earlier as illustrated in Figure 5.23 in Section 5.2.6.

Table 6.23 lists the prequalification questionnaires to register the contractor (Design-Bid-Build), Figure 6.27 is a sample bid clarification form and Figure 6.28 illustrates the contractor selection criteria

Based on the typical stages as per Figure 6.26, the audit methodologies can also be performed in four stages. Table 6.24 illustrates the audit methodology to assess/evaluate the process compliance at stage I, Table 6.25 illustrates the audit methodology to assess/evaluate the process compliance at stage II, Table 6.26 illustrates the audit methodology to assess/evaluate the process compliance at stage III, Table 6.27 illustrates the audit methodology to assess/evaluate the process compliance at stage IV and Figure 6.29 illustrates a logic flow diagram for the quality auditing of the bidding and tendering phase.

6.2.3.2 Auditing process for the contractor's tender submission process

Figure 6.30 illustrates a typical quality auditing process for the contractor's tender submission process.

The main objectives of performing a quality audit of the contractor's tender submission process are to:

1. Evaluate and assess process compliance with the organizational strategic policies and procedure

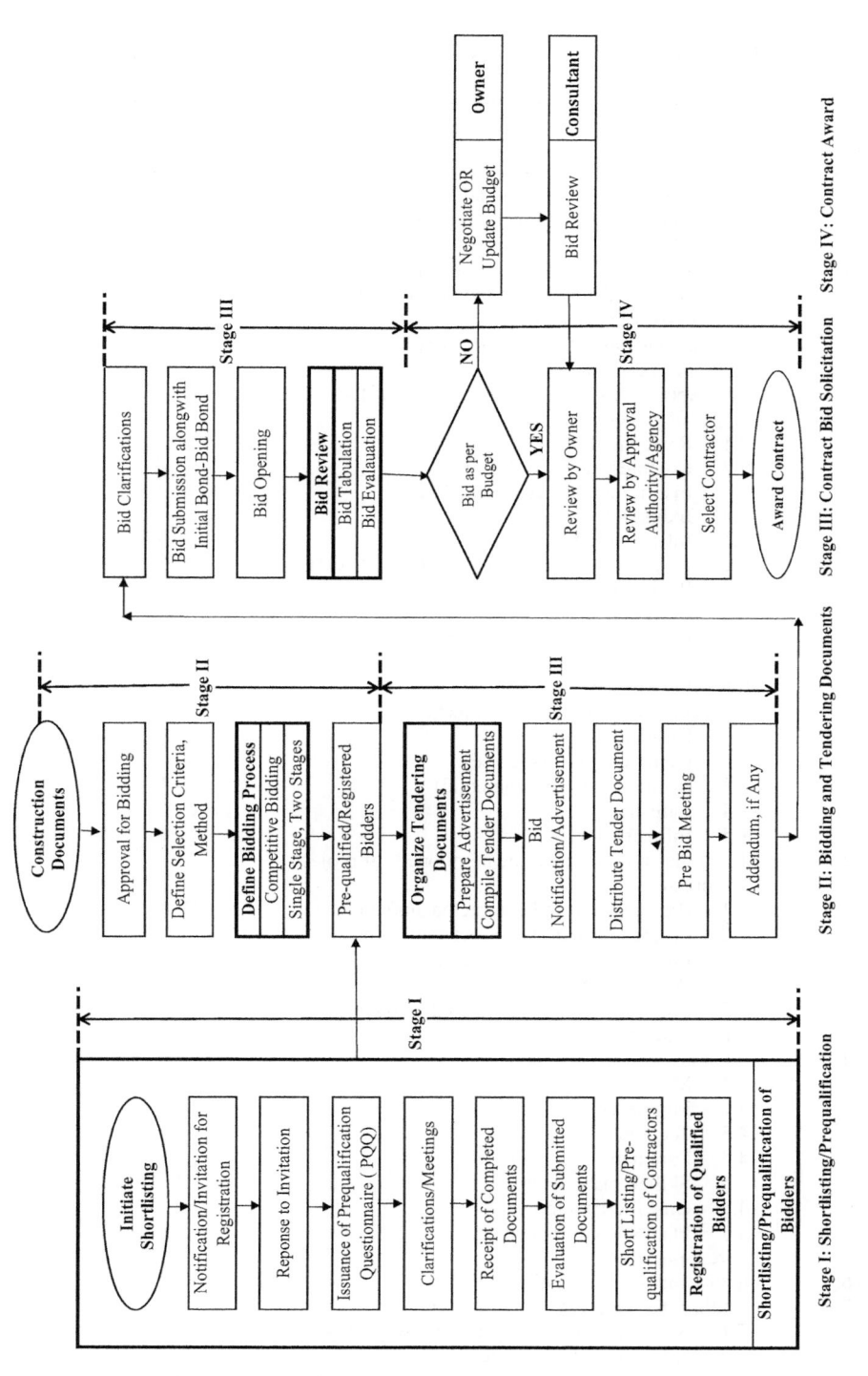

Figure 6.26 Bidding and tendering (procurement) process stages for the selection of the contractor (Design-Bid-Build)

Table 6.23 Prequalification questionnaires (PQQ) for selecting contractor (Design-Bid-Build)

Serial Number	Question	Answer
1	Name of the organization and address	
2	Organization's registration and license number	
3	ISO certification	
4	Registration/classification status of the organization	
5	Joint venture with any international contractor	
6	Total turnover last 5 years	
7	Audited financial report for last 3 years	
8	Insurance and bonding capacity	

9	Total experience (years) in construction of following types of project
	  9.1 Residential
	  9.2 Commercial (mixed use)
	  9.3 Institutional (governmental)
	  9.4 Industrial
	  9.5 Infrastructure	

10	Size of project (maximum amount single project)
	  10.1 Residential
	  10.2 Commercial (mixed use)
	  10.3 Institutional (governmental)
	  10.4 Industrial
	  10.5 Infrastructure	

11	List successfully completed projects
	  11.1 Residential
	  11.2 Commercial (mixed use)
	  11.3 Institutional (governmental)
	  11.4 Industrial
	  11.5 Infrastructure	

12	List similar types (type to be mentioned) of project completed
	  12.1 Project name and contracted amount
	  12.2 Project name and contracted amount
	  12.3 Project name and contracted amount
	  12.4 Project name and contracted amount
	  12.5 Project name and contracted amount	

13	List of subcontractors	

14	Resources
	  14.1 Management
	  14.2 Engineering
	  14.3 Technical
	  14.4 Foreman/supervisor
	  14.5 Skilled manpower
	  14.6 Unskilled manpower
	  14.7 Plant and equipment	

15	Current projects	
16	Quality management policy	

17	Health, safety and environment policy
	  17.1 Number of accidents during last 3 years
	  17.2 Number of fires on site	

Serial Number	Question	Answer
18	Staff development policy	
19	List of delayed projects	
20	List of failed contracts	
21	List of professional awards	
22	Litigation (dispute, claims) on earlier projects	

2. Assess conformance to QMS and that processes meet established quality system requirements
3. Check that the proposal preparation and submission comply with the organization's tender submission process. Figure 6.31 illustrates a typical tender submission process
4. Validate that the response to the contents of tender documents is complete in all respects. Table 6.21 illustrates the contents of tender documents (construction documents deliverable)
5. Verify that the tender documents are properly reviewed
6. Verify that the drawings match with the specifications and documents
7. Verify the correctness of BOQ with the working drawings and specifications
8. Review the process used to prepare the bid
9. Ensure that the prices of material, equipment and machinery considered in preparing the bid are valid for the entire duration of the construction phase
10. Check the availability of resources (manpower) for construction activities is properly considered
11. Validate the total cost of the bid
12. Validate that the construction schedule is practical and achievable
13. Review that the key risk factors are analyzed and managed
14. Check for compliance with regulatory requirements
15. Validate that the bid has been prepared by fully coordinating with all the stakeholders

The following audit tools are used to perform the quality audit of the tender submission process:

- Analytical tools
- Checklists
- Review of documents
- Interviews
- Questionnaires

The auditee team members are mainly from the following organizational groups:

- Construction/Project Manager
- Tendering Manager

Project Name
Project Number
Bid Clarification Form

Serial Number	Name of Contractor	Item No. and Clause Reference	Queries	Owner/Consultant's Clarification	Remarks

Figure 6.27 Bid clarification form

Evaluation Criteria	Weightage	Key Points for Consideration	Review Result
1. General Information			Yes OR No
1 Company information		Company's current position–a MUST information	
2. Financial	**25%**		
1 Total Turn Over (last 5 years)	25%	Sum of the turn over for the last five years	
2 Values of current work-in-hand	25%	Project value / Value of current work-in-hand	
3 Audit Financial Reports	10%	To confirm the ratio given in point three	
4 Financial Standing	30%		
33% Assets		Current Assets / Current Liabilities	
34% Liabilities		Total Liabilities – Total Equity / Total assets	
33% Profit/Loss		Net Profit before Tax / Total Equity	
5 Bonding and Insurance Limit	10%		
60% a. Performance & Bonding Capacity		Provided or Not provided	
40% b. Insurance		Provided or Not provided	
3. Organization Details			
3a. Business	**20%**		
1 Company's Core Area of Business	30%	Degree of satisfactory answer	
2 Experience of years in business	30%	No. of years	
3 ISO Certification	15%	Yes or No	
4 Registration/Classification Status	15%	Grade or Classification	
5 Organizational Chart	10%	Key staff indicated (Name/Title), Balanced Resources, Departmental(Specialization) Diversity, Lines Of Communication	
6 Dispute/Claims		Degree of satisfactory answer	
3b. Experience	**30%**		
50% a. Projects' value		No. of projects with comparable value	
50% b. Projects' type (similar type and complexity)		No. of projects with similar complexity	
4. Resources	**20%**		
1 Personnel	60%		
30% Management		No. of managerial staff	
30% Engineers		No. of engineers and project staff	
30% Technicians		No. of CAD technicians and foreman	
10% Staff Development		% turnover spent on training	
2 Technology	10%	% of turnover spent on acquiring latest construction technology	
3 Plant & Equipment	30%	List of plant and equipment	
5. GENERAL	**5%**		
1 Bank References	30%	Provided or Not provided	
2 Project References	30%	Provided or Not provided	
2 Health, Safety and Environment narration	40%	Degree of Satisfactory answer	

Figure 6.28 Contractor selection criteria

Table 6.24 Audit methodology for the assessment of the bidding and tender process for the selection of a contractor (Design-Bid-Build) – Stage I (shortlisting/registration of contractors)

Serial Number	Item To Be Assessed	Yes	No	Comments
A. Checklist				
1	Whether the notification for registration was announced in all technical newsletters and leading news media as per the organization's policy			
2	Whether there are registered designers to participate in the tender			
3	Whether prequalification questionnaires (PQQ) are issued to all the intending bidders (please refer to Table 6.23 for guidelines)			
4	Whether there are any requests for queries/bid clarification			
5	Whether the meeting for clarification is attended by all the intending bidders			
6	Whether the attendance sheet for the meeting is signed by all the attendees			
7	Whether minutes of the meeting are circulated to all the attendees/participating bidders			
8	Whether bid clarifications of the queries are sent to all the bidders (please refer to Figure 6.27 illustrative bid clarification form)			
9	Whether sufficient time is provided to prepare and submit completed PQQ			
10	Whether there are any requests for extension of submission date			
11	Whether all the intending bidders submitted the response to PQQ			
12	Whether received bid envelopes are placed in safe custody			
13	Whether there are any delayed submissions			
14	Whether received responses are acknowledged			
15	Whether received responses are clearly identified and recorded			
16	Whether responses are fairly evaluated as per the organization's selection policy			
17	Whether evaluation criteria, weightage and methodology are as specified (please refer to Figure 6.28 for guidelines for selection of contractor)			
18	Whether any of the submitted responses are found to be incomplete			
19	Whether there are any late submissions			
20	Whether risk factors are considered while evaluating the response and registration of designers			
21	Whether reasons for non-registration are conveyed to unsuccessful participants			
22	Whether shortlisted contractors are allotted a code and registration number			

B. *Interview questionnaires*

Serial Number	Description	Response	Comments
1	How responses were submitted: • Hand delivery • By post • Courier service		
2	Whether PQQ was available for electronic distribution		
3	How was the response for registration/shortlisting		

Table 6.25 Audit methodology for the assessment of the bidding and tender process for the selection of the contractor (Design-Bid-Build) – Stage II (bidding and tender documents)

Serial Number	Item To Be Assessed	Yes	No	Comments
A. Checklist				
1	Whether the tender documents include all the relevant information required to select the contractor (Design-Bid-Build)			
2	Whether the construction documents are approved by the project owner			
3	Whether bid/tender documents are prepared as per the procurement method and contract strategy adopted during the early stages of the project			
4	Whether all the information to enable bidders/contractors to submit the proposal is clear and unambiguous			
5	Whether documents are prepared taking into consideration the organization's Quality Management System (QMS)			
6	Whether regulatory/authority requirements are taken into consideration while preparing the tender documents			
7	Whether the selection criteria and selection method for contractor selection are clearly specified in the organization's QMS			
8	Whether budget approval is obtained for construction prior to the release of tender announcement			
9	Whether tender opening and closing dates are mentioned in the announcement			
10	Whether the schedule for completion of the construction phase is included in the tender documents			
11	Whether qualifications of personnel and core team members for construction are requested in the tender documents			
12	Whether project quality requirements are properly defined in the documents			
13	Whether the rights and liabilities of all the parties are mentioned in the documents			

(*Continued*)

Table 6.25 (Continued)

Serial Number	Item To Be Assessed	Yes	No	Comments
A. Checklist			
14	Whether change/variation clauses are included in the documents			
15	Whether cancellation/termination clauses are included in the documents			
16	Whether bond and insurance clauses are included in the documents			
17	Whether the tender documents are as per international/local standard documents used by the construction industry			
18	Whether evaluation and assessment criteria are well defined			
19	Whether review and analysis procedure of bid is included in the documents			
20	Whether the contract to be signed for construction is included in the documents			
21	Whether all the records are properly maintained			
22	Whether the request to provide the bidder's contact details is included in the documents			

B. Interview questionnaires

Serial Number	Description	Response	Comments
1	Whether any outside agency was involved in the preparation of the tender documents		
2	What guidelines were adopted to prepare the tender documents		
3	Whether any specific format/template was followed		
4	Whether all the requirements of the tender documents were coordinated and agreed by relevant stakeholders		
5	Whether risks of disputes and associated financial costs were considered while preparing the contract documents		
6	Whether problems and challenges were identified while preparing the bid/tender		
7	Who was responsible for authorizing the bidding requirements		

Table 6.26 Audit methodology for the assessment of the bidding and tender process for the selection of the contractor (Design-Bid-Build) – Stage III (contract bid solicitation)

Serial Number	Item To Be Assessed	Yes	No	Comments
A. Checklist			
1	Whether the tender documents are properly organized			

Serial Number	Item To Be Assessed	Yes	No	Comments

A. Checklist

2	Whether the tender documents are distributed to all the prequalified/shortlisted/registered bidders (contractors) to participate in the tender			
3	Whether the tender is notified to all the prequalified bidders			
4	Whether all the prequalified/shortlisted/registered bidders participated in the tender			
5	Whether the attendance sheet for the meeting was signed by all the attendees			
6	Whether minutes of the meeting were circulated to all the attendees/participating bidders/contractor			
7	Whether sufficient time is provided to prepare and submit the tender			
8	Whether the tender submission date is extended from the originally announced date			
9	Whether an addendum (if any) is issued/notified to all the participating bidders			
10	Whether there are any requests for extension of the submission date			
11	Whether the clarification meeting is attended by all participating bidders			
12	Whether any changes are made to the originally announced tender documents			
13	Whether received bid envelopes are placed in safe custody			
14	Whether technical and financial envelopes are submitted at the same time			
15	Whether there are any delayed submissions			
16	Whether the tenders are opened as per the announced date and time			
17	Whether the tenders are opened in the presence of all the bidders			
18	Whether received tenders are acknowledged			
19	Whether received tenders are clearly identified and recorded			
20	Whether applicable fees/bid bond are submitted by all the bidders who participated in the tender			
21	Whether tenders are fairly evaluated as per the announced policy and procedure			
22	Whether the comparison of bids is tabulated			
23	Whether evaluation criteria, weightage and methodology are as specified			
24	Whether the bid cost is higher than the budgeted cost			
25	Whether any submitted tender is found to be incomplete			
26	Whether there are any late submissions			
27	Whether all the scheduled pages are signed by the bidder			

(Continued)

Table 6.26 (Continued)

Serial Number	Item To Be Assessed	Yes	No	Comments

A. Checklist

28	Whether risk factors are considered while evaluating the bids			
29	Whether reasons for non-acceptance of the tender were conveyed to the unsuccessful bidders			

B. Interview Questionnaires

Serial Number	Description	Response	Comments
1	How the notification was announced		
2	How tenders were submitted: • Hand delivery • By post • Courier service		
3	Whether the tender was available for electronic distribution		
4	How was the response to participate in the tender		
5	Whether tender box(es) were placed at secured places and monitored through an electronic surveillance system		
6	How much time was allowed to open the bids after the notified tendering closing time		
7	Whether the tender submission date was extended and what was the reason for the extension		
8	Whether all the tender envelopes/packages were sealed with marking of the tenders and contain original tender documents and required number of sets as per the tender conditions		
9	Whether a bid review was established		
10	Whether the bid selection procedure was clearly defined		

Table 6.27 Audit methodology for the assessment of the bidding and tender process for the selection of the contractor (Design-Bid-Build) – Stage IV (contract award)

Serial Number	Item To Be Assessed	Yes	No	Comments

A. Checklist

1	Whether the selection of the contractor satisfies all the conditions to be a successful bidder			
2	Whether the selected bidder (contractor) is a legal entity			
3	Whether the selection was made by the selection committee			

Serial Number	Item To Be Assessed	Yes	No	Comments

A. Checklist

Serial Number	Item To Be Assessed	Yes	No	Comments
4	Whether the selection of the contractor is approved by the relevant applicable authority, if mandated by the organization's policy			
5	Whether the selected contractor is capable of carrying out the specified work and has the necessary resources			
6	Whether risk factors are considered prior to the signing of the contract			
7	Whether the standard contract format is used for signing the contract			
8	Whether the contract terms and conditions are clearly written and unambiguous			
9	Whether an addendum and minutes of the meeting are included in the contract documents			
10	Whether the contract period/schedule is properly described			
11	Whether staff qualifications (core staff) are properly defined			
12	Whether a performance bond is submitted by the consultant			
13	Whether variation/change management clauses are included in the contract agreement			
14	Whether dispute resolution and conflict resolution clauses are properly defined			
15	Whether contract documents are reviewed prior to signing			
16	Whether tender documents and contract are properly archived			
17	Whether a letter of award is sent			
18	Whether the final results are published and announced			

B. Interview Questionnaires

Serial Number	Description	Response	Comments
1	On what basis the bidder (contractor) was selected: low bid, competitive bid or quality based system		
2	Was there any tie between two or more bidders		
3	Was there any dispute or objection raised by any of the bidders before the award of the contract to the successful bidder		
4	Whether the final price was negotiated or as per the bid (quotation) received		
5	Whether the successful bidder failed to submit a performance bond		

(Continued)

Table 6.27 (Continued)

B. Interview Questionnaires

Serial Number	Description	Response	Comments
6	Whether there was any need to update the approved budget		
7	Whether the contractor declared any subcontractor for outsourcing (subcontracting) purposes		
8	Whether risks in the project execution were considered		
9	Whether the contract includes a penalty clause for delay in completion of the project		
10	Whether the contract includes a penalty clause for failure to provide the required resources (manpower) for the project		
11	Whether the contract includes a penalty clause for safety violation during the execution of the project		

- Planning Manager
- Contracts Manager/Contract Administrator
- Quality Manager
- Engineering Manager
- Procurement Manager
- Document Controller
- Safety Officer
- Finance Manager

Table 6.28 illustrates the audit methodology for the assessment of the contractor's tender submission process and Figure 6.32 illustrates a logic flow diagram for the quality auditing of the contractor's tender submission process.

6.2.3.3 Auditing process for the selection of the construction supervisor (consultant)

In the traditional type of contract, it is normal practice for the designer/consultant of the project to be contracted by the owner to supervise the construction process. The supervision firm is normally known as the "consultant" and is responsible for supervising the construction process and achieving project quality goals. The firm (consultant) appoints a representative, who is acceptable and approved by the owner/client, to be on site and is often called a resident engineer. The resident engineer, along with the supervision team members, is responsible for supervising, monitoring and controling, implementing the procedure specified in the contract documents and ensuring completion of the project within the specified time and budget and per the defined scope of work. The consultant is responsible for

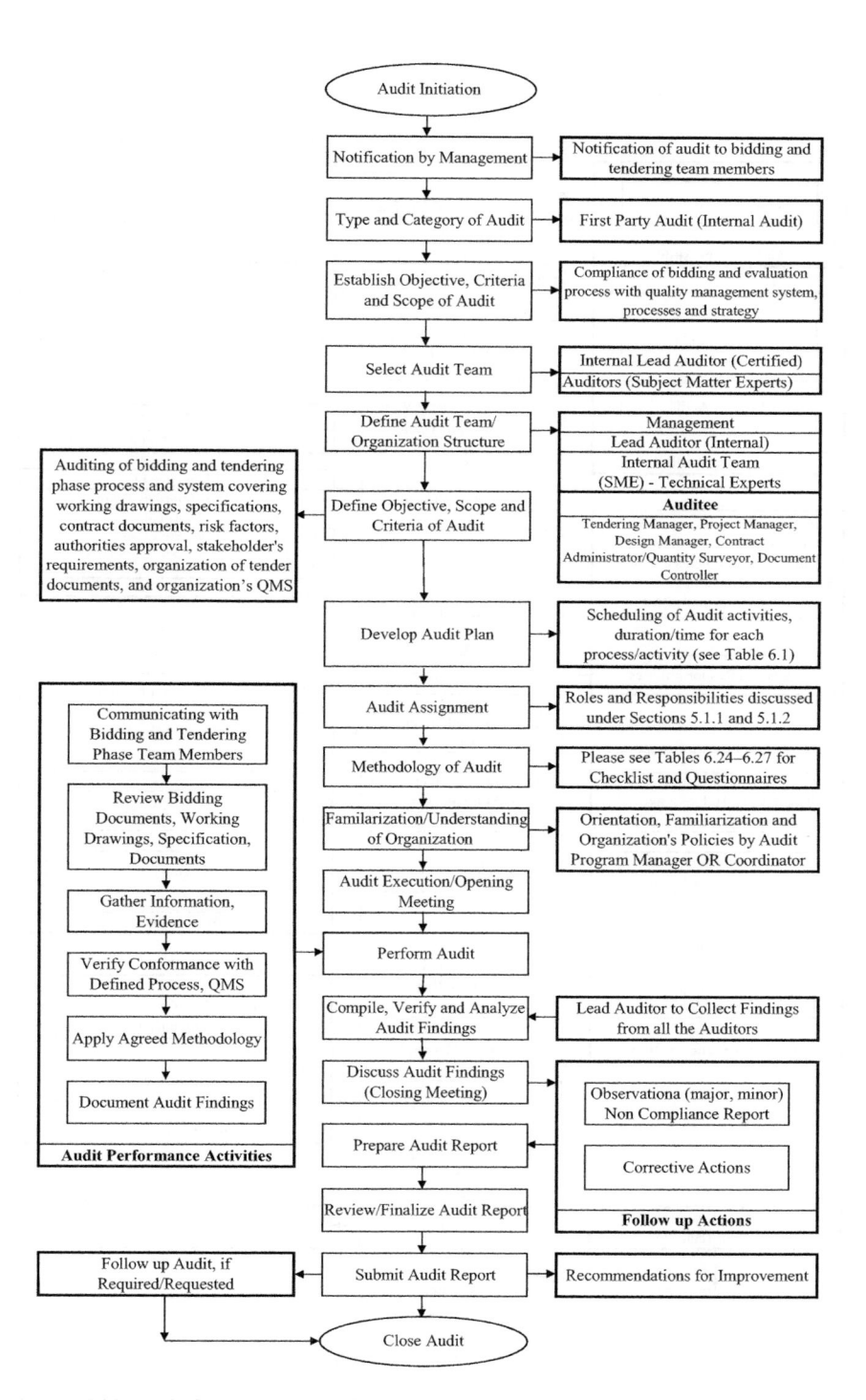

Figure 6.29 Logic flow diagram for the quality auditing of the bidding and tendering phase

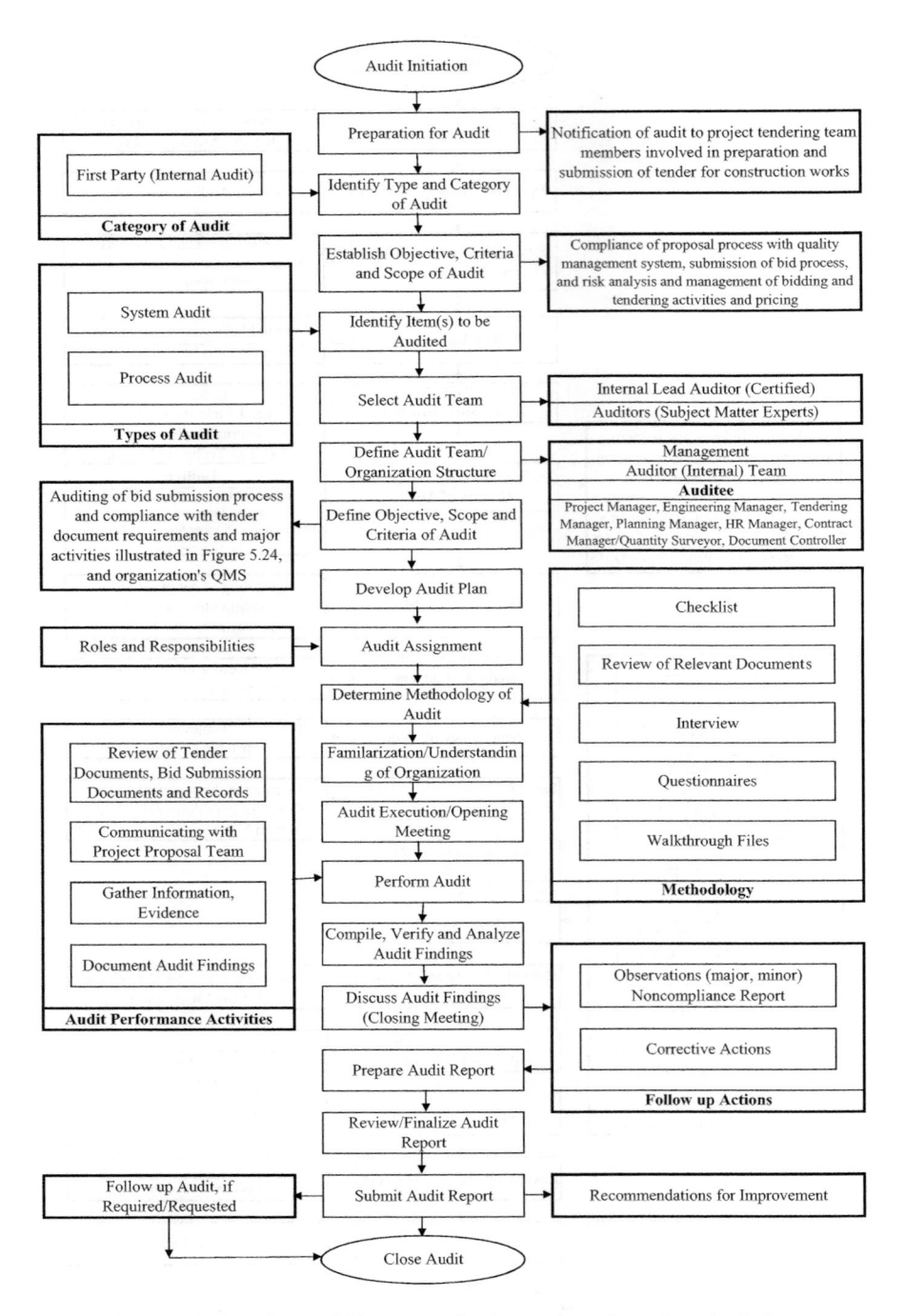

Figure 6.30 Typical quality auditing process for the contractor's tender submission process

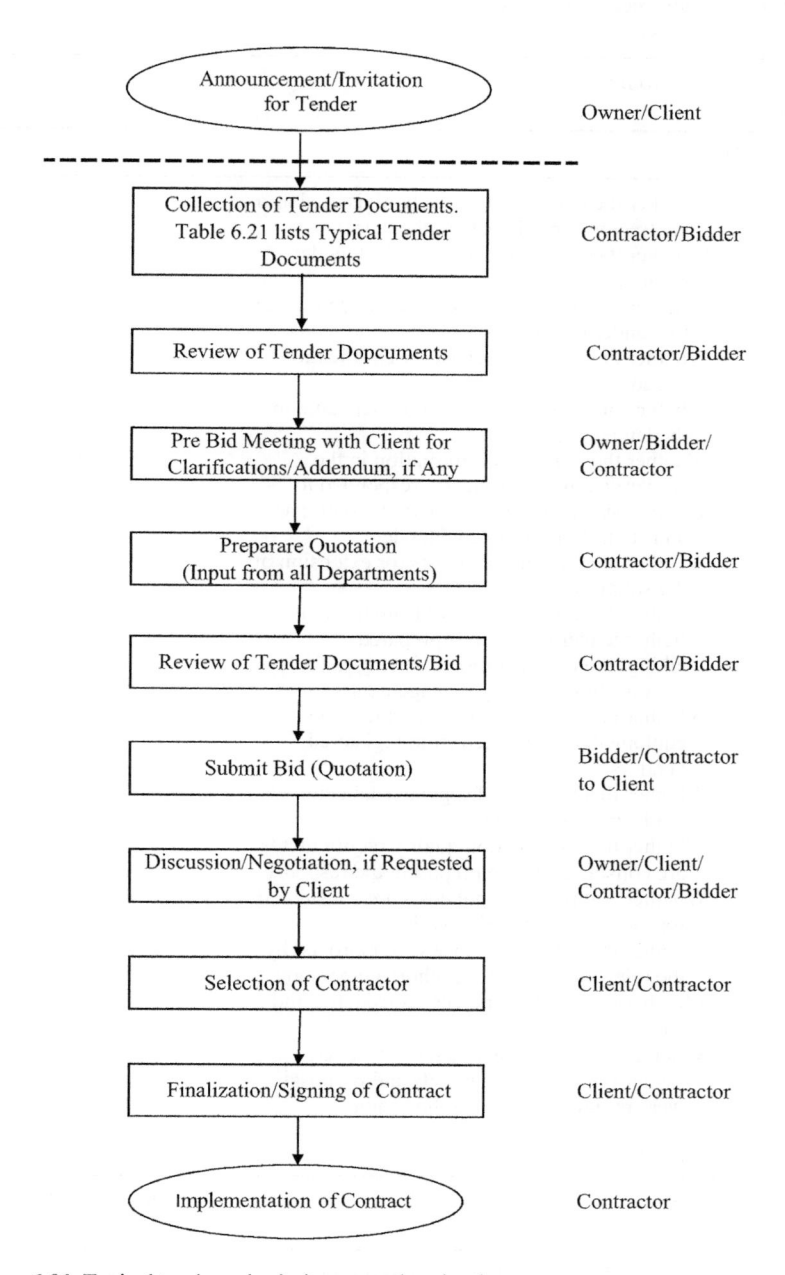

```
┌─────────────────────────┐
│ Announcement/Invitation │        Owner/Client
│      for Tender         │
└─────────────────────────┘
- - - - - - - - - - - - - - - - - - - -
┌─────────────────────────┐
│ Collection of Tender Documents. │
│ Table 6.21 lists Typical Tender │   Contractor/Bidder
│        Documents                │
└─────────────────────────┘
            │
┌─────────────────────────┐
│ Review of Tender Dopcuments │    Contractor/Bidder
└─────────────────────────┘
            │
┌─────────────────────────────┐
│ Pre Bid Meeting with Client for │   Owner/Bidder/
│ Clarifications/Addendum, if Any │   Contractor
└─────────────────────────────┘
            │
┌─────────────────────────┐
│   Preparare Quotation    │     Contractor/Bidder
│ (Input from all Departments) │
└─────────────────────────┘
            │
┌─────────────────────────┐
│ Review of Tender Documents/Bid │   Contractor/Bidder
└─────────────────────────┘
            │
┌─────────────────────────┐
│  Submit Bid (Quotation) │      Bidder/Contractor
└─────────────────────────┘       to Client
            │
┌─────────────────────────────┐
│ Discussion/Negotiation, if Requested │   Owner/Client/
│           by Client         │    Contractor/Bidder
└─────────────────────────────┘
            │
┌─────────────────────────┐
│  Selection of Contractor │     Client/Contractor
└─────────────────────────┘
            │
┌─────────────────────────┐
│ Finalization/Signing of Contract │  Client/Contractor
└─────────────────────────┘
            │
┌─────────────────────────┐
│ Implementation of Contract │    Contractor
└─────────────────────────┘
```

Figure 6.31 Typical tender submission procedure by the contractor

Table 6.28 Audit methodology for the assessment of the contractor's tender submission process

Serial Number	Item To Be Assessed	Yes	No	Comments
A. Checklist				
1	Whether the company is already registered as a contractor to participate in the tender			
2	Whether there are any fees to collect tender documents			
3	Whether the scope of work is clearly defined in the tender documents			
4	Whether there is any request for clarification raised			
5	Whether the response to the price schedule in the tender is complete			
6	Whether the requested information in the tender documents is clearly responded to			
7	Whether sufficient time is provided to prepare and submit the completed tender			
8	Whether there are any requests for extension of the submission date			
9	Whether there are any delayed submissions			
10	Whether tender response is prepared taking into consideration/referring to the organization's quality management system			
11	Whether prices for material, products and equipment are obtained from registered vendors			
12	Whether there are any designated subcontract works mentioned in the tender			
13	Whether the construction quality requirements are considered while preparing the response			
14	Whether equipment rental costs are considered for the execution of the works			
15	Whether the tender response was approved by the relevant committee/authorized person			
16	Whether the risk factors were considered and analyzed			
17	Whether all the relevant pages are signed			
18	Whether the tender documents and proposal were reviewed prior to submission			

Serial Number	Description	Response	Comments
B. Interview questionnaires			
1	Whether the notification for the proposal submission was announced in all technical newsletters and leading news media or it was only for registered bidders (contractors)		

B. Interview Questionnaires

Serial Number	Description	Response	Comments
2	How the tender was submitted: • Hand delivery • By post • Courier service		
3	Whether tender documents were available for electronic distribution		
4	What tools and techniques were used to price the tender		
5	Whether the organization's quality management plan was considered to evaluate the price schedule		
6	Whether the lack of resources/availability of resources (technical as well as skilled and semiskilled) was considered while preparing the price schedule		
7	Whether the availability of specified material and products was considered while preparing the price schedule		
8	Whether the availability of equipment and machinery for the construction period was considered while preparing the price schedule		
9	Whether any subcontracting work is from the registered vendors		
10	Whether the subcontractor cost was properly checked and evaluated		
11	Whether the construction schedule was properly reviewed by taking into consideration the availability of resources and their production output		
12	Whether the regulatory approval period was properly considered		
13	What risk factors were considered while evaluating the price schedule		
14	Whether risk factors were considered for outsourcing/subcontracting of works		
15	Whether any penalty due to non-conformance with the contract was considered in the tender price		
16	Whether purchasing prices for material, products and equipment are from the recommended/specified manufacturers		
17	Whether the tender prices were worked out as per 'rule of thumb' or actual cost records		
18	Whether any error in BOQ (Bill of Quantity) and contract drawings was observed		
19	Was there was any negotiation to the submitted quotation		

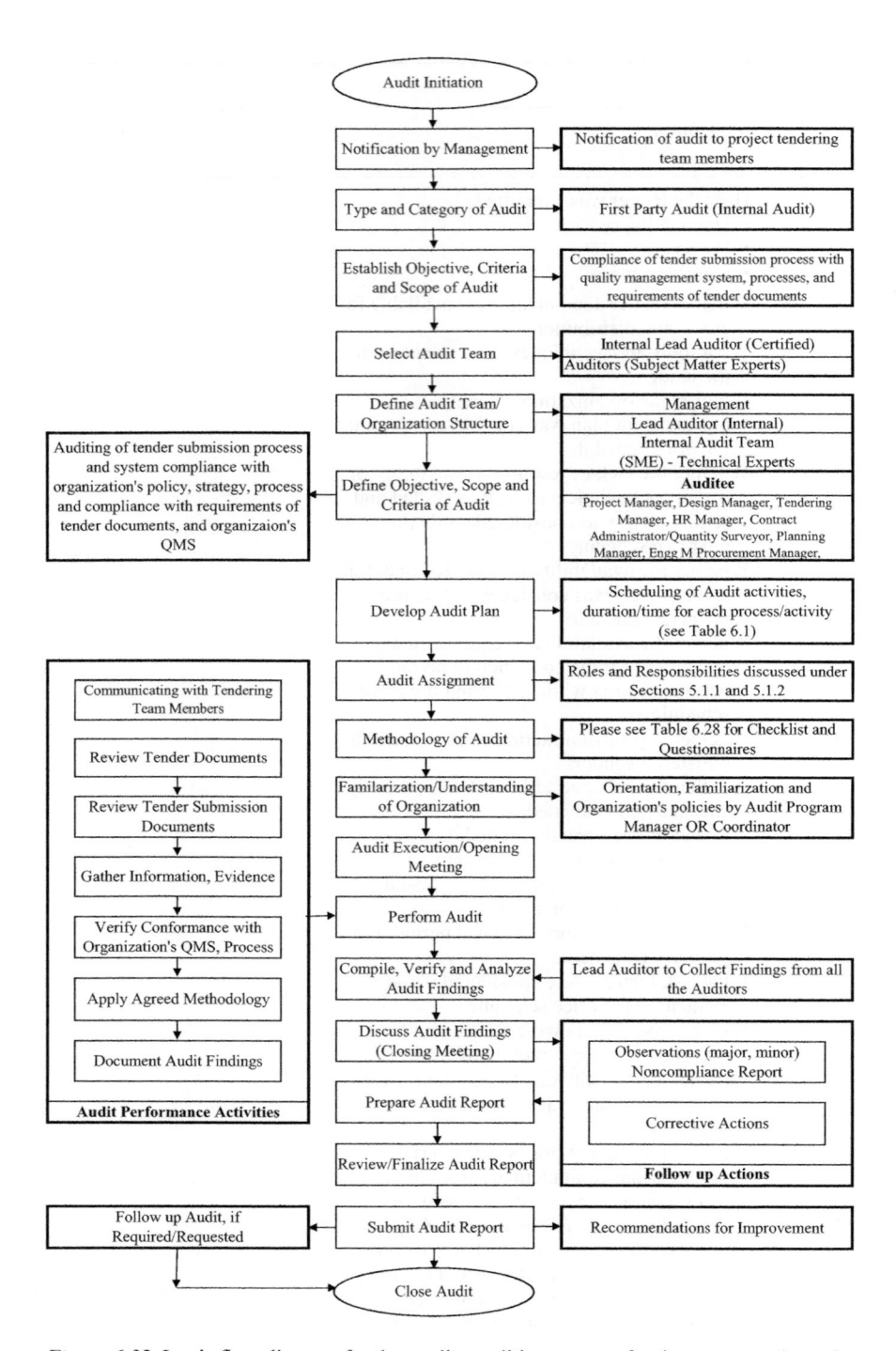

Figure 6.32 Logic flow diagram for the quality auditing process for the contractor's tender submission process

Auditing processes for project life cycle phases 299

achieving project quality goals and is also responsible for implementing the procedures specified in the contract documents. Table 6.29 lists the responsibilities the owner delegates to the engineer (consultant). The selected firm (consultant) and the team members should have the experience to supervise the construction activities listed in Table 6.29.

Figure 6.33 illustrates a typical quality auditing process for the selection of the construction supervisor (consultant).

The main objectives of performing a quality audit for the bidding and tendering process to select the supervisor (consultant) for a construction project are to:

1 Evaluate and assess process compliance with the organizational strategic policies and procedure
2 Assess conformance to QMS and that processes meet established quality system requirements
3 Assess that all the major activities are adequately performed and managed
4 Check that key risk factors are analyzed and managed to meet business requirements
5 Assess proper functioning of project management system

Table 6.29 Responsibilities of the construction supervisor (consultant)

Sr. No.	Description
1	Achieving the quality goal as specified
2	Review contract drawings and resolve technical discrepancies/errors in the contract documents
3	Review construction methodology
4	Approval of contractor's construction schedule
5	Regular inspection and checking of executed works
6	Review and approval of construction materials
7	Review and approval of shop drawings
8	Inspection of construction material
9	Monitoring and controlling construction expenditure
10	Monitoring and controlling construction time
11	Maintaining project record
12	Conduct progress and technical co-ordination meetings
13	Co-ordination of owner's requirements and comments related to site activities
14	Project related communication with contractor
15	Co-ordination with regulatory authorities
16	Processing of site work instruction for owner's action
17	Evaluation and processing of variation order/change order
18	Recommendation of contractor's payment to owner
19	Evaluating and making decisions related to unforeseen conditions
20	Monitor safety on site
21	Supervise testing, commissioning and handover of the project
22	Issue substantial completion certificate

Source: Abdul Razzak Rumane (2010). *Quality Management in Construction Projects*. Reprinted with permission of Taylor & Francis Group

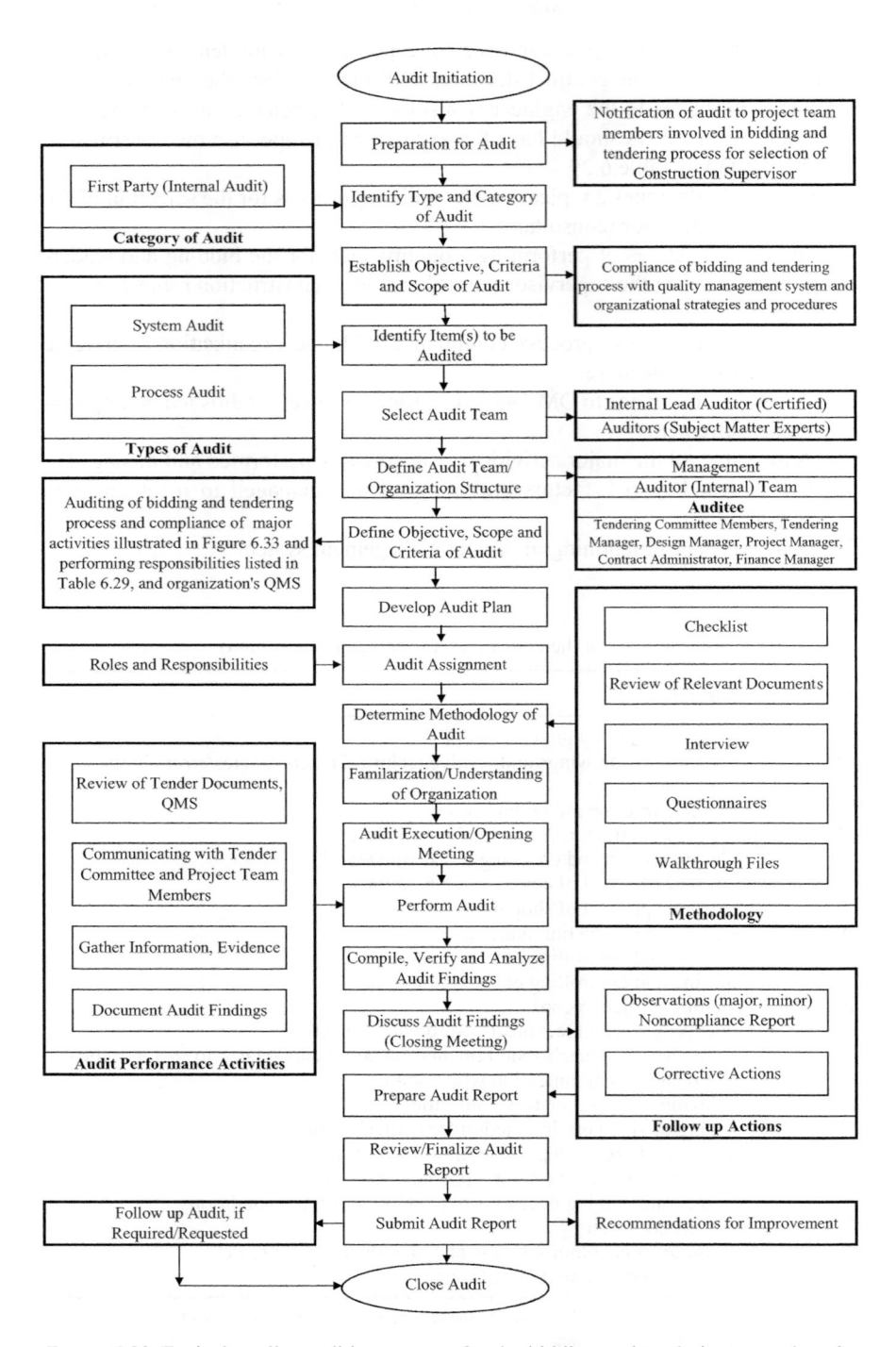

Figure 6.33 Typical quality auditing process for the bidding and tendering procedure for the selection of the construction supervisor (consultant)

Auditing processes for project life cycle phases 301

The following audit tools are used to perform a quality audit for the bidding and tendering process to select the construction supervisor:

- Checklists
- Review of documents
- Interviews
- Questionnaires
- Analysis of audit documentation

The auditee team members are mainly from the following organizational groups:

- Tendering Committee
- Tendering Manager
- Project Manager
- Contract Administrator
- Design Manager
- Document Controller
- Finance Manager

In order to manage and control this, the bidding and tendering process for the selection of the construction supervisor can be divided into four major stages. These stages are:

1 First stage: Shortlisting/Prequalification of supervisor
2 Second stage: Tender documents
3 Third stage: Contract bid solicitation
4 Fourth stage: Contract award

Figure 6.34 illustrates the typical quality auditing stages of the bidding and tendering process for the selection of the construction supervisor (consultant).

Table 6.30 lists the prequalification questionnaires to register the construction supervisor (consultant).

Based on the typical stages as per Figure 6.34, the audit methodologies can also be performed in four stages. Table 6.31 illustrates the audit methodology to assess/evaluate the process compliance at stage I, Table 6.32 illustrates the audit methodology to assess/evaluate the process compliance at stage II, Table 6.33 illustrates the audit methodology to assess/evaluate the process compliance at stage III, Table 6.34 illustrates the audit methodology to assess/evaluate the process compliance at stage IV and Figure 6.35 illustrates the logic flow diagram for the quality auditing of the selection of the construction supervisor.

6.2.3.4 Auditing process for the construction supervisor's tender submission process

Figure 6.36 illustrates a typical quality auditing process for the construction supervisor's tender submission process.

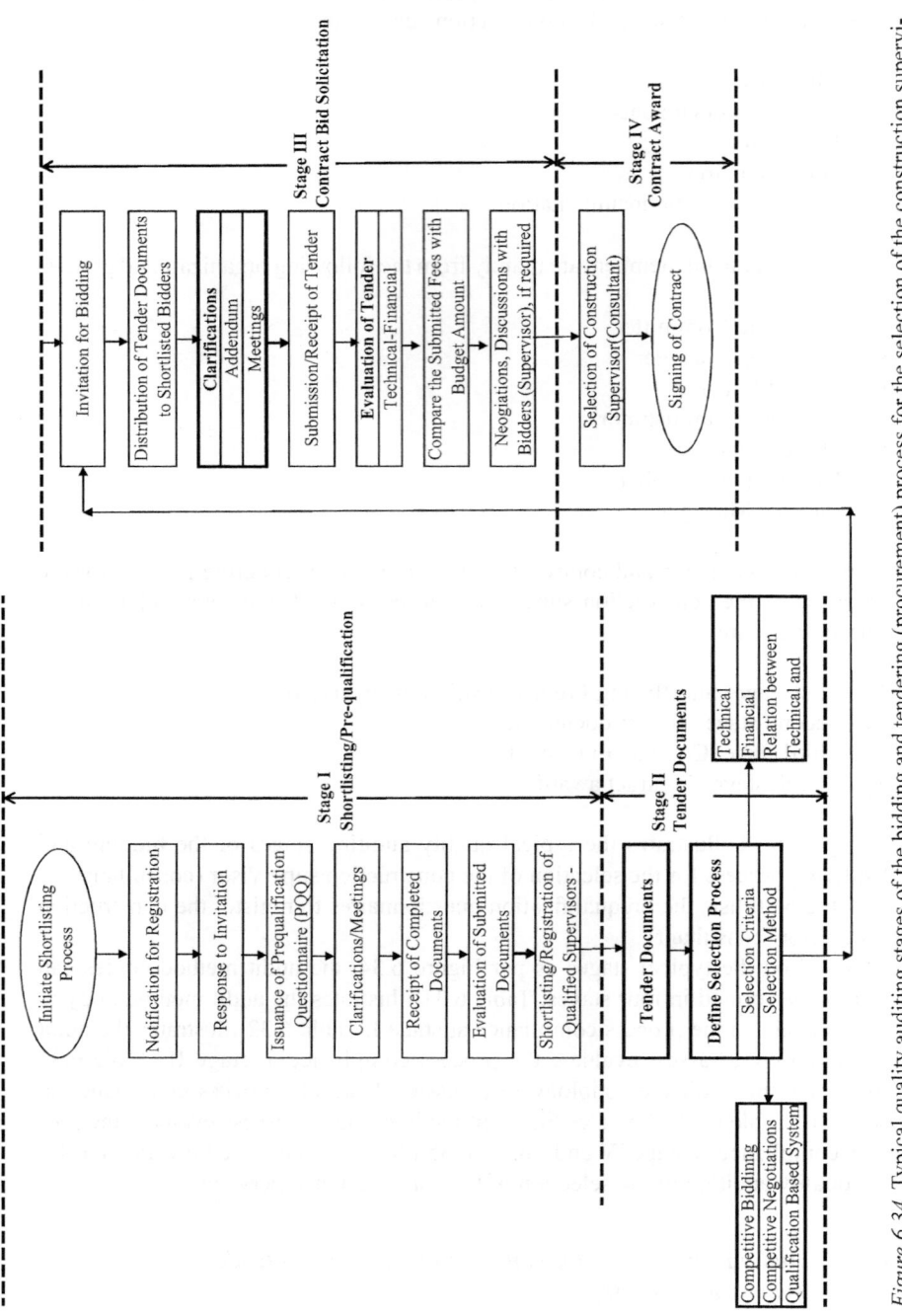

Figure 6.34 Typical quality auditing stages of the bidding and tendering (procurement) process for the selection of the construction supervisor (consultant)

Table 6.30 Prequalification questionnaires for the registration of the construction supervisor (consultant)

Serial Number	Element	Question	Response
1	General information	a Company name b Full address c Registration details/business permit d Management details e Membership of professional trade associations, if any f Award winning project, if any g Quality management certification	
2	Financial information	a Yearly turnover b Current workload c Audited financial report d Performance bonding capacity e Insurance limit	

3	a) Organization details (general)
	
	b) Organization details (experience)
		a Core area of business
		b How long in the same field of operation
		c Quality control/assurance organization
		a Number of years in the same business
		b Technical capability
		  i Supervision of construction projects
		  ii Construction quality audits
		  iii Monitoring and control
		  iv Testing and commissioning
		c List of previous contracts
		  i Name of project
		  ii Value of each contract
		  iii Contract period of each contract	

4	Resources	a Human resources
		  i Management
		  ii Construction supervision
		  iii Construction quality auditing
		  iv Project scheduling and planning
		  v Project monitoring and controlling
		  vi Contract administration
		  vii Conflict management
		  viii HSE
		b Human resource development plan	

5	Project reference		
6	Bank reference		

Table 6.31 Audit methodology for the assessment of the bidding and tender process for the selection of the construction supervisor (consultant) – Stage I (shortlisting/registration of construction supervisor)

Serial Number	Item To Be Assessed	Yes	No	Comments
A. Checklist				
1	Whether the notification for registration was announced in all technical newsletters and leading news media as per the organization's policy			
2	Whether there are registered construction supervisors to participate in the tender			
3	Whether prequalification questionnaires (PQQ) are issued to all the intending bidders (Please refer to Table 6.30 for guidelines)			
4	Whether there are any requests for queries/bid clarification			
5	Whether the meeting for clarification is attended by all the intending bidders			
6	Whether the attendance sheet for the meeting is signed by all the attendees			
7	Whether minutes of the meeting are circulated to all the attendees/participating bidders			
8	Whether sufficient time is provided to prepare and submit completed PQQ			
9	Whether there are any requests for extension for submission date			
10	Whether all the intending bidders submitted the response to PQQ			
11	Whether received bid envelopes are placed in safe custody			
12	Whether there are any delayed submissions			
13	Whether received responses are acknowledged			
14	Whether received responses are clearly identified and recorded			
15	Whether responses are fairly evaluated as per the organization's selection policy			
16	Whether any of the submitted response are found to be incomplete			
17	Whether there are any late submissions			
18	Whether risk factors are considered while evaluating the response and registration of construction supervisors			
19	Whether reasons for non-registration are conveyed to unsuccessful participants			
20	Whether shortlisted construction supervisors are allotted a code and registration number			

B. Interview Questionnaires

Serial Number	Description	Response	Comments
1	How responses were submitted: • Hand delivery • By post • Courier service		
2	Whether PQQ was available for electronic distribution		
3	How was the response for registration/shortlisting		
4	Whether tender was open for international bidders or only local companies		

Table 6.32 Audit methodology for the assessment of the bidding and tender process for the selection of the construction supervisor (consultant) – Stage II (bidding and tender documents)

Serial Number	Item To Be Assessed	Yes	No	Comments
A. Checklist				
1	Whether the tender documents includes all the relevant information required to select the construction supervisor (consultant)			
2	Whether the tender documents include all the related activities to be performed by the supervisor during the construction phase			
3	Whether activities listed in Table 6.29 have been considered while preparing the tender documents			
4	Whether the tender documents are approved by the project owner/tendering committee			
5	Whether the bid/tender documents are prepared as per the procurement method and contract strategy adopted during the early stages of the project (Design-Bid-Build)			
6	Whether all the information to enable bidders/supervisors to submit the proposal is clear and unambiguous			
7	Whether documents are prepared taking into consideration the organization's Quality Management System (QMS)			
8	Whether regulatory/authority requirements are taken into consideration while preparing the tender documents			
9	Whether the selection criteria and selection method for the contractor (supervisor) selection are clearly specified in the organization's QMS			

(*Continued*)

Table 6.32 (Continued)

Serial Number	Item To Be Assessed	Yes	No	Comments
A. Checklist				
10	Whether budget approval is obtained for the supervisor's fees prior to release of tender announcement			
11	Whether the tender opening and closing dates are mentioned in the announcement			
12	Whether the schedule for the supervision period is included in the tender documents			
13	Whether a staff deployment chart is included in the tender documents			
14	Whether the qualifications of supervision team members are prescribed in the tender documents			
15	Whether project quality requirements are properly defined in the tender documents			
16	Whether the rights and liabilities of all the parties are mentioned in the documents			
17	Whether change/variation clauses are included in the documents			
18	Whether cancellation/termination clauses are included in the documents			
19	Whether bond and insurance clauses are included in the documents			
20	Whether the tender documents are as per international/local standard documents used by the construction industry			
21	Whether evaluation and assessment criteria are well defined			
22	Whether review and analysis procedure of bid is included			
23	Whether the contract to be signed for the construction supervisor is included in the documents			
24	Whether all records are properly maintained			
25	Whether the request to provide the bidder's contact details are included in the documents			
B. Interview Questionnaires				

Serial Number	Description	Response	Comments
1	Whether any outside agency was involved in the preparation of the tender documents		
2	What guidelines were adopted to prepare the tender documents		
3	Whether any specific format/template was followed		

B. Interview Questionnaires

Serial Number	Description	Response	Comments
4	Whether all the requirements of tender documents were coordinated and agreed by relevant stakeholders		
5	Whether risks of disputes and associated financial costs were considered while preparing the contract documents		
6	Whether problems and challenges were identified while preparing the bid/tender		
7	Who was responsible for authorizing bidding requirements		
8	Whether supervision responsibilities are clearly defined in the documents		
9	Whether staff selection procedure is clearly defined in the documents		

Table 6.33 Audit methodology for the assessment of the bidding and tender process for the selection of the construction supervisor (consultant) – Stage III (contract bid solicitation)

Serial Number	Item To Be Assessed	Yes	No	Comments
A. Checklist				
1	Whether the tender documents are properly organized			
2	Whether the tender documents are distributed to all the prequalified/shortlisted/registered bidders (construction supervisors) to participate in the tender			
3	Whether the tender is notified to all the prequalified bidders			
4	Whether all the prequalified/shortlisted/registered bidders participated in the tender			
5	Whether the attendance sheet for the meeting was signed by all the attendees			
6	Whether minutes of the meeting were circulated to all the attendees/participating bidders/ construction supervisors			
7	Whether sufficient time is provided to prepare and submit the tender			
8	Whether the tender submission date is extended from the date originally announced			
9	Whether an addendum (if any) is issued/notified to all the participating bidders			
10	Whether there are any requests for extension of submission date			

(Continued)

Table 6.33 (Continued)

Serial Number	Item To Be Assessed	Yes	No	Comments
A. Checklist				
11	Whether the clarification meeting is attended by all participating bidders			
12	Whether any changes are made to the originally announced tender documents			
13	Whether received bid envelopes are placed in safe custody			
14	Whether there are any delayed submissions			
15	Whether tenders are opened as per the announced date and time			
16	Whether tenders are opened in the presence of all the bidders			
17	Whether received tenders are acknowledged			
18	Whether received tenders are clearly identified and recorded			
19	Whether applicable fees/bid bond are submitted by all the bidders who participated in the tender			
20	Whether tenders are fairly evaluated as per the announced policy and procedure			
21	Whether the comparison of bids is tabulated			
22	Whether the bid cost is higher than the budgeted cost			
23	Whether any submitted tender is found to be incomplete			
24	Whether there are any late submissions			
25	Whether all the scheduled pages are signed by the bidder			
26	Whether risk factors are considered while evaluating the bids			
27	Whether reasons for the non-acceptance of the tender were conveyed to the unsuccessful bidders			

Serial Number	Description	Response	Comments
B. Interview Questionnaires			
1	How the notification was announced		
2	How tenders were submitted: • Hand delivery • By post • Courier service		
3	Whether the tender was available for electronic distribution		
4	How was the response to participate in the tender		
5	Whether tender box(es) were placed in secured places and monitored through an electronic surveillance system		

B. *Interview Questionnaires*

Serial Number	Description	Response	Comments
6	How much time was allowed to open the bids after the notified tendering closing time		
7	Whether the tender submission date was extended and what was the reason for the extension		
8	Whether all the tender envelopes/packages were sealed with marking of the tenders and contain the original tender documents and required number of sets as per tender conditions		
9	Whether a bid review was established		
10	Whether the bid selection procedure was clearly defined		

Table 6.34 Audit methodology for the assessment of the bidding and tender process for the selection of the construction supervisor (consultant) – Stage IV (Contract Award)

Serial Number	Item To Be Assessed	Yes	No	Comments
A. Checklist				
1	Whether the selection of the construction supervisor satisfies all the conditions to be a successful bidder			
2	Whether the selected bidder (construction supervisor) is a legal entity			
3	Whether the selection was made by the selection committee			
4	Whether the selection of the construction supervisor is approved by the relevant applicable authority, if mandated by the organization's policy			
5	Whether the selected construction supervisor is capable of carrying out the specified work and has the necessary resources			
6	Whether risk factors are considered prior to the signing of the contract			
7	Whether the standard contract format is used for signing the contract			
8	Whether the contract terms and conditions are clearly written and unambiguous			
9	Whether an addendum and minutes of the meeting are included in the contract documents			
10	Whether the contract period/schedule is properly described			

(*Continued*)

Table 6.34 (Continued)

Serial Number	Item To Be Assessed	Yes	No	Comments
A. Checklist				
11	Whether the staff deployment schedule is included in the contract			
12	Whether supervision staff qualifications are properly defined			
13	Whether a performance bond is submitted by the consultant			
14	Whether variation/change management clauses are included in the contract agreement			
15	Whether dispute resolution and conflict resolution clauses are properly defined			
16	Whether contract documents are reviewed prior to signing			
17	Whether tender documents and contract are properly archived			
18	Whether a letter of award is sent			
19	Whether final results are published and announced			

B. Interview Questionnaires

Serial Number	Description	Response	Comments
1	Whether the construction supervisor is registered with relevant regulatory authorities for supervision of construction projects		
2	On what basis the bidder (construction supervisor) was selected: low bid, competitive bid or quality based system		
3	Was there any tie between two or more bidders		
4	Was there any dispute or objection raised by any of the bidders before the award of the contract to the successful bidder		
5	Whether the final price was negotiated or as per the bid (tender) received		
6	Whether the successful bidder failed to submit a performance bond		
7	Whether there was any need to update the approved budget		
8	Whether risks in non-deployment of any of the supervision team members was considered		
9	Whether the contract includes a penalty clause for delay in the deployment of supervision team members		
10	Whether the contract includes a penalty clause for failure to provide a required team member		
11	Who is responsible for selecting supervision team members		

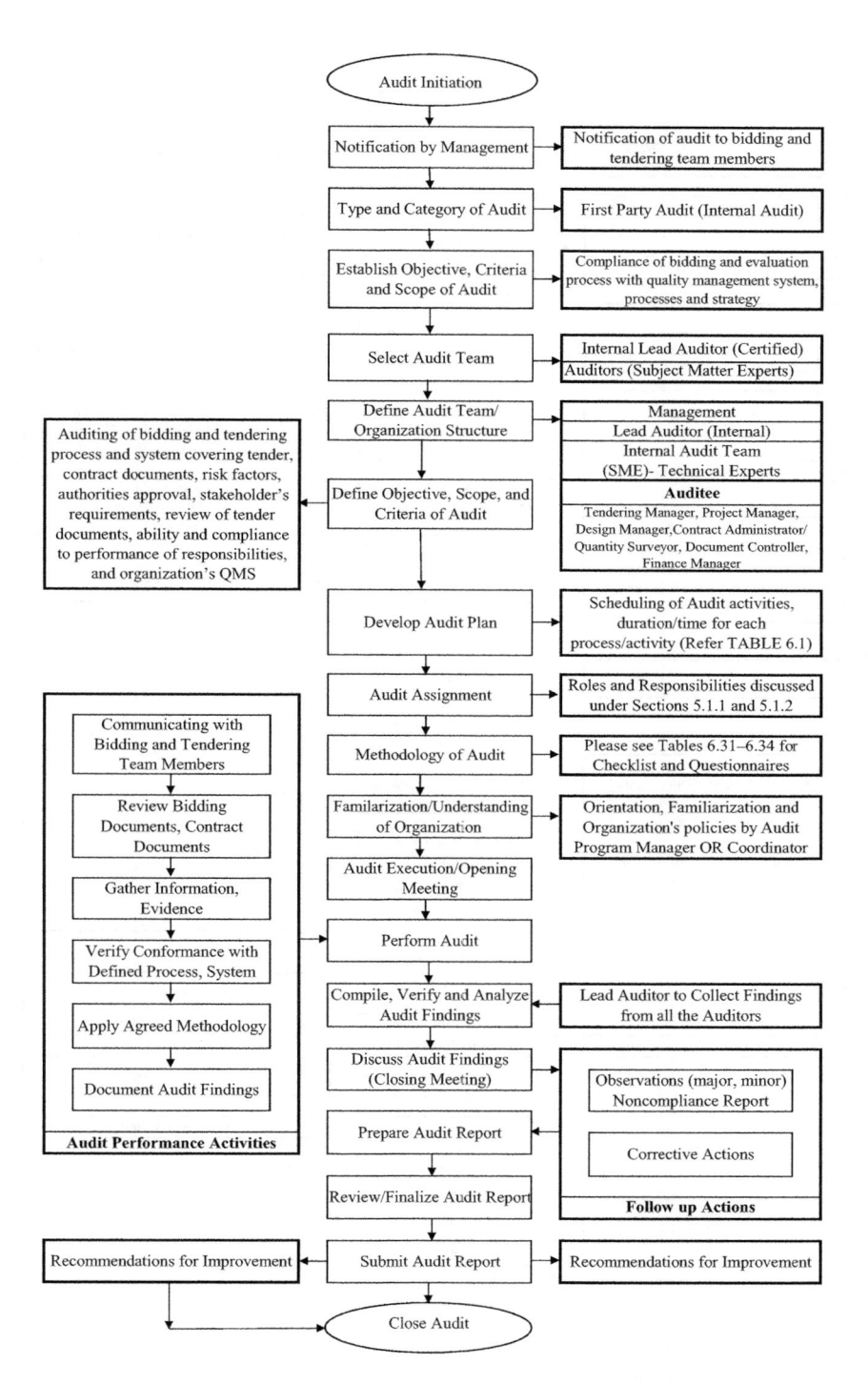

Figure 6.35 Logic flow diagram for the quality auditing of the selection of the construction supervisor (consultant)

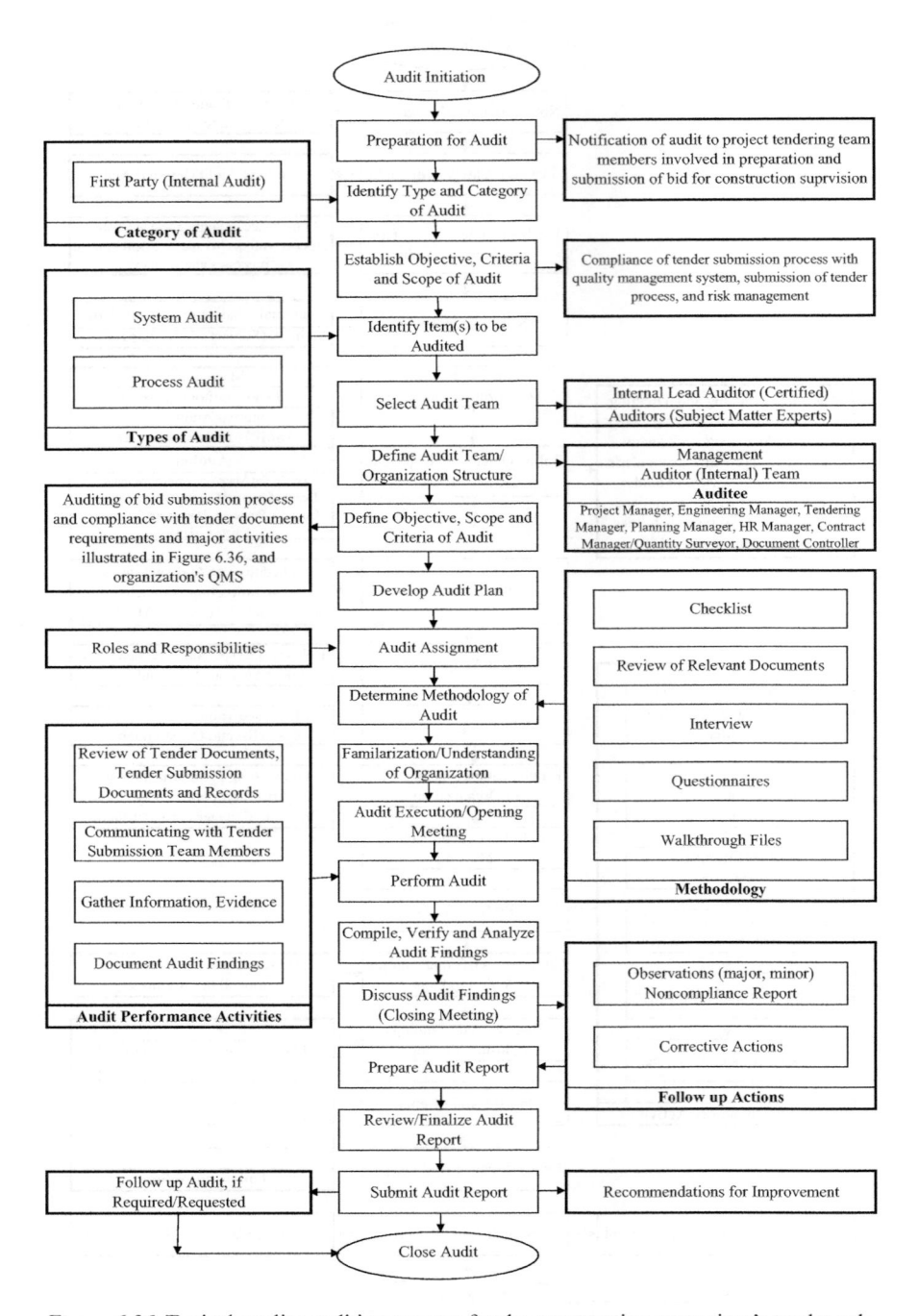

Figure 6.36 Typical quality auditing process for the construction supervisor's tender submission process

Auditing processes for project life cycle phases 313

The main objectives of performing a quality audit of the construction supervisor's proposal submission process are to:

1. Evaluate and assess process compliance with the organizational strategic policies and procedure
2. Assess conformance to QMS and that processes meet established quality system requirements
3. Review the process used to prepare the tender (quotation)
4. Verify that the proposal (tender) preparation and submission comply with the organization's tender (proposal) submittal process (Figure 6.37 illustrates the typical process)

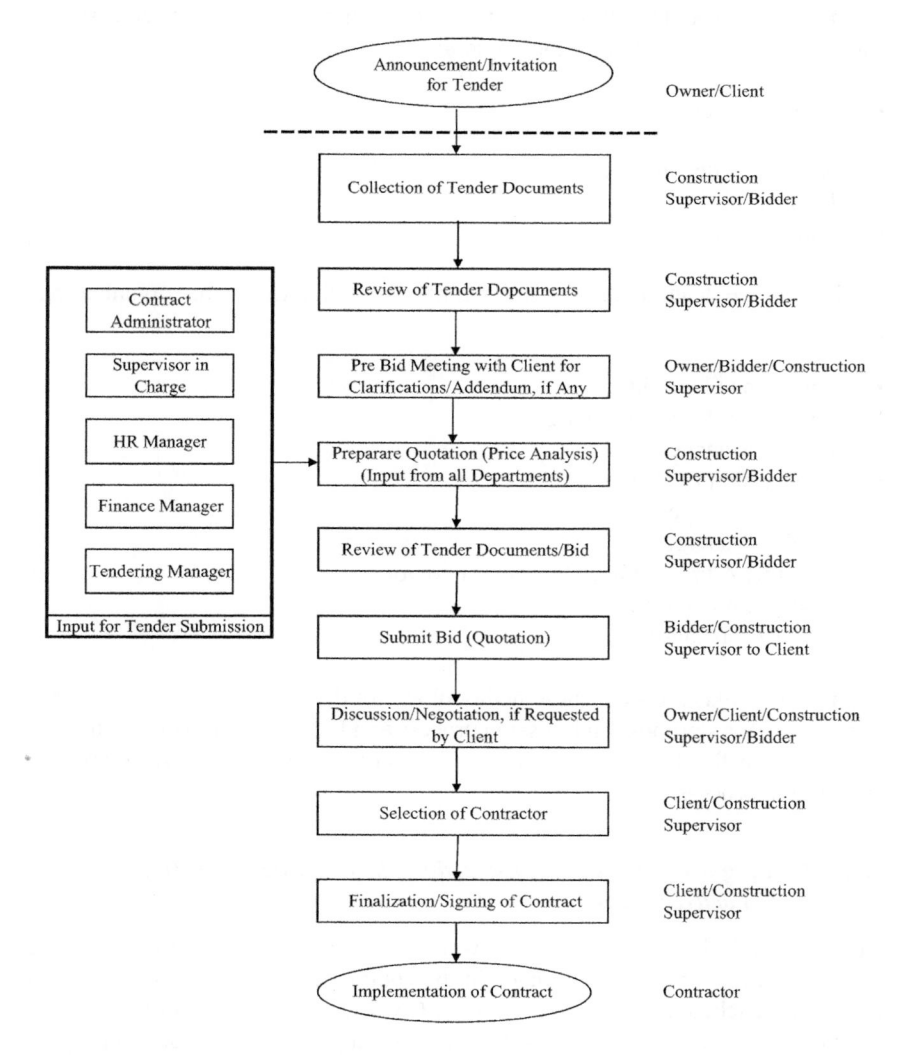

Figure 6.37 Typical tender submission procedure by the construction supervisor

5   Validate that the response to the contents of the tender documents is complete in all respects.
6   Check that availability of resources during the construction phase is considered
7   Review the process used to prepare the tender
8   Validate that the schedule for deployment of project staff complies with tender requirements
9   Validate the cost of the proposal (price analysis)
10 Review that the key risk factors are analyzed and managed
11 Check for compliance with regulatory requirements

The following audit tools are used to perform a quality audit of the construction supervisor's proposal submission process:

- Analytical tools
- Checklists
- Review of documents
- Interviews
- Questionnaires

The auditee team members are mainly from the following organizational groups:

- Tendering Committee
- Tendering/Proposal Manager
- Project Manager
- Supervision In Charge
- Design Manager
- Quality Manager
- Contract Administrator/Quantity Surveyor
- Document Controller
- Finance Manager

Table 6.35 illustrates the audit methodology for the assessment of the construction supervisor's proposal submission process and Figure 6.38 illustrates the logic flow diagram for the quality auditing of the construction supervisor's proposal submission process.

6.2.4 Auditing process for the construction phase (Design-Bid-Build project delivery system)

Construction is the translation of the owner's goals and objectives, by the contractor, to build the facility as stipulated in the contract documents, plans and specifications on schedule and within budget . Construction is the sixth phase of the construction project life cycle and is an important phase in construction projects.

Table 6.35 Audit methodology for the assessment of the construction supervisor's tender submission process

Serial Number	Item To Be Assessed	Yes	No	Comments
A. Checklist				
1	Whether the company is already registered as a construction supervisor (consultant) to participate in the tender			
2	Whether there are any fees to collect the tender documents			
3	Whether the scope of work is clearly defined in the tender documents			
4	Whether there is any request for clarification raised			
5	Whether the response to the price schedule/tender in pricing is complete			
6	Whether the requested information in the tender documents is clearly responded to			
7	Whether sufficient time is provided to prepare and submit the completed tender			
8	Whether there are any requests for extension of the submission date			
9	Whether there are any delayed submissions			
10	Whether the tender response is prepared taking into consideration/referring to the organization's quality management system			
11	Whether the availability of qualified personnel to supervise the construction is as per tender requirements			
12	Whether staff deployment costs include all the benefits and company overhead costs			
13	Whether the cost to review the Request for Information (RFI) during construction is considered while preparing the tender			
14	Whether the tender was reviewed prior to submission			
15	Whether the tender response was approved by the relevant committee/authorized person			
16	Whether the risk factors were considered and analyzed			
17	Whether all the relevant pages are signed			

Serial Number	Description	Response	Comments
B. Interview Questionnaires			
1	Whether the notification for tender submission was announced in all technical newsletters and leading news media or it was only for registered bidders (contractors)		

(Continued)

Table 6.35 (Continued)

B. Interview Questionnaires

Serial Number	Description	Response	Comments
2	How the tender was submitted: • Hand delivery • By post • Courier service		
3	Whether tender documents were available for electronic distribution		
4	What tools and techniques were used to price the tender		
5	Whether the organization's quality management plan was considered to evaluate the price schedule		
6	Whether the availability of resources (technical as well as skilled and semi-skilled) was considered while preparing the price schedule		
7	Whether overhead costs and leave salary of staff are taken into consideration while developing the monthly cost		
8	Whether the deployment schedule was properly reviewed by taking into consideration the availability of resources		
9	Whether the regulatory approval period was properly considered		
10	What risk factors were considered while evaluating the price schedule		
11	Whether any indirect costs are considered		
12	Was there any negotiation to the submitted quotation		

A majority of the total project budget and schedule is expended during construction. Similar to costs, the time required to construct the project is much higher than the time required for the preceding phases. Construction usually requires a large number of workforce and a variety of activities. Construction activities involve the erection, installation or construction of any part of the project. Construction activities are actually carried out by the contractor's own workforce or by subcontractors. Construction therefore requires more detailed attention to its planning, organization, monitoring and control of the project schedule, budget, quality, safety and environment concerns. The construction phase consists of various activities such as mobilization, submittals, planning and scheduling, management of resources/procurement, execution of works, control and monitoring, quality and inspection.

Figure 5.26 (discussed earlier in Section 5.2.7) illustrates the major activities relating to the construction phase developed based on the project management

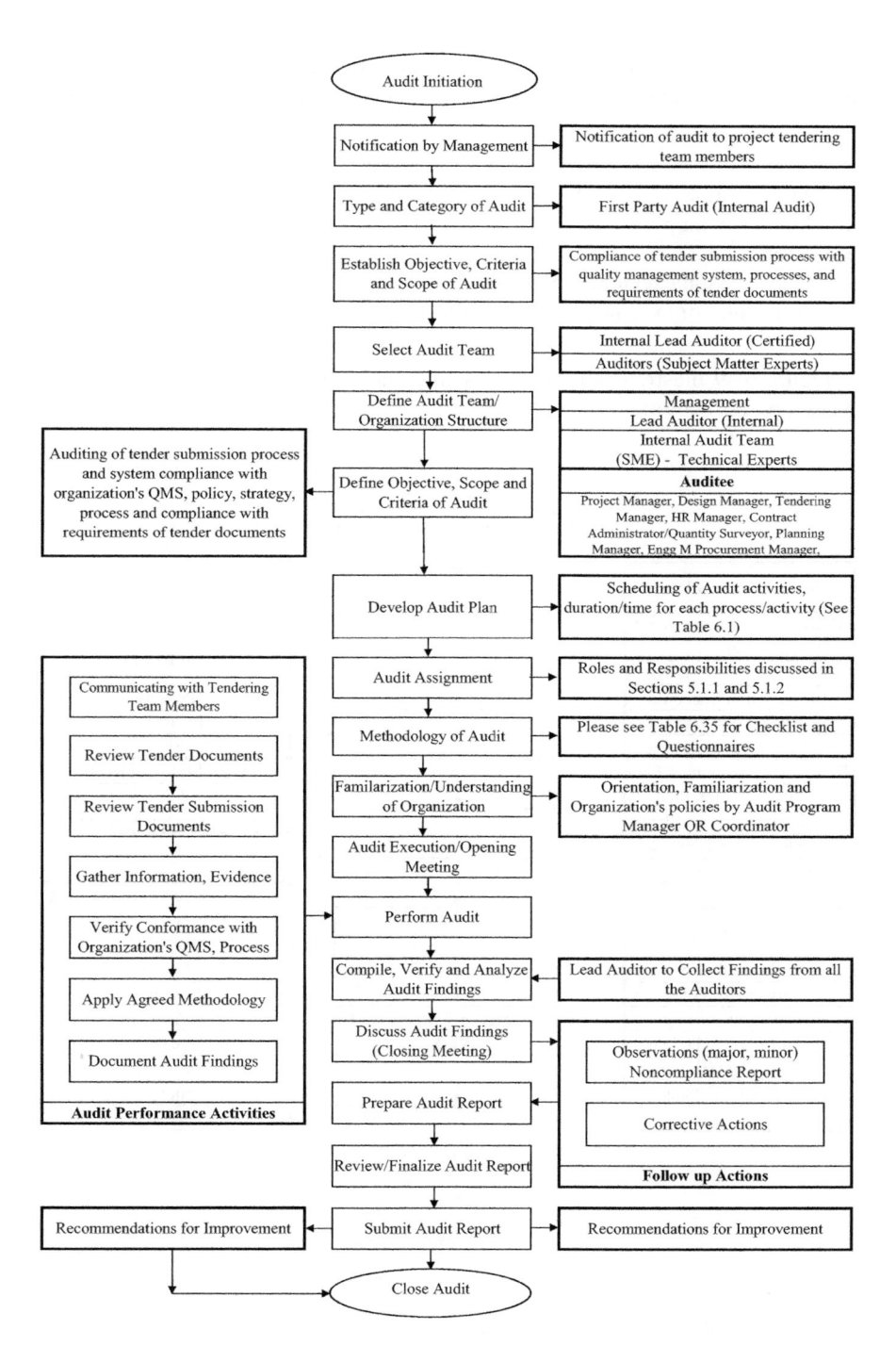

Figure 6.38 Logic flow diagram for the quality auditing process for the construction supervisor's tender submission process

process groups methodology. In order to manage and control the quality auditing of the construction process, these activities can be divided into the following stages:

1. Mobilization
2. Management plans
3. Submittals
4. Construction/execution of works
5. Inspection of executed works/installed systems

Figure 6.39 illustrates the typical quality auditing stages for the construction phase process activities.

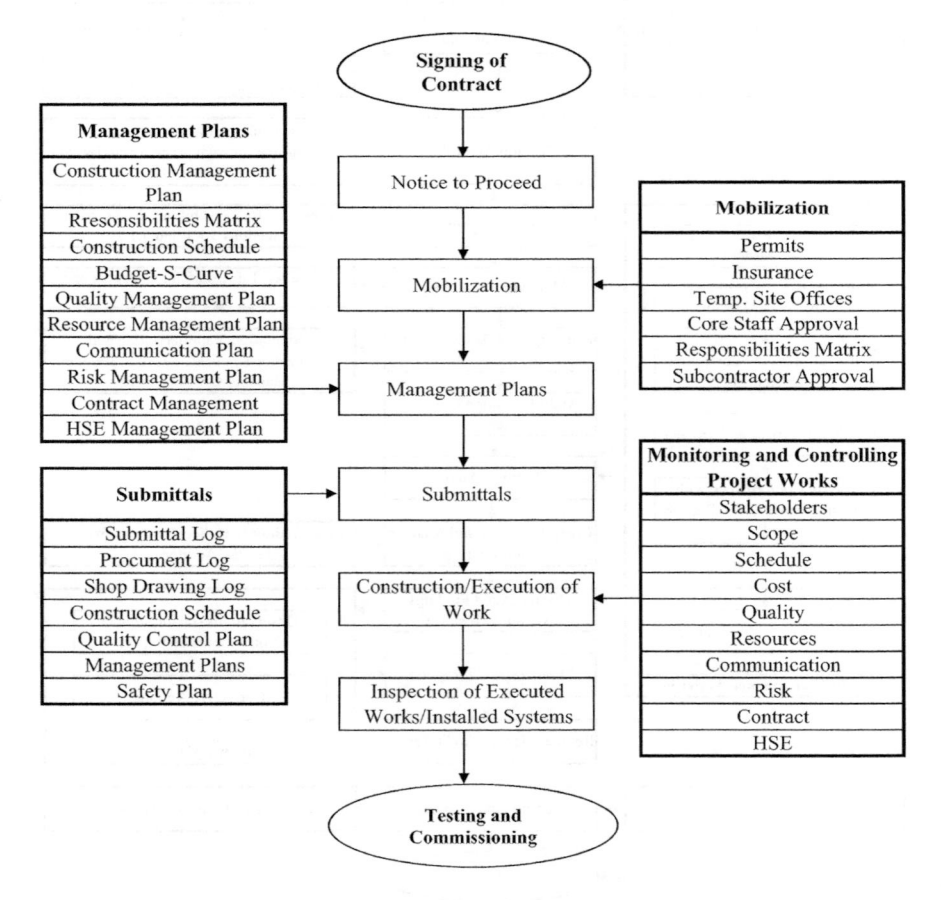

Figure 6.39 Typical quality auditing stages for the construction phase

Auditing processes for project life cycle phases 319

The main objectives of performing a quality audit of the construction phase (construction process/works) are to:

1. Determine whether appropriate systems, processes and control for managing construction are in place to achieve the contractual objectives and the owner/client's satisfaction
2. Evaluate and assess process compliance with the organizational strategic policies and procedure
3. Assess conformance to corporate QMS and that processes meet project specific quality system requirements
4. Review the process used for the construction/execution of works
5. Ensure that the construction of the project complies with the contract requirements
6. Ensure a responsibilities matrix is properly established taking into consideration all the stakeholders
7. Validate to what extent the project execution requirements (scope of work) are identified and implemented
8. Validate that the project schedule and cost are properly managed and controlled
9. Validate the compliance with quality standards and codes
10. Assess compliance with the contractor's quality control plan
11. Check for compliance with regulatory requirements
12. Verify the extent to which the resources are utilized
13. Verify that the selection of materials/systems complies with contract requirements and the organization's procurement policy
14. Verify that all the submittals are as per the approved schedule
15. Verify that project documents and logs are maintained
16. Review that the key risk factors are analyzed and managed during the construction phase for execution of the project/facility as per risk the management plan
17. Assess compliance to safety considerations as per the safety management plan
18. Review design drawings, specifications and contract documents
19. Assess that all the major activities are adequately performed and managed as listed in Figure 5.26
20. Assess the proper functioning of the project/construction management system
21. Verify the extent to which the project construction complied with rules and principles of the project/construction management system
22. Verify that the selection of subcontractors complies with contract documents and the organization's policy

The following audit tools are used to perform a quality audit of the construction phase activities:

- Analytical tools
- Checklists

- Review of documents
- Interviews
- Questionnaires
- Samples

The auditee team members are mainly from the following organizational groups:

- Construction/Project Manager
- Planning, Cost Control Manager
- Contract Administrator/Quantity Surveyor
- Engineering Manager
- Quality Engineer
- Material Engineer
- HSE Engineer
- Site Engineers (All Trades)
- Document Controller
- Subcontractors/Suppliers

Figure 6.40 illustrates the framework for the quality auditing of the construction phase.

6.2.4.1 Auditing process for the construction phase (contractor's works)

The quality auditing of the construction phase can be performed as follows:

1. During construction/execution of works
2. After completion of construction/execution of works

Figure 6.41 illustrates a typical quality auditing process for the construction phase.

6.2.4.1.1 AUDITING DURING THE CONSTRUCTION/EXECUTION OF WORKS

Quality audit during the construction phase can be performed at different stages of construction as it progresses and different activities/processes are performed or are under execution.

The audit objective is different for each of these activities/processes. Quality auditing during the construction phase can be performed for each stage as discussed in Figure 6.39.

The main objectives of performing a quality audit of mobilization activities during the construction phase are to:

1. Ascertain that contract documents are reviewed and the scope of work is properly identified

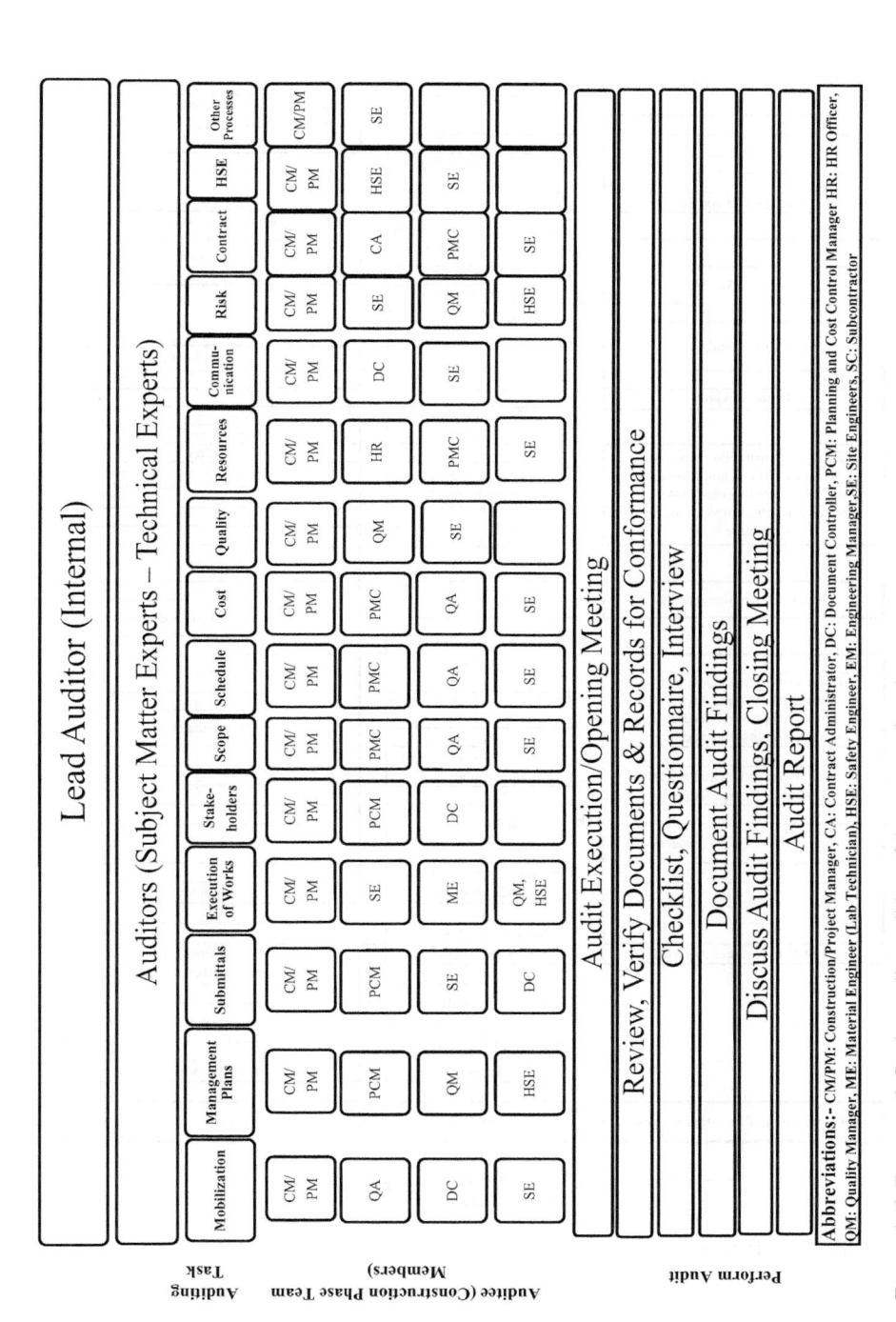

Figure 6.40 Framework for the quality auditing of the construction phase

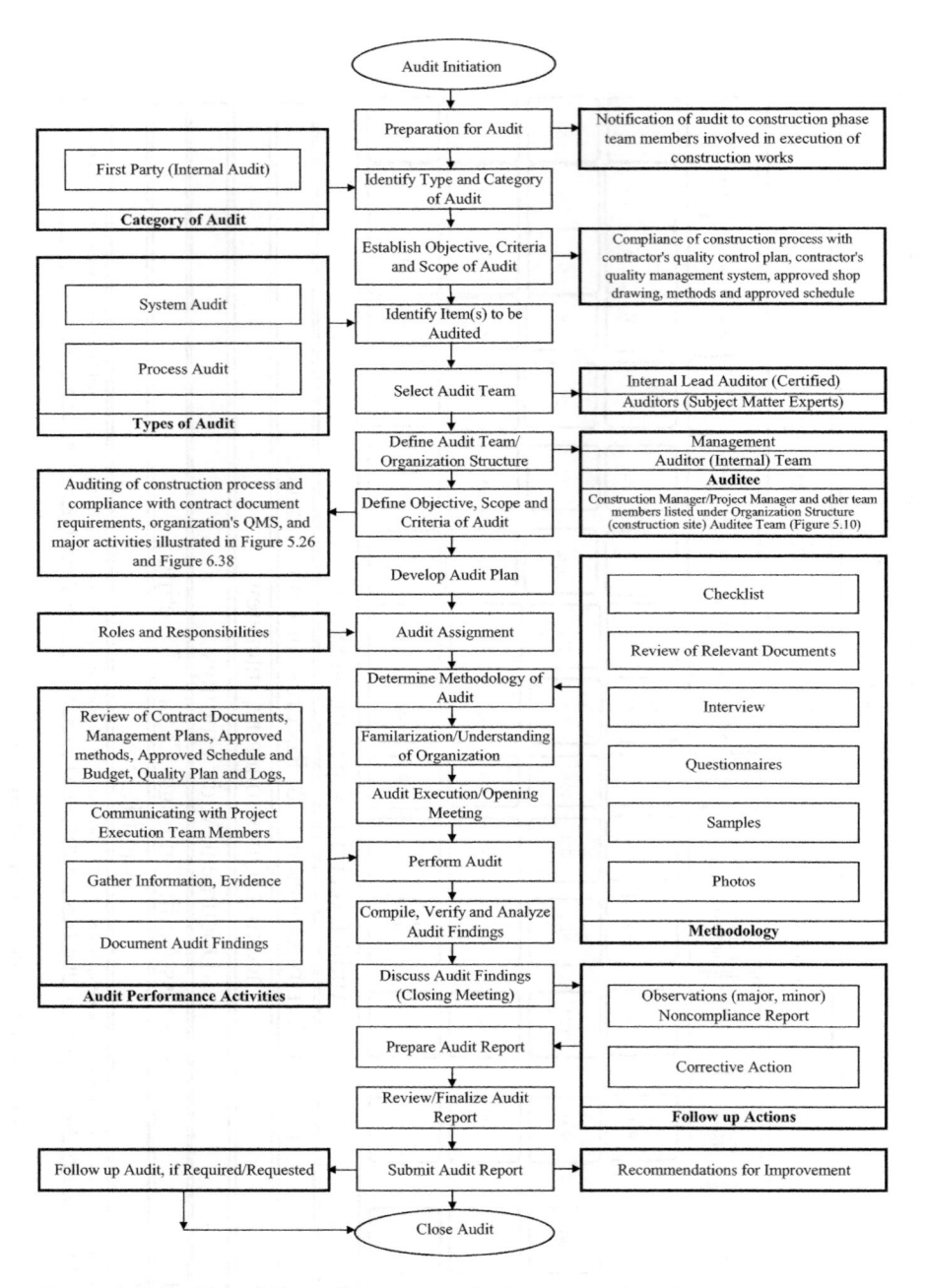

Figure 6.41 Typical quality auditing process for the construction phase

Auditing processes for project life cycle phases 323

2. Assess conformance to QMS and that processes meet project specific quality system requirements
3. Ascertain that all the permits from regulatory authorities are obtained to ensure mobilization works
4. Verify that the site is handed over as per the agreed-upon schedule by the owner/client
5. Verify that the construction work of temporary site offices is under progress and as per schedule
6. Verify that a responsibility matrix is established
7. Ensure that the core staff for the project are approved
8. Ensure that subcontractors are approved

Table 6.36 illustrates the audit methodology for the assessment of mobilization activities during the construction phase and Figure 6.42 illustrates the logic flow diagram for the quality auditing of mobilization activities during the construction phase.

Table 6.36 Audit methodology for the assessment of mobilization activities during the construction phase

Serial Number	Item To Be Assessed	Yes	No	Comments
A. Checklist				
1	Whether the contract documents are reviewed and contract execution requirements properly established			
2	Whether project manager/construction manager is assigned and approved by the project owner			
3	Whether a responsibilities matrix is established taking into consideration requirements of all the stakeholders			
4	Whether a project organization chart with responsibilities is established			
5	Whether the contract documents include all the construction documents that were part of the contract that was signed			
6	Whether the bonds and insurance are submitted on time			
7	Whether the necessary permits are obtained on time			
8	Whether the necessary survey records for boundaries are obtained from the municipality/authority			
9	Whether the site was handed over as per schedule			
10	Whether the works on temporary facilities are as per schedule			

(*Continued*)

Table 6.36 (Continued)

Serial Number	Item To Be Assessed	Yes	No	Comments
A. Checklist				
11	Whether sign boards and access road markings are completed			
12	Whether core staff are approved			
13	Whether the approval of subcontractors is in progress			
14	Whether necessary logs are in place			
B. Interview Questionnaires				

Serial Number	Description	Response	Comments
1	After how many days the notice to proceed was issued by the client after signing of the contract		
2	Who reviewed the contract documents and established the project execution requirements		
3	Was there any delay in getting authorities' approval		
4	Whether the construction equipment and tools are at the project site		
5	Whether the manpower requirements are established and there is a histogram for deployments of resources		
6	Whether authorities' approval is in progress and how long it will take		
7	Whether a temporary fire fighting system is installed and approved		
8	Whether a project sign board is installed		
9	Whether temporary access roads and fencing is in progress and when it will be completed		
10	Whether health and safety related works are complete		
11	Whether material testing laboratory was from the approved list		

The main objectives of performing a quality audit of the preparation of management plans activities during the construction phase are to:

1. Ascertain that contact documents are reviewed and required management plans are properly identified
2. Assess conformance to QMS and that processes meet project specific quality system requirements
3. Ascertain that all the management plans are prepared as per specification and contract requirements

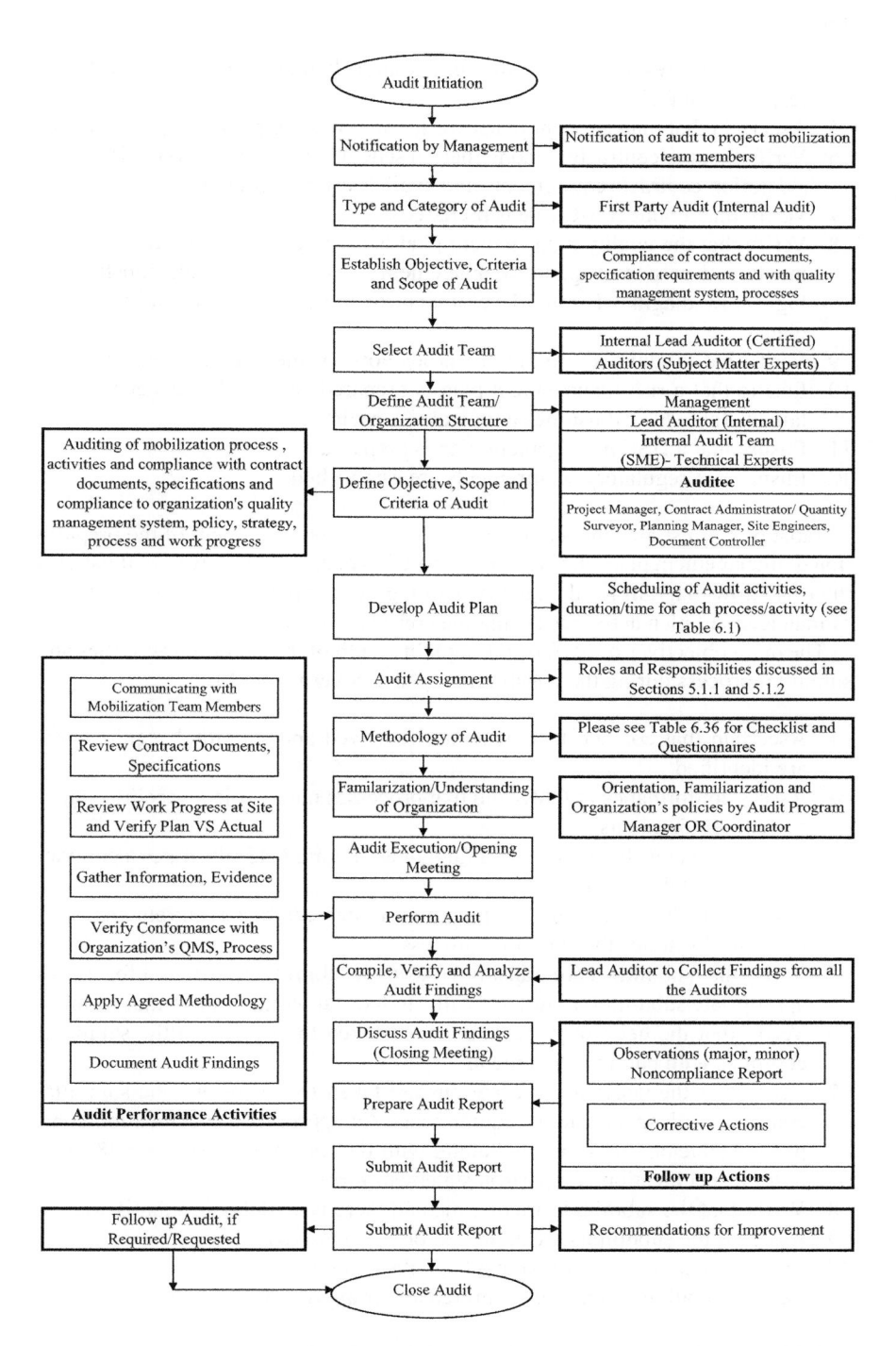

Figure 6.42 Logic flow diagram for the quality auditing process for mobilization activities during the construction phase

4   Verify that a responsibilities matrix is established and communicated to the respective stakeholders
5   Ascertain that management plans are submitted for approval as per schedule
6   Verify that the contractor's construction schedule is prepared taking into consideration contract requirements and availability of resources
7   Verify that a budget S-Curve is prepared
8   Verify that the quality management plan is prepared as per contract requirements and corporate quality management system. Figure 6.43 illustrates a logic flow diagram for the development of the contractor's quality control plan.
9   Verify that an administrative matrix is developed to meet contract requirements
10 Ensure that a risk management plan is prepared and a risk register is maintained. Table 6.37 illustrates a typical risk register.
11 Ensure that a HSE management plan is prepared
12 Ensure that regulatory requirements are established

Table 6.38 illustrates the audit methodology for the assessment of the preparation of management plans during the construction phase and Figure 6.44 illustrates the logic flow diagram for the quality auditing of the preparation and submission of management plan activities during the construction phase.

The main objectives of performing a quality audit of the submittal activities and submittal process during the construction phase covering are to:

1   Ascertain that contact documents are reviewed and submittal requirements are identified
2   Assess conformance to QMS and that processes meet project specific quality system requirements
3   Ascertain that different logs are maintained. Figure 6.45 illustrates a contractors' submittal status log.
4   Verify that the submittals are prepared as per contract requirements. Figure 6.46 illustrates the submittal process.
5   Verify that submittals are transmitted using a submittal transmittal form with appropriate submittal. Figure 6.47 illustrated a submittal transmittal form.
6   Verify that the necessary precautions are taken to ensure that the submittal complies with contract documents
7   Verify that the necessary precautions are taken to ensure that the submittal complies with specification requirements. An appropriate specification comparison statement is submitted along with the transmittal. Figure 6.48 illustrates a specification comparison statement form
8   Verify that the submittals are on time and there is no delay in submission
9   Assess if the submittals are approved or not approved
10 Verify if there are major comments on the submittals
11 Verify that all the plans are submitted as per approved schedule

Table 6.39 illustrates the audit methodology for the assessment of submittal activities during the construction phase and Figure 6.49 illustrates the logic flow

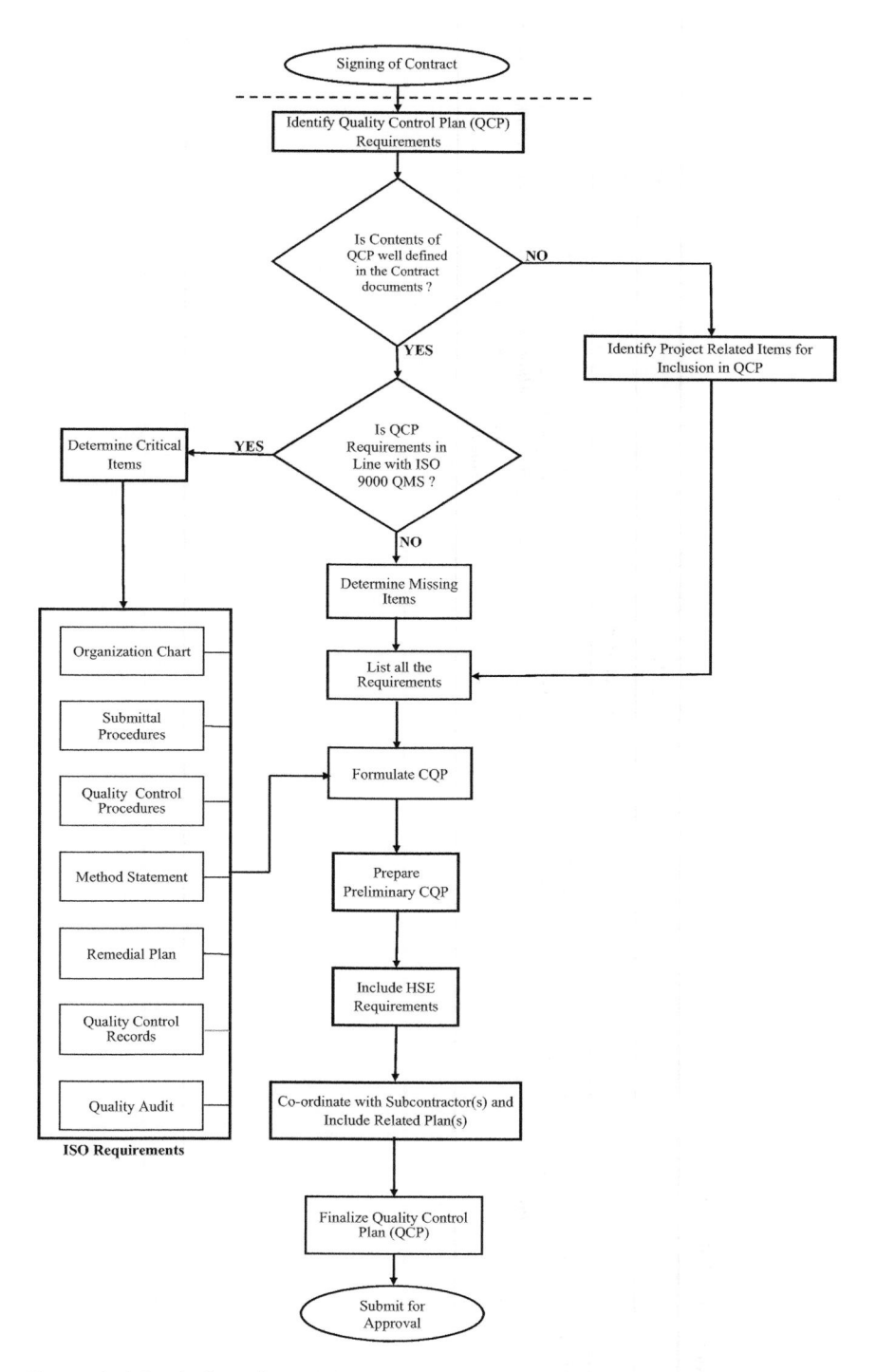

Figure 6.43 Logic flow diagram for the development of the contractor's quality plan

Table 6.37 Risk register

Project Name													
Risk Register													
Serial Number	Risk Identification (Risk ID)	Description of Risk	Owner of Risk	Estimated Likelihood of Risk	Impact	Estimated Severity	Prioritization	List of Activities Influenced for Risk	Leading Indicators	Risk Mitigation Plan	Risk Mitigation Plan on Leading Indicator	Timeline for Mitigation Action	Tracking of Leading Indicators	Date of Review/Update	Forecating Risk Happenings	Action to be Taken in Future

Table 6.38 Audit methodology for the assessment of the preparation of management plans during the construction phase

Serial Number	Item To Be Assessed	Yes	No	Comments
A. Checklist				
1	Whether contract requirements development of management plans is established			
2	Whether management plans are prepared properly identifying the contract requirements			
3	Whether the plans are prepared taking into consideration the corporate quality management system			
4	Whether stakeholders' requirements are fully incorporated in the management plans			
5	Whether a construction schedule is prepared and submitted			
6	Whether the construction schedule is prepared taking into consideration availability of manpower, material, machinery and equipment			
7	Whether a resource management plan is prepared			
8	Whether the contract administration plan is in place			
9	Whether a HSE plan is prepared and submitted			
10	Whether the risk factors are considered while developing the management plans			

Serial Number	Description	Response	Comments
B. Interview Questionnaires			
1	Who was responsible for reviewing the contract documents to identify the number of management plans to be submitted		
2	Were all the identified management plans kept in place for proper execution and monitoring and control of the project		
3	Whether a budget S-Curve is prepared		
4	Whether a contractor's quality plan is prepared and submitted		
5	Whether a communication plan is prepared		
6	Whether a risk management plan is prepared		
7	Whether a waste management plan is prepared		
8	Whether a finance management plan is prepared		
9	Whether all the regulatory approvals are as per schedule		
10	Whether all the plans are prepared taking into consideration subcontractors' requirements		

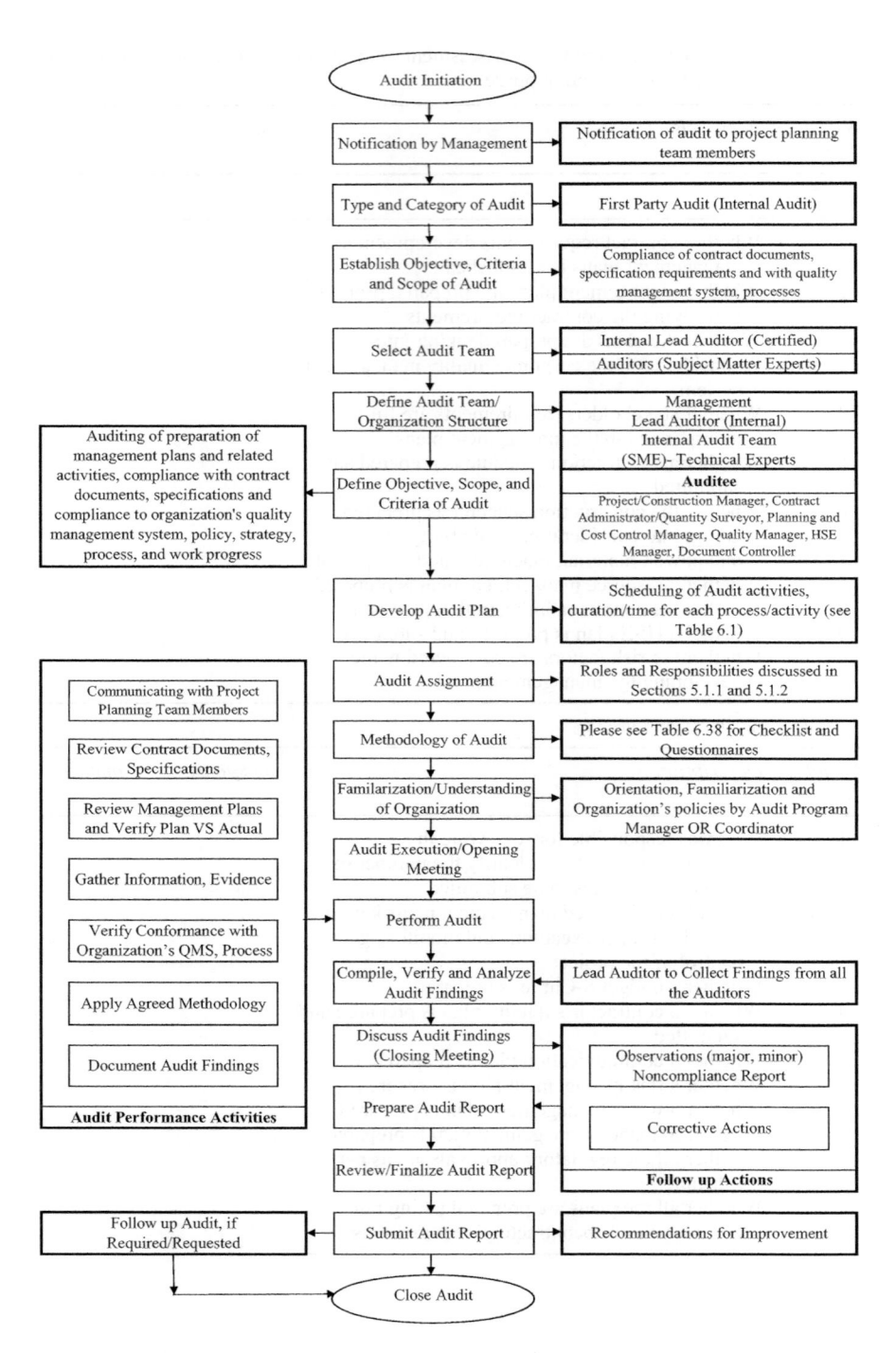

Figure 6.44 Logic flow diagram for the quality auditing process for the preparation of management plans activities during the construction phase

Project Name

Consultant Name

CONTRACTOR'S SUBMITTAL STATUS LOG (REPORT E-1)

CONTRACT NO. : ______________________ NO. : __________

CONTRACTOR : ______________________ DATE : __________

Document Specification or BOQ Page	Submittal Sequence No.	Rev	Description	Type of Submittal								Action Code : A = Approved as submitted B = Approved as noted C = Not Approved D = For Information						Notes
				Shop Drawing	Samples	Guarantee	Mf's Dats	Mt's Dats	Certificates	Test	Otr							
				SD	SM	GT	MF	MT	CT	TT	OT	Planned Sub. date	Submitted By Contrac.	Returned By A/E	Action	Approval Needed By	Required On Job	

Figure 6.45 Contractor's submittal status log

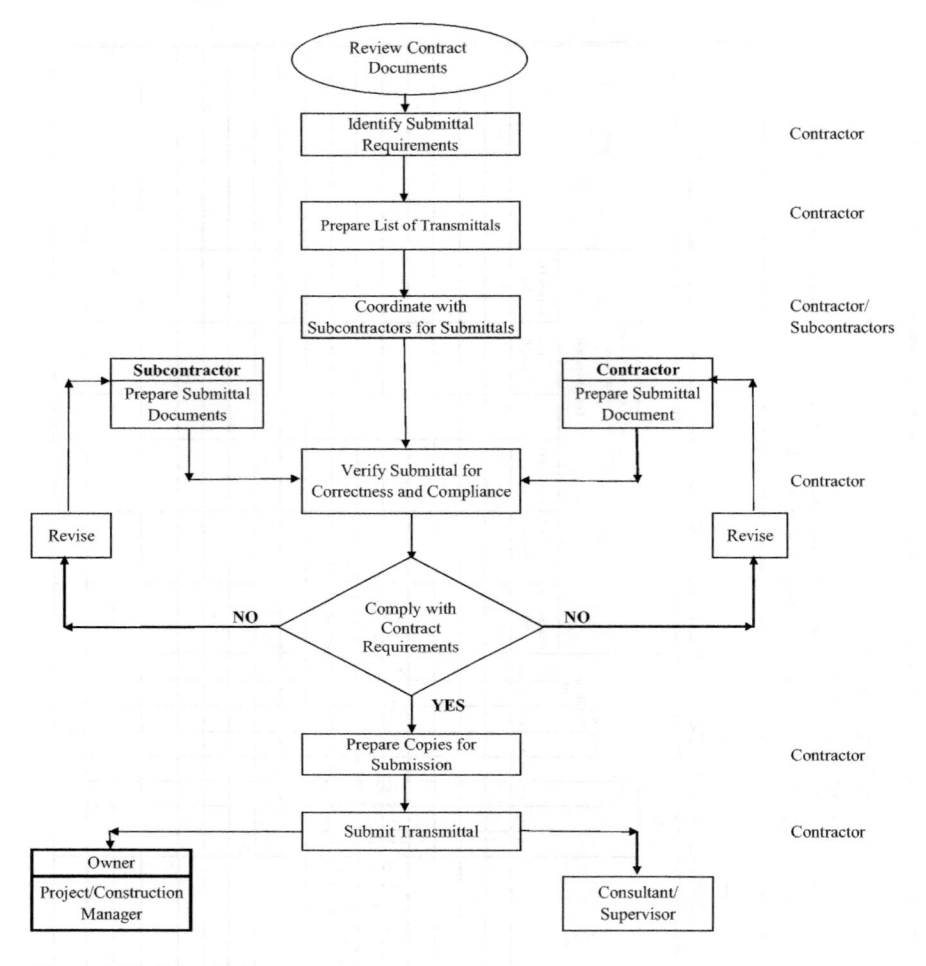

Figure 6.46 Submittal process

diagram for the quality auditing of submittal activities during the construction phase.

The main objectives of performing a quality audit of the construction/execution activities/process during the construction phase are to:

1 Ascertain that contact documents are reviewed and project execution requirements/scope are properly established
2 Assess conformance to QMS and that processes meet project specific quality system requirements
3 Ascertain that constructions activities and process are properly controlled, documented and analyzed

Auditing processes for project life cycle phases 333

Project Name
Consultant Name
SUBMITTAL TRANSMITTAL FORM

Contractor Name :	
Contract No. :	
To.	Resident Engineer
Transmittal No.:	Date :

Submittal Type:			Action Requested:	
DG	Shop Drawings	1	For Approval
SK	Sketches	2	For Review and Comment
PR	Material/Product/System	3	For Information
MD	Manufacturer's Data	4	For Construction
SM	Sample	5	For Incorporation Within the Design
MM	Minutes of Meeting	6	For Costing
RP	Reports	7	For Tendering
LG	Logs		
OT	Others (please specify)		

SAMPLE FORM

We are sending herewith the following:

			ENCLOSURES		
Item	Qty	Ref. No.	Description	Type	Action

Comments:
_____
_____
_____

Issued by:	Received by:
Signature:	Signature:
Date:	Date:

Figure 6.47 Submittal transmittal form

4 Verify that stakeholders' requirements are properly identified and informed as per agreed-upon requirements
5 Verify that the change management process is in place
6 Assess all the activities are performed as per approved construction schedule
7 Evaluate the project status/performance
8 Ensure a S-Curve is used to monitor cash flow and forecast
9 Verify that all the materials and system are approved
10 Verify that all the builders' shop drawings, coordination/composite drawings and shop drawings are approved

Project Name

Consultant Name

SPECIFICATION COMPARISON STATEMENT (SCS)

Contractor: ______________________ Date : ____________

Contract No. ______________________ A/S No.: ____________

Submittal No. : ____________ Revision: ________ Transmittal Ref.: ____________

Submittal Title: ____________ Specification Ref : ____________

S No.	SPECIFICATION REQUIREMENTS	CONTRACTOR'S PROPOSAL	REMARKS

SAMPLE FORM

Figure 6.48 Specification comparison statement

Table 6.39 Audit methodology for the assessment of submittal activities during the construction phase

Serial Number	Item To Be Assessed	Yes	No	Comments
A. Checklist				
1	Whether project submittal requirements are as per contract established			
2	Whether submittals are prepared properly identifying the contract requirements			
3	Whether the submittals are prepared taking into consideration the required number of copies			
4	Whether submittals are distributed to all the stakeholders as per contract documents			
5	Whether submittals are as per approved schedule			
6	Whether submittal logs are in place			
7	Whether submittals are submitted on a submittal transmittal form			
8	Whether submittals are submitted with all the required relevant documents and compliance statement			
9	Whether all the management plans are as per schedule			
10	Whether shop drawings submittals are as per schedule			
11	Whether coordination drawings submittals are as per schedule			
12	Whether composite drawings submittals are as per schedule			
13	Whether shop drawings are coordinated with all the relevant trades			
14	Whether drawings are prepared as per the specified format			
15	Whether drawings title, client name, logo, contract numbering and reference number are properly mentioned as per the specified format			
15	Whether material submittals are as per schedule			

Serial Number	Description	Response	Comments
B. Interview Questionnaires			
1	Who has reviewed the contract documents to identify the number of submittals to be prepared and submitted for approval		
2	Whether submittals are as per schedule		

(Continued)

Table 6.39 (Continued)

B. Interview Questionnaires

Serial Number	Description	Response	Comments
3	How many submittals are not approved and is there is any list of non-approved transmittals in place		
4	What are the reasons for non-approval or resubmission, if any, of transmittals		
5	Whether the reasons (causes) for non-approval, if any, are analyzed and corrective/preventive action taken		
6	What measures are taken to ensure that submittals comply with specifications		
7	Whether there is a list of the required number of shop drawings, coordination drawings and composite drawings		
8	Whether there is a list of materials and systems submittals to be submitted for approval		
9	Whether the required number of samples are submitted along with the transmittals		
10	Whether a method statement is required to be submitted		

11 Assess all the works are carried out as per approved shop drawings by installing approved material
12 Check that the execution/installation is carried out as per approved/recommended processes
13 Verify that product samples are inspected as per contract to ascertain compliance with project specifications
14 Check the concrete casting process is properly followed. Figure 6.50 illustrates a flow chart for concrete casting.
15 Verify casting samples are taken as per contract to ascertain compliance with project specifications
16 Ensure that the core staff are performing their assigned responsibilities
17 Ascertain the resources are properly utilized
18 Assess the productivity rate of workers
19 Ascertain the proper utilization of equipment and machinery
20 Verify that a risk register is maintained and a risk management plan is followed
21 Verify that a responsibility matrix is established. Table 6.40 illustrates the contractor's responsibilities to manage construction quality.
22 Ensure that all logs are maintained and updated
23 Ensure that subcontractors' works are properly monitored

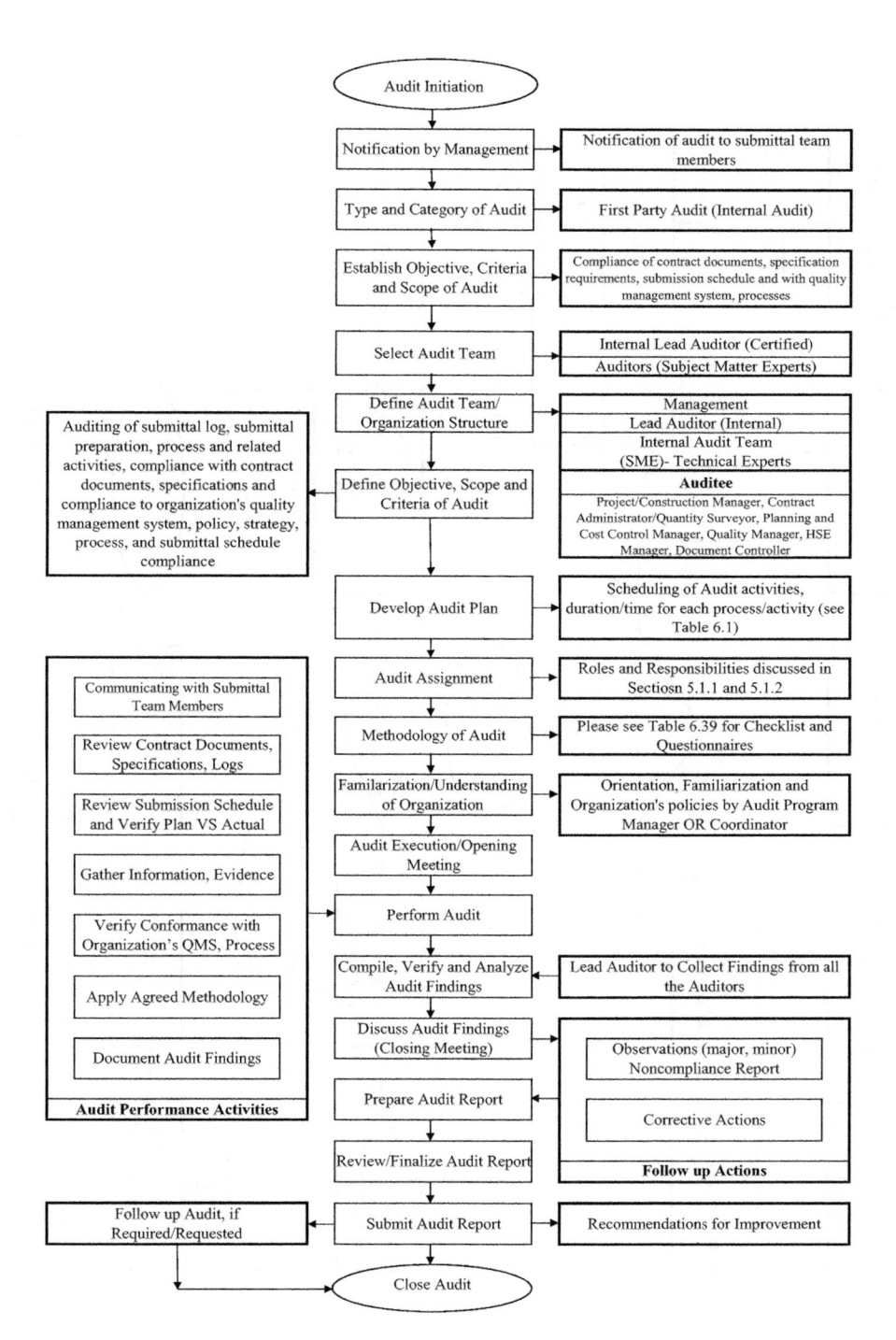

Figure 6.49 Logic flow diagram for the quality auditing process for submittal activities during the construction phase

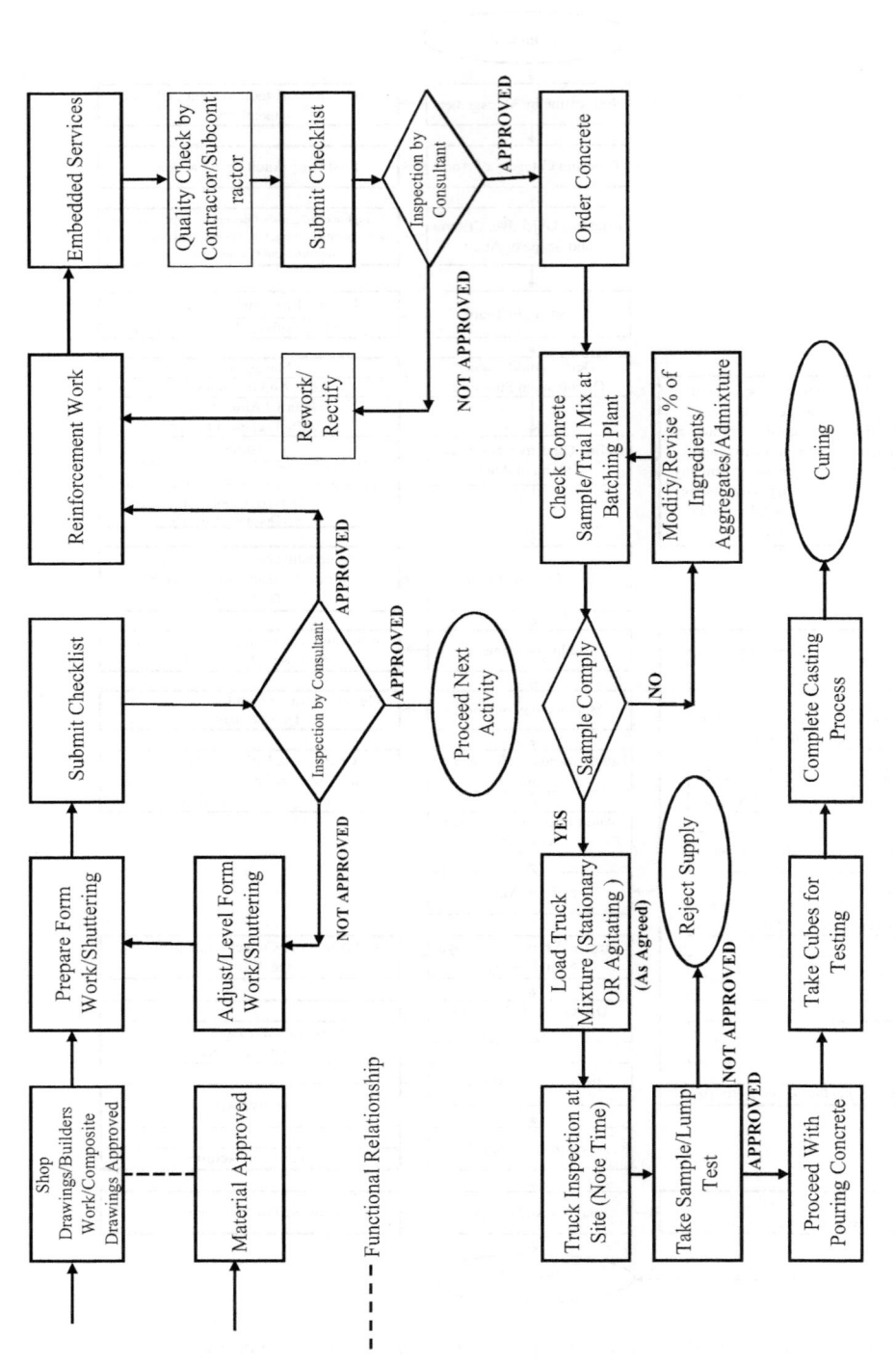

Figure 6.50 Flow chart for the concrete casting process for structural work

Source: Abdul Razzak Rumane (2013). *Quality Tools for Managing Construction Projects*. Reprinted with permission of Taylor & Francis Group

Table 6.40 Contractor's responsibilities to manage construction quality

		AREAS OF QUALITY CONTROL							
		MAIN CONTRACTOR		SUBCONTRACTORS					
Sr.No.	ACTIVITY	HEAD OFFICE/ QUALITY MANAGER	PROJECT SITE/ PROJECT MANAGER	STRUCTURAL	INTERIOR	MRCHANICAL (HVAC+PHFF)	ELECTRICAL	LANDSCAPE	EXTERNAL
1	Prepare Quality Control Plan	□	■	□	□	□	□	□	□
2	Construction Schedule	□	■	□	□	□	□	□	□
3	Staff Approval	□	■	□	□	□	□	□	□
4	Mobilization		■	□	□	□	□	□	□
5	Prepare Material Submittal		■	■	■	■	■	■	■
6	Submit Material Tranmittal		■	□	□	■	■	□	■
7	Prepare Shop Drawings		■	□	□	■	■	□	□
8	Submit Shop Draing Tranmittal		■	■	■	■	■	■	■
9	Material Sample	□	■	■	■	■	■	■	■
10	Receiving Material Inspection		■	■	■	■	■	■	■
11	Material Testing		■	■	■	■	■	■	■
12	Mock Up		■	■	■	■	■	■	■
13	Site Work Inspection		■	■	■	■	■	■	■
14	Quality of Work		■	■	■	■	■	■	■

(Continued)

Table 6.40 (Continued)

15	Prepare Checklist	■	■ □	■ □	■ □	■ □	■ □
16	Submit Checklist	■	□ □	□ ■	□ □	□ □	□ □
17	Corrective/Preventive Action	■	■ □	■ □	■ □	■ □	■ □
18	Daily Report	■	□ □	□ □	□ □	□ □	□ □
19	Monthly Progress Report	■	□ □	□ □	□ □	□ □	□ □
20	Progress Payment	□	□ □	□ □	□ □	□ □	□ □
21	Site Safety	■	■ □	■ □	■ □	■ □	■ □
22	Safety Report	■	□ □	□ ■	□ ■	□ ■	□ ■
23	Waste Disposal	■	■ □	■ □	■ □	■ □	■ □
24	Reply to Job Site Instruction	■	□ □	□ □	□ □	□ □	□ □
25	Reply to Nonconformance Report	■	□ □	□ □	□ □	□ □	□ □
26	Documentation	■	□ □	□ □	□ □	□ □	□ □
27	Testing and Commissioning	■	■ □	■ □	■ □	■ □	■ □
28	Project Closeout Documents	□	□ □	□ □	□ □	□ □	□ □
29	Punch List	■	■ □	■ □	■ □	■ □	■ □
30	Request for Issuance of Substantial Completion Letter	□	■ □	■ □	■ □	■ □	■ □

■ **Primary Responsibility**
□ **Advise/Assist**

Source: Abdul Razzak Rumane. (2013). Quality Tools for Managing Construction Projects
Reprinted with permission of Taylor & Francis Group

Auditing processes for project life cycle phases 341

24 Assess that the procurement procedure is followed as per the corporate quality management system. Figure 6.51 illustrates the material procurement procedure for construction projects.
25 Assess the implementation of site safety and environmental protection activities
26 Ascertain the timely submission of the payment certificate to the owner/client
27 Ascertain that subcontractors' payment is documented properly
28 Verify that the project monitoring and controlling process is properly followed. Figure 6.52 illustrates a logic flow diagram for the monitoring and control process.
29 Check what type of project monitoring system is followed: traditional or digitized. Figure 6.53 illustrates a traditional monitoring system whereas Figure 6.54 illustrates a digitized monitoring system.

Table 6.41 illustrates the audit methodology for the assessment of construction/execution activities/process during the construction phase and Figure 6.55 illustrates the logic flow diagram for quality auditing of the construction/execution activities/process during the construction phase.

The main objectives of performing a quality audit of the inspection of executed/installed works/process and systems during the construction phase are to:

1   Ascertain that contact documents are reviewed and an inspection and quality control procedure is properly established
2   Assess conformance to QMS and that processes meet project specific quality system requirements
3   Assess that a checklist is submitted for the completed work/work under progress as per the contract specifications
4   Verify that checklists are submitted after assessing that the checklist submittal procedure follows the sequence of execution of works as illustrated in Figure 6.56
5   Verify how many checklists are not approved and what the volume of rework is
6   Assess that on-site and off-site inspection is carried out as per contract requirements
7   Verify that preventive action is carried out
8   Verify that corrective action is carried out
9   Ensure that rejection analysis is done for the not approved/rejected work
10 Ensure that action is taken against remedial notes
11 Ensure that action is taken against non-conformance reports
12 Ensure that material inspection is carried out as per contract requirements
13 Ensure that work inspection and testing is carried out on regular basis

Table 6.42 illustrates the audit methodology for the assessment of inspection activities related to execution/installation works during the construction phase and Figure 6.57 illustrates the logic flow diagram for the quality auditing of the inspection of the executed/installed works/process and systems during the construction phase.

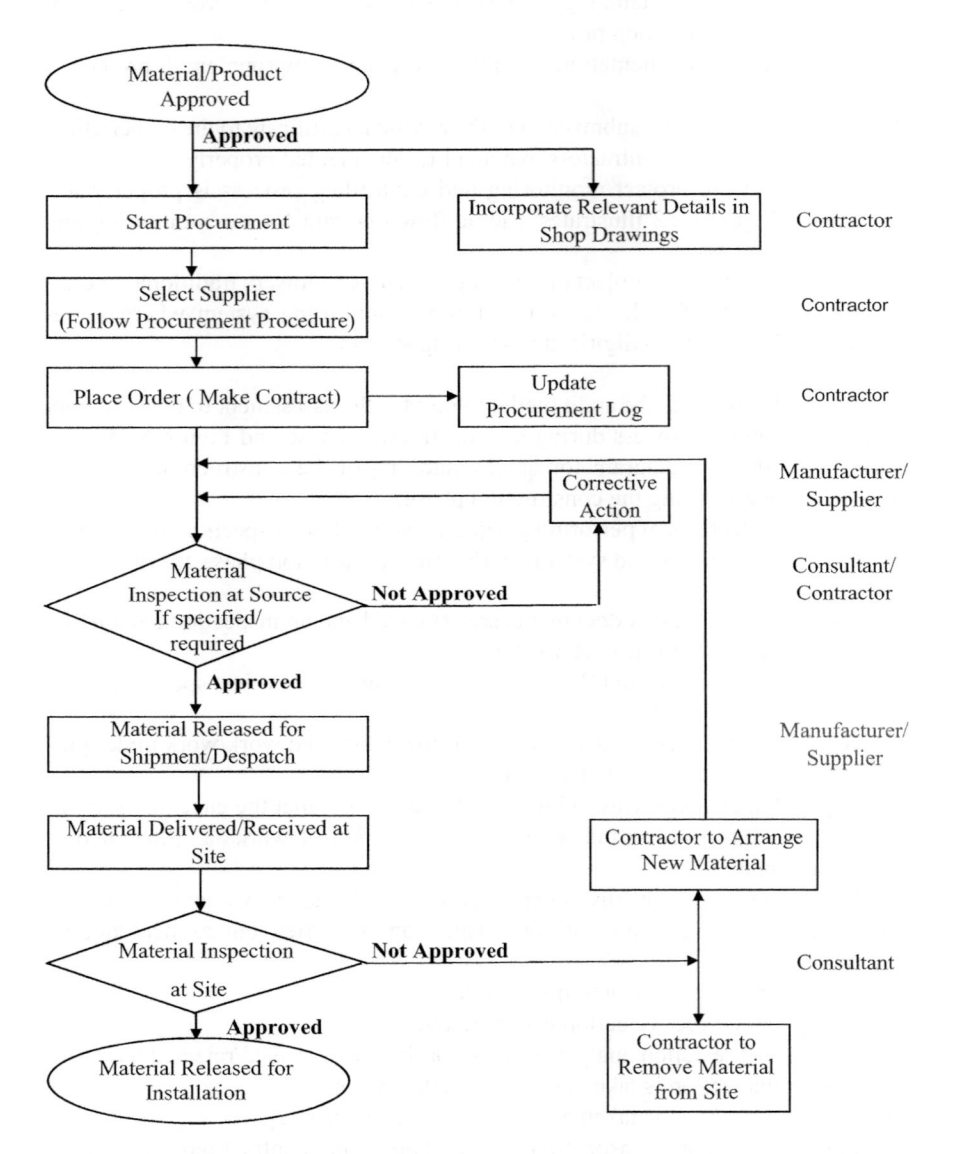

Figure 6.51 Material procurement procedure

Source: Abdul Razzak Rumane (2013). *Quality Tools for Managing Construction Projects.* Reprinted with permission of Taylor & Francis Group

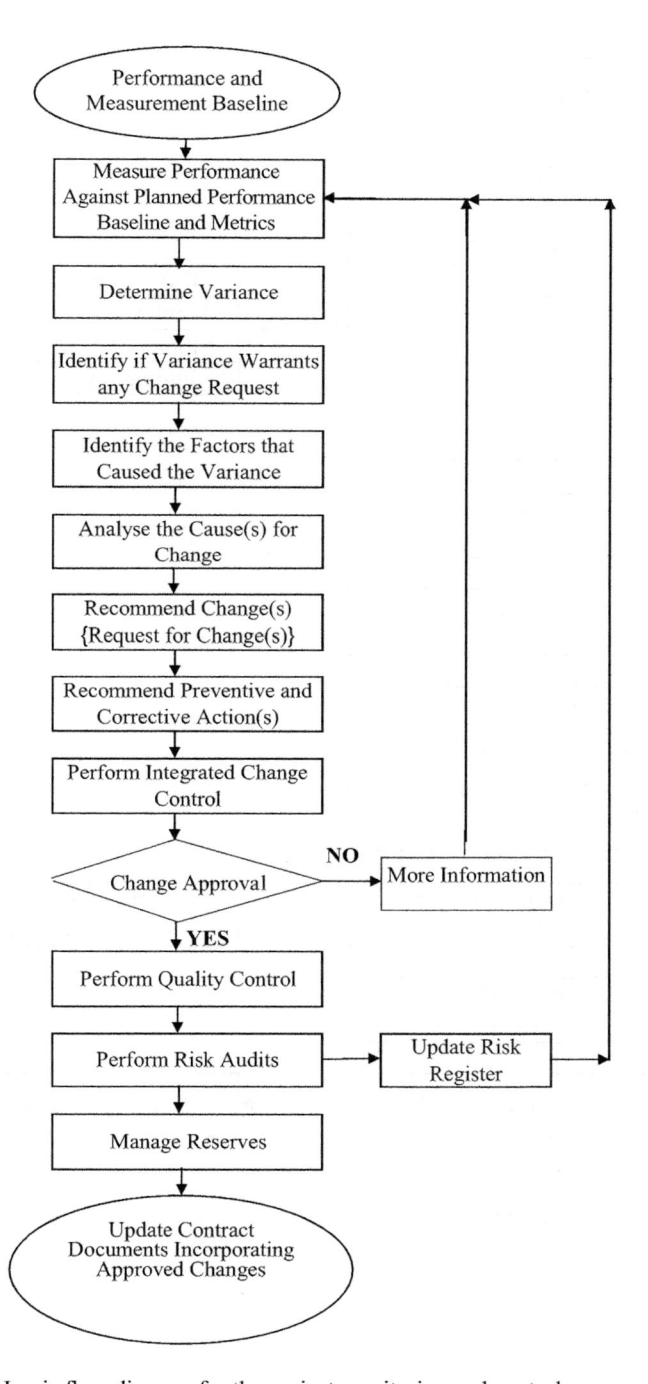

Figure 6.52 Logic flow diagram for the project monitoring and control process

Source: Abdul Razzak Rumane (2013). *Quality Tools for Managing Construction Projects.* Reprinted with permission of Taylor & Francis Group

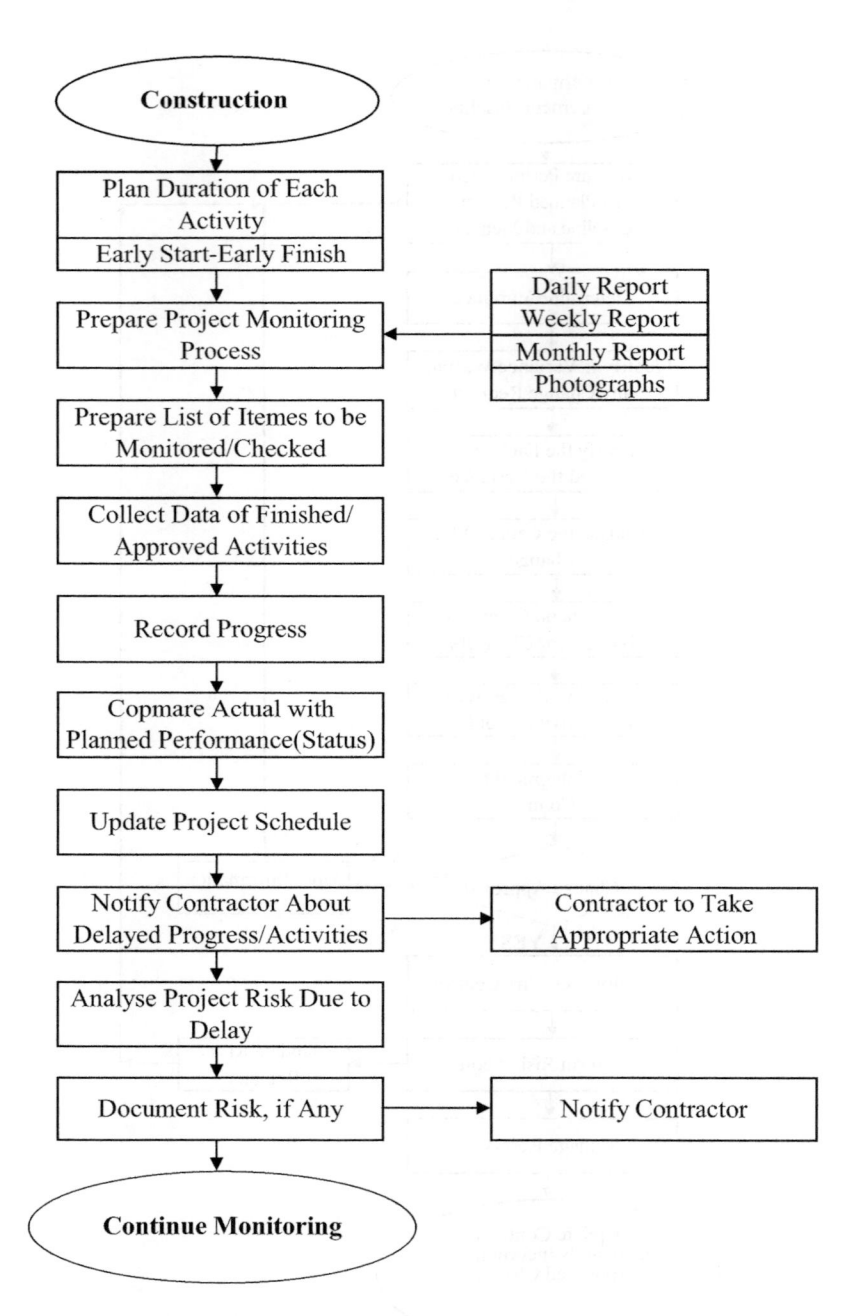

```
                    ┌─────────────────┐
                    │  Construction   │
                    └────────┬────────┘
                             ▼
              ┌──────────────────────────┐
              │  Plan Duration of Each   │
              │        Activity          │
              │ Early Start-Early Finish │
              └──────────────┬───────────┘
                             ▼                    ┌──────────────────┐
              ┌──────────────────────────┐        │   Daily Report   │
              │ Prepare Project Monitoring│◄──────│   Weekly Report  │
              │         Process           │       │   Monthly Report │
              └──────────────┬───────────┘        │    Photographs   │
                             ▼                    └──────────────────┘
              ┌──────────────────────────┐
              │ Prepare List of Itemes to be │
              │    Monitored/Checked     │
              └──────────────┬───────────┘
                             ▼
              ┌──────────────────────────┐
              │ Collect Data of Finished/│
              │    Approved Activities   │
              └──────────────┬───────────┘
                             ▼
              ┌──────────────────────────┐
              │     Record Progress      │
              └──────────────┬───────────┘
                             ▼
              ┌──────────────────────────┐
              │   Copmare Actual with    │
              │ Planned Performance(Status)│
              └──────────────┬───────────┘
                             ▼
              ┌──────────────────────────┐
              │  Update Project Schedule │
              └──────────────┬───────────┘
                             ▼                    ┌──────────────────────┐
              ┌──────────────────────────┐        │ Contractor to Take   │
              │  Notify Contractor About │───────►│  Appropriate Action  │
              │ Delayed Progress/Activities│      └──────────────────────┘
              └──────────────┬───────────┘
                             ▼
              ┌──────────────────────────┐
              │ Analyse Project Risk Due to│
              │           Delay          │
              └──────────────┬───────────┘
                             ▼                    ┌──────────────────┐
              ┌──────────────────────────┐        │ Notify Contractor│
              │  Document Risk, if Any   │───────►└──────────────────┘
              └──────────────┬───────────┘
                             ▼
                    ┌─────────────────┐
                    │Continue Monitoring│
                    └─────────────────┘
```

Figure 6.53 Traditional monitoring system

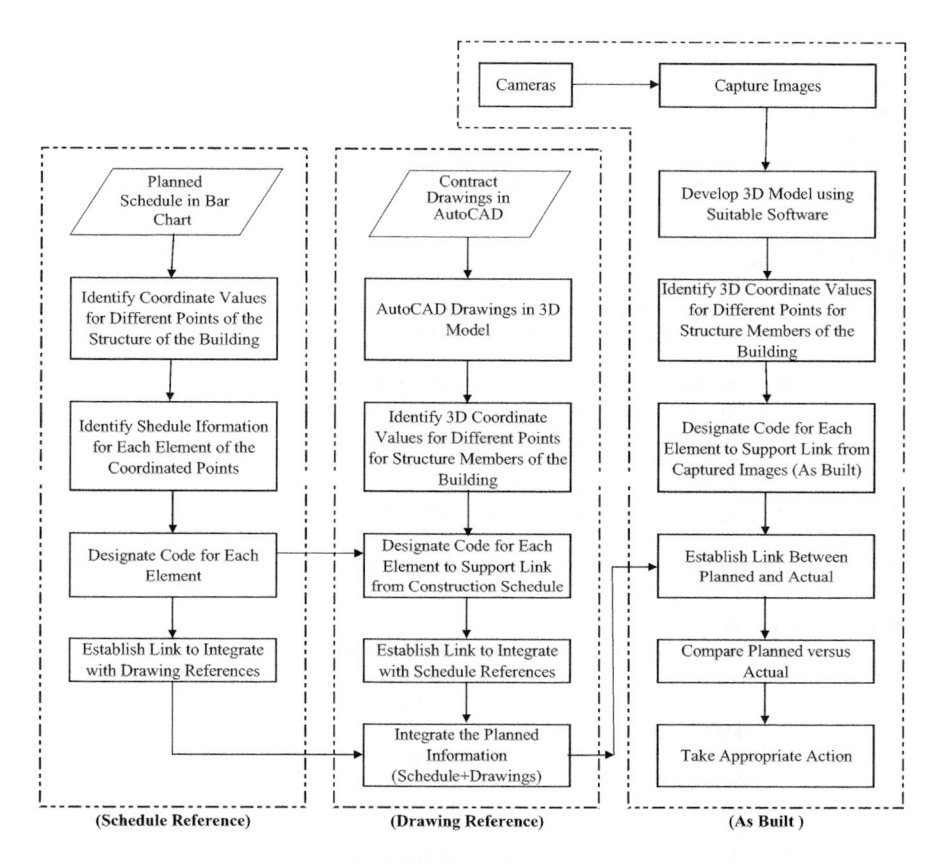

Figure 6.54 Digitized monitoring system

Source: Abdul Razzak Rumane (2013). *Quality Tools for Managing Construction Projects.* Reprinted with permission of Taylor & Francis Group

Table 6.41 Audit methodology for the assessment of the construction/execution activities/processes during the construction phase

Serial Number	Item To Be Assessed	Yes	No	Comments

A. Checklist

A-1 Construction/execution of works
1. Whether the project execution requirements are properly established
2. Whether the shop drawings are approved
3. Whether the materials/systems are approved
4. Whether the method of installation (method statement) is approved
5. Whether execution/installation is carried out as per approved/recommended processes

(*Continued*)

Table 6.41 (Continued)

Serial Number	Item To Be Assessed	Yes	No	Comments
A. Checklist				

B-1 Monitoring and Controlling
1. Whether the project monitoring and control process is as illustrated in Figure 6.52
2. Whether a change management system is in place
3. Whether project progress is monitored by a traditional monitoring system (please refer to Figure 6.53 for guidance purposes) or by digitized monitoring system (Please refer to Figure 6.54 for guidance purposes)
4. Whether critical activities that may affect the project are identified
5. Whether progress photos are taken to monitor work progress

B-1.1 Stakeholders
1. Whether project stakeholders are identified and listed
1. Whether a responsibilities matrix of all the stakeholders is established
2. Whether reports are distributed to all the stakeholders
3. Whether a meeting invitation is sent to all the stakeholders
4. Whether core staff are approved and available at the project
5. Whether all the approvals from the authorities are obtained

B-1.2 Scope
1. Whether work is executed as per defined scope and agreed methodology
2. Is there any change in the contracted scope of work and specifications
3. Whether preventive and corrective measure are being taken to control the scope
4. Is there any Request for Information (RFI) raised
5. Is there any Request for Variation submitted
6. Is there any request for alternate materials
7. Is there any site work instruction (SWI)

B-1.3 Schedule
1. Whether contractor's construction schedule is approved
2. Whether sequence of activities properly coordinated
3. Whether performance baseline for all the activities identified

Serial Number	Item To Be Assessed	Yes	No	Comments

A. Checklist

4	Whether all the submittals are as per approved schedule and monitored			
5	Whether project progress status are prepared to know the actual progress versus planned progress			
6	Whether project forecasting is done to predict future progress of activities to be performed			
7	Whether percentage of completion of each activity is based on approved checklist			
8	Whether schedule is updated based on approved changes			
9	Whether subcontractor's work progress is monitored			
10	Whether contractor was submitting daily progress report			
11	Whether contractor was submitting daily manpower, equipment report available on site			

B-1.4 Cost

1	Whether there is an approved construction budget			
2	Whether there is approved S-Curve			
3	Whether cash flow is properly maintained			
4	Whether Earned Value Management in place			
5	Whether the project cost is exceeding the planned budget			
6	Whether forecasted cost is as planned or likely to exceed			
7	Whether there is any approved variation cost to the contracted project cost			
8	Whether S-Curve is updated based on approved variation orders			

B-1.5 Quality

1	Whether contractor's quality control plan is approved			
2	Whether builders workshop drawings, coordinated drawings, composite drawings, shop drawings are approved			
3	Is there are any comments on the shop drawing approval transmittals			
4	Whether all the materials/systems approved			
5	Is there are any comments on the submitted material transmittals			
6	Whether incoming material is inspected upon receipt on site			

(Continued)

Table 6.41 (Continued)

Serial Number	Item To Be Assessed	Yes	No	Comments
A. Checklist				
7	Whether test certificates, country of origin submitted along with the material incoming checklist			
8	Is there any rejection to the material received on site			
9	Is there any factory inspection carried out for the material/system			
10	Whether works are being carried out as per approved builders work shop drawings, coordinated drawings, composite drawings, and shop drawings			
11	Whether works are carried out as per approved methods			
12	Whether approved materials/systems are installed			
13	Is there any variation to the approved methods			
14	Whether quality meetings are conducted			
15	Whether works are executed to comply with regulatory requirements			
16	Whether subcontractor's quality management system is properly monitored to comply with contract documents			
B-1.6 Resources				
1	Whether project staff, and other workforce requirements were established			
2	Whether project staff and workforce is available on site as per the approved manpower/staff deployment chart			
3	Is there any conflict among project team members			
4	Whether sufficient workforce is available to meet the project execution plan Whether procurement log is properly maintained			
5	Whether equipment, machinery is available on site to execute the construction works			
6	Whether subcontractors have sufficient resources to execute the works as per schedule			
7	Whether work is assigned on daily basis			
8	Whether attendance sheet is maintained			
9	Whether subcontractor management is in place			
B-1.7 Communication				
1	Whether site administration matrix is in place			
2	Whether logs are maintained			
3	Whether communication plan is in place			

Serial Number	Item To Be Assessed	Yes	No	Comments

A. Checklist

4	Whether correspondence procedure between all the stakeholders is established			
5	Whether communication method is established			
6	Whether Not Approved transmittals are addressed promptly			
7	Whether logs are updated on regular basis			
8	Whether progress meetings, coordination meetings are held as scheduled			
9	What is the frequency of progress meeting, Quality meeting, HSE meeting			
10	Whether daily, weekly, and monthly reports are properly prepared and circulated to the relevant stakeholders			
11	Whether project documents are kept in safe and accessible location			

B-1.8 Risk

1	Whether risk management plan is in place			
2	Whether risk register is maintained and updated			
3	Whether occurrence of new risk is reported to update project progress and performance			
4	Whether risk response plan is prepared to address the identified and assessed risks			
5	Whether effectiveness of risk response plan is regularly evaluated and effective measures taken			
6	Whether financial risk is reported to the owner/client			

B-1.9 Contract

1	Whether contract management process is in place			
2	Whether project procurement requirements are identified			
3	Whether bidding and tendering procedure is followed to procure the material, system, machinery, and equipment			
4	Whether subcontractors list is finalized			
5	Whether procurement follow-up system is in place			
6	Whether cost of subcontracted items is as per estimated value			
7	Whether there is any delay in receipt of material, system, equipment that may affect project execution schedule			
8	Whether subcontractor's works are as per schedule			

(*Continued*)

Table 6.41 (Continued)

Serial Number	Item To Be Assessed	Yes	No	Comments

A. Checklist

B-1.10 HSE
1. Whether HSE management plan is approved
2. Whether safety meetings are conducted
3. Whether safety drills are conducted
4. Whether temporary fire fighting system is operative
5. Whether accident preventive measure are followed
6. Whether first aid and emergency medical services are available on site
7. Whether accident reporting system is followed properly
8. Whether accident records are maintained
9. Whether hazardous area are marked
10. Whether the personnel are using protective equipment and other safety gears
11. Whether periodic safety inspection is carried out and what is the frequency of such inspection
12. Whether safety violation report is issued
13. Whether safety disciplinary notice is issued
14. Whether access and escape route signs are installed on site
15. Whether machinery, equipment, and tools are regularly inspected for safe usage
16. Whether crane and hoist have valid third party certification
17. Whether protective clothing and equipment are used while concreting
18. Whether sirens and bells are in working conditions
19. Whether barricade are installed to avoid accidents
20. Whether all formwork and scaffoldings are checked as per permissible load prior to casting

B-1.11 Finance
1. Whether there is contractor's finance plan in place
2. Whether progress payment is submitted on time
3. Whether cash flow is as per approved S-Curve
4. Whether funding limit reconciliation is done periodically
5. Whether the project progress payment is received on time as per contract

Serial Number	Item To Be Assessed	Yes	No	Comments

A. Checklist

6	Whether there is any deduction from the approved progress payment			
7	Whether subcontractors are paid as per contract			
8	Is there interim payment, advance payment for procurement of material			
9	Whether staff salaries are paid on regular basis			
10	Whether payment to the suppliers is as per the agreed-upon terms			
11	Whether there is any funding agency to finance the project			
12	Whether finance records and logs are maintained properly			

B-1.12 Claim

1	Whether any claim is identified, recorded and informed to the client/owner			
2	Whether there is any dispute noticed related to contract documents			
3	Is there any claim that may affect project progress			

B. Interview Questionnaires

Serial Number	Description	Response	Comments
1	Whether the project team members know the scope of work to executed		
2	Whether the project team members know the construction phase deliverables		
3	Whether all the related data of each activities are collected and recorded for project monitoring purpose		
4	Whether there is an organized and efficient means for measuring, collecting, verifying and quantifying data to monitor progress of the project		
5	Whether project deliverables are as defined in the scope baseline or there are any changes in the scope baseline		
6	Whether there is any approved procedure for integrating the changes		
7	Was there any system to initiate the changes/ variation and if there are any standard forms		
8	What tools and techniques are used to improve schedule and recover delays		

(*Continued*)

Table 6.41 (Continued)

B. *Interview Questionnaires*

Serial Number	Description	Response	Comments
9	What is the frequency of progress status reporting		
10	Whether Earned Value Management (EVM) methodology is used to measure and evaluate project performance		
11	Whether there are not approved checklists and how rework is managed		
12	Whether productivity rate and performance of workers is as estimated during tendering stage		
13	Whether material management process is properly followed		
14	Whether identified risks are used to update project schedule and cost		
14	Whether risk management process cycle is properly followed		
15	Whether risk probability level is being followed to determine risk impact on the project		
16	What is the procedure followed to select the subcontractor		
17	Whether there are any items expensive than the estimated cost		
18	Whether supply chain management system is followed while selecting the supplier		
19	Whether safety awareness/training programs are conducted		
20	Whether emergency evacuation plan is prepared		
21	Whether evacuation routes are displayed		
22	What method is followed to resolve the claim		
23	Whether the reasons (causes) for comments on shop drawings, if any, analyzed and corrective/preventive action taken		
24	Whether the reasons (causes) for rejection of material, if any, analyzed and corrective/preventive action taken		
25	Whether the reasons (causes) for rejection/non-approval of executed works, if any, analyzed and corrective/preventive action taken		

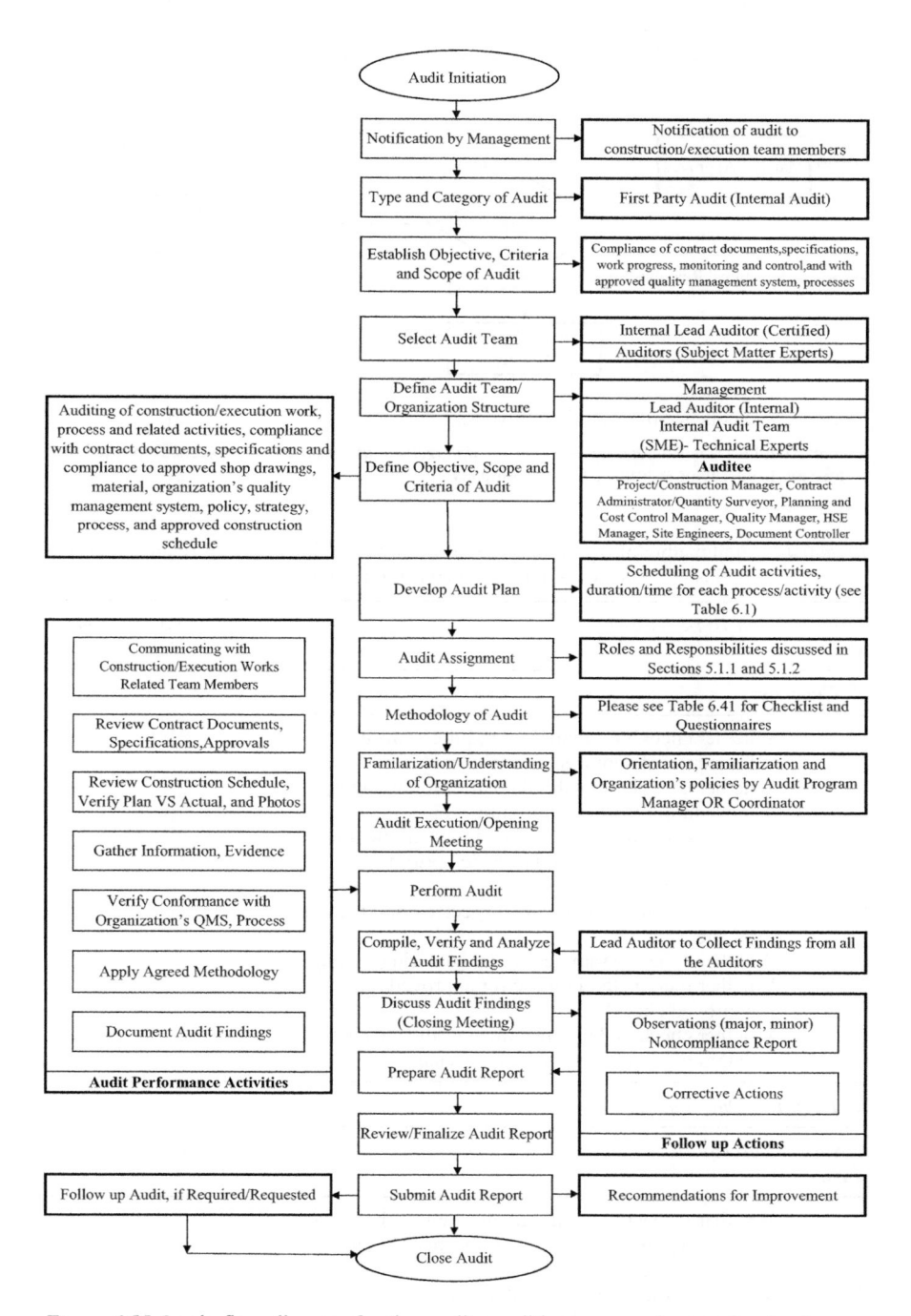

Figure 6.55 Logic flow diagram for the quality auditing process for construction/execution activities during the construction phase

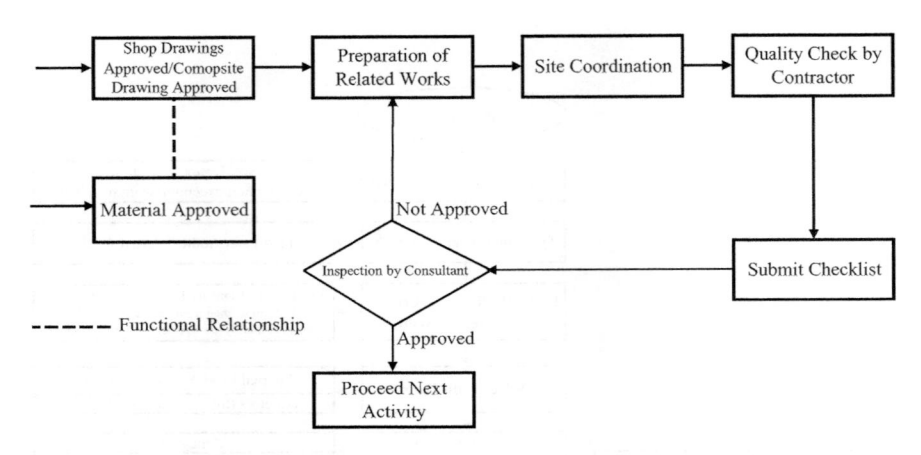

```
┌──────────────────┐     ┌──────────────┐     ┌──────────────┐     ┌──────────────┐
│ Shop Drawings    │     │ Preparation  │     │ Site         │     │ Quality Check│
→│ Approved/Composite├───→│ of Related   ├───→│ Coordination ├───→│ by Contractor│
│ Drawing Approved │     │ Works        │     │              │     │              │
└──────────────────┘     └──────────────┘     └──────────────┘     └──────────────┘
         ╎                     ▲                                            │
         ╎                     │                                            ▼
┌──────────────────┐            Not Approved                        ┌──────────────┐
→│ Material Approved│            ◇                        ◀─────────│Submit Checklist│
└──────────────────┘   Inspection by Consultant                     └──────────────┘
  - - - Functional Relationship  │
                              Approved
                                 ▼
                         ┌──────────────┐
                         │ Proceed Next │
                         │   Activity   │
                         └──────────────┘
```

Figure 6.56 Sequence of execution of works

Source: Abdul Razzak Rumane (2010). *Quality Management in Construction Projects*. Reprinted with permission from Taylor & Francis Group Company

Table 6.42 Audit methodology for the assessment of inspection activities related to execution/installation works during the construction phase

Serial Number	Item To Be Assessed	Yes	No	Comments
A. Checklist				
1	Whether checklist submission requirements and frequency of submission established			
2	Whether checklist is submitted after following the procedure illustrated in Figure 6.56			
3	Whether checklist is coordinated with all the trades			
4	Whether material sample is verified prior to submitting for approval			
5	Whether incoming materials inspected upon receiving on site			
6	Whether concrete samples are taken during casting			
7	Whether there is log in place for approved/not approved checklists			
8	Whether checklist has attachment of the location/area diagram of the works to be inspected			

B. Interview Questionnaires

Serial Number	Description	Response	Comments
1	Whether inspection requirements are identified and frequency of inspection is established through Inspection and Test Plan (IPT)		
2	Whether inspection is carried out continually as per the progress of works		
3	Whether laboratory tests are carried out on concrete		
4	Whether the inspection is visual or non-destructive test is also carried out?		
5	What is the frequency of submitting checklists?		

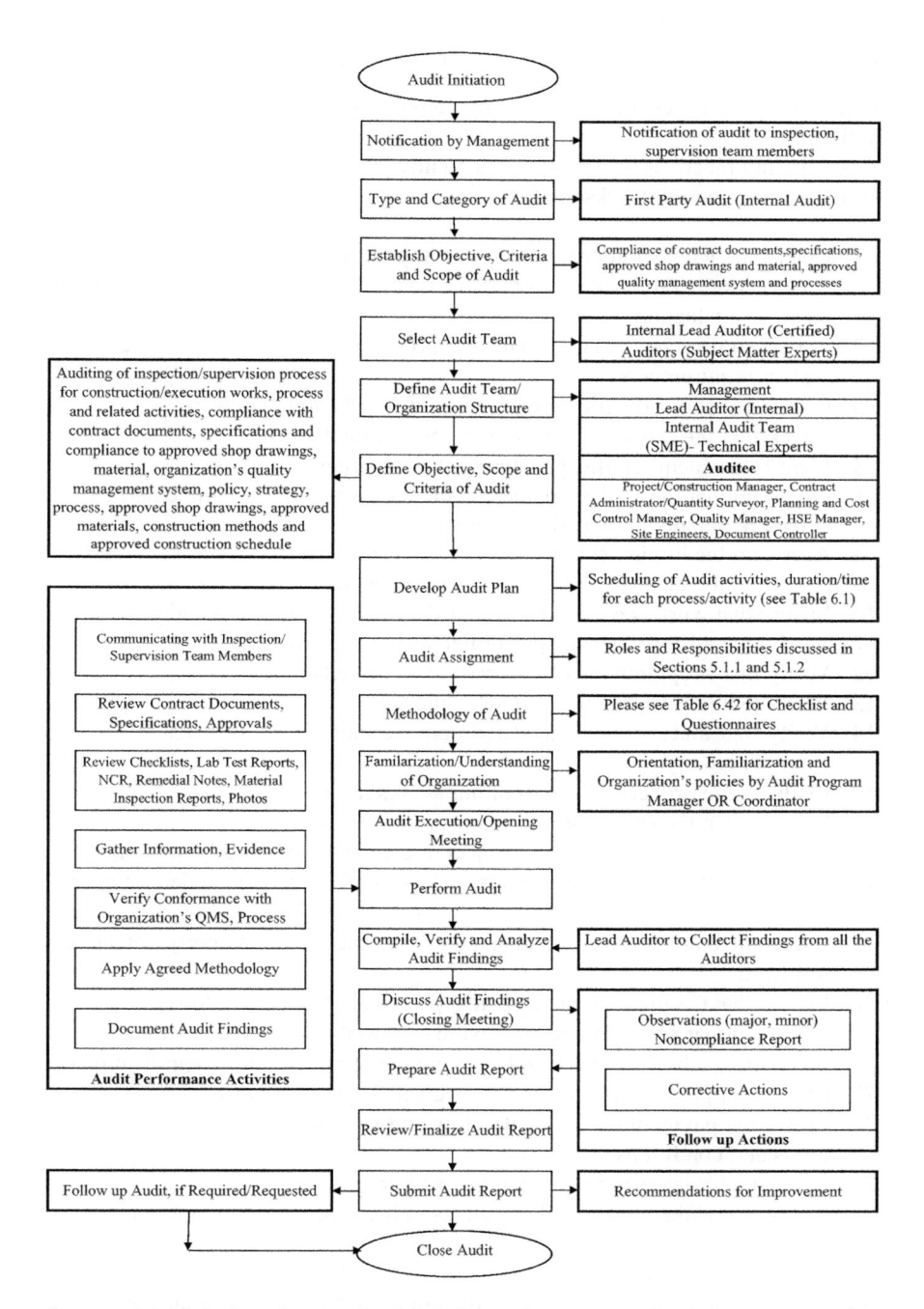

Figure 6.57 Logic flow diagram for the quality auditing process for the inspection of executed/installed activities during the construction phase

6.2.4.1.2 AUDITING AFTER COMPLETION OF CONSTRUCTION/EXECUTION OF WORKS

Figure 6.58 illustrates the typical quality auditing process for construction phase activities.

The main objectives of performing a quality audit after completion of the construction/execution of works are to:

1. Determine whether appropriate systems, processes and control for managing construction were in place to achieve contractual objectives and the owner/client's satisfaction
2. Assess conformance to QMS and that processes meet project specific quality system requirements
3. Check that stakeholders were properly identified and managed
4. Check that the scope of work was properly established
5. Identify problems/challenges during the construction process
6. Ascertain that the change management process is properly implemented
7. Compare planned vs actual performance
8. Ascertain the implementation of cost control activities
9. Ascertain the quality management plan/processes
10. Ensure that work is carried out as per the approved methods and procedures
11. Verify that all activities are performed as per approved methods
12. Verify that shop drawings are approved as per schedule
13. Verify material approval as per schedule
14. Verify that material was inspected upon receipt on site or prior to dispatch from the factory
15. Evaluate the proper utilization of human resources
16. Validate that the responsibility matrix is followed
17. Ascertain that all the logs and project documents are properly maintained
18. Check that the risk management process is in place and the update of the risk register
19. Ascertain that the contract management process was properly set up and followed
20. Ensure that the material procurement process is properly followed
21. Ascertain that proper procedures were followed to select subcontractor(s)
22. Ensure that a health, safety and environmental protection system is in place
23. Check that progress payments are received as per cash flow
24. Check the payments to subcontractors and suppliers are as per agreed-upon terms and conditions
25. Check that salaries are paid on regular basic
26. Verify that claims towards the changes in the contract are submitted

Table 6.43 illustrates the audit methodology for the assessment of construction/execution activities/processes as illustrated in Figure 5.26 after completion of construction and Figure 6.59 illustrates the logic flow diagram for the quality auditing process for construction phase activities.

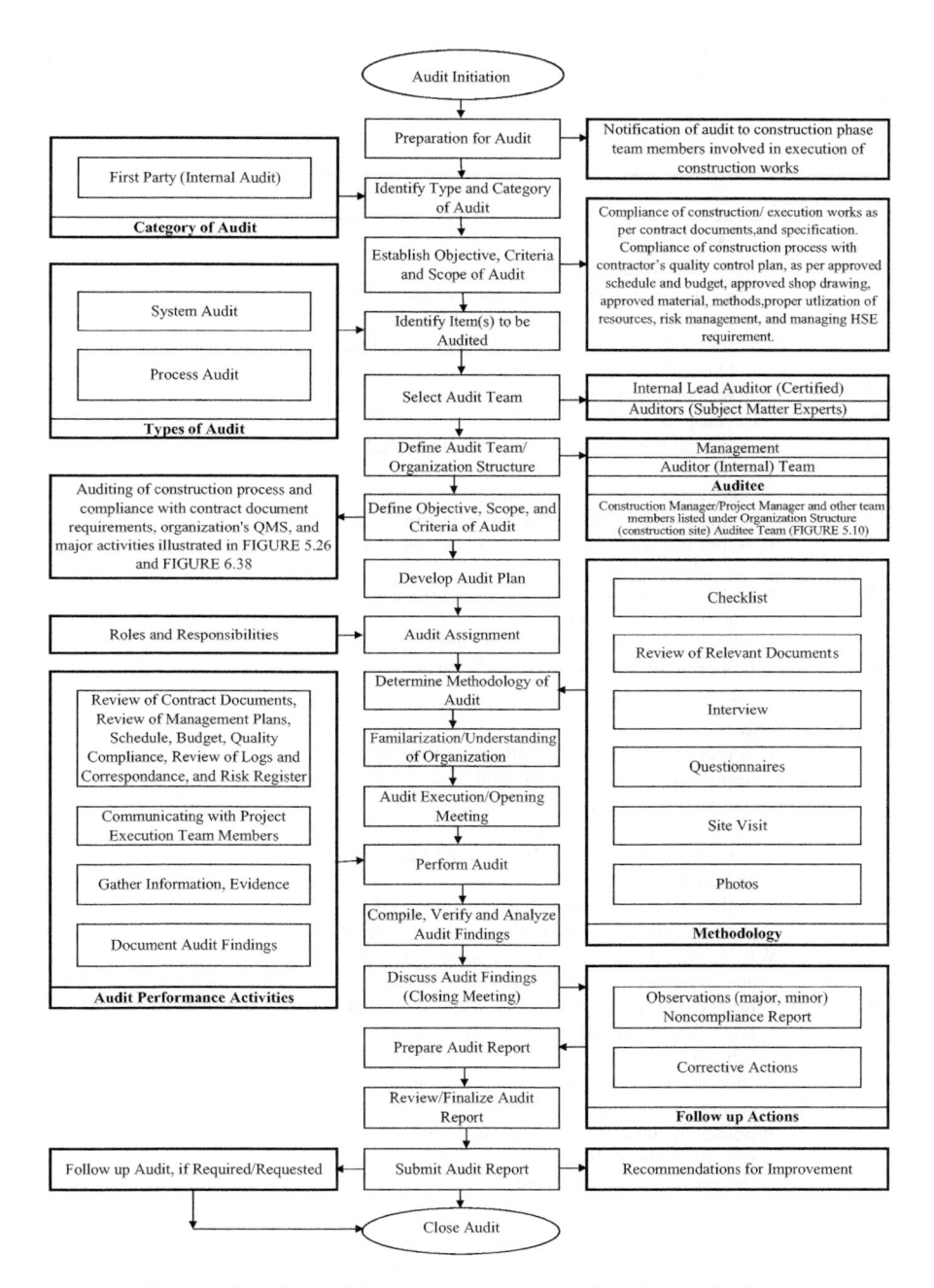

Figure 6.58 Typical quality auditing process for construction phase activities

Table 6.43 Audit methodology for the assessment of construction phase activities/processes after completion of the construction

Serial Number	Item To Be Assessed	Yes	No	Comments

A. Checklist

A-1 Integration Management
1. Whether contract documents reviewed and project execution requirements/construction phase deliverables properly established
2. Whether any discrepancies, conflicts identified between drawings and documents
3. Whether all the management plans are developed
4. Validate all the management plans approved before start of related activities
5. Whether necessary permits obtained on time
6. Whether project site handed over on time
7. Whether mobilization started and completed and schedule
8. Whether list of submittals as per contract documents prepared
9. Whether there was any delay in submittals
10. Whether change management plan was established and followed

A-2 Stakeholders
1. Whether project staff, and other workforce requirements were established
2. Whether responsibilities matrix of all the stakeholders is established
3. Whether reports are distributed to all the stakeholders
4. Whether core staff submitted and approved as per schedule and was available at the project
5. Whether all the subcontractors were selected and approved as scheduled
6. Whether regulatory authorities were identified and relevant submittals were made as per schedule
7. Is there nominated subcontractors listed in the contract documents
8. Whether the concerned stakeholders were informed about the changes in the baseline (scope, schedule, and cost)

A-3 Scope
1. Whether work is executed as per defined scope
2. Whether scope baseline is correctly identified and regularly updated for any changes
3. Is there any change in the contracted scope of work and specifications

Serial Number	Item To Be Assessed	Yes	No	Comments

A. Checklist

4	Whether preventive and corrective measure taken to control the scope			
5	Is there any Request for Information (RFI) raised			
6	Is there any Request for Variation submitted			
7	Is there any request for alternate material			
8	Is there any site work instruction (SWI)			
9	Whether subcontractors' scope of works properly defined and monitored			

A-4 Schedule

1	Whether the contractor's construction schedule is approved at first submission			
2	Whether the construction schedule is prepared by identifying the contract requirements			
3	Whether the construction schedule is prepared taking into consideration availability of all the resources			
4	Whether the sequence of activities is properly coordinated and in an organized and structured manner			
5	Whether the schedule is prepared taking into consideration the logical relationship among the activities for a smooth flow of work			
6	Whether the performance baseline for all the activities is identified			
	Whether engineering categories and respective activities were properly listed in the schedule			
7	Whether all the submittals are as per approved schedule and monitored			
8	Whether the project progress status is prepared to know the actual progress versus planned progress			
9	Whether project forecasting is done to predict the future progress of activities to be performed			
10	Whether the percentage of completion of each activity is based on the approved checklist			
11	Whether the schedule is updated based on approved changes			
12	Whether subcontractors' work progress is monitored			
13	Whether resources (equipment and manpower) are submitted with the schedule			
14	Whether a cost loaded schedule was prepared			
15	Whether work progress was properly monitored and the schedule updated			

(Continued)

Table 6.43 (Continued)

Serial Number	Item To Be Assessed	Yes	No	Comments
A. Checklist				
16	Whether any delayed activities were noticed			
17	Whether project progress photographs are used to monitor work progress			
A-5 Cost				
1	Whether there is an approved construction budget			
2	Whether there is an approved S-Curve			
3	Whether cash flow is properly maintained			
4	Whether the project cost is exceeding the planned budget			
5	Whether a cost loaded curve is prepared for the project			
6	Whether there is any approved variation cost to the contracted project value			
7	Whether the S-Curve is updated based on approved variation orders			
8	Whether a cost control process was in place			
9	Whether there was a need to change the cost baseline			
10	Whether there was cost overrun			
11	Whether all the changes were recorded and cash flow updated			
A-6 Quality				
1	Whether the contractor's quality control plan was approved as per schedule			
2	Whether builders' workshop drawings, coordinated drawings, composite drawings and shop drawings are approved before the start of construction activities			
3	Whether there are any comments on the shop drawing approval transmittals			
4	Whether all the materials/systems were approved as per schedule			
5	Whether there are any comments on the submitted material transmittals			
6	Whether incoming material is inspected upon receiving it on site			
7	Whether there is any rejection to the material received on site			
8	Whether there are any factory inspections/ visits carried out for the material/system			
9	Whether test certificates are received with the material supplied			
10	Whether laboratory tests are carried out for concrete works			

Serial Number	Item To Be Assessed	Yes	No	Comments

A. Checklist

11	Whether all the related shop drawings and material were approved before the start of the particular work			
12	Whether checklists are coordinated with all the trades before submission to the supervising engineer			
13	Whether there is any variation to the approved methods			
14	Whether inspection of works was continually carried out			
15	Whether quality meetings are conducted			
16	Whether there is rejection of works for not complying with regulatory requirements			
17	Whether the subcontractors' quality management system is properly monitored to comply with the contract documents			

A-7 Resources

1	Whether project staff/manpower are deployed as per project manpower plan			
2	Whether sufficient workforce was available to meet the project execution plan			
3	Whether workers were assigned daily activities			
4	Whether project equipment was available on site as per schedule			
5	Whether there was any conflict recorded among project team members			
6	Whether the procurement log is properly maintained			
7	Whether there was any delay in execution of works due to non-availability of resources			
8	Whether subcontractors have sufficient resources to execute the works as per schedule			
9	Whether work is assigned on a daily basis			
10	Whether an attendance sheet is maintained			
11	Whether subcontractor management is in place			

A-8 Communication

1	Whether the site administration matrix was followed properly			
2	Whether a communication plan was developed			
3	Whether a correspondence procedure between all the stakeholders was established			
4	Whether a communication method was established			
5	Whether submittals were distributed using the submittal transmittal form			

(*Continued*)

Table 6.43 (Continued)

Serial Number	Item To Be Assessed	Yes	No	Comments

A. Checklist

6	Whether logs are maintained			
7	Whether logs are updated on a regular basis			
8	Whether the project performance/status is regularly distributed to the stakeholders			
9	Whether a meeting invitation is sent to all the stakeholders			
10	Whether progress meetings and coordination meetings are held as scheduled on a regular basis			
11	Whether daily, weekly and monthly reports are properly prepared and circulated to the relevant stakeholders			
12	Whether project documents are kept in a safe and accessible location			

A-9 Risk

1	Whether a risk management plan is prepared and followed			
2	Whether high risk items are identified and a response plan prepared			
3	Whether the risk register is maintained and updated			
4	Whether project risks are documented and recorded			
5	Whether the occurrence of new risk is reported to update project progress and performance			
6	Whether financial risk is reported to the owner/client			

A-10 Contract

1	Whether a contract management process was established			
2	Whether project procurement requirements are identified			
3	Whether the bidding and tendering procedure is followed to procure the material, system, machinery and equipment			
4	Whether the subcontractors' list was finalized and subcontractors were selected as per organizational procurement strategy and contract requirements			
5	Whether the procurement follow-up system was in place and followed up			
6	Whether the cost of subcontracted items is as per estimated value			
7	Whether there is any delay in receipt of material, system and equipment that may affect the project execution schedule			

Serial Number	Item To Be Assessed	Yes	No	Comments

A. Checklist

8	Whether the subcontractors' workforce was available as per contract requirements			
9	Whether the subcontractors' performance was analyzed and documented			

A-11 HSE

1	Whether the HSE management plan is approved			
2	Whether safety meetings are conducted			
3	Whether safety drills were conducted			
4	Whether a temporary fire fighting system was operative			
5	Whether any major accidents were recorded			
6	Whether the accident reporting system was followed properly			
7	Whether there were safety violation reports issued			
8	Whether safety disciplinary notices were issued			
9	Whether periodic safety inspection is carried out			
10	Whether machinery, equipment and tools were regularly inspected for safe usage			
11	Whether the crane and hoist had valid third party certification			
12	Whether there is any report that sirens and bells were not in working conditions			
13	Whether all formwork and scaffoldings are checked prior to casting			
14	Whether any insurance was claimed for accidents			
15	Whether there was any waste management plan operative			
16	Whether safety violations and accidents are recorded and documented			
17	Whether there was an accident investigation carried out and recorded			
18	Whether safety training was conducted			
19	Whether roles and responsibilities of all individuals were clearly defined			

A-12 Finance

1	Whether there is a contractor's finance plan in place			
2	Whether progress payment is submitted on time			
3	Whether cash flow is as per approved S-Curve			
4	Whether the project progress payment was received on time as per contract			

(*Continued*)

Table 6.43 (Continued)

Serial Number	Item To Be Assessed	Yes	No	Comments
A. Checklist				
5	Whether there was any deduction from the approved progress payment			
6	Whether there is interim payment and advance payment for procurement of material			
7	Whether staff salaries are paid on a regular basis			
8	Whether subcontractors are paid as per contract on a regular basis			
9	Whether payment to the suppliers was as per the agreed-upon terms			
10	Whether finance records and logs are maintained properly			
11	Whether there was delay in receipt of progress payments			
12	Whether an outside financial agency was involved to finance the contractor's payment to its subcontractors, suppliers and salaries			
A-13 Claim				
1	Whether any claim was identified, recorded and informed to the client/owner			
2	Whether there is any dispute noticed related to contract documents			
3	Was there any claim that affected project progress			
4	Was there any claim from subcontractors			
5	Whether all the claims were resolved/settled			

B. Interview Questionnaires

Serial Number	Description	Response	Comments
1	Whether contract requirements (scope of works) were established after proper review of particular specifications, general specifications, BOQ, contract drawings and condition of contract		
2	Whether the project team members know the construction phase deliverables		
3	Whether all the related data of each activity are collected and recorded for project monitoring purposes		
4	Whether there is an organized and efficient means for measuring, collecting, verifying and quantifying data to monitor the progress of the project		

B. Interview Questionnaires

Serial Number	Description	Response	Comments
5	Whether project deliverables are as defined in the scope baseline or there are any changes in the scope baseline		
6	What tools and techniques are used to improve the schedule and recover delays		
7	What was the frequency of progress status reporting		
8	Whether Earned Value Management (EVM) methodology was used to measure and evaluate project performance		
9	Whether a cost control structure and policy was established		
10	Whether all the changes were promptly managed as they occurred		
11	Whether there are not-approved (rejection) checklists and how rework is managed		
12	Whether the quality management responsibilities matrix was being properly followed as per Table 6.40 and who was responsible for monitoring		
13	What method was followed for selection of materials and systems		
14	Whether there were any factory visits as per the contract		
15	Whether the reasons (causes) for comments on shop drawings, if any, were analyzed and corrective/preventive action taken		
16	Whether the reasons (causes) for rejection of material, if any, were analyzed and corrective/preventive action taken		
17	Whether the reasons (causes) for rejection/non-approval of executed works, if any, were analyzed and corrective/preventive action taken		
18	Whether there was any problem in getting approval of core staff and other project team members		
19	Whether the productivity rate and performance of workers was as estimated during the tendering stage		
20	Was there any overcharging for labor and material as was estimated		
21	Whether any training and development program was conducted for project team members		
22	Whether the project team acquisition process was properly followed		

(*Continued*)

Table 6.43 (Continued)

B. *Interview Questionnaires*

Serial Number	Description	Response	Comments
23	Whether equipment and machinery were available on site to execute the construction works without any interruption		
24	Whether the material management process is properly followed		
25	What methods were used to distribute correspondence to the stakeholders		
26	Whether all the documents and correspondences, transmittals, are saved electronically		
27	Whether identified risks were used to update the project schedule and cost		
28	Whether the risk management process cycle was properly followed		
29	Whether the risk probability level was followed to determine the risk impact on the project		
30	Whether there was a team/project team member assigned to identify project risk		
31	Whether different categories of risks under each of the construction process were regularly monitored		
32	Whether risk assessment was being done for each of the identified risks		
33	Whether all the risk factors were considered while selecting the subcontractors and material suppliers		
34	What was the procedure followed to select the subcontractors		
35	Was there any delay in getting approval of subcontractor(s)		
36	Whether there are any items expensive than the estimated cost		
37	Whether the supply chain management system was followed while selecting the supplier		
38	Whether subcontractor(s) were selected on low bid basis or quality based system		
39	Whether safety awareness/training programs were conducted		
40	Whether an emergency evacuation plan was prepared		
41	Whether evacuation routes were displayed		
42	Whether access and escape route signs were installed on site		
43	Whether protective clothing and equipment were used during the execution of work and in particular during concrete work		

B. Interview Questionnaires

Serial Number	Description	Response	Comments
44	Whether health surveillance was carried out		
45	Was there any site visit by the regulatory authority		
46	Whether emergency evacuation plans were displayed at various locations		
47	Whether hazardous areas were properly monitored		
48	Whether a safe working environment was observed during execution of works		
49	Whether a safety audit was performed		
50	What method was followed to resolve/settle claims		
51	Whether any claim is pending to be settled		
52	Whether the payments are received/paid for all the resolved/settled claims		

6.2.4.2 Auditing process for the construction phase (construction supervisor's activities)

Complex and major construction projects have many challenges such as delays, changes, disputes, accidents on site, etc. and therefore need efficient management of the construction phase of the facility/project to meet the intended use and owner's expectations. The owner/client may not have the necessary staff/resources in-house to manage the design, planning, monitoring and controlling, and construction of the construction project to achieve the desired results. Therefore, in such cases, the owner engages a professional firm which is trained and has expertise in the management of construction processes. It is a normal practice for the designer/consultant of the project to be contracted by the owner to supervise the construction process. The supervision firm is responsible for achieving the project quality goals and is also responsible for implementing the procedures specified in the contract documents. Table 6.44 lists the responsibilities the owner delegates to the engineer (consultant).

The project owner appoints an engineer's representative to supervise the project construction process. The engineer's representative is supported by a supervision team consisting of professionals with experience and expertise in the supervision and administration of similar construction projects. The engineer's representative is also called the resident engineer. Depending on the type and size of the project, the supervision team usually consists of the following personnel:

1　Resident engineer
2　Contract administrator/quantity surveyor
3　Planning/scheduling engineer

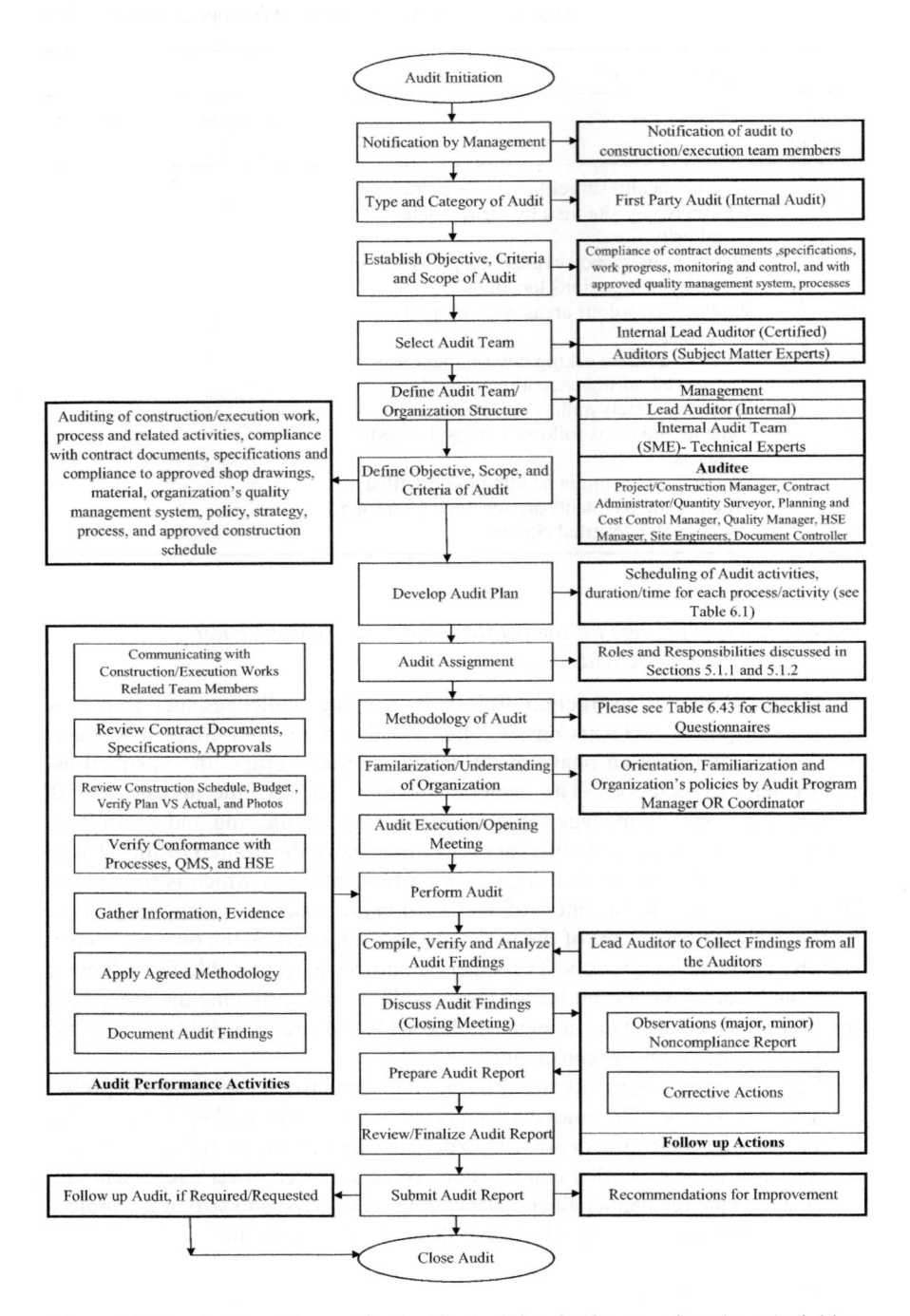

Figure 6.59 Logic Flow Diagram for Quality Auditing for Construction Phase Activities

Auditing processes for project life cycle phases 369

Table 6.44 Responsibilities of the construction supervisor (consultant)

Sr. No.	Description
1	Achieving the quality goal as specified
2	Review contract drawings and resolve technical discrepancies/errors in the contract documents
3	Review construction methodology
4	Approval of contractor's construction schedule
5	Regular inspection and checking of executed works
6	Review and approval of construction materials
7	Review and approval of shop drawings
8	Inspection of construction material
9	Monitoring and controlling construction expenditure
10	Monitoring and controlling construction time
11	Maintaining project record
12	Conduct progress and technical co-ordination meetings
13	Co-ordination of owner's requirements and comments related to site activities
14	Project related communication with contractor
15	Co-ordination with regulatory authorities
16	Processing of site work instruction for owner's action
17	Evaluation and processing of variation order/change order
18	Recommendation of contractor's payment to owner
19	Evaluating and making decisions related to unforeseen conditions
20	Monitor safety on site
21	Supervise testing, commissioning and handover of the project
22	Issue substantial completion certificate

4   Engineers from different trades such as architectural, structural, mechanical, HVAC, electrical, low voltage system, landscape, infrastructure
5   Inspectors from different trades
6   Interior designer
7   Document controller
8   Office secretary

The construction phase consists of various activities such as mobilization, execution of work, planning and scheduling, control and monitoring, management of resources/procurement, quality and inspection. Figure 6.39 discussed earlier illustrates the construction phase stages for the quality auditing of construction process activities.

Figure 6.60 illustrates the typical quality auditing process for the construction supervisor's activities related to the construction phase.

The main objectives of performing a quality audit of the construction supervisor's activities are to:

1   Evaluate and assess process compliance with the organizational strategic policies and procedure
2   Assess conformance to QMS and that processes meet established quality system requirements

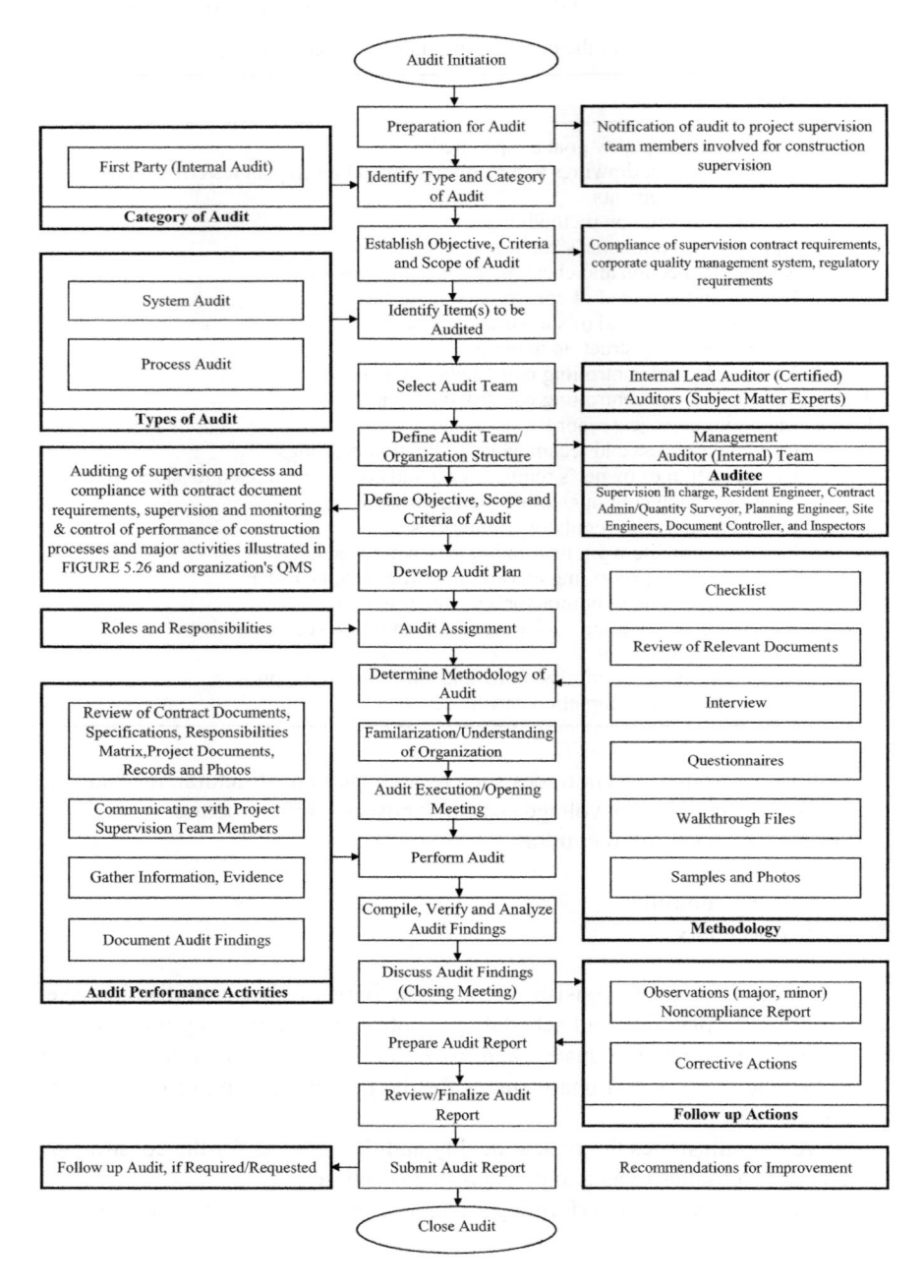

Figure 6.60 Typical quality auditing process for the construction supervisor's activities related to the construction phase

Auditing processes for project life cycle phases 371

3 Review the process used to perform the activities listed in the contract. The main responsibilities are listed in Table 6.44
4 Validate the continued supervision/inspection of construction process activities
5 Check project performance as per schedule is properly monitored and controlled
6 Check the project cost as per the budget/contracted value is properly monitored and monitored
7 Check if there are any changes in the contract documents and reasons for changes
8 Check that any variation order is properly evaluated and monitored
9 Check if there was any delay in getting approval of project staff
10 Check that quality management process was properly followed
11 Check if the resident engineer developed a checklist listing all the activities as per the supervision manual and the supervisor's responsibilities as per the contract. Table 6.45 illustrates an example checklist listing the items to be verified by the resident engineer to ensure availability of all the necessary documents and information to facilitate the smooth flow of supervision work

Table 6.45 Consultant's checklist for the smooth functioning of the project

Serial Number		Items To Be Checked/Verified
I		**Project details**
	I.1	Scope of work
	I.2	Project objectives
	I.3	Project deliverables
II		**Project Organization**
	II.1	Organization chart and roles and responsibilities of defined supervision staff
	II.2	Supervision staff deployment matching with project requirements
	II.3	Contractor's staff deployment plan approved as per contract requirements
	II.4	Responsibility matrix prepared and approved by the client and distributed among all project parties
	II-5	Project directory
III		**Mobilization**
	III.1	Site permit from authorities available
	III.2	Project plot boundaries are marked as per the permit
	III.3	Project commencement order issued
	III.4	Copy of permit issued to the contractor
	III.5	Temporary site offices drawings approved
	III.6	Temporary firefighting plan approved by respective authority
	III.7	Copies of contractor's performance bond, guarantees, insurance policies and licenses available on site
	III.8	Copies of consultant's performance bond, guarantees, insurance policies and licenses available on site
	III.9	Pre-construction meeting conducted and submittal and approval procedures discussed and agreed

(*Continued*)

Table 6.45 (Continued)

Serial Number	Items To Be Checked/Verified
IV	**Project Administration**
	III-1 Contract documents
	IV-1.1 Signed copy of contract between owner and contractor available on site
	IV-1.2 Copies of contract documents available on site
	IV-1.3 Contracted Bill of Quantity (BOQ) is available
	IV-1.4 All volumes of particular specifications available
	IV-1.5 Contracted drawings are available
	IV-1.6 Authority approved drawings, duly stamped, available
	IV-1.7 Addendum (if any) to the contract available
	IV-1.8 Replies to tender queries available
	IV-1.9 Copy of signed contract documents and drawings handed over to contractor and has acknowledged the same
	IV-1.10 Log for codes and standards available
	IV-2 Document management
	IV-2.1 Document control system is in place
	IV-2.2 Filing index is available
	IV-2.3 Material submittal log is available
	IV-2.4 Shop drawing submittal log is available
	IV-2.5 Logs for correspondence between various parties available
	IV-2.6 Log for checklist (request for inspection) available
	IV-2.7 Log for JSI (job site instruction) available
	IV-2.8 Log for SWI (site work instruction) available
	IV-2.9 Log for RFI (request for information) available
	IV-2.10 Log for VO (variation order) available
	IV-2.11 Log for NCR (non-conformance report) available
	IV-2.12 Material sample log and place identified
	IV-2.13 Log for equipment test certificate available
	IV-2.14 Log for visitors on site
	IV-2.15 Contractor's staff approval log in place
	IV-2.16 Subcontractor's approval log in place
	IV-2.17 Consultant's staff approval in place
	IV-2.18 Overtime request log available
V	**Communication**
	V-1 Communication matrix established and agreed by all the parties
	V-2 Distribution system for transmittals/submittals agreed
VI	**Project Monitoring and Control**
	VI-1 Daily report log in place
	VI-2 Weekly report log in place
	VI-3 Monthly report log in place
	VI-4 Progress meetings log in place
	VI-5 Minutes of meetings log in place
	VI-6 Progress payment log in place
	VI-7 Construction schedule log in place
VII	**Construction**
	VII-1 Quality control plan log in place
	VII-2 Safety management plan log in place
	VII-3 Risk management plan log in place
	VII-4 Method statement submittal log in place

Serial Number	Items To Be Checked/Verified	
	VII-5	Accident and fire report
	VII-6	Off-site inspection visits
	VII-7	Location of gathering point established
VIII	**General**	
	VIII-1	Correspondence between site and head office
	VIII-2	Staff related matters
	VIII-3	Copy of supervision manual available
	VIII-4	Emergency contact telephones and contact details displayed on site

12. Check all the project control documents are in place as illustrated in Table 6.46
13. Verify that project team members are available as per deployment schedule
14. Check if any conflict is reported among team members
15. Check that all the submittals are as per schedule
16. Check that all logs illustrated in Table 6.47 were maintained
17. Validate that all the submittals are properly reviewed and actioned
18. Check the RFI and scope change request are actioned/responded to within the stipulated time
19. Review that the key risk factors are analyzed and managed
20. Check that payment applications are properly reviewed and submitted within the stipulated time
21. Verify if there is any claim by the contractor
22. Check that regulatory requirements are considered while approving the executed/installed works

The following audit tools are used to perform a quality audit of the construction supervisor's activities related to the construction phase:

- Analytical tools
- Checklists
- Review of documents
- Interviews
- Questionnaires

The auditee team members are mainly from the following organizational groups:

- Supervisor in Charge
- Resident Engineer
- Contract Administrator/Quantity Surveyor
- Planning/Scheduling Engineer
- Engineers from different trades such as architectural, structural, mechanical, HVAC, electrical, low voltage system, landscape, infrastructure

Table 6.46 List of project control documents

Serial Number	Document Name
I	**Administrative**
I-1	Material Entry Permit
I-2	Material Removal Permit
I-3	Vehicular Entry Permit
I-4	Site Entry Permit
I-5	Visitor Entry Permit
I-6	Municipality Permit
I-7	Request for Overtime
I-8	Theft & Damage Report
I-9	Performance Bonds
I-10	Advance Payment Guarantee
I-11	Insurance
I-12	Accident Report
I-13	Sample Tag
II	**Contracts Related**
II-1	Notice to Proceed
II-2	Job Site Instruction
II-3	Site Works Instruction
II-4	Attachment to Site Works
II-5	Request for Staff Approval
II-6	Request for Subcontractor Approval
II-7	Variation Order
II-8	Attachment to Variation Order
II-9	Material Delivered on site
II-10	Baseline Change Request Form
II-11	Extension of Time
II-12	Suspension of Work
II-13	Attendees
II-14	Minutes of Meeting
II-15	Transmittal for Minutes of Meeting
II-16	Submittal Form
III	**Engineering Submittal**
III-1	Master Schedule
III-2	Cost Loaded Schedule
III-3	Material Approval
III-4	Specification Comparison Statement
III-5	Product Data
III-6	Product Sample
III-7	Workshop Drawings
III-8	Builders Drawings
III-8	Composite Drawings
III-10	Method Statement
III-11	Request for Information
III-12	Request for Modification
III-13	Variation Order (Proposal)
III-14	Request for Alternative or Substitution

Serial Number	Document Name
IV	**PCS Reporting Forms**
IV-1	Contractor's Submittal Status Log E-1
IV-2	Contractor's Procurement Log E-2
IV-3	Contractor's Shop Drawing Status Log
IV-4	Daily Progress Report
IV-5	Weekly Progress Report
IV-6	Look Ahead Schedule
IV-7	Monthly Progress Report
IV-8	Progress Photographs
IV-9	Daily Checklist Status
IV-10	Progress Payment Request
IV-11	Payment Certificate
IV-12	Submittal Schedule
IV-13	Schedule Update Report
V	**Management Plans**
V-1	Quality Control Plan
V-2	Safety Plan
V-3	Environmental Protection Plan
VI	**Quality Control Forms**
VI-1	Checklist (Request for Inspection)
VI-2	Checklist for Form Work
VI-3	Notice for Daily Concrete Casting
VI-4	Checklist for Concrete Casting
VI-5	Quality Control of Concreting
VI-6	Report on Concrete Casting
VI-7	Notice for Testing at Lab
VI-8	Concrete Quality Control Form
VI-9	Checklist for Mechanical Work
VI-10	Checklist for Electrical Work
VI-11	Checklist for Finishing
VI-12	Checklist for External Work
VI-13	Checklist for Landscape
VI-14	Remedial Note
VI-15	Non-Conformance/Non-compliance Report
VI-16	Material Inspection Report
VI-17	Safety Violation Notice
VI-18	Notice of Commencement of New Activity
VI-19	Removal of Rejected Material
VI-20	Testing and Commissioning
VII	**Closeout Forms**
VII-1	As Built Drawings
VII-2	Substantial Completion Certificate
VII-3	Handing-Over Certificate
VII-4	Taking Over Certificate
VII-5	Manuals
VII-6	Handing Over of Spare Parts
VII-7	Defect Liability Certificate

Source: Abdul Razzak Rumane (2013). *Quality Tools for Managing Construction Projects*. Reprinted with permission of Taylor & Francis Group

Table 6.47 List of logs

Section	Log
1	Incoming and Outgoing Letters (Owner, Contractor)
2	Staff Approval
3	Subcontractor Approval
4	Transmittal for Shop Drawing
5	Material Source Approval
6	Transmittal for Material
7	Transmittal for Sample
8	Request for Alternative
9	Request for Information
10	Request for Modification
11	Job Site Instruction
12	Variation Order
13	Request for Substitution
14	Non-Compliance Report
15	Remedial Note
16	Payment Request/Certificate
17	Daily Report
18	Checklist
19	Daily checklist status
20	Checklist for concrete casting
21	Material delivered on site
22	Material Inspection Report
23	Concrete Test Report
24	Weekly Progress Report
25	Monthly Progress Report
26	Photographs
27	Notice of Meeting
28	Minutes of Meeting
29	Safety Violation Report
30	Accident Report
31	Request for Proposal
32	Construction Schedule
33	Safety Management Plan
34	Quality Control Plan
35	Authority Approved drawings
36	Correspondence with Authorities
37	Issue Log
38	Inter Office Memo
39	Video Recording

- Inspectors from different trades
- Interior Designer
- Document Controller
- Office Secretary

Table 6.48 illustrates the audit methodology for the assessment of the construction supervisor's activities related to the construction phase and Figure 6.61

Table 6.48 Audit methodology for the assessment of the construction supervisor's activities related to the construction phase

Serial Number	Item To Be Assessed	Yes	No	Comments
A. Checklist				
A-1 Integration management				
1	Whether contract documents were reviewed and project execution requirements/ construction phase deliverables properly established			
2	Whether any discrepancies, conflicts were identified between drawings and documents			
3	Whether all the management plans were developed by the contractor prior to the start of related activities and approved			
4	Whether the necessary permits were obtained by contractor prior to the start of the construction works			
5	Whether the supervision license was valid and had regulatory approval (if applicable)			
6	Whether bonds and insurance were submitted by the contractor on time			
7	Whether the project site was handed over on time			
8	Whether mobilization was started and completed on schedule			
9	Whether the project control documents listed in Table 6.47 were in place			
10	Whether a change management plan was established and followed			
A-2 Stakeholders				
1	Whether a responsibilities matrix of all the stakeholders was established and circulated			
2	Whether project staff were submitted and approved as per schedule and were available at the project site			
3	Whether the regulatory authorities were identified and the contractor was informed of relevant submittals as per schedule			
4	Whether the concerned stakeholders were informed about the changes in the baseline (scope, schedule and cost)			
A-3 Scope				
1	Whether the scope baseline was correctly identified and regularly updated for any changes			
2	Whether there was any change in the contracted scope of work and specifications			
3	Whether there was any Request for Information (RFI) raised			

(*Continued*)

Table 6.48 (Continued)

Serial Number	Item To Be Assessed	Yes	No	Comments
A. Checklist				
4	Whether there was any request for variation submitted and the reason for variation			
5	Whether there was any request for alternate material submitted and approved			
6	Whether there was any site work instruction (SWI)			
7	Whether there was any Request for Proposal submitted by the contractor for additional/new works			
A-4 Schedule				
1	Whether the contractor's construction schedule was approved at first submission			
2	Whether the construction schedule was prepared by identifying the contract requirements			
3	Whether the construction schedule was prepared taking into consideration the availability of all the resources			
4	Whether the sequence of activities was properly coordinated and in an organized and structured manner			
5	Whether the performance baseline for all the activities was identified			
6	Whether the engineering categories and respective activities were properly listed in the schedule			
7	Whether all the submittals were submitted as per approved schedule			
8	Whether the project progress status was monitored and updated to reflect actual progress versus planned progress			
9	Whether project forecasting was done to predict future progress and delay of activities to be performed			
10	Whether the percentage of completion of each activity was based on the approved checklist			
11	Whether the schedule was updated based on approved changes			
12	Whether the subcontractors' work progress was included in the progress report			
13	Whether resources (equipment and manpower) were included in the schedule			
14	Whether a cost loaded schedule was prepared by the contractor and submitted for approval			
15	Whether work progress was properly monitored and the schedule updated			

Serial Number	Item To Be Assessed	Yes	No	Comments

A. Checklist

16	Whether any delayed activities were noticed and remedial action was proposed by the contractor			

A-5 Cost

1	Whether the project was completed as per approved construction budget			
2	Whether there was an approved S-Curve to monitor the expenses			
3	Whether cash flow was properly maintained			
4	Whether the project cost is exceeding the planned budget			
5	Whether a cost loaded curve was prepared for the project			
6	Whether there is any approved variation cost to the contracted project value			
7	Whether the S-Curve was regularly updated based on approved variation orders			
8	Whether the cost control process was in place			
9	Whether there was a need to change the cost baseline			
10	Whether there was cost overrun			
11	Whether all the changes were recorded and cash flow updated			

A-6 Quality

1	Whether the contractor's quality control plan was approved as per schedule			
2	Whether builders' workshop drawings, coordinated drawings, composite drawings and shop drawings are approved before the start of construction activities			
3	Whether there are any comments on the shop drawing approval transmittals			
4	Whether all the materials/systems were approved as per schedule			
5	Whether there are any comments on the submitted material transmittals			
6	Whether all incoming material was inspected upon receipt on site			
7	Whether there is any rejection to the material received on site			
8	Whether there are any factory inspections/ visits carried out for the material/system and what was the outcome			
9	Whether test certificates were received with the material supplied			
10	Whether laboratory tests were carried out for concrete works			

(Continued)

Table 6.48 (Continued)

Serial Number	Item To Be Assessed	Yes	No	Comments
A. Checklist				
11	Whether checklists are coordinated with all the trades before submission to the supervising engineer			
12	Whether there was any variation to the approved methods			
13	Whether quality meetings were conducted			
14	Whether there was rejection of works for not complying with regulatory requirements			
15	Whether the subcontractor's quality management system was properly monitored to comply with contract documents			
16	Whether a method statement was submitted by the contractor for approval			
17	Whether test reports were properly documented and records were maintained			
A-7 Resources				
1	Whether project staff/manpower requirements were established as per project contract			
2	Whether project staff/manpower were deployed as per project manpower plan			
3	Whether the contractor had sufficient workforce available to meet the project execution plan			
4	Whether the contractor's project equipment were available on site as per schedule			
5	Whether all the equipment had a valid license and approval from the appropriate agencies			
6	Whether there was any conflict recorded among project team members			
7	Whether there was any delay in execution of works due to non-availability of resources			
8	Whether there was any excess material available than what was required			
9	Whether subcontractors had sufficient resources to execute the works as per schedule			
10	Whether the subcontractors' workforce was available as per contract requirements			
11	Whether the requisite daily labor/manpower was available on site and daily records were maintained			
12	Whether a staff attendance sheet is maintained			
13	Whether the owner-supplied material (if applicable) was available on site as per schedule			

Serial Number	Item To Be Assessed	Yes	No	Comments

A. Checklist

14	Whether there was any delay in receipt of material, system and equipment that affected the project execution schedule			

A-8 Communication

1	Whether the site administration matrix was followed properly			
2	Whether a communication plan was developed			
3	Whether the correspondence procedure between all the stakeholders was established			
4	Whether a communication method was established			
5	Whether submittals were distributed using the submittal transmittal form			
6	Whether logs were maintained			
7	Whether logs were updated on regular basis			
8	Whether project performance/status was regularly distributed to the stakeholders			
9	Whether a meeting invitation was sent to all the stakeholders			
10	Whether progress meetings and coordination meetings were held as scheduled on a regular basis			
11	Whether daily, weekly and monthly reports were submitted by the contractor and circulated to the relevant stakeholders			
12	Whether checklists were properly maintained			
13	Whether the daily checklist status log was maintained			
14	Whether project documents were kept in a safe and accessible location			

A-9 Risk

1	Whether a risk management plan was prepared and followed			
2	Whether high risk items were identified and a response plan prepared			
3	Whether the risk register was maintained and updated			
4	Whether project risks were documented and recorded			
5	Whether occurrence of new risks was reported to update the project progress and performance			
6	Whether financial risk was reported to the owner/client			

(Continued)

Table 6.48 (Continued)

Serial Number	Item To Be Assessed	Yes	No	Comments

A. Checklist

A-10 Contract
1. Whether a contract management process was established
2. Whether supervision requirements were properly identified
3. Whether subcontractors were approved and selected as per specifications and contract requirements
4. Whether there was major changes or additions to the signed contract

A-11 HSE
1. Whether the HSE management plan was approved
2. Whether safety meetings were conducted
3. Whether safety drills were conducted
4. Whether a temporary fire fighting system was operative
5. Whether any major accidents were recorded
6. Whether the accident reporting system was followed properly
7. Whether there were safety violation reports issued
8. Whether safety disciplinary notices were issued
9. Whether periodic safety inspection was carried out
10. Whether machinery, equipment and tools were regularly inspected for safe usage
11. Whether the crane and hoist had valid third party certification
12. Whether there was any report that sirens and bells were not in working conditions
13. Whether all formwork and scaffoldings were checked prior to casting
14. Whether any insurance was claimed for accidents
15. Whether there was any waste management plan operative
16. Whether safety violations and accidents were recorded and documented
17. Whether there were accident investigations carried out and recorded
18. Whether safety training was conducted
19. Whether the roles and responsibilities of all individuals towards safety were clearly defined
20. Whether there was any non-compliance to HSE requirements

Serial Number	Item To Be Assessed	Yes	No	Comments

A. Checklist

A-12 Finance

1	Whether progress payment was being submitted by the contractor on time			
2	Whether cash flow was as per the approved S-Curve			
3	Whether the project progress payment was received on time as per contract			
4	Whether there was any deduction from the approved progress payment			
5	Whether there was interim payment and advance payment for the procurement of materials			
6	Whether the contractor's staff/laborers' salaries were paid on a regular basis			
7	Whether subcontractors were paid as per contract on a regular basis			
8	Whether there was any complaint for payment to the subcontractor not paid as per the agreed-upon terms			
9	Whether there was any complaint for payment to the suppliers not paid as per the agreed-upon terms			
10	Whether finance records and logs were maintained properly			
11	Whether there was delay in receipt of progress payments by the contractor			
12	Whether supervision payment was on a regular basis			

A-13 Claim

1	Whether any claim was identified, recorded and informed to the client/owner			
2	Whether there was any dispute noticed related to the contract documents			
3	Whether there was any claim that affected the project progress			
4	Whether there was any claim from subcontractors			
5	Whether all the claims were resolved/settled			

B. Interview Questionnaires

Serial Number	Description	Response	Comments
1	Whether supervision responsibilities were clearly defined in the contract		
2	Whether the project team members know their responsibilities		

(*Continued*)

Table 6.48 (Continued)

B. Interview Questionnaires

Serial Number	Description	Response	Comments
3	Whether all the related data of each activities were collected and recorded for project monitoring purposes		
4	Was there any change request by the project owner		
5	Whether all the changes were promptly managed as they occur		
6	What was the frequency of progress status reporting		
7	Whether Earned Value Management (EVM) methodology was used to measure and evaluate project performance		
8	Whether a cost control structure and policy was established		
8	Whether there are not-approved (rejection) checklists and how rework is managed		
9	Whether there were any factory visits as per the contract		
10	Whether there was any problem in getting the approval of project staff and other project team members		
11	Whether there wasany unauthorized material stored on site		
12	Whether any training and development program was conducted for project team members		
13	Whether the project team acquisition process was properly followed		
14	Whether the material management process was properly followed by the contractor		
15	What methods were used to distribute correspondence to the stakeholders		
16	Whether all the documents and correspondences, transmittals, were saved electronically		
17	Whether identified risks were used to update the project schedule and cost		
18	Whether the risk management process cycle was properly followed		
19	Whether the risk probability level was followed to determine the risk impact on the project		
20	Whether there was a team/project team member assigned to identify project risk		
21	Whether different categories of risks under each of the construction processes were regularly monitored		

B. Interview Questionnaires

Serial Number	Description	Response	Comments
22	Whether risk assessment was being done for each of the identified risks		
23	Whether all the risk factors were considered while selecting the subcontractor and material supplier		
24	Whether there was any delay in getting approval of the subcontractor(s) from the project owner		
25	Whether safety awareness/training programs were conducted		
25	Whether an emergency evacuation plan was prepared		
26	Whether evacuation routes were displayed		
27	Whether access and escape route signs were installed on site		
28	Whether protective clothing and equipment were used during the execution of work and in particular during concrete work		
29	Whether health surveillance was carried out		
30	Whether there was any site visit by the regulatory authority		
31	Whether emergency evacuation plans were displayed at various locations		
32	Whether hazardous areas were properly monitored		
33	What method was followed to resolve/settle claims		
34	Whether any claim is pending to be settled		
35	Whether the payments were received/paid for all the resolved/settled claims		

illustrates a logic flow diagram for the quality auditing process for the construction supervisor's activities related to the construction phase.

6.2.5 Auditing process for the testing, commissioning and handover phase

Testing, commissioning and handover is the last phase of the construction project life cycle. This phase involves the testing of electro-mechanical systems, commissioning of the project, obtaining authorities' approval, training of user's personnel and handing over of technical manuals, documents and as-built drawings to the owner/owner's representative. During this period the project is transferred/handed over to the owner/end user for their use and the substantial completion certificate is issued to the contractor. Figure 5.28 discussed in Section 5.2.8 illustrate the

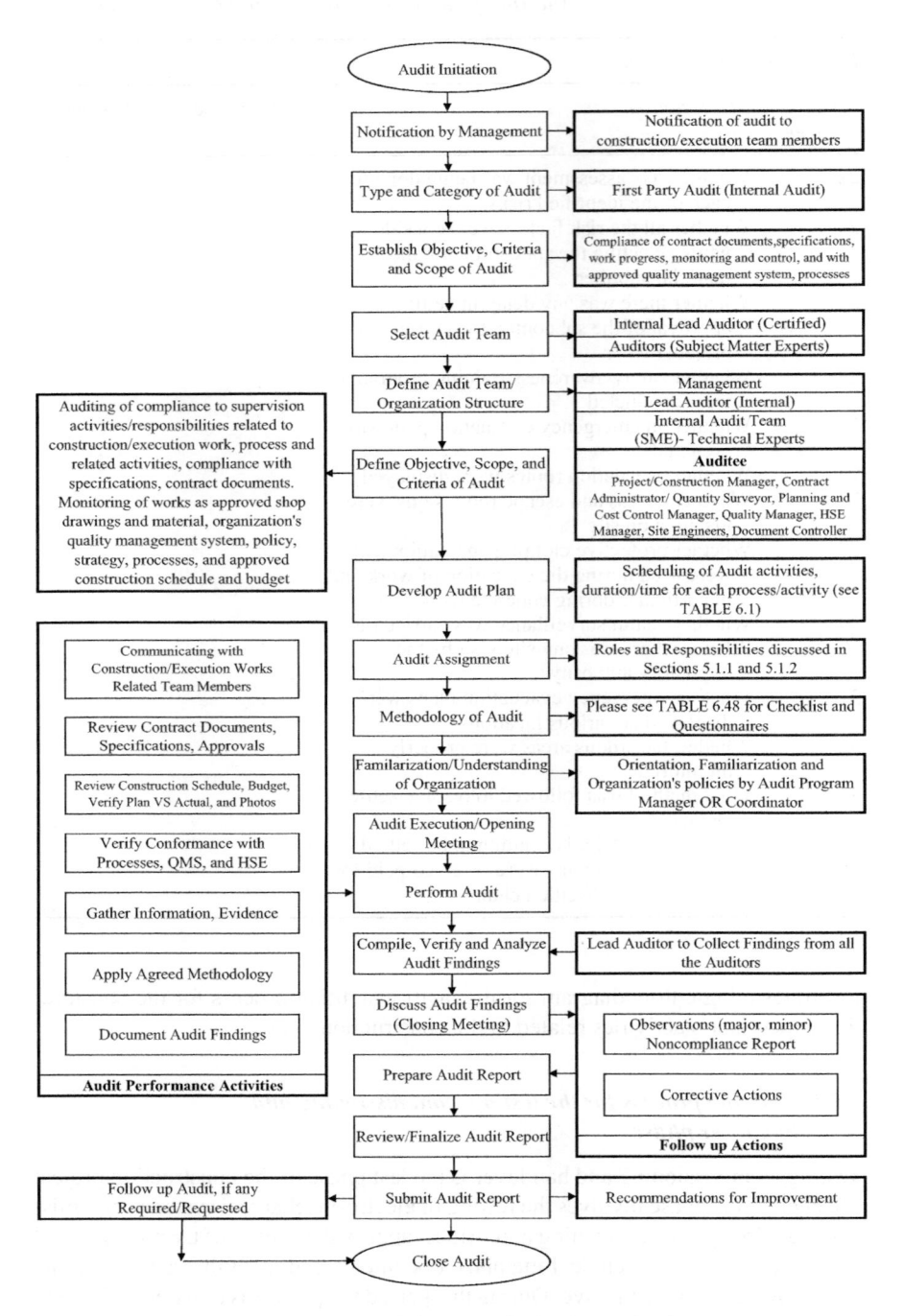

Figure 6.61 Logic flow diagram for the quality auditing process for the construction supervisor's activities related to the construction phase

major activities relating to the testing, commissioning and handover phase developed based on the project management process groups methodology.

6.2.5.1 Contractor's work

Figure 6.62 illustrates the typical quality auditing process for the contractor's activities related to the testing, commissioning and handover phase.

The main objectives of performing quality audit activities related to the testing, commissioning and handover phase are to:

1. Ascertain that contract documents are reviewed and testing, commission and handover activities are properly identified
2. Assess conformance to QMS and that processes meet project specific quality system requirements
3. Validate that the stakeholders' requirements were properly documented
4. Validate that an inspection and test plan was developed as illustrated in Figure 6.63
5. Verify that the testing and commissioning schedule was prepared and approved
6. Verify that the handover schedule (move-in plan) was prepared and approved
7. Verify that authorities' approval requirements were identified and complied
8. Check that project quality requirements for testing, commissioning and handover phase were properly followed
9. Check that punch list/snag list works are completed
10. Check that test results and other related documents were handed over to the stakeholders as per contract requirements
11. Check that a risk management plan was developed
12. Verify that the project closeout documents as listed in Table 6.49 were prepared and approved.
13. Verify that project closeout documents was as per contract documents
14. Ensure that the selection of subcontractors (testing, balancing and adjusting) was as per corporate policy taking into consideration contract documents
15. Check all the project completion was as per approved schedule
16. Check that the project handover/acceptance certificate was issued by the owner/construction supervisor
17. Check that payments were made to all the subcontractors
18. Check that payments were made for all the purchases
19. Check that a HSE management plan was prepared and followed
20. Check all the claims are settled
21. Check if there is any "Lessons Learned" list prepared
22. Check that project records are archived as per the organization's policy

The following audit tools are used to perform a quality audit of the testing, commissioning and handover phase activities:

- Analytical tools
- Checklists

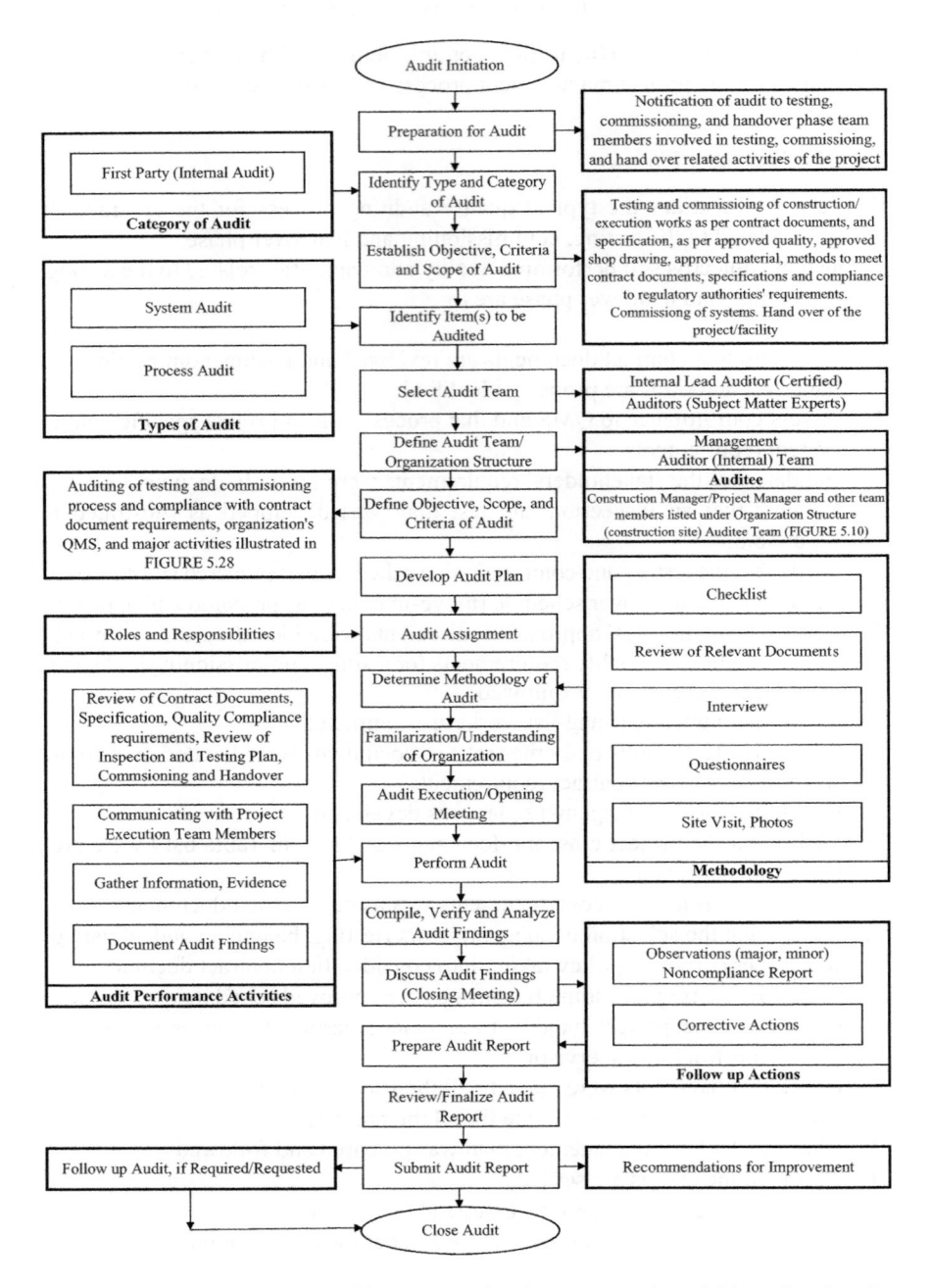

Figure 6.62 Typical quality auditing process for the contractor's activities related to the testing, commissioning and handover phase

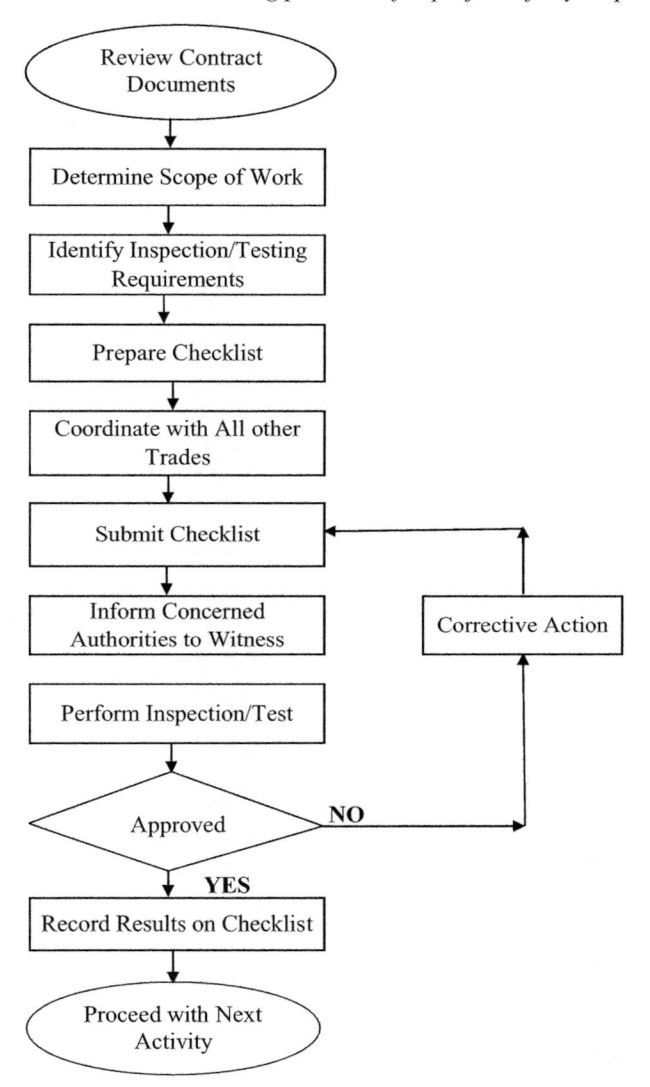

Figure 6.63 Development of the inspection and test plan

- Review of documents
- Interviews
- Questionnaires

The auditee team members are mainly from the following organizational groups:

- Construction/Project Manager
- Planning, Cost Control Manager

Table 6.49 Contractor's responsibilities to manage construction quality

Description	Testing and Commissioning	As-Built Drawings	Operation and Maintenance Manuals	Guarantees	Warranties	Government Authorities Approvals	Record Documents	Test Certificates	Samples	Spare Parts	Punch Lists	Final Cleaning	Training	Taking Over Certificate	Remarks

ARCHITECTURAL WORKS

CIVIL WORKS

MECHANICAL WORKS

HVAC WORKS

ELECTRICAL WORKS (LIGHT & POWER)

SAMPLE FORM

ELECTRICAL WORKS (LOW VOLTAGE)

FINISHES

EXTERNAL WORKS

- Contract Administrator/Quantity Surveyor
- Quality Engineer
- HSE Engineer
- Site Engineers (All Trades)
- Document Controller

Table 6.50 illustrates the audit methodology for the assessment of the contractor's activities related to the testing, commissioning and handover phase and Figure 6.64 illustrates a logic flow diagram for the quality auditing process for the contractor's activities related to the testing, commissioning and handover phase.

Table 6.50 Audit methodology for the assessment of the contractor's activities related to the testing, commissioning and handover phase

Serial Number	Item To Be Assessed	Yes	No	Comments
A. Checklist				
1	Whether the scope of works was established taking into consideration contract documents and specifications			
2	Whether testing and startup requirements were identified as per contract documents			
3	Whether the testing, commissioning and handover plan was developed taking into consideration all the executed/installed works/systems			
4	Whether all stakeholders involved during the testing, commissioning and handover phase were properly identified			
5	Whether a testing schedule was developed and approved			
6	Whether a commissioning schedule was developed and approved			
7	Whether quality procedures for testing, commissioning and handover were developed and approved			
8	Whether a punch list/snag list was prepared			
9	Whether the necessary testing equipment was available as per schedule			
10	Whether a startup risk management plan was established			
11	Whether the project closeout documents as listed in Table 6.49 were submitted and approved			
12	Whether a project handing-over schedule was developed and was approved			
13	Whether a "Lessons Learned" list was prepared			
14	Whether payments were made to subcontractors			
15	Whether the claims were resolved and settled			

Auditing processes for project life cycle phases 393

B. *Interview Questionnaires*

Serial Number	Description	Response	Comments
1	Whether testing and commissioning requirements were identified		
2	Whether authorities' approval was obtained as per schedule		
3	Whether handing-over certificate (substantial completion certificate) was issued by the owner/construction supervisor		
4	Whether project team members were terminated/demobilized or assigned to new project		
5	Whether there were any incomplete works that are not executed/installed		
6	Whether payments towards all the purchases were made		
7	Whether any claim is not resolved/settled		

6.2.5.2 Construction supervisor's activities

Figure 6.65 illustrate a typical quality auditing process for the construction supervisor's activities related to the testing, commissioning and handover phase.

The main objectives of performing quality audit activities related to the testing, commissioning and handover phase are to:

1 Check that contract documents were reviewed and the scope of work for testing, commissioning and handover are properly identified
2 Assess conformance to QMS and that processes meet established quality system requirements
3 Validate that the team members were informed of typical responsibilities as listed in Table 6.51
4 Verify that all the stakeholders and team members were identified and a responsibility matrix was established
5 Ascertain that all requirements of the regulatory authorities were identified to ensure the smooth handing-over of the project/facility
6 Verify that the testing and commissioning schedule was approved
7 Check that the punch list and snag list were prepared and actioned/responded to
8 Verify that closeout documents as per Table 6.49 were submitted by the contractor and were reviewed and actioned
9 Check that all the relevant documents were distributed to the concerned stakeholders
10 Check a risk management plan was developed
11 Check that the project completion was as per approved schedule

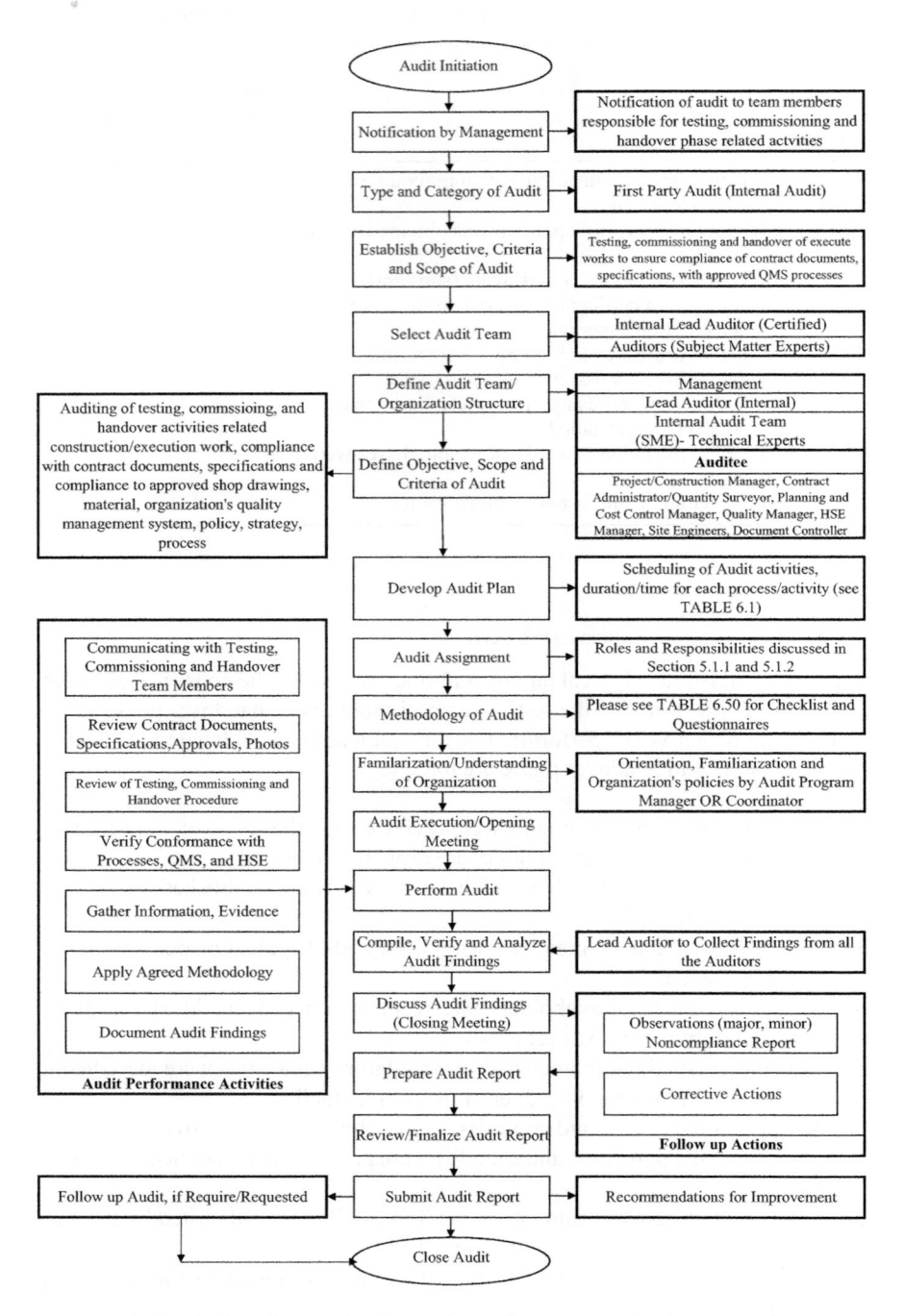

Figure 6.64 Logic flow diagram for the quality auditing process for the contractor's activities related to the testing, commissioning and handover phase

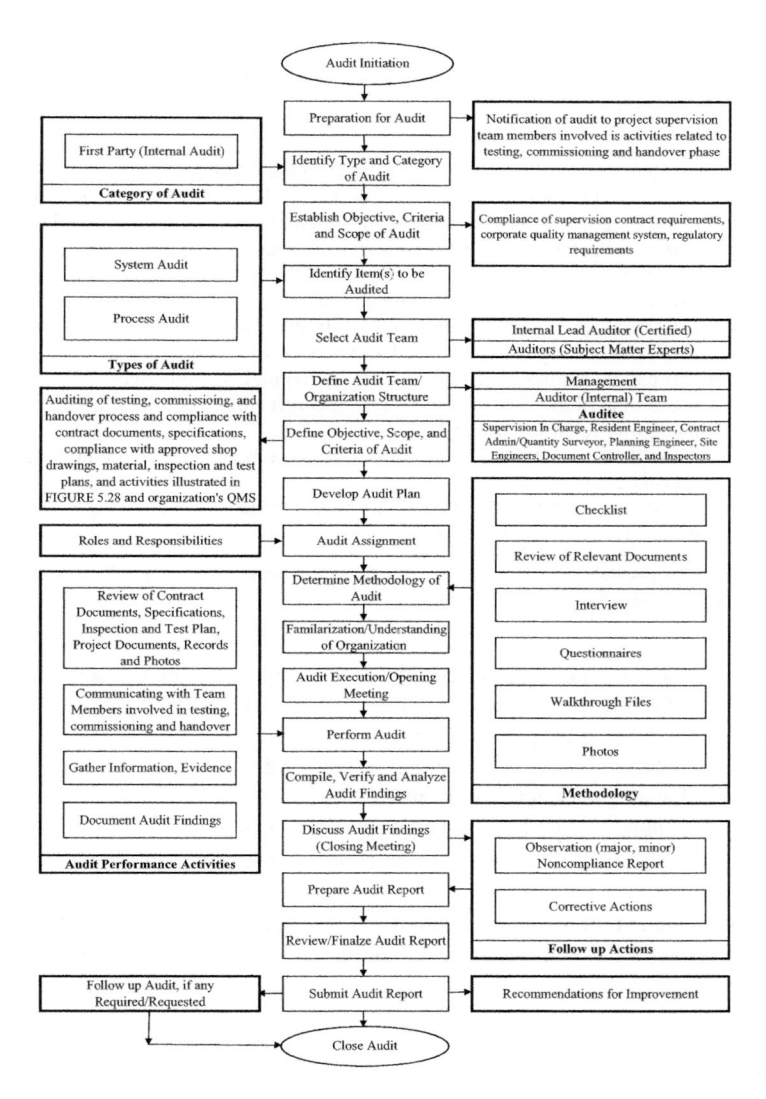

Figure 6.65 Typical quality auditing process for the construction supervisor's activities related to the testing, commissioning and handover phase

Table 6.51 Typical responsibilities of the construction supervisor during the testing, commissioning and handover phase

Serial Number	Responsibilities
1	Ensure that occupancy permit from respective authorities is obtained
2	Ensure that all the systems are functioning and operative
3	Ensure that Job Site Instruction (JSI) and Non-Conformance Report (NCR) are closed
4	Ensure that site is cleaned and all the temporary facilities and utilities are removed

(*Continued*)

Table 6.51 (Continued)

Serial Number	Responsibilities
5	Ensure that master keys are handed over to the owner/end user
6	Ensure that guarantees, warrantees, bonds are handed over to the client
7	Ensure that operation and maintenance manuals are handed over to the client
8	Ensure that test reports, test certificates, inspection reports are handed over to the client
9	Ensure as-built drawings are handed over to the client/end user
10	Ensure that spare parts are handed over to the client
11	Ensure that a snag list is prepared and handed over to the client
12	Ensure that training for client/end user personnel is completed
13	Ensure that all the dues of suppliers, subcontractors, contractor are paid
14	Ensure that retention money is released
15	Ensure that a substantial completion certificate issued and a maintenance period commissioned
16	Ensure that the supervision completion certificate from the owner is obtained
17	Lessons learned documented

Source: Abdul Razzak Rumane (2013). *Quality Tools for Managing Construction Projects*. Reprinted with permission of Taylor & Francis Group

12 Check that the handing-over certificate was signed by the concerned parties. Figure 6.66 illustrates a typical handing-over certificate
13 Check the HSE management plan was followed
14 Check that all the payments were approved and paid
15 Check that all the claims were settled
16 Check if there is any "Lesson Learned" list prepared
17 Check that project records are archived as per the organization's policy

The following audit tools are used to perform a quality audit of the construction supervisor's activities related to the testing, commissioning and handover phase:

- Analytical tools
- Checklists
- Review of documents
- Interviews
- Questionnaires

The auditee team members are mainly from the following organizational groups:

- Supervisor in Charge
- Resident Engineer
- Contract Administrator/Quantity Surveyor
- Planning/Scheduling Engineer
- Engineers from different trades such as architectural, structural, mechanical, HVAC, electrical, low voltage system, landscape, infrastructure
- Inspectors from different trades

Auditing processes for project life cycle phases 397

```
                        Project Name
                        Project Name
                    HANDING OVER CERTIFICATE
CONTRACTOR :  _____        CERTIFICATE No. : [____]
SUBCONTRACTOR: _____                   DATE [____]

SPECIFICATION NO : _____   DIVISION : _____   SECTION : _____
DRAWING No.                     BOQ REF:  _____
AREA :   [ ]  Building Works    [ ]  Electrical Works   [ ]  Mechanical Works
         [ ]  HVAC Works        [ ]  Finishes Works     [ ]

Description of Work/System:          SAMPLE FORM
_____
_____
_____

    The work/system mentioned above is completed by the contractor as specified and has been inspected and tested as per contract
documents. The work/system is fully functional to the satisfaction of owner/end user. The contractor hand over the said work/system to
the owner/end user as on ---------. The guarantee/warranty of work/system shall start as of --------- and shall be valid for a
period of ---------- years(duration) from the date of issuance of the substantial completion certificate. The contractor shall be liable
contractually till the end of warranty/guarantee period.


SIGNED BY:

OWNER/END USER: _____      CONTRACTOR: _____

CONSULTANT: _____          SUBCONTRACTOR: _____
```

Figure 6.66 Handing over certificate

Source: Abdul Razzak Rumane (2010). *Quality Management in Construction Projects.* Reprinted with permission of Taylor & Francis Group

- Interior Designer
- Document Controller
- Office Secretary

Table 6.52 illustrates the audit methodology for the assessment of the construction supervisor's activities related to the testing, commissioning and handover phase and Figure 6.67 illustrates a logic flow diagram for the quality auditing process for the construction supervisor's activities related to the testing, commissioning and handover phase.

Table 6.52 Audit methodology for the assessment of the construction supervisor's activities related to the testing, commissioning and handover phase

Serial Number	Item To Be Assessed	Yes	No	Comments
A. Checklist				
1	Whether the scope of works was established taking into consideration contract documents and specifications			
2	Whether testing and startup requirements were identified as per contract documents			
3	Whether a testing, commissioning and handover plan was developed taking into consideration all the executed/installed works/systems			
4	Whether all stakeholders involved during the testing, commissioning and handover phase were properly identified			
5	Whether contract closeout documents were identified as per contract documents and specifications			
6	Whether the testing and commissioning works/systems were identified and listed and distributed to all the team members			
7	Whether a testing schedule was developed and approved			
8	Whether a commissioning schedule was developed and approved			
9	Whether a punch list/snag list was prepared			
10	Whether a startup risk management plan was established			
11	Whether project closeout documents as listed in Table 6.49 were submitted by the contractor			
12	Whether the project handing-over schedule was developed and monitored			
13	Whether a "Lesson Learned" list was prepared			
14	Whether payments were made to subcontractors			
15	Whether the claims were resolved and settled			

B. *Interview Questionnaires*

Serial Number	Description	Response	Comments
1	Whether testing and commissioning requirements were identified		
2	Whether project team members were notified about their responsibilities as listed in Table 6.51		
3	Whether authorities' approval was obtained as per schedule		
4	Whether a handing-over certificate (substantial completion certificate) was issued		
5	Whether the project/facility was accepted/handed over to the client/end user		
6	Whether project team members were terminated/demobilized or assigned to new project		
7	Whether there is any incomplete works that are not executed/installed		
8	Whether the contractor settled all the payments		
9	Whether supervision fees were paid by the owner/client		
10	Whether any claim was not resolved/settled		

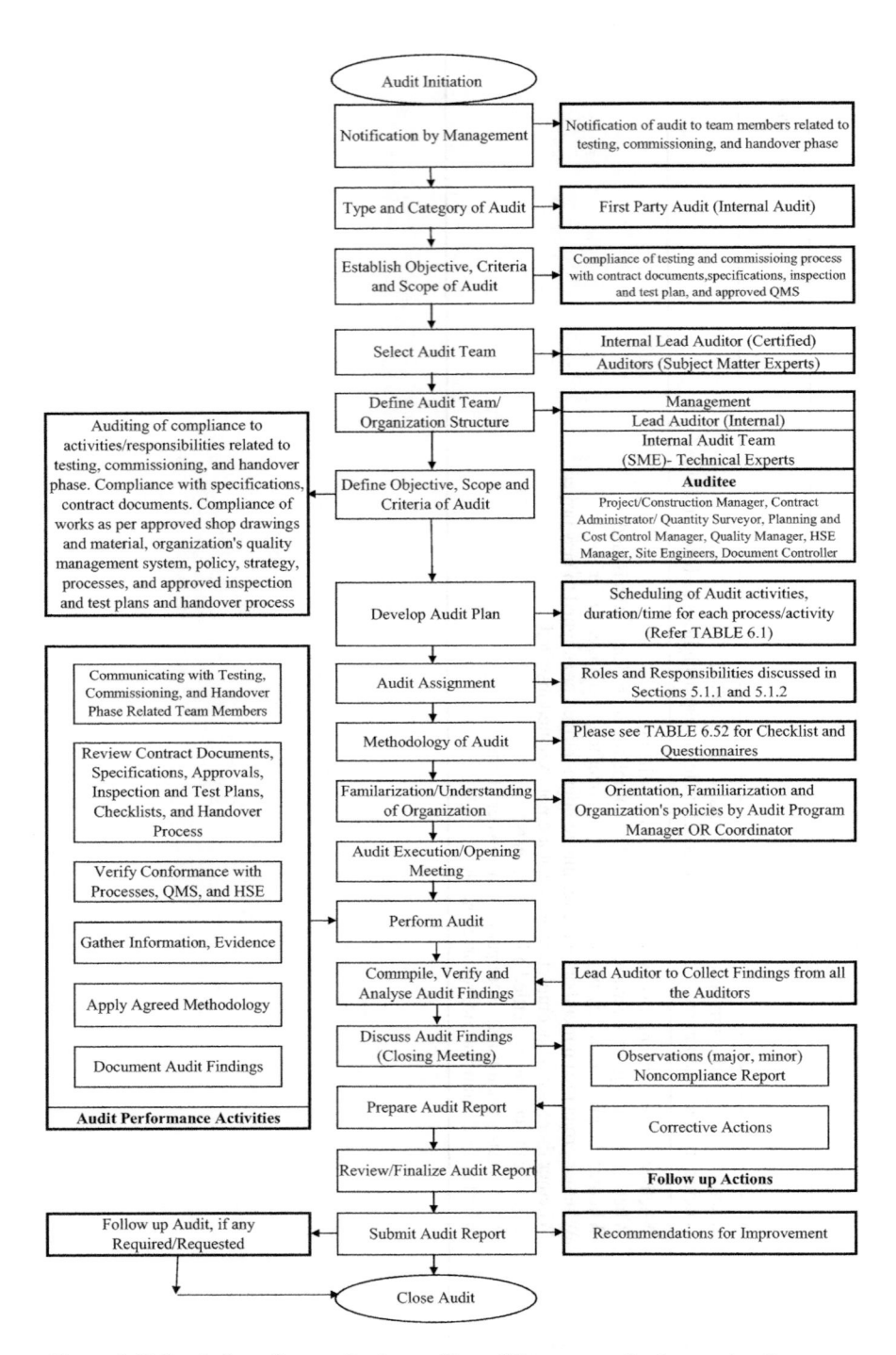

Figure 6.67 Logic flow diagram for the quality auditing process for the construction supervisor's activities related to the testing, commissioning and handover phase

Auditing processes for project life cycle phases 401

6.3 Auditing process for the construction project (Design-Build type of project delivery system) phases

In the Design-Build type of contract, the owner contracts a firm (contractor) solely responsible for designing and building the project. In this type of contracting system the contractor is appointed based on an outline design or design brief to understand the owner's intent for the project. The owner has to clearly and explicitly define their needs, performance specifications/requirements and comprehensive scope of works (also called statement of requirements) prior to the signing of the contract. It is a must that the project definition is understood by the contractor to avoid any conflicts in future as the contractor is responsible for the detailed design and construction of the project. The concept design should comply with project goals and objectives that fully meet the owner's requirements. The bidding and tendering documents to select the Design-Build contractor should clearly define the statement of requirements, instructions and evaluation criteria. A Design-Build type of contract is a contractually defined project and often used to shorten the time required to complete a project. Since the contract with the Design-Build firm is awarded before starting any design or construction, a cost plus contract, guaranteed maximum price or reimbursable arrangement is normally used instead of a lump-sum, fixed-cost arrangement. This type of contract requires the extensive involvement of the owner during the entire life cycle of project. The owner/client has to be involved in taking decisions during the selection of design alternatives and the monitoring of the schedule and cost during construction and therefore the owner has to maintain/hire a team of qualified professionals or hire a project management expertise personnel/firm to perform these activities. The degree of overlap between design and construction activities can vary substantially depending on the specific project requirements. Design-Build contracts are used for relatively straightforward work, where no significant risk or change is anticipated and when the owner is able to specify precisely what is required.

In the case of the Design-Bid-Build type of project delivery system, the contractor is contracted after the completion of design and construction documents based on successful bidding whereas in the Design-Build type of deliverable system, the contractor is contracted right from the early stages of the construction project and is responsible for the design development of the project as well as construction. The Design-Build contractor is responsible for coordinating design development activities.

Figure 6.68a illustrates a typical logic flow diagram for the Design-Build type of project delivery system and Figure 6.68b illustrates the quality auditing categories in the construction project life cycle (Design-Build type of project delivery system) phases.

The quality auditing categories applied to the Design-Bid-Build type of project delivery system are discussed earlier, as illustrated in Figure 6.5a.

The quality auditing categories applied in this book for the Design-Build type of project delivery system are as illustrated in Figure 6.68b and are mainly to understand the quality auditing processes using these types of categories for different phases/stages.

In general the quality auditing process and methodology discussed earlier for The Design-Bid-Build type of project delivery system can be used for the Design-Build type of delivery system. However, the aims, objectives and audit

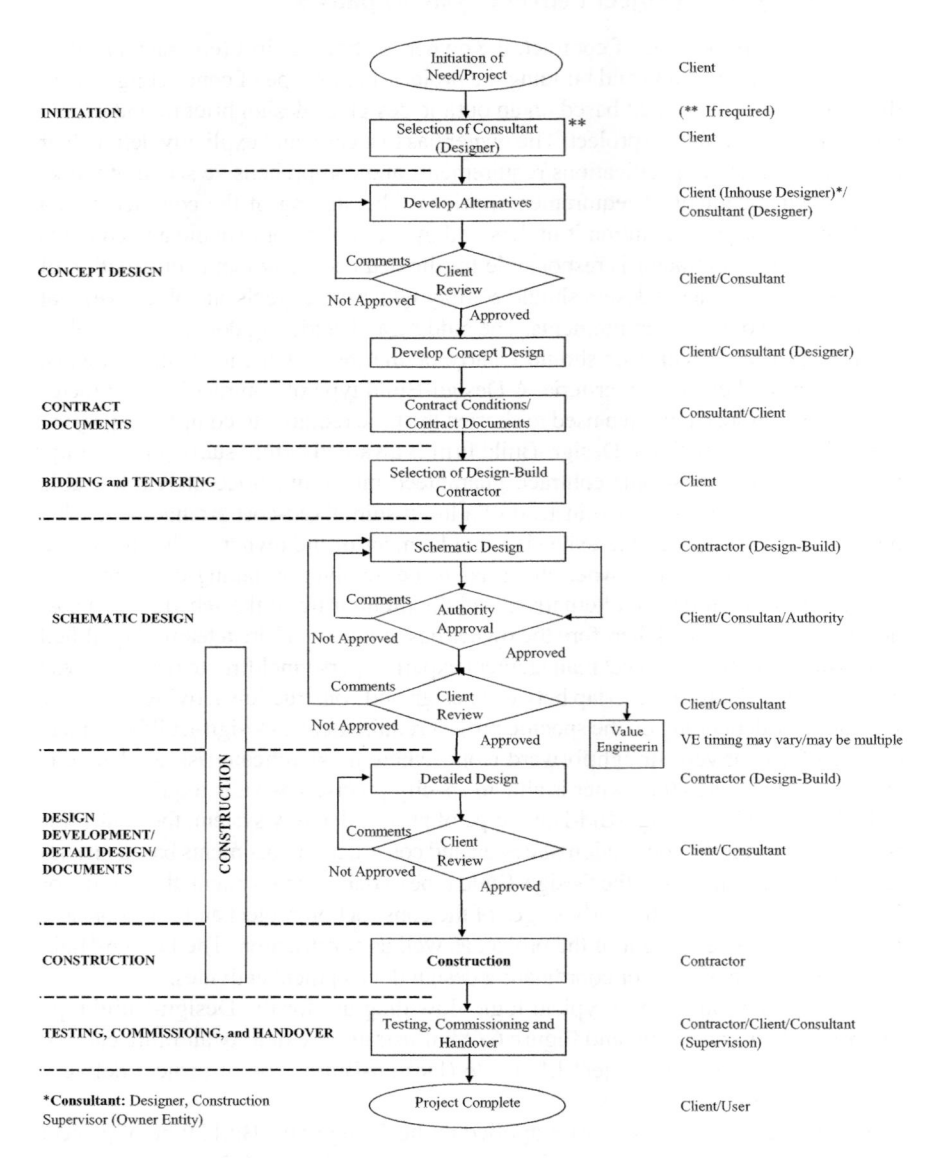

Figure 6.68a Logic flow diagram for construction projects – Design-Build project delivery system

Source: Abdul Razzak Rumane (2010). *Quality Management in Construction Projects.* Reprinted with permission from Taylor & Francis Group Company

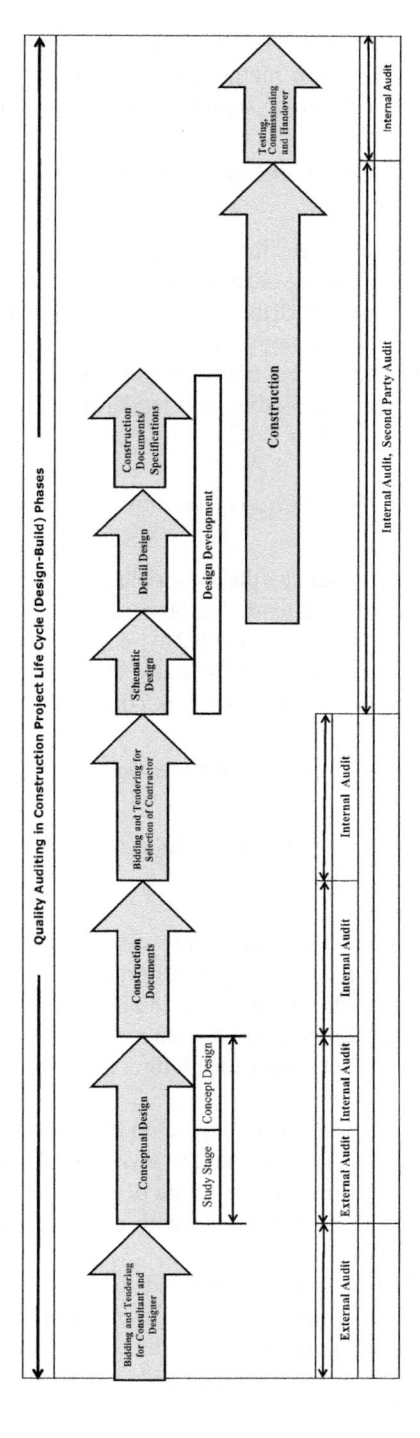

Figure 6.68b Quality auditing categories in the construction project life cycle (Design-Build) phases

methodology of certain stages will differ due to the concurrent engineering approach used in the Design-Build type of project delivery system in order to expedite project delivery. The audit methodology applied in this book for the Design-Build type of project delivery system is as shown in Figure 6.68b.

6.3.1 Auditing process for bidding and tendering (Design-Build project delivery system)

Tendering and bidding documents for the Design-Build type of project delivery system are prepared based on the organizational strategy to procure the contract. Bid and tendering documents are distributed to shortlisted contractors. The bidding and tendering process/methodology illustrated in Figure 6.6 and discussed earlier in Section 6.2.1 to select the contractor/consultant/supplier can be used to perform the quality auditing of the bidding and tendering process of the Design-Build type of project delivery system.

6.3.1.1 Auditing process for the selection of the consultant for the study stage (Design-Build)

The quality auditing process that includes the aims, objectives and audit methodology for the assessment of the bidding and tendering process for the selection of the consultant for the study stage discussed in section 6.2.1.1 can be used to perform quality auditing process for selection of consultant for study stage of Design-Build type of project. Please refer to Figure 6.7 and Tables 6.3–6.5 discussed in Section 6.2.1.1.

6.3.1.2 Auditing process for the selection of the concept designer (Design-Build)

The designer's responsibilities in the Design-Build type of project delivery system consist of:

1 Development of the concept design. In the case of projects that are more complex, the owner may use the designer to prepare 15–30% documents in order to provide enough definition to select the Design-Build contractor
2 Development of contract/construction documents and specifications
3 Coordinating with the project owner for bidding and tendering

It is the responsibility of the designer to develop the concept design (preliminary design, if applicable) that clearly defines the owner needs, performance specifications/requirements and comprehensive scope of works. The designer should have the ability to develop the project design with constructability in mind and design management experience.

Unlike the Design-Bid-Build type of project delivery system, the designer's roles and responsibilities are the development of the concept design preliminary design (if applicable) as design development is the responsibility of the Design-Build contractor. The Design-Build contractor either uses their in-house design team if available or hires an A/E firm to develop the design. The conflict between

Auditing processes for project life cycle phases 405

project professionals is an internal matter for the Design-Build contractor team and may not involve the owner.

The auditing process that include the aims, objectives, audit tools and audit methodology for the assessment of the selection of the designer discussed in Section 6.2.1.2 can be applied for the auditing process for the selection of the designer (Design-Build).

The bidding and tendering process stages illustrated in Figure 6.9 and discussed earlier in Section 6.2.1.2 can be applied to the Design-Build project delivery system. Table 6.53 lists the prequalification questionnaire to register the designer for the Design-Build type of project delivery system.

Table 6.53 Prequalification questionnaires (PQQ) for the registration of the designer (A/E) for Design-Build projects

Serial Number	Question	Answer
1	Name of the organization and address	
2	Organization's registration and license number	
3	ISO certification	
4	LEED or similar certification	
5	Total experience (years) in designing the following types of project for the Design-Build type of project delivery system 5.1 Residential 5.2 Commercial (mixed use) 5.3 Institutional (governmental) 5.4 Industrial 5.5 Infrastructure	
6	Size of project (maximum amount single project) 6.1 Residential 6.2 Commercial (mixed use) 6.3 Institutional (governmental) 6.4 Industrial 6.5 Infrastructure 6.6 Design-Build (specify type)	
7	List successfully completed projects (Design-Build) 7.1 Residential 7.2 Commercial (mixed use) 7.3 Institutional (governmental) 7.4 Industrial 7.5 Infrastructure	
8	List similar types (type to be mentioned) of projects completed 8.1 Project name and contracted amount 8.2 Project name and contracted amount 8.3 Project name and contracted amount 8.4 Project name and contracted amount 8.5 Project name and contracted amount	

(*Continued*)

Table 6.53 (Continued)

Serial Number	Question	Answer
9	Total experience in green building design	
10	Joint venture with any international organization	
11	Resources	
	11.1 Management	
	11.2 Engineering	
	11.3 Technical	
	11.4 Design equipment	
	11.5 Latest software	
12	Design production capacity	
13	Design standards	
14	Present work load	
15	Experience in value engineering (list projects)	
16	Financial capability (turnover for last 5 years)	
17	Financial audited report for last 3 years	
18	Insurance and bonding capacity	
19	Organization details	
	19.1 Responsibility matrix	
	19.2 CVs of design team members	
20	Design review system (quality management during design) for conceptual design specific for Design-Build type of projects	
21	Experience in preparation of contract documents	
22	Knowledge about regulatory procedures and requirements	
23	Experience in training of owner's personnel	
24	List of professional awards	
25	Litigation (dispute, claims) on earlier projects	

Based on the typical stages as discussed in Figure 6.9, the audit methodology can be performed in a similar manner to that listed in Tables 6.7, 6.8, 6.9 and 6.10. However, any specific/additional audit requirement, aims and objectives by the management for the Design-Build type of project delivery system can be considered in the checklist and questionnaire in the audit methodology for the assessment of the selection of the designer.

6.3.1.3 Auditing process for the Design-Build concept designer's proposal submission process

The auditing process that includes the aims, objectives and audit methodology for the assessment of the Design-Build designer's proposal submission process discussed in Section 6.2.1.3 can be used to perform the quality auditing process for the designer's proposal submission process for the designer (concept design) of the Design-Build type of project. Please refer to Figure 6.10–6.12 and Tables 6.11–6.12 discussed in Section 6.2.1.3.

6.3.2 Auditing process for contract/construction documents

The contract/construction documents in the Design-Bid type of project delivery system differ to those of the Design-Bid-Build type of delivery system. The contract/construction documents in the case of the Design-Build type of project delivery system are prepared taking into consideration that the project definition is understood by the contractor to avoid any conflict in the future as the contractor is responsible for the detailed design and construction of the project to performance specifications/requirements and the comprehensive scope of works (also called statement of requirements). Table 6.54 illustrates the list of contract/construction documents that are prepared by the owner/designer for the bidding and tendering of the Design-Build type of construction projects.

Table 6.54 Construction documents deliverables for Design-Build types of project

Serial Number	Deliverables	
1	Document I	
1.1	Tendering procedure	
	I-1 Invitation to tender	
	I-2 Instructions to bidders	
	I-3 Forms for tender and appendix	
	I-4 List of equipment and machinery	
	I-5 List of contractor's staff	
	I-6 Contractor's certificate of work statement	
	I-7 List of subcontractor(s) or specialist(s)	
	I-8 Initial bond	
	I-9 Final bond	
	I-10 Forms of agreement	
2	Document II	
2.1	Conditions of contract	
	II-1	General conditions, legal clauses
	II-2	Particular conditions
	II-3	Public tender laws
3	Document III	
	III-1	General specifications
	III-2	Particular specifications
	III-3	Conceptual drawings
	III-4	Schedule of rates and bill of quantities
	III-5	Analysis of prices
	III-6	Addenda
	III-7.1	Contractor's scope of works
	III-7.2	Soil investigation report
	III-7.3	Tender requirements (if any) and any other instructions issued by the owner

Figure 6.69 illustrates a typical quality auditing process for the contract/construction documents phase.

The main objectives of performing a quality audit of the contract/construction documents phase are to:

1. Evaluate and assess process compliance with the organizational strategic policies and procedure
2. Assess conformance to QMS and that processes meet established quality system requirements
3. Review the process used to prepare construction documents
4. Verify that the construction documents comply with requirements for the Design-Build type of project delivery system
5. Verify that the construction documents comply with the requirements for the bidding and tendering process
6. Verify that construction documents deliverables listed in Table 6.54 are developed and reviewed.
7. Validate that the project schedule is meeting the project requirements
8. Validate the cost of the project meeting the project requirements
9. Validate that concept design drawings and documents comply with quality standards and codes
10. Assess the compliance of the concept design drawings with the design quality management system
11. Validate that performance specifications are properly described in the contract documents and there is no ambiguity in the statement of requirements
12. Check for compliance with regulatory requirements
13. Check for sustainability considerations in the development of the concept design
14. Check that conservation of energy requirements is considered
15. Check that facility management requirements are considered
16. Review that the key risk factors are analyzed and managed while preparing construction documents
17. Review concept drawings, report, specifications and contract documents
18. Assess that all major activities related to the Design-Build project are adequately performed and managed as listed in Figure 5.22

The following audit tools are used to perform a quality audit of the design development phase:

- Analytical tools
- Checklists
- Review of documents
- Interviews
- Questionnaires

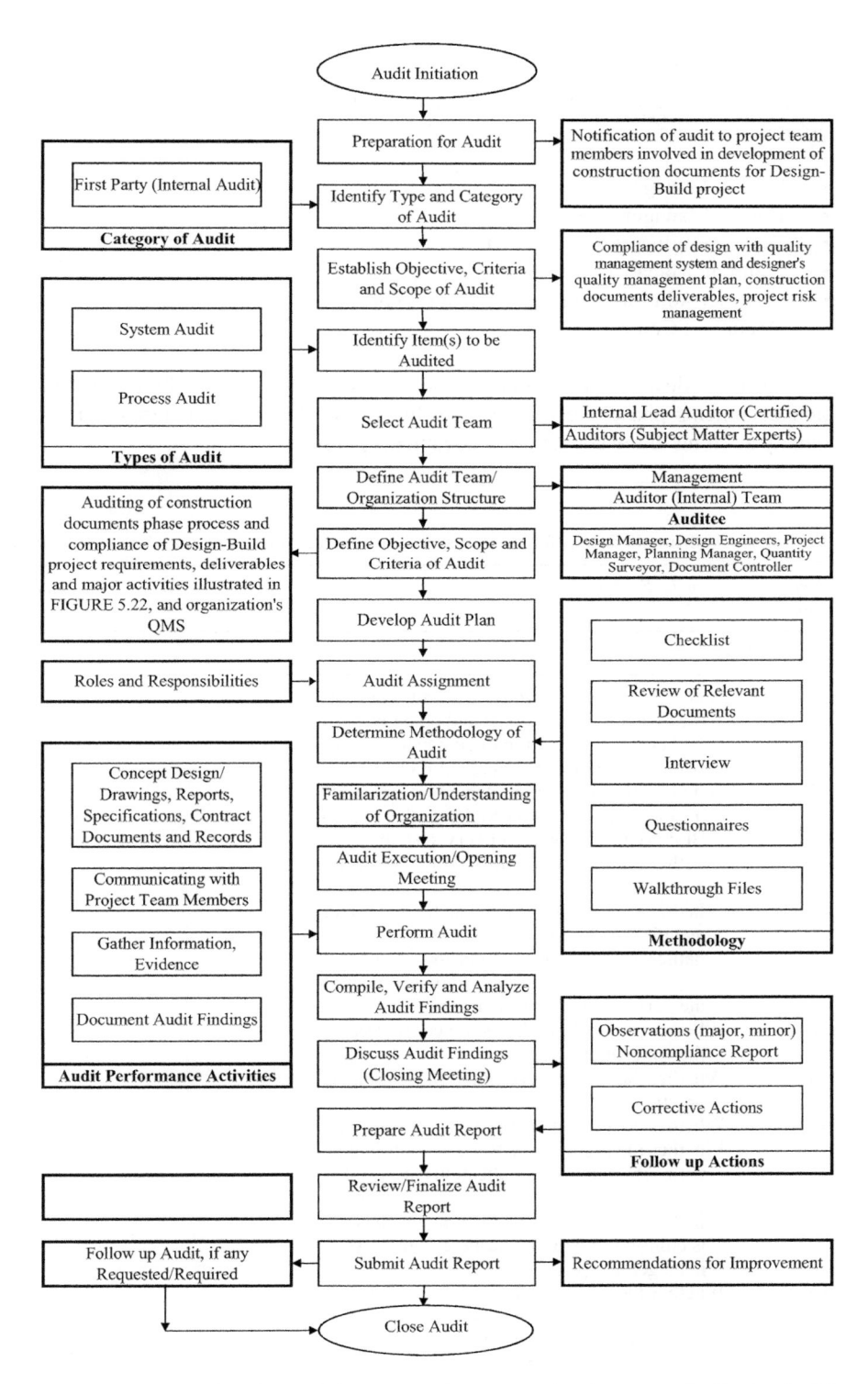

Figure 6.69 Typical quality auditing process for construction documents for Design-Build projects

The auditee team members are mainly from the following organizational groups:

- Project Manager
- Design Manager
- Design Engineers (All Trades)
- Planning Manager
- Contract Administrator/Quantity Surveyor
- Document Controller

Table 6.55 illustrates the audit methodology for the assessment of the construction documents phase and Figure 6.70 illustrates a logic flow diagram for the quality auditing of the construction documents phase

Table 6.55 Audit methodology for the assessment of construction documents (Design-Build)

Serial Number	Item To Be Assessed	Yes	No	Comments
A. Checklist				
1	Whether the construction documents support the owner's project goals and objectives			
2	Whether the construction documents comply with organizational strategic policies			
3	Whether the construction documents meet all the elements necessary to define the Design-Build type of project delivery system			
4	Whether regulatory/statutory requirements are considered			
5	Whether the concept design has functional and technical compatibility			
6	Whether the concept design is constructible			
7	Whether the concept design drawings meet all the performance requirements			
8	Whether project boundaries are considered in the concept drawings			
9	Whether the project schedule is achievable in practice			
10	Whether the project cost is properly estimated			
11	Whether the preliminary BOQ is coordinated with performance specifications and contract documents			
12	Whether the construction documents phase is complying with the designer's quality management plan			
13	Whether the concept drawings are based on specified quality standards and codes			
14	Whether all the drawings are numbered.			

Serial Number	Item To Be Assessed	Yes	No	Comments

A. Checklist

15	Whether following information is included on all the drawings: • Client name • Client logo • Location map • North orientation • Project name • Drawing title • Drawing number • Date of drawing • Revision number • Drawing scale • Contract reference number • Signature block • Signed by the designer for check and approval			
16	Whether the designer has considered the availability of resources during construction and the testing and commissioning phase			
17	Whether the recommended materials meet the owner's objectives			
18	Whether the construction document conforms with the project delivery system			
19	Whether the construction documents conform with the type of contract/pricing with the adopted methodology			
20	Whether the following documents are included along with the concept drawings: • Existing site conditions/site plan • Site surveys • Design calculations • Studies and reports			
21	Whether the specification is prepared taking into consideration international construction specifications and division numbers			
22	Whether construction documents are prepared taking into consideration international construction contract documents			
23	Whether risks have been identified and analyzed, and responses planned for mitigation of risk while preparing the concept drawings			
24	Whether health and safety requirements in the concept drawings are considered			
25	Whether the design conforms with fire and egress requirements			

(*Continued*)

Table 6.55 (Continued)

Serial Number	Item To Be Assessed	Yes	No	Comments
A. Checklist				
26	Whether environmental constraints are considered			
27	Whether the reports are complete and include adequate information about the project			
28	Whether the report is properly formatted and there is a table of contents for each report			
29	Whether there is any approved review procedure mentioned in the contract documents			
30	Whether the documents include a penalty for delay in the completion of the project			
31	Whether the documents include a penalty for failing to provide the required resources (manpower) for the project			
32	Whether the documents include a penalty for delay and violation of safety during the execution of the project			|//

Serial Number	Description	Response	Comments
B. Interview Questionnaires			
1	What procedure was followed to coordinate between the concept design and specifications		
2	How were the availability of resources during the construction phase ascertained		
3	Whether authorities' approval was obtained to initiate the project		
4	Whether construction documents were prepared within the agreed-upon schedule		
5	Whether construction documents were submitted in time for the owner's review		
6	What method was used to prepare the project schedule		
7	What methodology was used to estimate the project cost		
8	Whether numbers of sets were prepared to meet the bidding and tendering requirements		
9	Whether the construction documents were reviewed prior to release for tendering		
10	Whether the owner issued a construction documents acknowledgement letter		
11	Whether the reasons for comments on the construction documents, if any, were analyzed and corrective/preventive action taken		

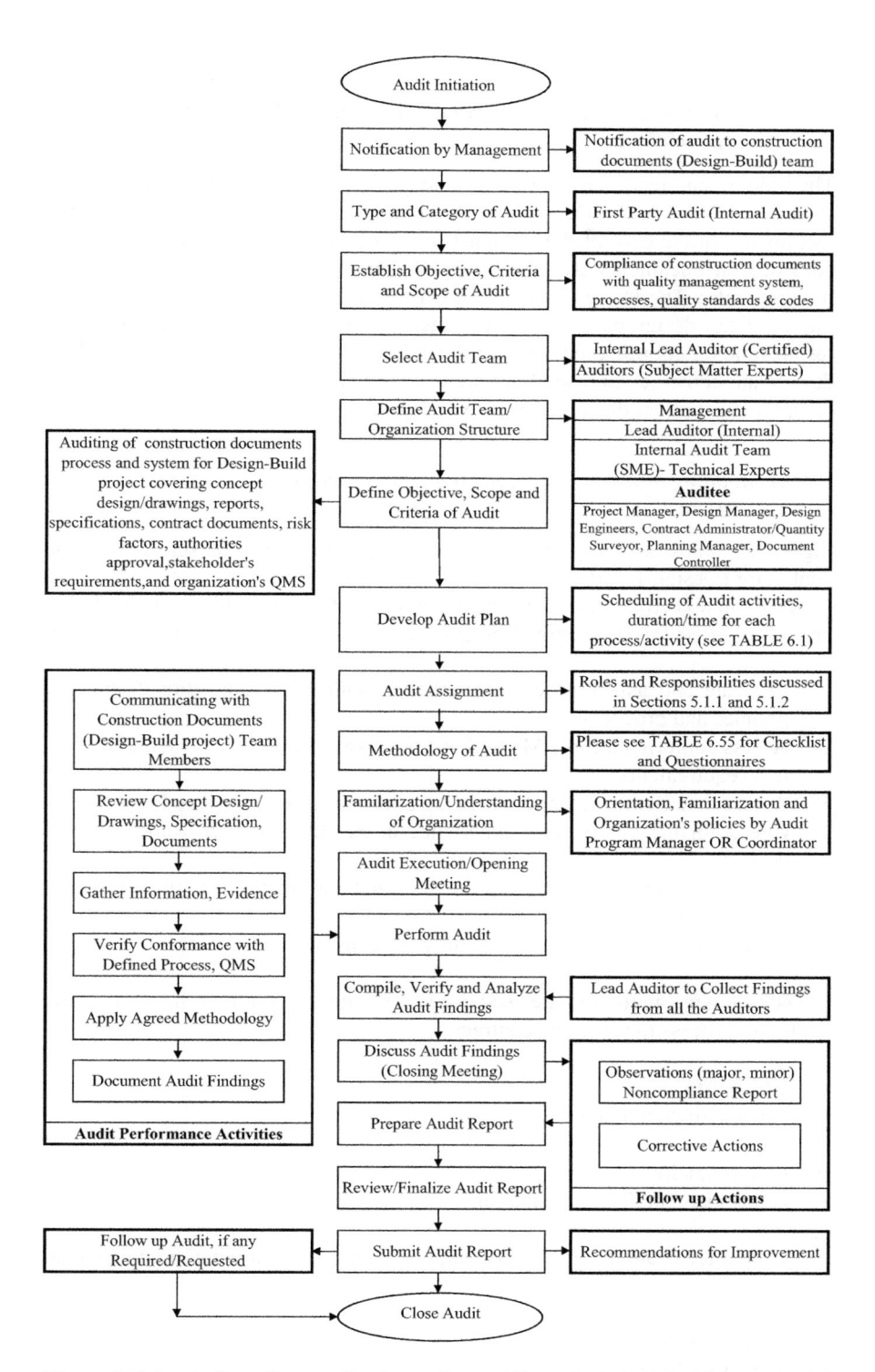

Figure 6.70 Logic flow diagram for the quality auditing of construction documents for Design-Build projects

6.3.3 Auditing process for the bidding and tendering (Design-Build) contractor

The bidding and tendering process for the Design-Build type of project delivery system is similar to the Design-Bid-Build contract strategy applied by the project owner. In the case of the Design-Build type of project delivery system, the construction documents are based on performance specifications (statement of requirements). The Design-Build contractor should have experience in developing the detail design based on the contractual concept design. The contractor is responsible for coordination between the design and construction disciplines. The design and construction disciplines are contractually obliged to work together to complete the project that meets the owner's needs.

6.3.3.1 Auditing process for the selection of the contractor (Design-Build)

Figure 6.71 illustrates a typical quality auditing process for the selection of the contractor (Design-Build).

The main objectives of performing a quality audit for the bidding and tendering process to select the contractor (Design-Build) for the construction project are to:

1. Evaluate and assess process compliance with the organizational strategic policies and procedure
2. Assess conformance to QMS and that processes meet established quality system requirements
3. Assess that all the major activities are adequately performed and managed
4. Check for the contractor's experience in executing Design-Build types of project
5. Check that the key risk factors are analyzed and managed to meet business requirements
6. Assess the proper functioning of the project management system

The following audit tools are used to perform a quality audit for the bidding and tendering process to select the contractor:

- Checklists
- Review of documents
- Interviews
- Questionnaires
- Analysis of audit documentation

The auditee team members are mainly from the following organizational groups:

- Tendering Committee
- Tendering Manager

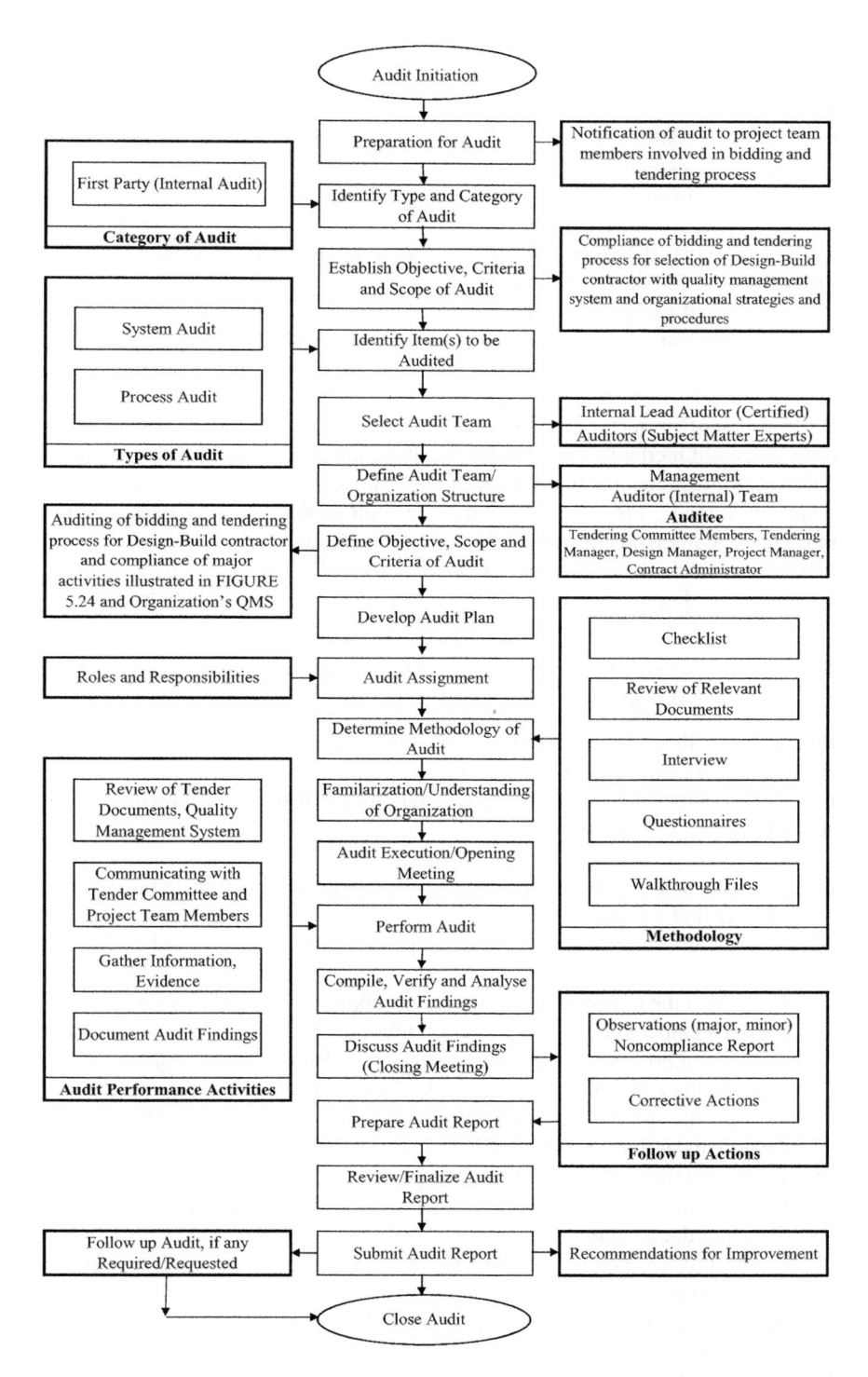

Figure 6.71 Typical quality auditing process for the bidding and tendering procedure for the selection of the contractor (Design-Build projects)

416   *Abdul Razzak Rumane*

- Project Manager
- Contract Administrator
- Design Manager
- Finance Manager

Table 6.56 lists the prequalification questionnaires to register the contractor for the Design-Build type of project.

As discussed earlier in Section 6.2.3.1, Figure 6.26 illustrates four stages that can be applied to select the contractor for Design-Build types of project. Contractor selection criteria can be evolved as per the organization's strategy. Based on the typical stages as per Figure 6.26, the audit methodologies for Design-Build types of project can also be performed in four stages. Table 6.57 illustrates the audit methodology to assess/evaluate the process compliance at stage I, Table 6.58 illustrates the audit methodology to assess/evaluate the process compliance at stage II, Table 6.59 illustrates the audit methodology to assess/evaluate the process compliance at stage III, Table 6.60 illustrates the audit methodology to assess/evaluate the process compliance at stage IV and Figure 6.72 illustrates the logic flow diagram for the quality auditing of the bidding and tendering phase.

6.3.3.2 *Auditing process for the Design-Build contractor's tender submission process*

Figure 6.73 illustrate a typical quality auditing process for the contractor's (Design-Build) tender submission process.

The main objectives of performing a quality audit of the contractor's tender submission process are to:

1. Evaluate and assess process compliance with the organizational strategic policies and procedure
2. Assess conformance to QMS and that processes meet established quality system requirements
3. Check that the proposal preparation and submission complies with the organization's tender submission process. Figure 6.74 illustrates the typical tender submission process by the contractor (Design-Build type of projects)
4. Validate the response to the contents of tender documents is complete in all respects. Table 6.54 illustrates the contents of tender documents (construction documents deliverable) for the Design-Build)type of project
5. Verify that the tender documents are properly reviewed
6. Verify that the concept design/drawings match the performance specifications and documents
7. Verify the correctness of BOQ with the concept design/drawings and specifications
8. Review the process used to prepare the bid
9. Ensure that the design development cost is included in the bid
10. Ensure that the prices of material, equipment and machinery considered to prepare the bid are valid for the entire duration of construction (Design-Build)

Table 6.56 Prequalification questionnaires (PQQ) for selecting the Design-Build contractor

Serial Number	Question	Answer
1	Name of the organization and address	
2	Organization's registration and license number	
3	ISO certification	
4	Registration/classification status of the organization	
5	Joint venture with any international contractor	
6	Total turnover last 5 years	
7	Audited financial report for last 3 years	
8	Insurance and bonding capacity	
9	Total experience (years) as Design-Build contractor	
10	Total experience (years) as contractor	
Design- Build contracts information		
11	Total experience (years) in construction of following types of project 11.1 Residential 11.2 Commercial (mixed use) 11.3 Institutional (governmental) 11.4 Industrial 11.5 Infrastructure	
12	Size of project (maximum amount single project) 12.1 Residential 12.2 Commercial (mixed use) 12.3 Institutional (governmental) 12.4 Industrial 12.5 Infrastructure	
13	List successfully completed projects 13.1 Residential 13.2 Commercial (mixed use) 13.3 Institutional (governmental) 13.4 Industrial 13.5 Infrastructure	
14	List similar types (type to be mentioned) of project completed 14.1 Project name and contracted amount 14.2 Project name and contracted amount 14.3 Project name and contracted amount 14.4 Project name and contracted amount 14.5 Project name and contracted amount	
15	Resources 15.1 Management 15.2 Engineering 15.3 Technical 15.4 Foreman/supervisor 15.5 Skilled manpower 15.6 Unskilled manpower 15.7 Plant and equipment	

(*Continued*)

Table 6.56 (Continued)

Serial Number	Question	Answer
16	Quality management policy	
17	Health, safety and environment policy	
	17.1 Number of accidents during last 3 years	
	17.2 Number of fires on site	
18	Current projects	
19	Staff development policy	
20	List of delayed projects	
21	List of failed contracts	
Designer's information		
22	Total years of experience in Design-Build type of projects	
23	Size of project (maximum value)	
24	List similar type of successfully completed projects	
25	List successfully Design-Build projects	
26	Resources	
	26.1 Architect	
	26.2 Structural engineer	
	26.3 Civil engineer	
	26.4 HVAC engineer	
	26.5 Mechanical engineer	
	26.6 Electrical engineer	
	26.7 Low voltage engineer	
	26.8 Interior designer	
	26.9 Landscape engineer	
	26.10 CAD technicians	
	26.11 Quantity surveyor	
	26.12 Equipment	
	26.13 Design software	
27	LEED or similar certification	
28	Total experience in green building design	
29	Design philosophy/methodology	
30	Quality management system	
31	HSE considerations in design	
32	List of professional awards	
33	Litigation (dispute, claims) on earlier projects	

Table 6.57 Audit methodology for the assessment of the bidding and tender process for the selection of contractor (Design-Build) – Stage I (shortlisting/registration of contractors)

Serial Number	Item To Be Assessed	Yes	No	Comments
A. Checklist				
1	Whether the notification for registration was announced in all technical newsletters and leading news media as per the organization's policy			

Serial Number	Item To Be Assessed	Yes	No	Comments
A. Checklist				
2	Whether there are registered designers to participate in the tender			
3	Whether prequalification questionnaires (PQQ) are issued to all the intending bidders (please refer to Table 6.56 for guidelines)			
4	Whether there are any requests for queries/bid clarifications			
5	Whether the meeting for clarifications is attended by all the intending bidders			
6	Whether the attendance sheet for the meeting is signed by all the attendees			
7	Whether minutes of the meeting are circulated to all the attendees/participating bidders			
8	Whether bid clarification of the queries is sent to all the bidders (please refer to Figure 6.27)			
9	Whether sufficient time is provided to prepare and submit completed PQQ			
10	Whether there are any requests for extension of the submission date			
11	Whether all the intending bidders submitted the response to PQQ			
12	Whether received bid envelopes are placed in safe custody			
13	Whether there are any delayed submissions			
14	Whether received responses are acknowledged			
15	Whether received responses are clearly identified and recorded			
16	Whether responses are fairly evaluated as per the organization's selection policy			
17	Whether evaluation criteria, weightage and methodology are as per the organization's strategy			
18	Whether any of the submitted responses are found to be incomplete			
19	Whether there are any late submissions			
20	Whether risk factors are considered while evaluating the response and registration of designers			
21	Whether reasons for non-registration are conveyed to unsuccessful participants			
22	Whether shortlisted contractors are allotted a code and registration number			

(*Continued*)

Table 6.57 (Continued)

B. Interview Questionnaires

Serial Number	Description	Response	Comments
1	How responses were submitted: • Hand delivery • By post • Courier service		
2	Whether PQQ was available for electronic distribution		
3	How was the response for registration/shortlisting		

Table 6.58 Audit methodology for the assessment of the bidding and tender process for the selection of the contractor (Design-Build) – Stage II (bidding and tender documents)

Serial Number	Item To Be Assessed	Yes	No	Comments
A. Checklist				
1	Whether the tender documents include all the relevant information required to select the contractor (Design-Build)			
2	Whether the concept design and performance documents properly define the owner needs			
3	Whether the construction documents are approved by the project owner			
4	Whether the bid/tender documents are prepared as per the procurement method and contract strategy adopted during the early stages of the project			
5	Whether all the information to enable the bidders/contractors to submit the proposal is clear and unambiguous			
6	Whether documents are prepared taking into consideration the organization's Quality Management System (QMS)			
7	Whether regulatory/authority requirements are taken into consideration while preparing the tender documents			
8	Whether the selection criteria and selection method for contractor selection are clearly specified in the organization's QMS			
9	Whether budget approval is obtained for construction prior to release of the tender announcement			
10	Whether the tender opening and closing dates are mentioned in the announcement			
11	Whether the schedule for completion of the construction phase is included in the tender documents			

Serial Number	Item To Be Assessed	Yes	No	Comments
A. Checklist				
12	Whether the qualifications of personnel and core team members for construction are requested in the tender documents			
13	Whether project quality requirements are properly defined in the documents			
14	Whether rights and liabilities of all the parties are mentioned in the documents			
15	Whether change/variation clauses are included in the documents			
16	Whether cancellation/termination clauses are included in the documents			
17	Whether bond and insurance clauses are included in the documents			
18	Whether the tender documents are as per international/local standard documents used by the construction industry			
19	Whether evaluation and assessment criteria are well defined			
20	Whether review and analysis procedure of bid is included in the documents			
21	Whether the contract to be signed for construction is included in the documents			
22	Whether all the records are properly maintained			
23	Whether the request to provide the bidder's contact details are included in the documents			

B. Interview Questionnaires

Serial Number	Description	Response	Comments
1	Whether any outside agency was involved in the preparation of the tender documents		
2	What guidelines were adopted to prepare the tender documents for Design-Build type of project		
3	Whether any specific format/template was followed		
4	Whether all the requirements of tender documents were coordinated and agreed by relevant stakeholders		
5	Whether risks of disputes and associated financial costs were considered while preparing the contract documents		
6	Whether problems and challenges were identified while preparing bid/tender		
7	Who was responsible for authorizing bidding requirements		

Table 6.59 Audit methodology for the assessment of the bidding and tender process for the selection of the contractor (Design-Build) – Stage III (contract bid solicitation)

Serial Number	Item To Be Assessed	Yes	No	Comments
A. Checklist				
1	Whether the tender documents are properly organized			
2	Whether the tender documents are distributed to all the prequalified/shortlisted/registered bidders (contractors) to participate in the tender			
3	Whether the tender is notified to all the prequalified bidders			
4	Whether all the prequalified/shortlisted/ registered bidders participated in the tender			
5	Whether the attendance sheet for the meeting was signed by all the attendees			
6	Whether minutes of the meeting were circulated to all the attendees/participating bidders/contractor (Design-Build)			
7	Whether sufficient time is provided to prepare and submit the tender			
8	Whether the tender submission date is extended from the date originally announced			
9	Whether an addendum (if any) is issued/ notified to all the participating bidders			
10	Whether there are any requests for extension of the submission date			
11	Whether the clarification meeting is attended by all participating bidders			
12	Whether any changes are made to the originally announced tender documents			
13	Whether the received bid envelopes are placed in safe custody			
14	Whether technical and financial envelopes are submitted at the same time			
15	Whether there are any delayed submissions			
16	Whether the tenders are opened as per the announced date and time			
17	Whether the tender is opened in the presence of all the bidders			
18	Whether received tenders are acknowledged			
19	Whether received tenders are clearly identified and recorded			
20	Whether applicable fees/bid bond are submitted by all the bidders who participated in the tender			
21	Whether tenders are fairly evaluated as per the announced policy and procedure			
22	Whether the comparison of bids is tabulated			

Serial Number	Item To Be Assessed	Yes	No	Comments

A. Checklist

23	Whether evaluation criteria, weightage and methodology are as specified			
24	Whether the bid cost is higher than the budgeted cost			
25	Whether any submitted tender is found to be incomplete			
26	Whether there are any late submissions			
27	Whether all the scheduled pages are signed by the bidder			
28	Whether risk factors are considered while evaluating the bids			
29	Whether reasons for non-acceptance of tender were conveyed to the unsuccessful bidders			

B. Interview Questionnaires

Serial Number	Description	Response	Comments
1	How the notification was announced		
2	How tenders were submitted: • Hand delivery • By post • Courier service		
3	Whether the tender was available for electronic distribution		
4	How was the response to participate in the tender		
5	Whether tender box(es) were placed at secured places and monitored through an electronic surveillance system		
6	How much time was allowed to open the bids after the notified tendering closing time		
7	Whether the tender submission date was extended and what was the reason for extension		
8	Whether all the tender envelopes/packages were sealed with marking of the tenders and contain the original tender documents and required number of sets as per the tender conditions		
9	Whether the bid review was established		
10	Whether the bid selection procedure was clearly defined		

Table 6.60 Audit methodology for the assessment of the bidding and tender process for the selection of the contractor (Design-Build) – Stage IV (contract award)

Serial Number	Item To Be Assessed	Yes	No	Comments
A. Checklist				
1	Whether the selection of the contractor satisfies all the conditions to be a successful bidder			
2	Whether the selected bidder (Design-Build contractor) is a legal entity			
3	Whether the contractor has their own design team or they will hire a design firm			
4	Whether the selection was made by the selection committee			
5	Whether the selection of the contractor is approved by the relevant applicable authority, if mandated by the organization's policy			
6	Whether the selected contractor is capable of carrying out the specified work and has the necessary resources for the design and construction of the project			
7	Whether risk factors are considered prior to the signing of the contract			
8	Whether the standard contract format is used for signing the contract			
9	Whether the contract terms and conditions are clearly written and unambiguous			
10	Whether an addendum and minutes of the meeting are included in the contract documents			
11	Whether the contract period/schedule is properly described			
12	Whether staff qualifications (core staff) are properly defined for design works			
13	Whether staff qualifications (core staff) are properly defined for construction works			
14	Whether a performance bond is submitted by the consultant			
15	Whether variation/change management clauses are included in the contract agreement			
16	Whether dispute resolution and conflict resolution clauses are properly defined			
17	Whether contract documents are reviewed prior to signing			
18	Whether tender documents and contract are properly archived			
19	Whether a letter of award is sent			
20	Whether final results are published and announced			

B. Interview Questionnaires

Serial Number	Description	Response	Comments
1	On what basis the bidder (contractor) was selected: low bid, competitive bid or quality based system		
2	Was there any tie between two or more bidders		
3	Was there any dispute or objection raised by any of the bidders before the award of the contract to the successful bidder		
4	Whether the final price was negotiated or as per the bid (quotation) received		
5	Whether the successful bidder failed to submit a performance bond		
6	Whether there was any need to update the approved budget		
7	Whether the contractor declared any subcontractor for outsourcing for design purposes		
8	Whether the contractor declared any subcontractor for outsourcing (construction subcontracting) purposes		
9	Whether risks in the project execution (design as well as construction works) were considered		
10	Whether the contract includes a penalty clause for delay in completion of the project		
11	Whether the contract includes a penalty clause for failure to provide the required resources (manpower) for the project		
12	Whether the contract includes a penalty clause for safety violation during the execution of the project		

11 Check that the availability of resources (manpower) for construction activities is properly considered
12 Validate the total cost of the bid
13 Validate that the construction schedule is practical and achievable
14 Review that the key risk factors are analyzed and managed
15 Check for compliance with regulatory requirements
16 Validate that bid has been prepared by fully coordinating with all the stakeholders

The following audit tools are used to perform the quality audit of the tender submission process:

- Analytical tools
- Checklists

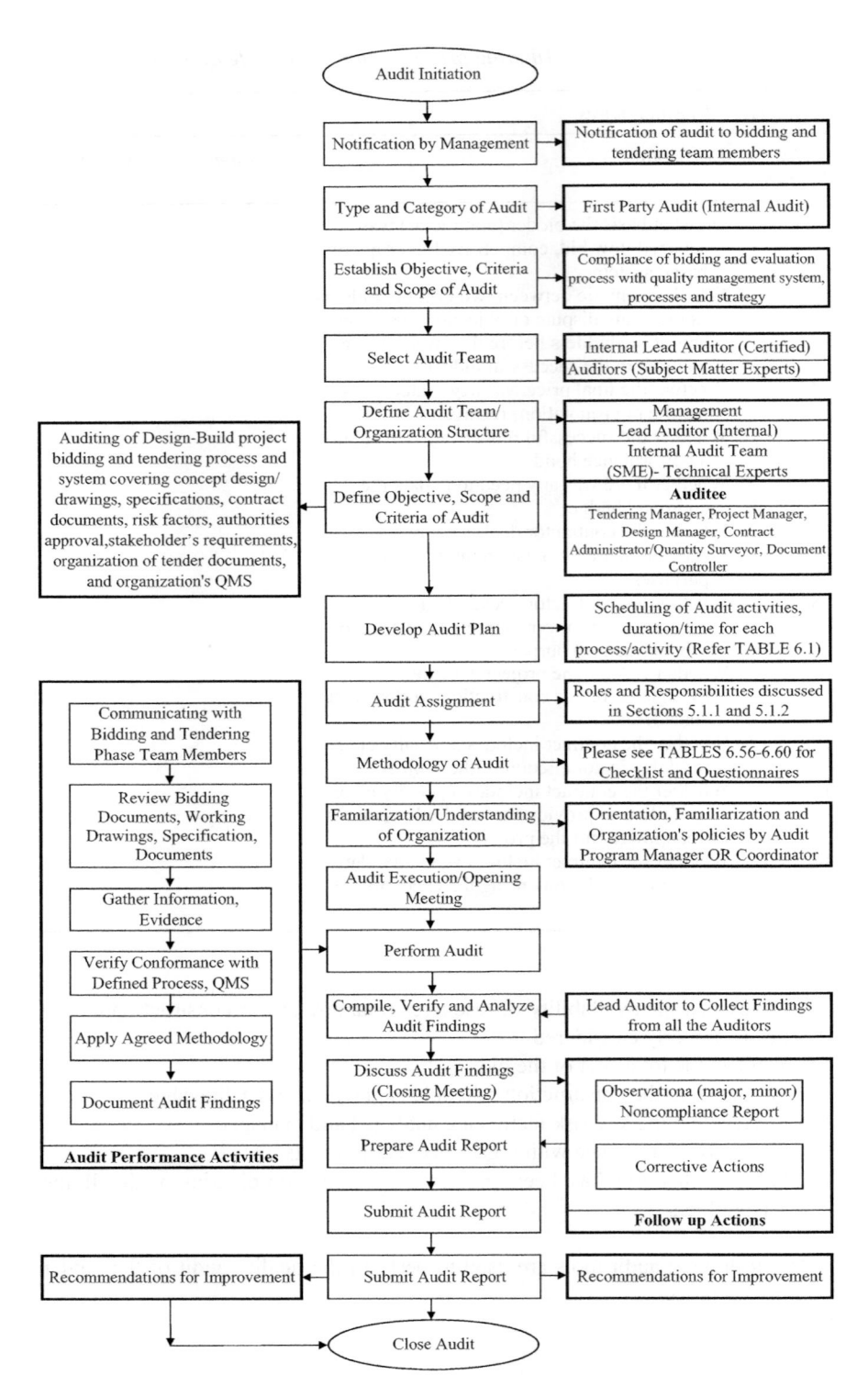

Figure 6.72 Logic flow diagram for the quality auditing of the bidding and tendering for the selection of the contractor (Design-Build projects)

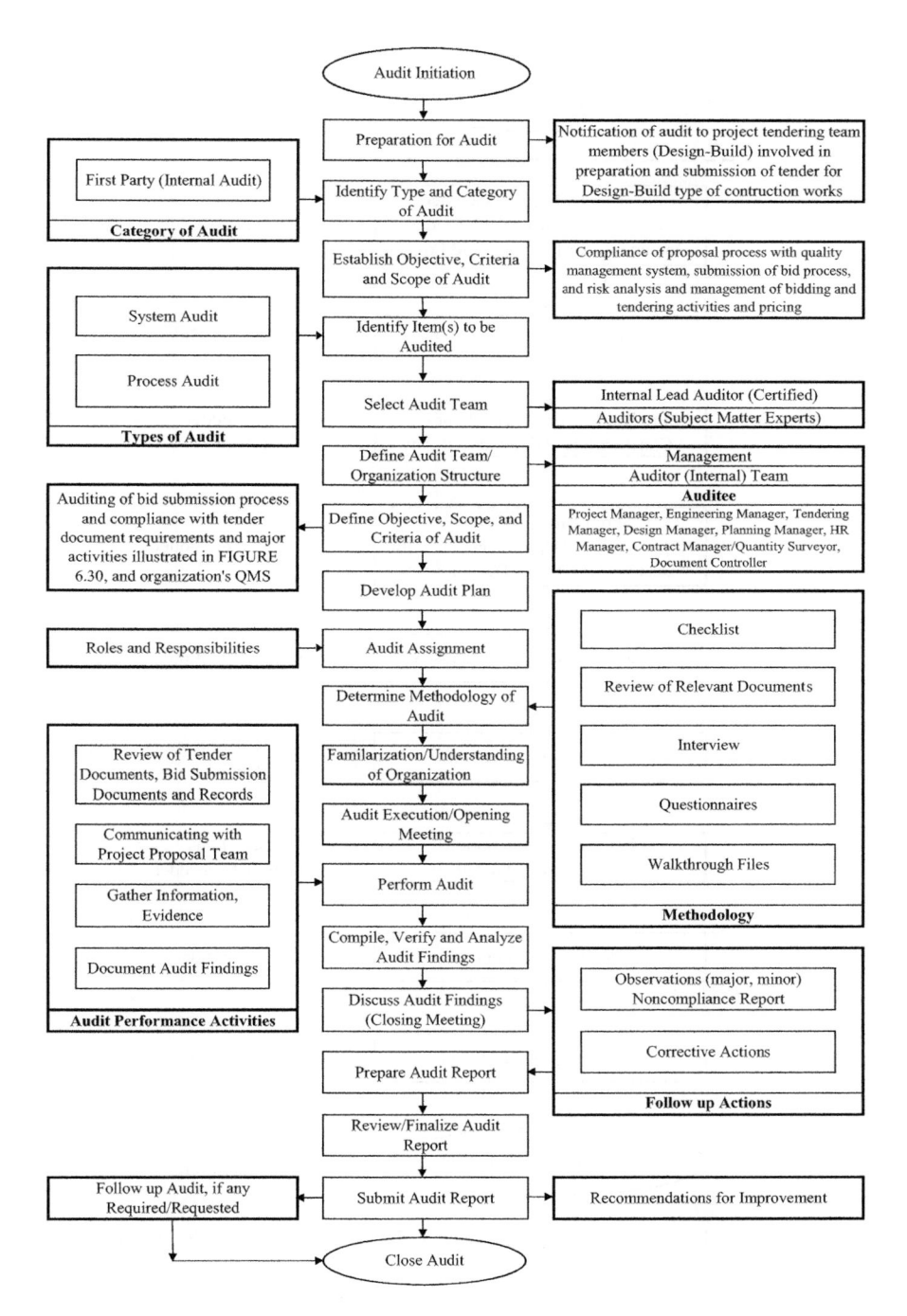

Figure 6.73 Typical quality auditing process for the contractor's (Design-Build) tender submission process

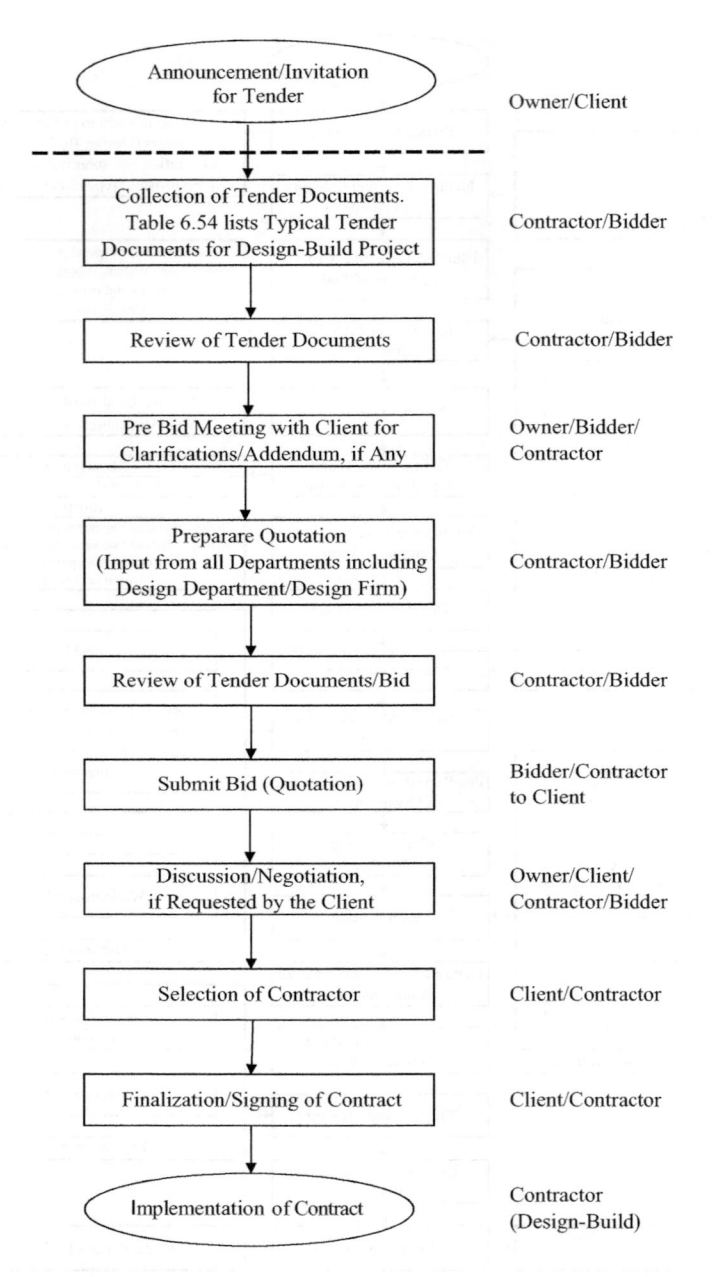

```
┌─────────────────────────┐
│ Announcement/Invitation │        Owner/Client
│      for Tender         │
└─────────────────────────┘
- - - - - - - - - | - - - - - - - - -
                  ▼
┌─────────────────────────────────┐
│ Collection of Tender Documents. │
│ Table 6.54 lists Typical Tender │   Contractor/Bidder
│ Documents for Design-Build Project │
└─────────────────────────────────┘
                  ▼
┌─────────────────────────┐
│ Review of Tender Documents │      Contractor/Bidder
└─────────────────────────┘
                  ▼
┌─────────────────────────────────┐
│ Pre Bid Meeting with Client for │   Owner/Bidder/
│ Clarifications/Addendum, if Any │   Contractor
└─────────────────────────────────┘
                  ▼
┌─────────────────────────────────┐
│       Preparare Quotation       │
│ (Input from all Departments including │ Contractor/Bidder
│ Design Department/Design Firm)  │
└─────────────────────────────────┘
                  ▼
┌─────────────────────────────┐
│ Review of Tender Documents/Bid │   Contractor/Bidder
└─────────────────────────────┘
                  ▼
┌─────────────────────────┐        Bidder/Contractor
│   Submit Bid (Quotation) │        to Client
└─────────────────────────┘
                  ▼
┌─────────────────────────────┐    Owner/Client/
│   Discussion/Negotiation,    │    Contractor/Bidder
│ if Requested by the Client  │
└─────────────────────────────┘
                  ▼
┌─────────────────────────┐
│  Selection of Contractor │        Client/Contractor
└─────────────────────────┘
                  ▼
┌─────────────────────────────┐
│ Finalization/Signing of Contract │ Client/Contractor
└─────────────────────────────┘
                  ▼
┌─────────────────────────┐        Contractor
│ Implementation of Contract │      (Design-Build)
└─────────────────────────┘
```

Figure 6.74 Typical tender submission procedure by the contractor (Design-Build)

Auditing processes for project life cycle phases 429

- Review of documents
- Interviews
- Questionnaires

The auditee team members are mainly from the following organizational groups:

- Construction/Project Manager
- Tendering Manager
- Planning Manager
- Contracts Manager/Contract Administrator
- Quality Manager
- Engineering Manager
- Design Manager
- Procurement Manager
- Document Controller
- Safety Officer
- Finance Manager

Table 6.61 illustrates the audit methodology for the assessment of the contractor's (Design-Build) tender submission process and Figure 6.75 illustrates the logic flow diagram for the quality auditing of the contractor's (Design-Build) tender submission process.

Table 6.61 Audit methodology for the assessment of the contractor's (Design-Build) tender submission process

Serial Number	Item To Be Assessed	Yes	No	Comments
A. Checklist				
1	Whether the company is already registered as a contractor to participate in the tender			
2	Whether there are any fees to collect tender documents			
3	Whether the scope of work (statement of requirements) is clearly defined in the tender documents			
4	Whether there is any request for clarification raised			
5	Whether the response to price schedule in the tender is complete			
6	Whether the requested information in the tender documents is clearly responded to			
7	Whether sufficient time is provided to prepare and submit the completed tender			
8	Whether there are any requests for extension of submission date			

(*Continued*)

Table 6.61 (Continued)

Serial Number	Item To Be Assessed	Yes	No	Comments
A. Checklist				
9	Whether there are any delayed submissions			
10	Whether the tender response is prepared taking into consideration/referring to the organization's quality management system			
11	Whether the price for design development is included in the quotation			
12	Whether prices for material, products and equipment are obtained from registered vendors			
13	Whether there is any designated subcontract works mentioned in the tender			
14	Whether the construction quality requirements are considered while preparing the response			
15	Whether there is equipment rental cost considered for the execution of works			
16	Whether the tender response was approved by the relevant committee/authorized person			
17	Whether the risk factors were considered and analyzed			
18	Whether all the relevant pages are signed			
19	Whether tender documents, proposal was reviewed prior to submission			

B. Interview Questionnaires

Serial Number	Description	Response	Comments
1	Whether the notification for the proposal submission was announced in all technical newsletters and leading news media or it was only for registered bidders (Design-Build contractors)		
2	How the tender was submitted: • Hand delivery • By post • Courier service		
3	Whether tender documents were available for electronic distribution		
4	What tools and techniques were used to price the tender		
5	Whether the organization's quality management plan was considered to evaluate the price schedule		
6	Whether the design team members are internal or an outside firm is contracted		

B. Interview Questionnaires

Serial Number	Description	Response	Comments
7	Whether a lack of resources/availability of resources (technical as well as skilled and semiskilled) was considered while preparing the price schedule		
8	Whether the availability of specified material and products was considered while preparing the price schedule		
9	Whether the availability of equipment and machinery for the construction period was considered while preparing the price schedule		
10	Whether any subcontracting work is from the registered vendors		
11	Whether the subcontractor cost was properly checked and evaluated		
12	Whether the construction schedule was properly reviewed by taking into consideration the availability of resources and their production output		
13	Whether the regulatory approval period was properly considered		
14	What risk factors were considered while evaluating the price schedule		
15	Whether risk factors were considered for outsourcing/subcontracting of works		
16	Whether any penalty due to non-conformance with the contract was considered in the tender price		
17	Whether purchasing prices for material, products and equipment are from the recommended/specified manufacturers		
18	Whether the tender prices were worked out as per 'Rule of Thumb' or actual cost records		
19	Whether any error in BOQ (Bill of Quantity) and contract drawings (concept design/drawings) was observed		
20	Whether there was any negotiation to the submitted quotation		

6.3.3.3 Auditing process for the selection of the construction supervisor (Design-Build)

For the Design-Build type of construction projects, the supervision team should have expertise in both:

1. Reviewing and monitoring the design development by the Design-Build contractor
2. Supervision of construction activities

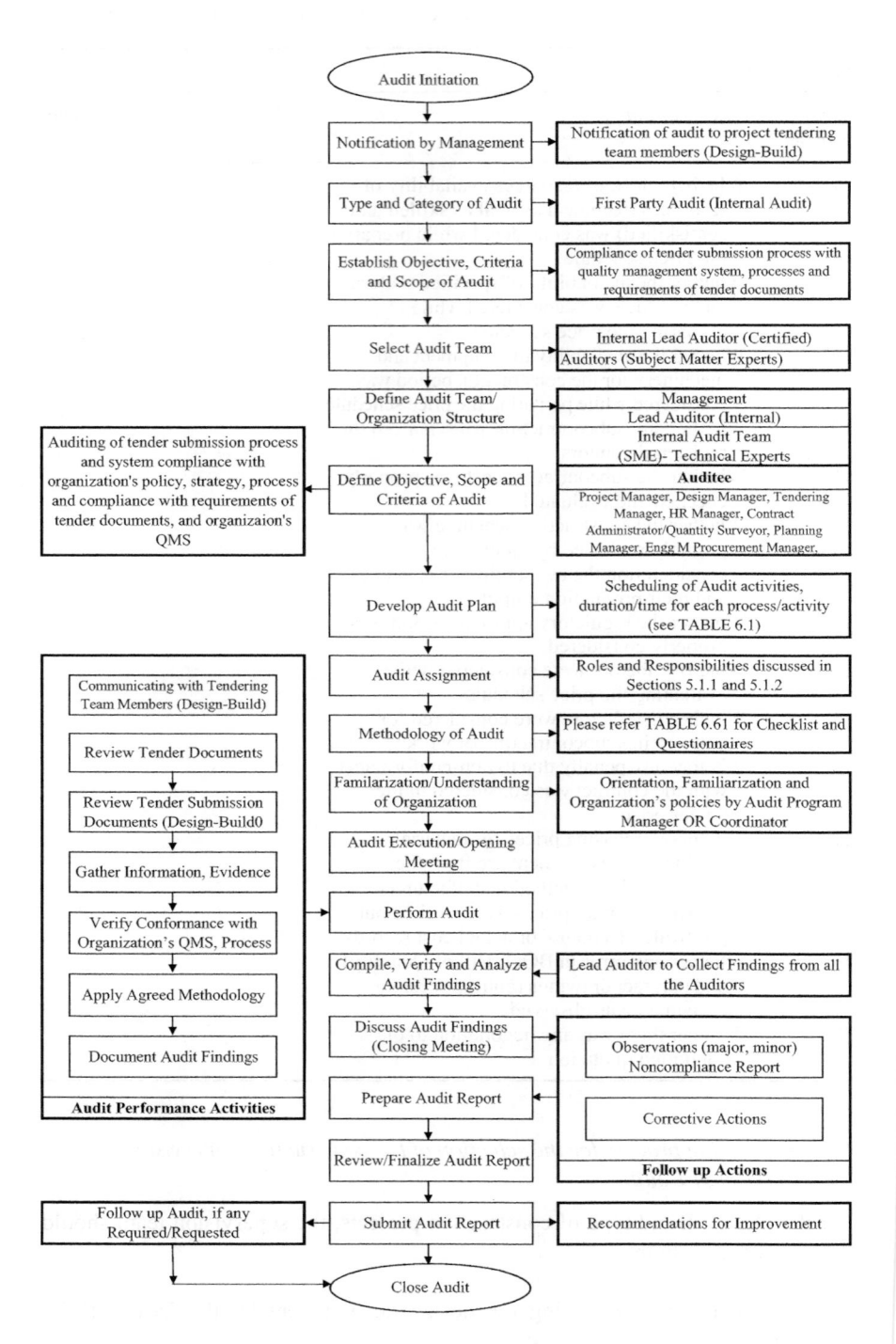

Figure 6.75 Logic flow diagram for the quality auditing process for the contractor's (Design-Build) tender submission process

Auditing processes for project life cycle phases 433

Table 6.62 lists the responsibilities of the supervision team members (firm) who have to supervise the design and construction activities.

Figure 6.76 illustrates a typical quality auditing process for the selection of the construction supervisor (consultant) for a Design-Build project.

The main objectives of performing a quality audit for the bidding and tendering process to select the supervisor (consultant) for the construction project are to:

1 Evaluate and assess process compliance with the organizational strategic policies and procedure
2 Assess conformance to QMS and that processes meet established quality system requirements
3 Assess that all the major activities are adequately performed and managed
4 Check that the key risk factors are analyzed and managed to meet business requirements
5 Assess the proper functioning of the project management system
6 Check the selected supervisor has experience in both design and construction supervision activities

Table 6.62 Responsibilities of the construction supervisor (Design-Build)

Sr. No.	Description
A. Design Management Activities	
A.1	Review contracted concept design and performance specifications for Design-Build construction
A.2	Review preliminary design to ensure compliance with owner requirements
A.3	Validate that design is developed to suit simultaneous construction works
A.4	Validate that design is developed to facilitate construction of project to meet performance specifications
A.5	Ensure that design development is fully coordinated to meet construction schedule
A.6	Validate compliance with quality standards and codes
A.7	Check that design complies with regulatory requirements
A.8	Review that risk factors are considered during design development
A.9	Check that constructability is considered
A.10	Verify that design complies with project goals and objectives
A.11	Assess compliance to safety considerations in the design
A.12	Achieve the quality goals to meet owner requirements
A.13	Ensure that overall project schedule (design and construction) is maintained
B-Construction Work (Phase)	
B.1	Review construction methodology
B.2	Approval of contractor's construction schedule
B.3	Regular inspection and checking of executed works
B.4	Review and approval of construction materials
B.5	Review and approval of shop drawings
B.6	Inspection of construction material
B.7	Monitoring and controlling construction expenditure
B.8	Monitoring and controlling construction time
B.9	Maintaining project record

(*Continued*)

434   Abdul Razzak Rumane

Table 6.62 (Continued)

Sr. No.	Description
B.10	Conduct progress and technical co-ordination meetings
B.11	Co-ordination of owner's requirements and comments related to site activities
B.12	Project related communication with contractor
B.13	Co-ordination with regulatory authorities
B.14	Processing of site work instruction for owner's action
B.15	Evaluation and processing of variation order/change order
B.16	Recommendation of contractor's payment to owner
B.17	Evaluating and making decisions related to unforeseen conditions
B.18	Monitor safety on site
B.10	Supervise testing, commissioning and handover of the project
B.20	Issue substantial completion certificate

Source: Abdul Razzak Rumane. (2010). *Quality Management in Construction Projects*. Reprinted with permission of Taylor & Francis Group

The following audit tools are used to perform a quality audit for the bidding and tendering process to select the construction supervisor:

- Checklists
- Review of documents
- Interviews
- Questionnaires
- Analysis of audit documentation

The auditee team members are mainly from the following organizational groups:

- Tendering Committee
- Tendering Manager
- Project Manager
- Contract Administrator
- Design Manager
- Document Controller
- Finance Manager

In order to manage and control this, the bidding and tendering process for the selection of the Design-Build construction supervisor can be divided into four major stages. These stages are:

1   First stage: Shortlisting/prequalification of supervisor
2   Second stage: Tender documents
3   Third stage: Contract bid solicitation
4   Fourth stage: Contract award

Figure 6.77 illustrates the typical quality auditing stages of the bidding and tendering process for the selection of the Design-Build construction supervisor (consultant).

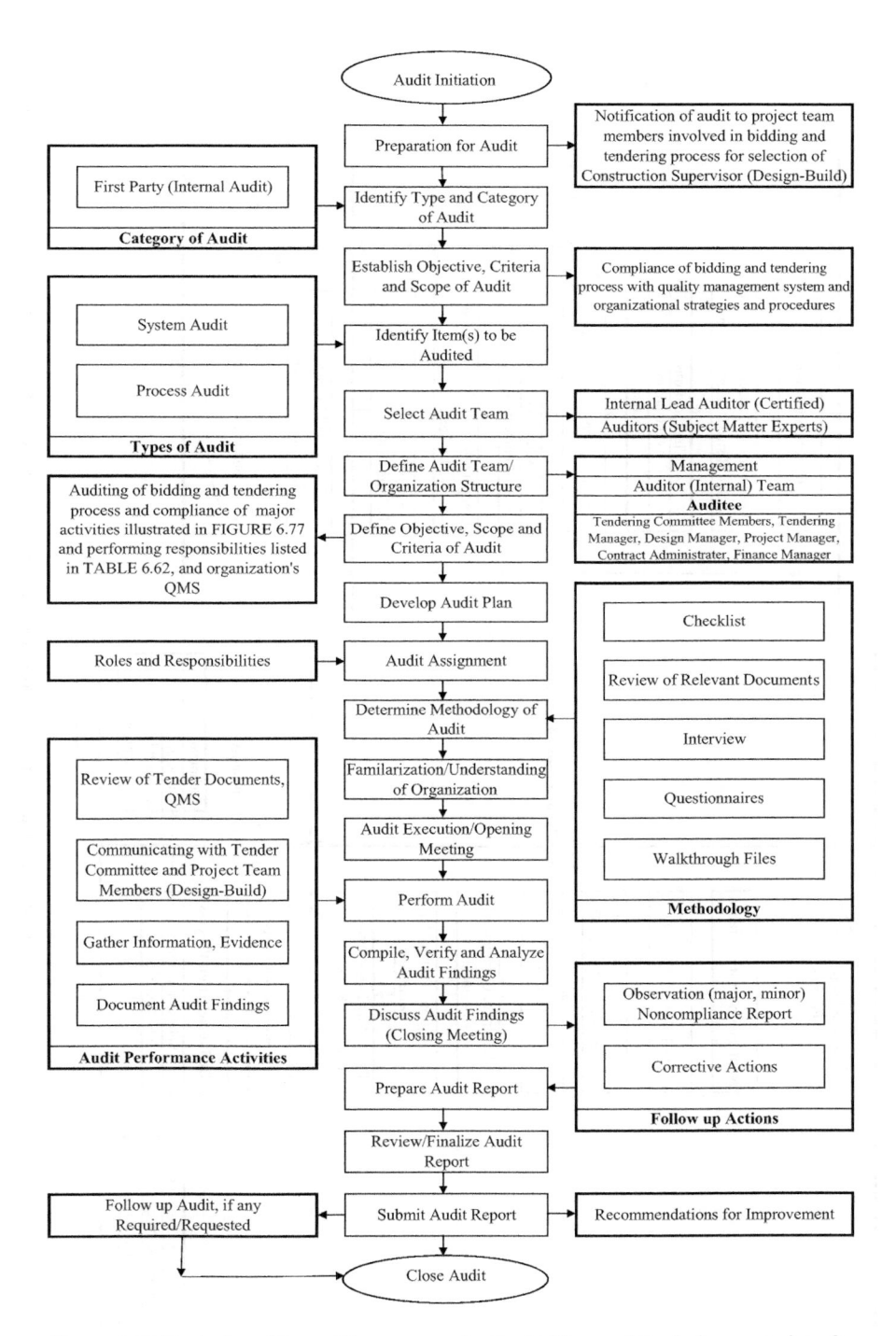

Figure 6.76 Typical quality auditing process for the bidding and tendering procedure for the selection of the construction supervisor (Design-Build)

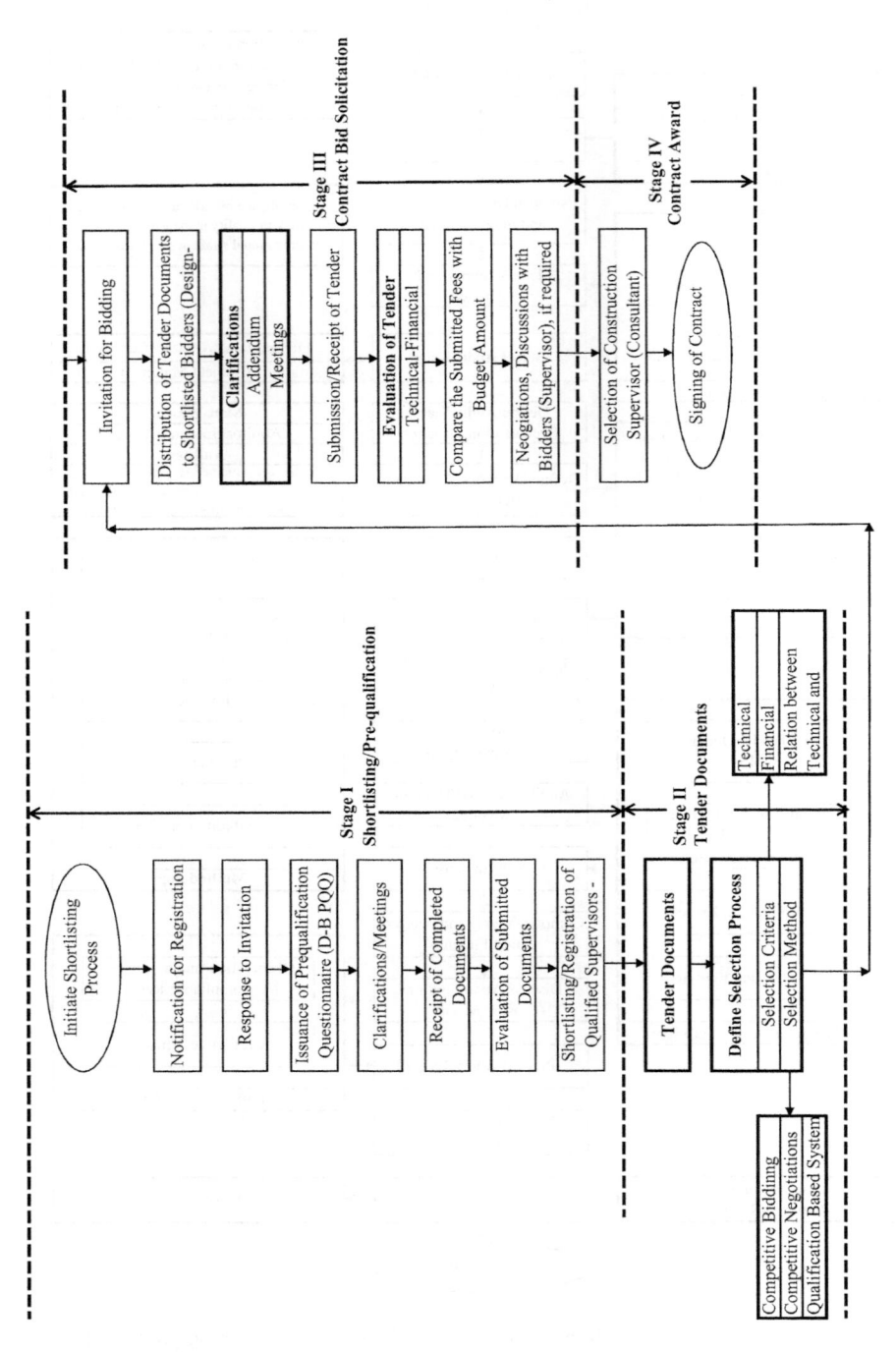

Figure 6.77 Typical quality auditing stages of the bidding and tendering process for the selection of the construction supervisor (Design-Build)

Auditing processes for project life cycle phases 437

Table 6.63 lists the prequalification questionnaires to register the Design-Build construction supervisor (consultant).

Based on the typical stages as per Figure 6.77, the audit methodologies can also be performed in four stages. Table 6.65 illustrates the audit methodology to assess/evaluate the process compliance at stage I, Table 6.66 illustrates the audit methodology to assess/evaluate the process compliance at stage II, Table 6.67 illustrates the audit methodology to assess/evaluate the process compliance at stage III, Table 6.68 illustrates the audit methodology to assess/evaluate the process compliance at stage IV and Figure 6.78 illustrates the logic flow diagram for the quality auditing of the selection of the construction supervisor.

6.3.3.4 Auditing process for the Design-Build construction supervisor tender submission process

Figure 6.79 illustrates a typical quality auditing process for the construction supervisor's (Design-Build) tender submission process.

The main objectives of performing a quality audit of the construction supervisor's proposal submission process are to:

1. Evaluate and assess process compliance with the organizational strategic policies and procedure
2. Assess conformance to QMS and that processes meet established quality system requirements
3. Review the process used to prepare the tender (quotation)
4. Verify that the proposal (tender) preparation and submission comply with the organization's tender (proposal) submittal process (Figure 6.80 illustrates the typical process)
5. Validate that the response to the contents of the tender documents is complete in all respects
6. Check that the availability of resources for the Design-Build (construction) is considered
7. Review the process used to prepare the tender
8. Validate that the schedule for deployment of project staff during design as well as construction complies with tender requirements
9. Validate the cost of the proposal (price analysis)
10. Ensure cost towards design development stage is considered
11. Review that the key risk factors are analyzed and managed
12. Check for compliance with regulatory requirements

The following audit tools are used to perform a quality audit of the construction supervisor's proposal submission process:

- Analytical tools
- Checklists
- Review of documents
- Interviews
- Questionnaires

Table 6.63 Prequalification questionnaires for the registration of the construction supervisor (Design-Build)

Serial Number	Element	Question	Response
1	General information	a Company name b Full address c Registration details/business permit d Management details e Membership of professional trade associations, if any f Award winning project, if any g Quality management certification	
2	Financial information	a Yearly turnover b Current workload c Audited financial report d Performance bonding capacity e Insurance limit	
3	a Organization details (general)	a Core area of business b How long in the same field of operation c Quality control/assurance organization	

	b Organization details (experience)	a Number of years in the same business
		b Technical capability
		   i Design development and review
		   ii Supervision of construction projects
		   iii Coordination between design and construction activities
		   iv Construction quality audits
		   v Monitoring and control
		   vi Testing and commissioning
		c List of previous contracts
		   i Name of project
		   ii Value of each contract
		   iii Contract period of each contract
		   iv List of Design-Build contracts	

4	Resources	a Human resources
		   i Management
		   ii Design development and review
		   iii Construction supervision
		   iv Design quality auditing
		   v Construction quality auditing
		   vi Project scheduling and planning
		   vii Project monitoring and controlling
		   viii Contract administration
		   ix Conflict management
		   x HSE
		b Human resource development plan	

5	Project reference		
6	Bank reference		

Table 6.64 Audit methodology for the assessment of the bidding and tender process for the selection of the construction supervisor (Design-Build) – Stage I (shortlisting/registration of construction supervisor)

Serial Number	Item To Be Assessed	Yes	No	Comments
A. Checklist				
1	Whether the notification for registration was announced in all technical newsletters and leading news media as per the organization's policy			
2	Whether there are registered construction supervisors (Design-Build type of projects) to participate in the tender			
3	Whether prequalification questionnaires (PQQ) are issued to all the intending bidders (please refer to Table 6.63 for guidelines)			
4	Whether there are any requests for queries/bid clarification			
5	Whether the meeting for clarification is attended by all the intending bidders			
6	Whether the attendance sheet for the meeting is signed by all the attendees			
7	Whether minutes of the meeting are circulated to all the attendees/participating bidders			
8	Whether sufficient time is provided to prepare and submit completed PQQ			
9	Whether there are any requests for extension of the submission date			
10	Whether all the intending bidders submitted the response to PQQ			
11	Whether received bid envelopes are placed in safe custody			
12	Whether there are any delayed submissions			
13	Whether received responses are acknowledged			
14	Whether received responses are clearly identified and recorded			
15	Whether responses are fairly evaluated as per the organization's selection policy			
16	Whether any of the submitted responses are found to be incomplete			
17	Whether there are any late submissions			
18	Whether risk factors are considered while evaluating the response and registration of construction supervisors			
19	Whether reasons for non-registration are conveyed to unsuccessful participants			
20	Whether shortlisted construction supervisors are allotted a code and registration number			

(*Continued*)

Table 6.64 (Continued)

B. Interview Questionnaires

Serial Number	Description	Response	Comments
1	How responses were submitted: • Hand delivery • By post • Courier service		
2	Whether PQQ was available for electronic distribution		
3	How was the response for registration/shortlisting		
4	Whether tender was open for international bidders or only local companies		

Table 6.65 Audit methodology for the assessment of the bidding and tender process for the selection of the construction supervisor (Design-Build) – Stage II (bidding and tender documents)

Serial Number	Item To Be Assessed	Yes	No	Comments
A. Checklist				
1	Whether the tender documents includes all the relevant information required to select the construction supervisor (Design-Build)			
2	Whether the tender documents include all the related activities to be performed by the supervisor during the construction phase			
3	Whether the activities listed in Table 6.62 have been considered while preparing the tender documents for the Design-Build project			
4	Whether the tender documents are approved by the project owner/tendering committee			
5	Whether the bid/tender documents are prepared as per the procurement method and contract strategy adopted during the early stages of the projectc(Design-Build)			
6	Whether all the information to enable bidders/supervisors to submit the proposal is clear and unambiguous			
7	Whether documents are prepared taking into consideration the organization's Quality Management System (QMS)			
8	Whether regulatory/authority requirements are taken into consideration while preparing the tender documents			

Serial Number	Item To Be Assessed	Yes	No	Comments
A. Checklist				
9	Whether the selection criteria and selection method for contractor (supervisor) selection is clearly specified in the organization's QMS			
10	Whether budget approval is obtained for supervisor's fees prior to the release of the tender announcement			
11	Whether the tender opening and closing dates are mentioned in the announcement			
12	Whether the schedule for the supervision period is included in the tender documents			
13	Whether the staff deployment chart is included in the tender documents			
14	Whether the qualifications of supervision team members are prescribed in the tender documents			
15	Whether project quality requirements are properly defined in the documents			
16	Whether the rights and liabilities of all the parties are mentioned in the documents			
17	Whether change/variation clauses are included in the documents			
18	Whether cancellation/termination clauses are included in the documents			
19	Whether bond and insurance clauses are included in the documents			
20	Whether the tender documents are as per international/local standard documents used by the construction industry			
21	Whether evaluation and assessment criteria are well defined			
22	Whether review and analysis procedure of bid is included			
23	Whether the contract to be signed for the construction supervisor is included in the documents			
24	Whether all the records are properly maintained			
25	Whether the request to provide the bidder's contact details are included in the documents			

B. Interview Questionnaires

Serial Number	Description	Response	Comments
1	Whether any outside agency was involved in the preparation of the tender documents		

(*Continued*)

Table 6.65 (Continued)

B. Interview Questionnaires

Serial Number	Description	Response	Comments
2	What guidelines were adopted to prepare the tender documents		
3	Whether any specific format/template was followed		
4	Whether all the requirements of tender documents were coordinated and agreed by relevant stakeholders		
5	Whether risks of disputes and associated financial costs were considered while preparing the contract documents		
6	Whether problems and challenges were identified while preparing the bid/tender		
7	Who was responsible for authorizing the bidding requirements		
8	Whether supervision responsibilities are clearly defined in the documents		
9	Whether staff selection procedure is clearly defined in the documents		
10	Whether the design team is an internal entity of the Design-Build contractor or an outside firm		

Table 6.66 Audit methodology for the assessment of the bidding and tender process for the selection of the construction supervisor (Design-Build) – Stage III (contract bid solicitation)

Serial Number	Item To Be Assessed	Yes	No	Comments
A. Checklist				
1	Whether the tender documents are properly organized			
2	Whether the tender documents are distributed to all the prequalified/shortlisted/registered bidders (construction supervisors) to participate in the tender			
3	Whether the tender is notified to all the prequalified bidders			
4	Whether all the prequalified/shortlisted/registered bidders participated in the tender			
5	Whether the attendance sheet for the meeting was signed by all the attendees			
6	Whether minutes of the meeting were circulated to all the attendees/participating bidders/construction supervisors			

Serial Number	Item To Be Assessed	Yes	No	Comments

A. Checklist

7	Whether sufficient time is provided to prepare and submit the tender			
8	Whether the tender submission date is extended from the date originally announced			
9	Whether an addendum(if any) is issued/ notified to all the participating bidders			
10	Whether there are any requests for extension of the submission date			
11	Whether the clarification meeting is attended by all participating bidders			
12	Whether any changes are made to the originally announced tender documents			
13	Whether the received bid envelopes are placed in safe custody			
14	Whether there are any delayed submissions			
15	Whether the tenders are opened as per the announced date and time			
16	Whether the tenders are opened in the presence of all the bidders			
17	Whether received tenders are acknowledged			
18	Whether received tenders are clearly identified and recorded			
19	Whether applicable fees/bid bond are submitted by all the bidders who participated in the tender			
20	Whether tenders are fairly evaluated as per the announced policy and procedure			
21	Whether the comparison of bids is tabulated			
22	Whether the bid cost is higher than the budgeted cost			
23	Whether any submitted tender is found to be incomplete			
24	Whether there are any late submissions			
25	Whether all the scheduled pages are signed by the bidder			
26	Whether risk factors are considered while evaluating the bids			
27	Whether reasons for the non-acceptance of the tender was conveyed to the unsuccessful bidders			

B. Interview Questionnaires

Serial Number	Description	Response	Comments
1	How the notification was announced		
2	How tenders were submitted: • Hand delivery • By post • Courier service		

(*Continued*)

Table 6.66 (Continued)

B. Interview Questionnaires

Serial Number	Description	Response	Comments
3	Whether the tender was available for electronic distribution		
4	How was the response to participate in the tender		
5	Whether tender box(es) were placed at secured places and monitored through an electronic surveillance system		
6	How much time was allowed to open the bids after the notified tendering closing time		
7	Whether the tender submission date was extended and what was the reason for extension		
8	Whether all the tender envelopes/packages were sealed with marking of tenders and contain the original tender documents and required number of sets as per the tender conditions		
9	Whether the bid review was established		
10	Whether the bid selection procedure was clearly defined		

Table 6.67 Audit methodology for the assessment of the bidding and tender process for the selection of the construction supervisor (Design-Build) – Stage IV (contract award)

Serial Number	Item To Be Assessed	Yes	No	Comments
A. Checklist				
1	Whether the selection of the construction supervisor (Design-Build) satisfies all the conditions to be a successful bidder			
2	Whether the selected bidder (construction supervisor) is a legal entity			
3	Whether the selection was made by the selection committee			
4	Whether the selection of the construction supervisor was approved by the relevant applicable authority, if mandated by the organization's policy			
5	Whether the selected construction supervisor (Design-Build) is capable of carrying out the specified work and has the necessary resources			

Serial Number	Item To Be Assessed	Yes	No	Comments

A. Checklist

6	Whether risk factors are considered prior to the signing of the contract			
7	Whether the standard contract format is used for signing the contract			
8	Whether the contract terms and conditions are clearly written and unambiguous			
9	Whether an addendum and minutes of the meeting are included in the contract documents			
10	Whether the contract period/schedule is properly described			
11	Whether the staff deployment schedule is included in the contract			
	Whether the design team/design firm qualifications are properly defined			
12	Whether the supervision staff qualifications are properly defined			
13	Whether a performance bond is submitted by the consultant			
14	Whether variation/change management clauses are included in the contract agreement			
15	Whether dispute resolution and conflict resolution clauses are properly defined			
16	Whether contract documents are reviewed prior to signing			
17	Whether tender documents and contract are properly archived			
18	Whether a letter of award is sent			
19	Whether final results are published and announced			

B. Interview Questionnaires

Serial Number	Description	Response	Comments
1	Whether the construction supervisor is registered with the relevant regulatory authorities for supervision of construction projects		
2	On what basis the bidder (construction supervisor) was selected: low bid, competitive bid or quality based system		
3	Was there any tie between two or more bidders		
4	Was there any dispute or objection raised by any of the bidders before the award of the contract to the successful bidder		

(*Continued*)

Table 6.67 (Continued)

B. Interview Questionnaires

Serial Number	Description	Response	Comments
5	Whether the final price was negotiated or as per the bid (tender) received		
6	Whether the successful bidder failed to submit a performance bond		
7	Whether there was any need to update the approved budget		
8	Whether risks in non-deployment of any of the supervision team members were considered		
9	Whether the contract includes a penalty clause for delay in the deployment of supervision team members		
10	Whether the contract includes a penalty clause for failure to provide the required team member		
11	Who is responsible for selecting supervision team members		

Table 6.68 Audit methodology for the assessment of the construction supervisor's (Design-Build) tender submission process

Serial Number	Item To Be Assessed	Yes	No	Comments
A. Checklist				
1	Whether the company is already registered as a construction supervisor (consultant) to participate in the tender			
2	Whether there are any fees to collect tender documents			
3	Whether the scope of work is clearly defined in the tender documents			
4	Whether there is any request for clarification raised			
5	Whether the response to price schedule/tender in pricing is complete			
6	Whether the requested information in the tender documents is clearly responded to			
7	Whether sufficient time is provided to prepare and submit the completed tender			
8	Whether there are any requests for extension of the submission date			
9	Whether there are any delayed submissions			
10	Whether the tender response is prepared taking into consideration/referring to the organization's quality management system			

Serial Number	Item To Be Assessed	Yes	No	Comments

A. Checklist

11	Whether the availability of qualified personnel to supervise the Design-Build construction is as per tender requirements			
12	Whether the availability of qualified personnel to review design development and coordination with construction activities is as per tender requirements			
13	Whether the staff deployment costs include all the benefits and company overhead cost			
14	Whether the cost to review RFI during construction is considered while preparing the tender			
15	Whether the tender was reviewed prior to submission			
16	Whether the tender response was approved by the relevant committee/authorized person			
17	Whether the risk factors were considered and analyzed			
18	Whether all the relevant pages are signed			

B. Interview Questionnaires

Serial Number	Description	Response	Comments
1	Whether the notification for tender submission was announced in all technical newsletters and leading news media or it was only for registered bidders (contractors)		
2	How the tender was submitted: • Hand delivery • By post • Courier service		
3	Whether the tender documents were available for electronic distribution		
4	What tools and techniques were used to price the tender		
5	Whether the organization's quality management plan was considered to evaluate the price schedule		
6	Whether the availability of resources (technical as well as skilled and semi-skilled) was considered while preparing the price schedule		
7	Whether the overhead cost of vacation pay of staff is taken into consideration while developing monthly cost		

(Continued)

Table 6.68 (Continued)

B. Interview Questionnaires

Serial Number	Description	Response	Comments
8	Whether the deployment schedule was properly reviewed by taking into consideration the availability of resources		
9	Whether the regulatory approval period was properly considered		
10	What risk factors were considered while evaluating the price schedule		
11	Whether any indirect costs are considered		
12	Was there any negotiation to the submitted quotation		
13	Whether design team is an internal entity or outside design firm		

The auditee team members are mainly from the following organizational groups:

- Tendering Committee
- Tendering/Proposal Manager
- Project Manager
- Supervisor in Charge
- Design Manager
- Quality Manager
- Contract Administrator/Quantity Surveyor
- Document Controller
- Finance Manager

Table 6.68 illustrates the audit methodology for the assessment of the construction supervisor's (Design-Build) proposal submission process and Figure 6.81 illustrates a logic flow diagram for the quality auditing of the construction supervisor's (Design-Build) proposal submission process.

6.3.4 *Auditing process for construction (Design-Build) activities*

In the Design-Build type of contract, the owner contracts a firm (contractor) solely responsible for designing and building the project. A Design-Build type of contract is a contractually defined project often used to shorten the time required to complete a project. The degree of overlap between design and construction activities can vary substantially depending on specific project requirements. In the Design-Build type of deliverable system, the contractor is contracted right from the early stages of the construction project and is responsible for the design development of the project as well as construction. The Design-Build contractor

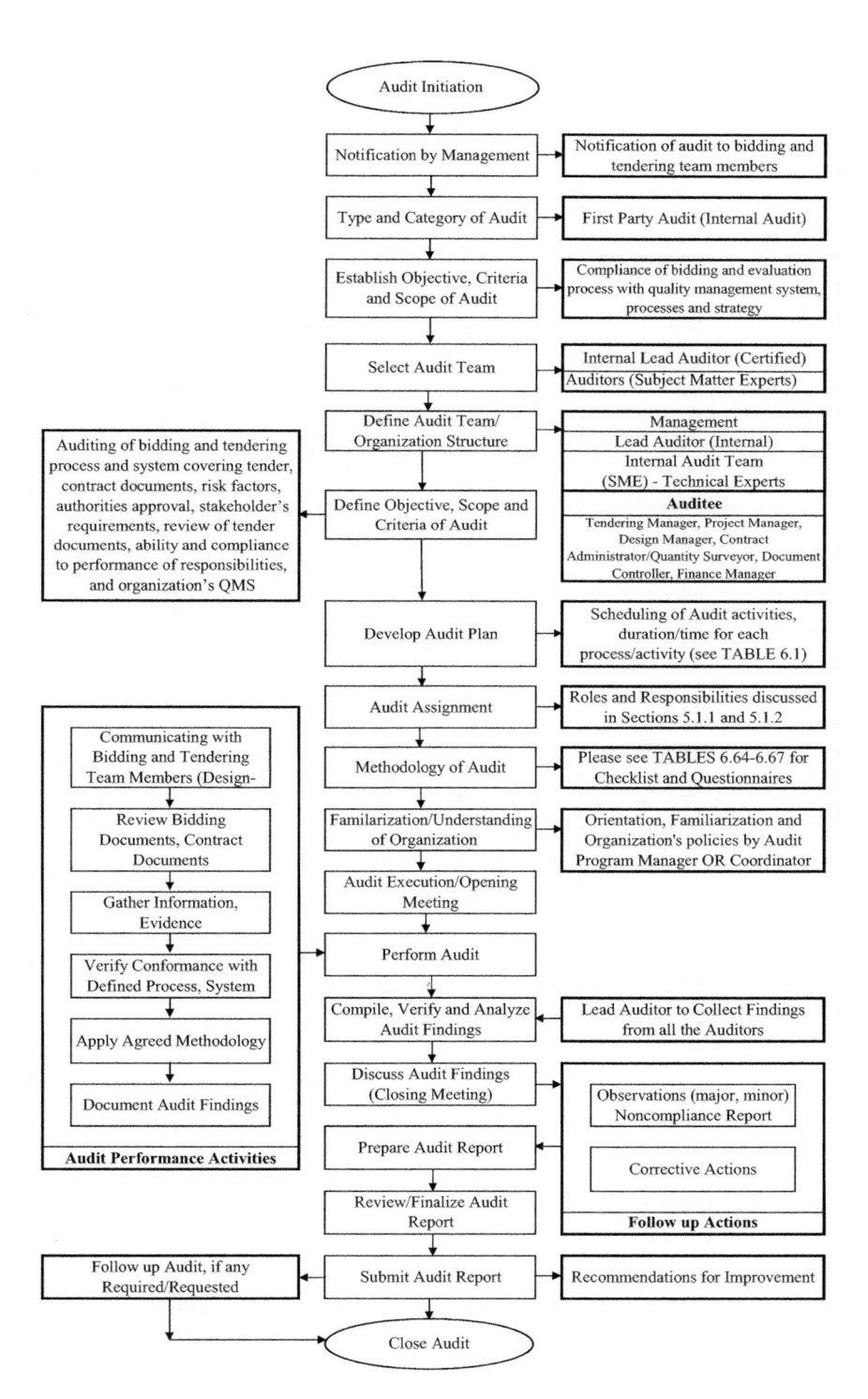

Figure 6.78 Logic flow diagram for the quality auditing of the selection of the construction supervisor (Design-Build)

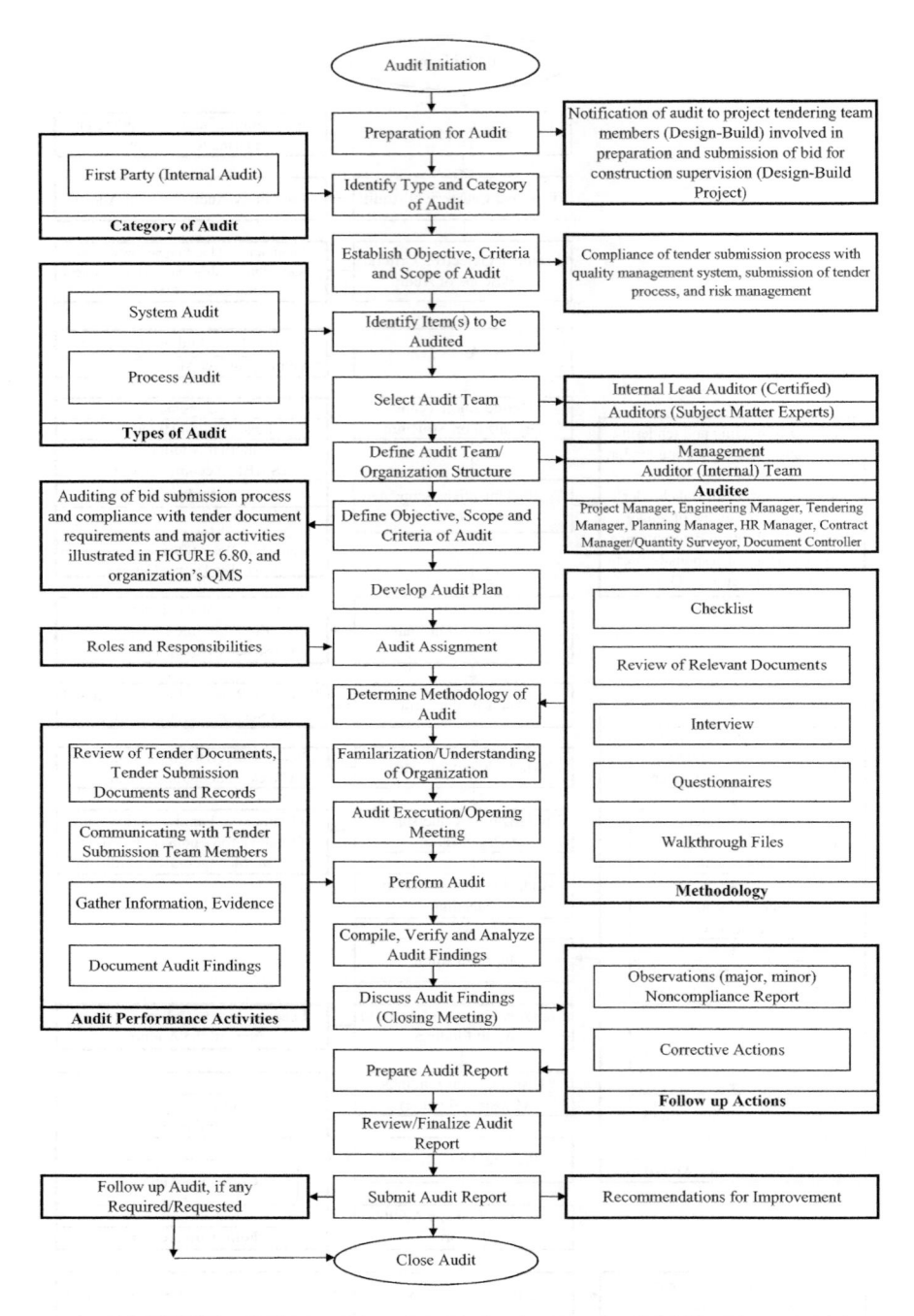

Figure 6.79 Typical quality auditing process for the construction supervisor's (Design-Build) tender submission process

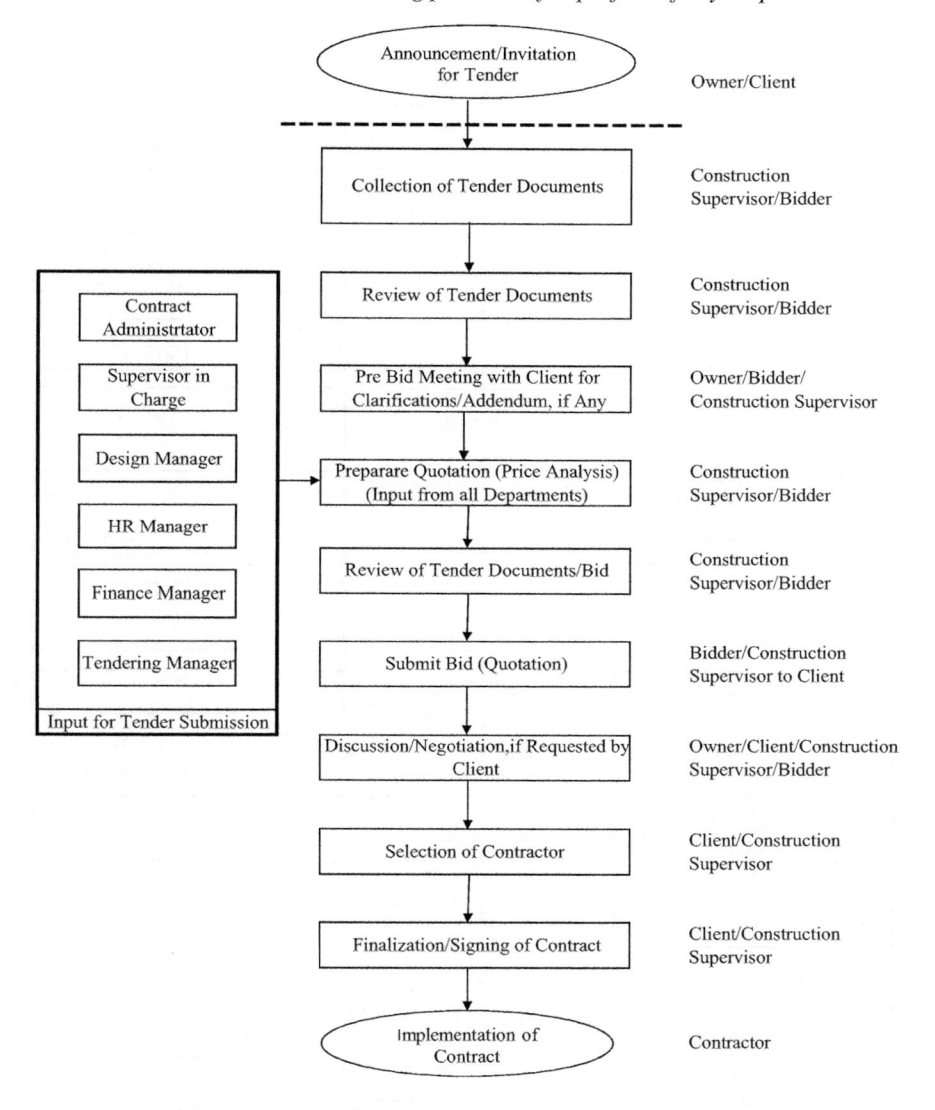

Auditing processes for project life cycle phases 451

Figure 6.80 Typical tender submission procedure by the construction supervisor (Design-Build)

is responsible for coordinating design development activities. The Design-Build contractor coordinates with the project architect (either Design-Build contractor entity or specialist A/E firm) and endeavors to further the interest of the project owner and the project. The Design-Build contractor provides pre-construction phase services (design development) and construction phase services (construction) and completes the project in an expeditious manner to make the project qualitative, competitive and economical to satisfy the owner's objectives.

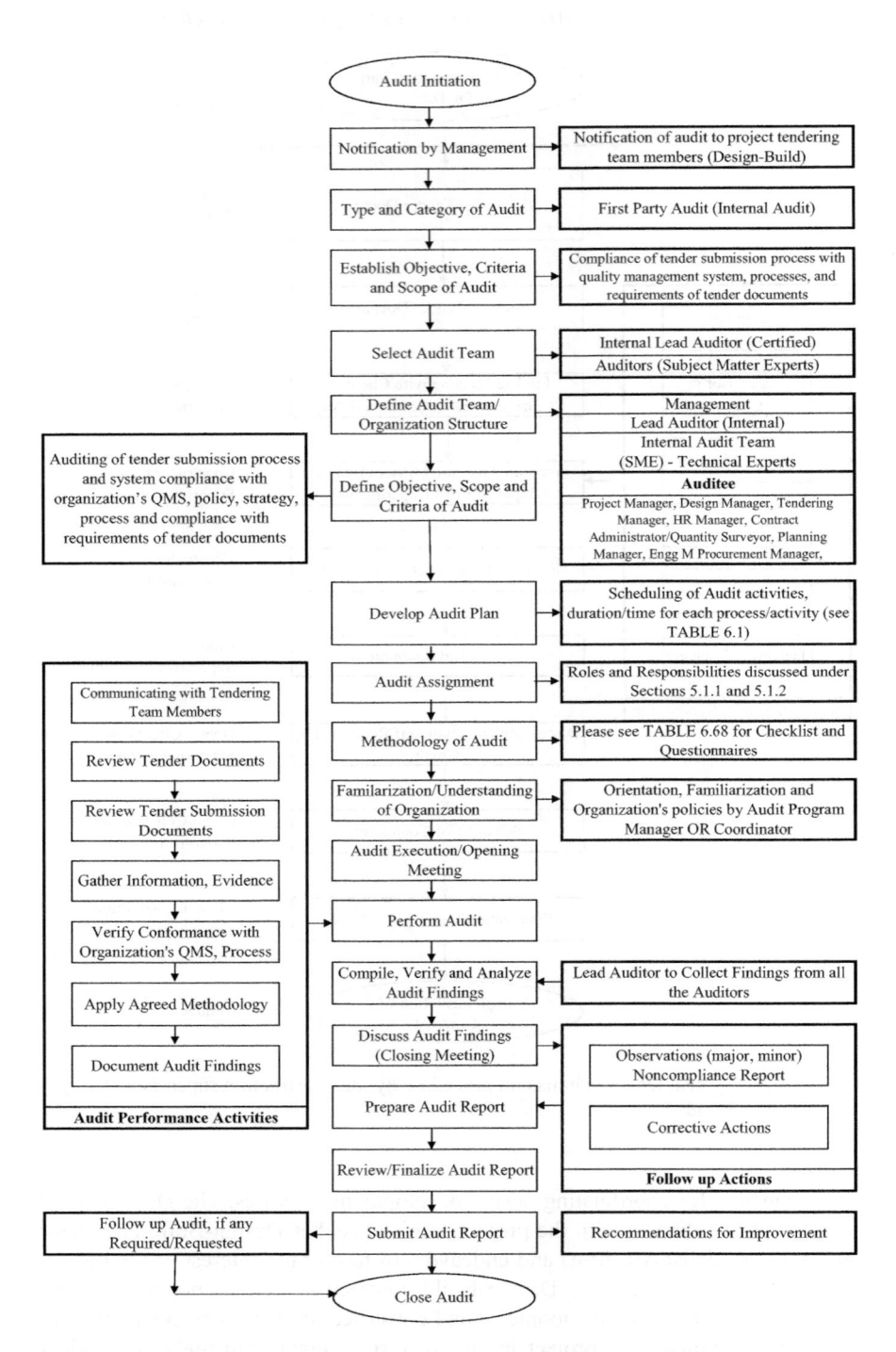

Figure 6.81 Logic flow diagram for the quality auditing process for the construction supervisor's (Design-Build) tender submission process

The quality auditing of the Design-Build type of project can be performed as follows:

1. During implementation of design development and construction works
2. After completion of design and construction works

However, this section discusses quality auditing after the completion of:

1. Design development activities
2. Construction works

6.3.4.1 Contractor's work

The quality auditing of the contractor's work for the Design-Build type of project can be performed in two stages. These are as follows:

1. Design development activities
2. Construction works

6.3.4.1.1 DESIGN DEVELOPMENT ACTIVITIES

The design developments activities commence upon the date specified in a notice to proceed issued by the project owner and continue through completion of the construction documents and specifications. Design development activities may overlap construction works. Design development activities mainly include:

1. Review of the owner's design criteria and construction requirements described in the contract documents
2. Development of the schedule for design development activities including reasonable time for the owner's (construction supervisor's) review and approval of design drawings and specifications
3. Development of schematic design based on the contracted conceptual design and performance specifications and requirements
4. Development of detail design and construction drawings
5. Development of construction drawings, documents and specifications
6. Coordination with project owner and construction team to ensure the owner's project objectives are taken care

The main objectives of performing the quality auditing of design development activities of the Design-Build project are:

1. Ascertain that contract/construction documents, conceptual design and performance specifications are reviewed and understood to start the project design (schematic design)

2   Ascertain the project schedule for design development stages and construction phase is prepared and approved
3   Ascertain that communication and coordination procedures among project stakeholders are established
4   Validate that contract requirements comply with the applicable laws, rules and regulations, and codes and standards
5   Assess conformance to QMS and that processes meet established quality system requirements

The quality audit aims and objectives of design development activities of the Design-Build type of project may vary as per the Design-Build contractor's requirements and to what extent the details are to be assessed and verified. Normally the quality audit of the design development activities of the Design-Build project can be performed at the following stages:

1   Schematic design
2   Detail design
3   Construction documents, specifications

In general the quality auditing processes for the schematic design phase and detail design phase discussed earlier in Sections 6.2.2.2 and 6.2.2.3 for the Design-Bid-Build type of project can be used to perform quality auditing for the schematic design and detail design of the Design-Build project. The aims and objectives can be modified to include Design-Build project requirements.

As regards the quality auditing process of construction documents and specifications, it slightly differs to the quality auditing process for construction documents discussed in Section 6.2.2.4 as it may not be necessary to develop all the construction document deliverables in the Design-Build type of project. The quality auditing process discussed in Section 6.2.2.4 can be modified by considering the construction document deliverables that will be different to those of the deliverables listed in Table 6.21 and used to perform the quality auditing.

6.3.4.1.2 CONSTRUCTION WORKS

Figure 6.82 illustrates the typical quality auditing stages for the construction work (phase) process activities of the Design-Build project.

The quality auditing of the construction work (phase) of the Design-Build type of project can be performed as follows:

1   During construction/execution of works
2   After completion of construction/execution of works

Figure 6.41 discussed earlier in Section 6.2.4 illustrates the typical quality auditing process for construction work (phase).

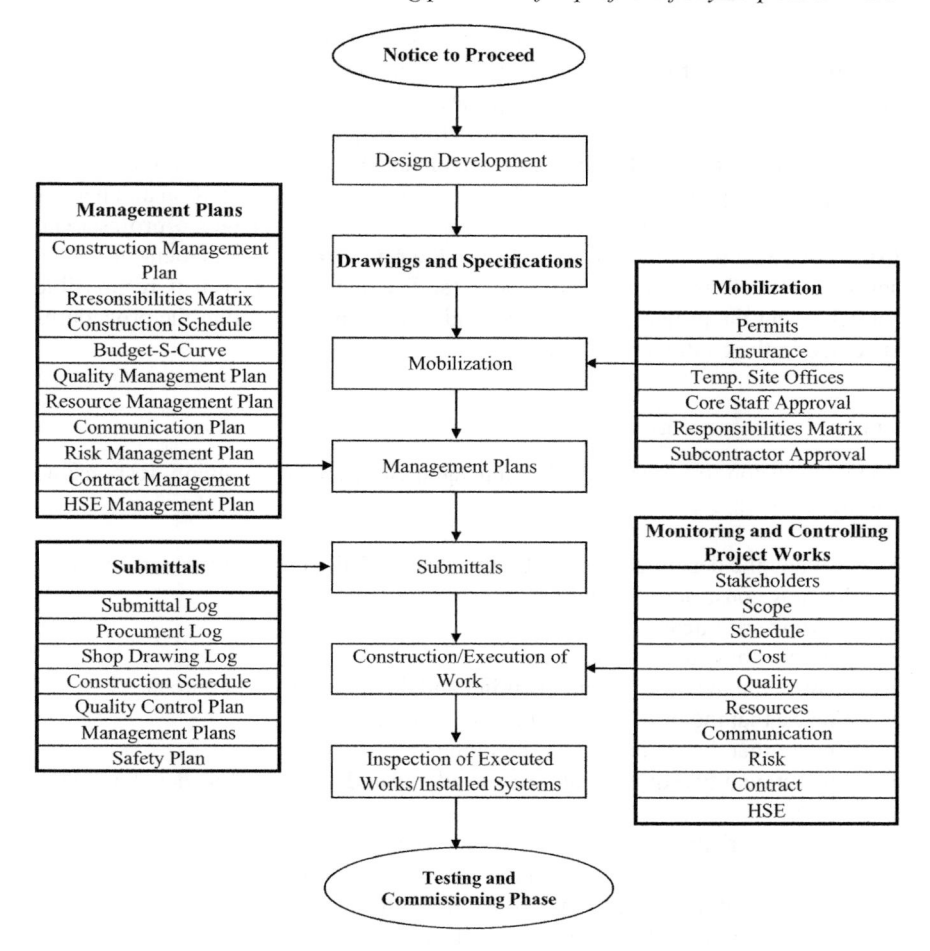

Figure 6.82 Typical quality auditing stages for the construction work (phase) of the Design-Build project

The auditing process that includes the aims, objectives and audit methodology for the assessment of construction work (phase) activities of the Design-Build type of project discussed earlier in Sections 6.2.4.1 and 6.2.4.2 can be used to perform the quality auditing process for the construction work (phase) of the Design-Build type of project. Please refer to Figures 6.42–6.59 and Tables 6.36–6.43 discussed in Sections 6.2.4.1 and 6.2.4.2.

6.3.4.2 Construction supervisor's activities

Complex and major construction projects have many challenges such as delays, changes, disputes, accidents on site, etc. and therefore need efficient management of the construction phase of the facility/project to meet the intended use and the owner's expectations. The owner/client may not have the necessary staff/ resources in-house to manage the design, planning, monitoring and controlling and construction of the construction project to achieve the desired results. In the case of the Design-Build type of construction project, the owner has to monitor the project activities right from the design management, as design development is the responsibility of the Design-Build contractor up to the completion and handover of the project. Therefore, in such cases, the owner engages a professional firm which is trained and has expertise in the review and management of design services as well as construction processes. The supervision firm is responsible for achieving the project quality goals and also for implementing the procedures specified in the contract documents.

6.3.4.2.1 DESIGN ACTIVITIES

The supervision firm during design management activities is mainly responsible for the following:

- Review the design developed by the Design-Build contractor's design team
- Review construction drawings
- Review specifications
- Review construction documents
- Advise the owner about any conflict and deviations to the contract
- Ensure that the project schedule is maintained
- Ensure that the project cost is maintained
- Project quality requirements are considered and assured while developing the design
- Validate that the regulatory requirements are taken into consideration
- Check that applicable codes and standards are followed
- Validate that the designer has taken into consideration the project risk
- Coordinate with the Design-Build design team to ensure that the project design is as per the owner's design guidelines and design criteria
- Validate that value methodology is applied
- Approve the design documents on behalf of the owner

Figure 6.83 illustrates a typical quality auditing process for the construction supervisor's activities related to the design development stages of the Design-Build project.

The main objectives of performing the quality audit of the construction supervisor's activities related to the design development stages of the Design-Build project are to:

1. Evaluate and assess process compliance with the contract requirements as per concept design and performance requirements (statement of requirements)
2. Assess conformance to QMS and that processes meet established quality system requirements

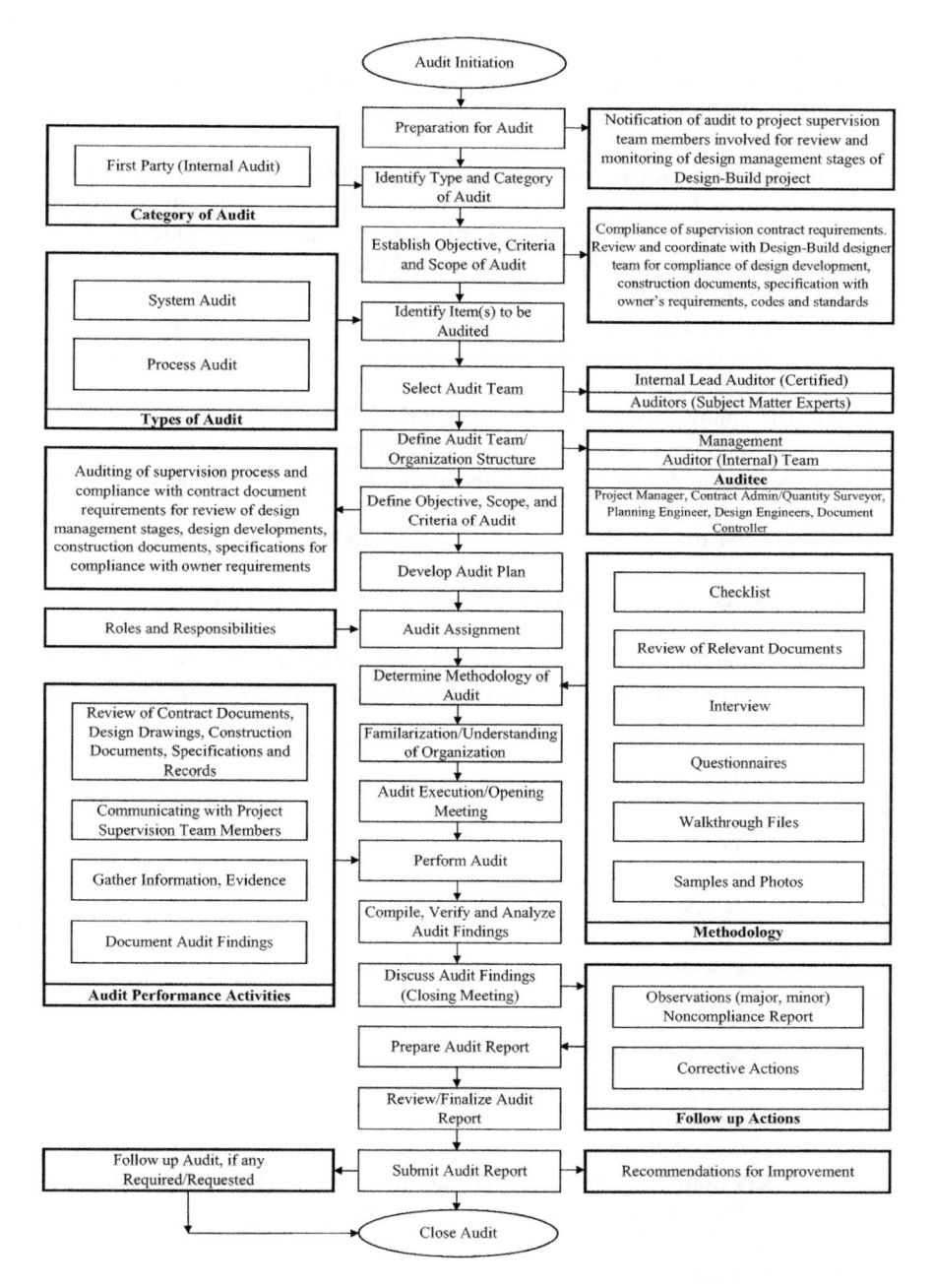

Figure 6.83 Typical quality auditing process for the construction supervisor's activities related to the design management stages of the Design-Build project

3. Check the design complies with project goals and objectives
4. Review the process used to perform the activities listed in the contract. The main responsibilities are listed in Table 6.62 (A – Design development activities)
5. Validate that the design progress by the Design-Build design team was properly monitored to develop the project design as per the owner's design guidelines
6. Check if there are any changes/variation to the contract requirements
7. Check the design development is as per schedule
8. Check for project cost variance
9. Check that design deliverables were produced as per contract documents
10. Check that construction documents and specifications were produced as per contract requirements
11. Validate compliance with quality standards and codes
12. Check for compliance with regulatory requirements
13. Check that the designer has considered the risk factors while developing the project design
14. Review construction documents and specifications for compliance with contract documents
15. Check for compliance to safety considerations in the design

The following audit tools are used to perform a quality audit of the construction supervisor's activities related to the design management stages of the Design-Build project:

- Analytical tools
- Checklists
- Review of documents
- Interviews
- Questionnaires

The auditee team members are mainly from the following organizational groups:

- Project Manager
- Design Manager
- Contract Administrator/Quantity Surveyor
- Planning/Scheduling Engineer
- Engineers from different trades such as architectural, structural, mechanical, HVAC, electrical, low voltage system, landscape, infrastructure
- Document Controller
- Office Secretary

Table 6.69 illustrates the audit methodology for the assessment of the construction supervisor's activities related to the design management stages of the Design-Build project and Figure 6.84 illustrates a logic flow diagram for the quality

Table 6.69 Audit methodology for the assessment of the construction supervisor's activities related to the design development stages of the Design-Build project

Serial Number	Item To Be Assessed	Yes	No	Comments
A. Checklist				
A-1 Design and Drawings				
1	Whether contract documents were reviewed and design management requirements were properly established			
2	Whether design management deliverable were established and listed to ensure compliance with contract requirements			
3	Whether the design drawings at all stages were reviewed to ensure that they support the owner's project goals and objectives			
4	Whether the design management documents were reviewed to ensure that the designer developed the design, construction documents and specifications to meet all the elements specified in the contract documents			
5	Whether the owner's requirements are fully considered			
6	Whether the contracted concept design has all the necessary detail to proceed to the next stage of design development			
7	Whether there were comments on schematic design			
8	Whether there were comments on detail design			
9	Whether the designer has considered regulatory/statutory requirements			
10	Whether the designer developed the design based on specified quality standards and codes			
11	Whether the designer has developed the design to meet all the performance requirements			
12	Whether the designer has taken care of the constructability factor while developing the design			
13	Whether sustainability is considered in developing the design			
14	Whether the design meets LEED requirements			
15	Whether energy conservation is considered			
16	Whether the designer has considered the facility management requirements			
17	Whether the designer has taken into consideration project boundaries and assumptions			
18	Whether value engineering study recommendations are considered while developing the design			

(*Continued*)

Table 6.69 (Continued)

Serial Number	Item To Be Assessed	Yes	No	Comments
A. Checklist				
19	Whether project objectives in respect of time, cost and quality are considered while developing the design			
20	Whether design is developed within agreed-upon schedule			
A-2 Specifications and Documents				
1	Whether the designer prepared the specification taking into consideration international construction contract documents			
2	Whether the specifications and construction documents meet performance specifications as per contract documents			
3	Whether division numbers in the specifications match with applicable international standards			
4	Whether the documents are developed within the agreed-upon schedule			
A-3 Quality				
1	Whether design is fully coordinated for conflict between different trades			
2	Whether the designer has considered the availability of resources during the entire project life cycle			
3	Whether all the drawings are numbered			
4	Whether the design matches with property limits			
5	Whether drawings are reviewed prior to submission			
6	Whether the design was coordinated for construction activities			
A-4 Risk				
1	Whether the designer has considered project risks while developing the design and documents			
2	Whether valid assumptions and constraints are considered			
A-5 Safety				
1	Whether the designer considered health and safety requirements while developing the design			
2	Whether environmental constraints and requirements are considered by the designer			
3	Whether the design confirms with fire and egress requirements			

Serial Number	Item To Be Assessed	Yes	No	Comments

A. Checklist

4	Whether environmental compatibility is considered while developing the design			

A-6 General

1	Whether all the related data and information was collected by the designer to develop the design			
2	Whether site investigations were done by the designer prior to the development of the design and documents			
3	Whether the design team members were approved for the design development			
4	Whether a responsibilities matrix was developed			
5	Whether regulatory approval was obtained and comments, if any, incorporated and all review comments responded to			
6	Whether the design and documents were approved by all the concerned stakeholders			

B. Interview Questionnaires

Serial Number	Description	Response	Comments
1	Was there any design approval submission plan		
2	Whether construction documents were fully coordinated with design drawings and specifications		
3	Whether coordination meetings were held among all the team members		
4	Whether the availability of resources during the construction phase was considered by the designer		
5	Was there any approved design, specifications, construction review procedure		
6	Whether the design, documents and specification were developed within the agreed-upon time		
7	Was there any delay to the start of construction due to non-completion of design documents		

auditing process for the construction supervisor's activities related to the design management stages of the Design-Build project.

6.3.4.2.2 CONSTRUCTION WORKS

The auditing process that includes the aims, objectives and audit methodology for the assessment of the construction supervisor's activities related to the

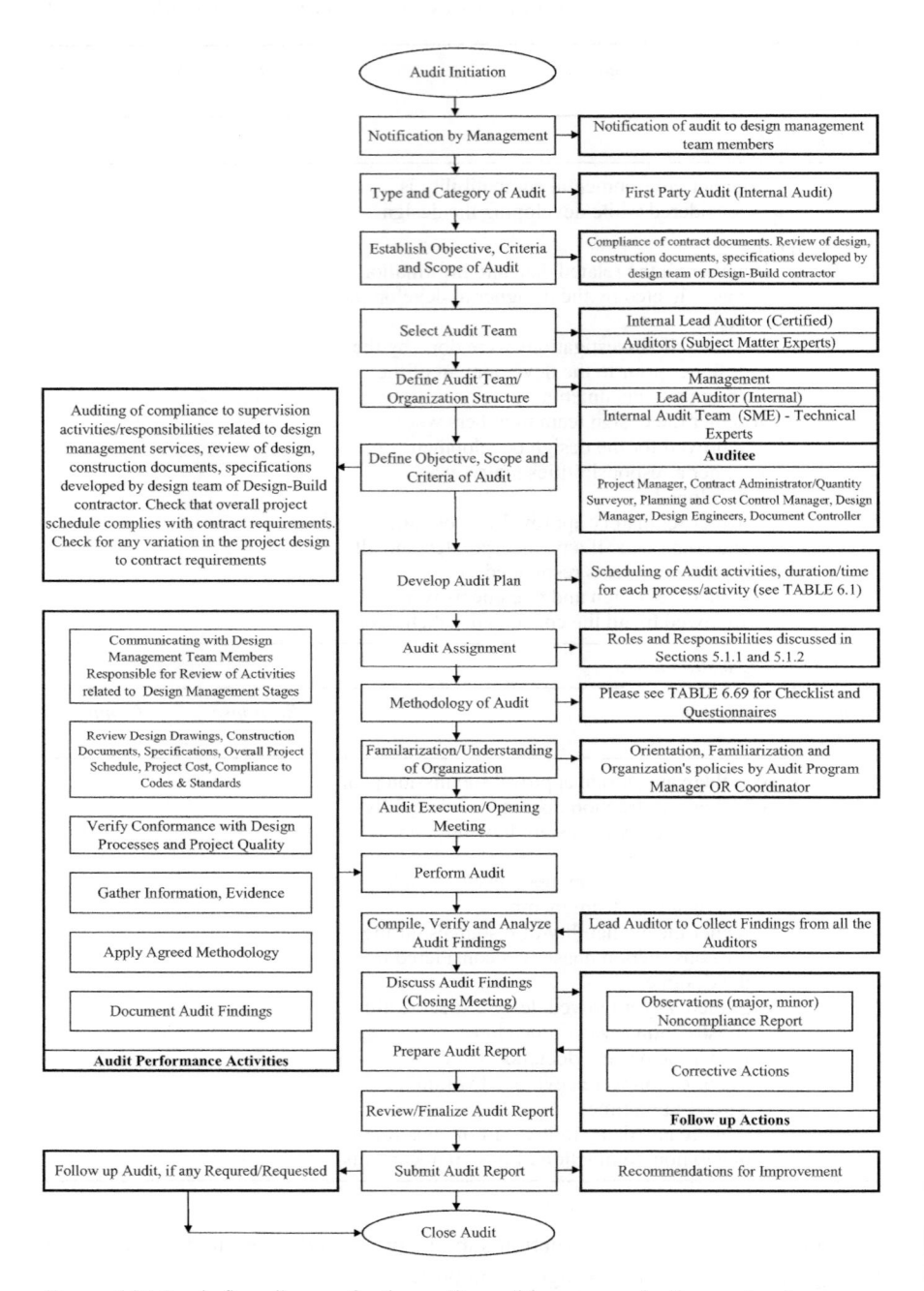

Figure 6.84 Logic flow diagram for the quality auditing process for the construction supervisor's activities related to the design management stages of the Design-Build project

construction work (phase) of the Design-Bid-Build type of project discussed earlier in Section 6.2.4.2 can be used to perform the quality auditing process for the construction supervisor's activities related to the construction work (phase) of the Design-Build type of project. Please refer to Figures 6.60–6.61 and Tables 6.44–6.48 discussed in Section 6.2.4.2.

6.3.5 Auditing process for the testing, commissioning and handover phase

Testing, commissioning and handover is the last phase of the construction project life cycle. This phase involves the testing of electro-mechanical systems, commissioning of the project, obtaining authorities' approval, training of user's personnel and handing over of technical manuals, documents and as-built drawings to the owner/owner's representative. During this period the project is transferred/handed over to the owner/end user for their use and a substantial completion certificate is issued to the contractor.

The quality auditing process discussed in Section 6.2.5 for the testing, commissioning and handover phase of the Design-Bid-Build type of project can be used for performing the quality audit of Design-Build projects as all the activities in both types of project delivery system are the same.

6.3.5.1 Contractor's work

The quality auditing process that includes the aims, objectives and audit methodology for the assessment of the testing, commissioning and handover process discussed in Section 6.2.5.1 can be used for the auditing process for the testing, commission and handover of the Design-Build type of project. Please refer to Figures 6.62, 6.63, Table 6.49, Figure 6.64, and Table 6.50 discussed in Section 6.2.5.1.

6.3.5.2 Construction supervisor's activities

The quality auditing process that includes the aims, objectives and audit methodology for the assessment of the testing, commissioning and handover process discussed in Section 6.2.5.2 can be used for the auditing process for the testing, commission and handover of the Design-Build type of project. Please refer to Figure 6.65, Table 6.51, Figures 6.66 and 6.67 and Table 6.52 discussed in Section 6.2.5.2.

6.4 Auditing process for the project manager type of project delivery system

A project manager contract is used by the owner when the owner decides to turn over the entire project management to a professional project manager. In the project manager type of contract, the project manager is the owner's representative, and is directly responsible to the owner. The project manager is responsible for the planning, monitoring and management of the project. In its broadest sense, the project manager has responsibility for all phases of the project from inception of the project to the completion and handing-over of the project to the owner/end user. The project manager is involved in giving advice to the owner and is responsible for appointing the design professional(s), consultant and supervision firm, and selecting the contractor to construct the project. Figure 6.85 illustrates the contractual relationship between the owner, project manager, designer (A/E) and contractor.

6.4.1 Auditing process for the selection of the project manager

Normally, the project manager firm is hired on the basis of **qualifications**. The basis of selection is solely based on demonstrated competence, professional qualifications and experience for the type of services required. In quality based selection, the contract price is negotiated after the selection of the best qualified firm.

The project manager firm has a direct contractual relationship with the designer (A/E) and contractor.

The project manager firm is responsible for the administration of the construction project on behalf of the owner. It has responsibilities for the following:

- Selection of project teams
- Define project scope and baseline
- Review project design/design management
- Managing project

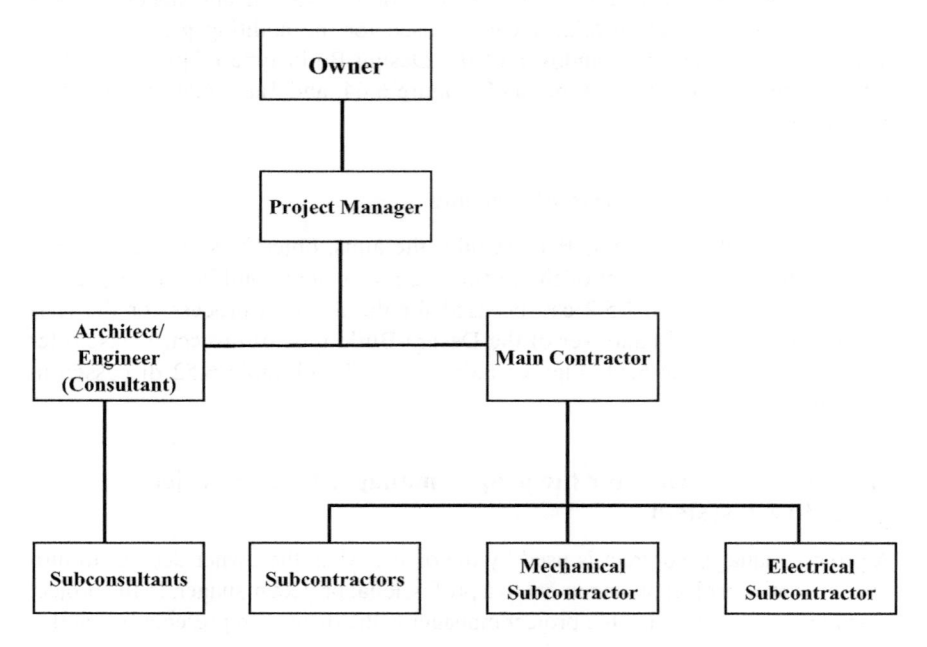

Figure 6.85 Project manager type delivery system contractual relationship

Auditing processes for project life cycle phases 465

- Develop project schedule
- Monitor project progress
- Reporting project status
- Monitoring project cost
- Managing project quality
- Resource management
- Update project owner/client about project progress
- Coordinate with various project teams and stakeholders
- Managing project risk
- Administration of contracts
- Manage variation/changes
- Project safety
- Managing project payments
- Claim management

Figure 6.86 illustrates the typical quality auditing process for the bidding and tendering for the selection of the project manager firm.

The main objectives of performing a quality audit for the bidding and tendering process to select the project manager to manage the construction project on behalf of the owner/client are to:

1. Evaluate and assess process compliance with the organizational strategic policies and procedure
2. Assess conformance to QMS and that processes meet established quality system requirements
3. Verify that all the major activities are adequately performed and managed
4. Ensure that key risk factors are analyzed and managed to meet business requirements
5. Check the proper functioning and implementation of the project management system

The following audit tools are used to perform a quality audit for the bidding and tendering process to select the project manager:

- Checklists
- Review of documents
- Interviews
- Questionnaires
- Analysis of audit documentation

The auditee team members are mainly from the following organizational groups:

- Business Manager
- Tendering Manager
- Project Manager
- Financial Manager

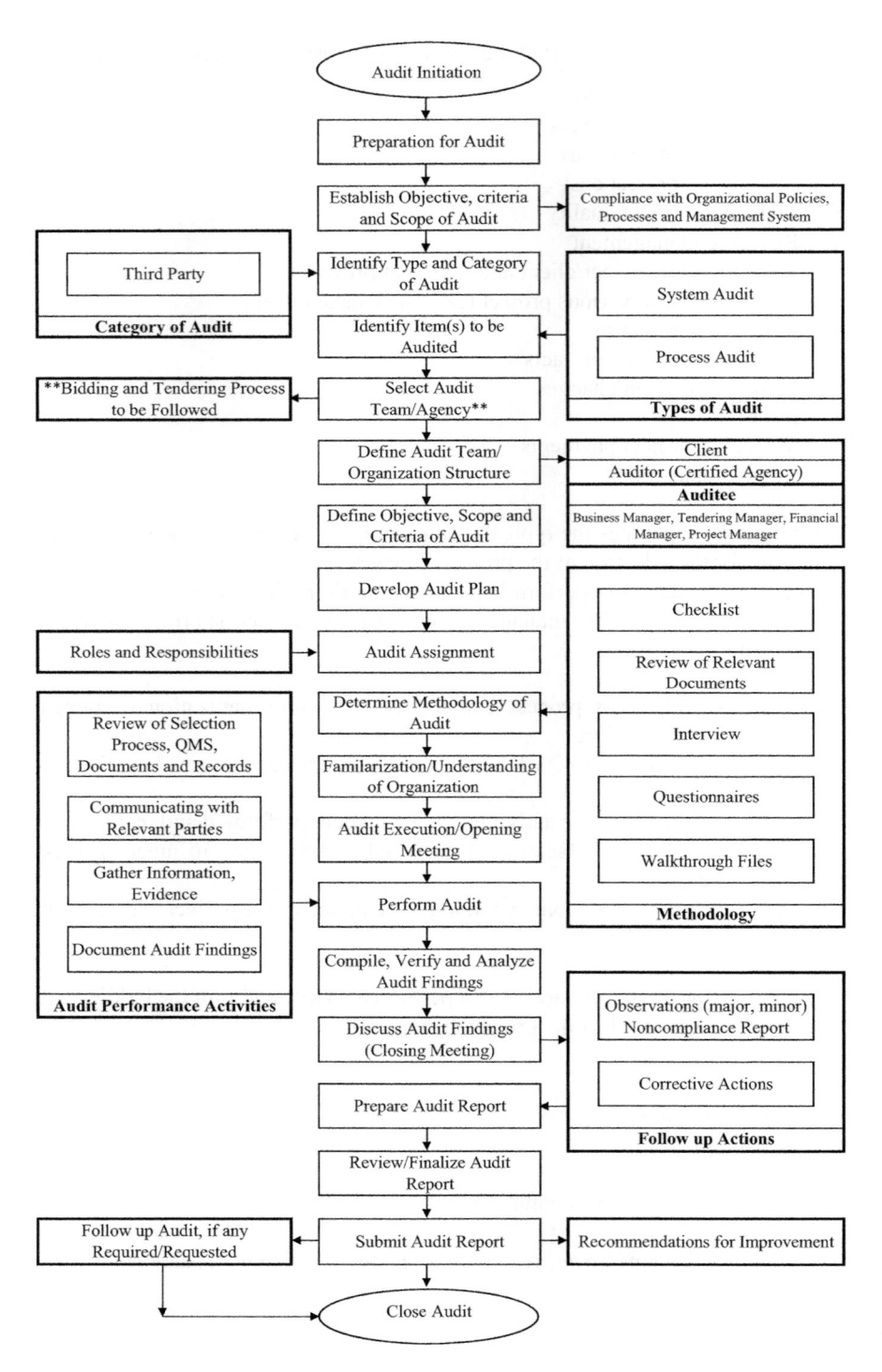

Figure 6.86 Typical quality auditing process for the bidding and tendering procedure for the selection of the project manager

Auditing processes for project life cycle phases 467

The bidding and tendering process stages discussed in Section 6.2.1.2 as illustrated in Figure 6.9 can be used to assess/evaluate the process compliance of the selection of the project manager.

Table 6.70 illustrates the audit methodology to assess/evaluate the process compliance at stage I, Table 6.71 illustrates the audit methodology to assess/evaluate the process compliance at stage II, Table 6.72 illustrates the audit methodology to assess/evaluate the process compliance at stage III, Table 6.73 illustrates the audit methodology to assess/evaluate the process compliance at stage IV and Figure 6.87 illustrates the logic flow diagram for the quality auditing process for the selection of the project manager.

Table 6.70 Audit methodology for the assessment of the bidding and tender process for the selection of the project manager firm – Stage I (shortlisting/registration of designers)

Serial Number	Item To Be Assessed	Yes	No	Comments
A. Checklist				
1	Whether the notification for registration was announced in all technical newsletters and leading news media as per the organization's policy			
2	Whether there are registered project manager firms to participate in the tender			
3	Whether prequalification questionnaires(PQQ) are issued to all the intending bidders			
4	Whether tender documents clearly indicate the responsibilities of the project manager for all relevant activities such as selection of designer (A/E), consultant, design management, selection of contractor and subcontractor and project management of construction			
5	Whether there are any requests for bid clarification			
6	Whether the meeting for bid clarification is attended by all the intending bidders			
7	Whether the attendance sheet for the meeting is signed by all the attendees			
8	Whether minutes of the meeting are circulated to all the attendees/participating bidders			
9	Whether sufficient time is provided to prepare and submit completed PQQ			
10	Whether there are any requests for extension to submission date			
11	Whether all the intending bidders submitted the response to PQQ			

(*Continued*)

Table 6.70 (Continued)

Serial Number	Item To Be Assessed	Yes	No	Comments
A. Checklist				
12	Whether received bid envelopes are placed in safe custody			
13	Whether there are any delayed submissions			
14	Whether received responses are acknowledged			
15	Whether received responses are clearly identified and recorded			
16	Whether responses are fairly evaluated as per the organization's selection policy			
17	Whether evaluation criteria, weightage and methodology are specified for selection of project manager			
18	Whether any of the submitted responses are found to be incomplete			
19	Whether there are any late submissions			
20	Whether risk factors are considered while evaluating the response and registration of designers			
21	Whether reasons for non-registration are conveyed to unsuccessful participants			
22	Whether shortlisted project manager firms are allotted a code and registration number			

B. Interview Questionnaires

Serial Number	Description	Response	Comments
1	How responses were submitted: • Hand delivery • By post • Courier service		
2	Whether PQQ was available for electronic distribution		
3	How was the response for registration/shortlisting		
4	Whether the quoted price was as per the allocated budget for project management		

Table 6.71 Audit methodology for the assessment of the bidding and tendering process for the selection of the project manager firm – Stage II (proposal documents)

Serial Number	Item To Be Assessed	Yes	No	Comments
A. Checklist				
1	Whether the Request for Proposal (RFP) includes all the relevant information required to select the project manager firm			

Serial Number	Item To Be Assessed	Yes	No	Comments
A. Checklist				
2	Whether the RFP properly defines the project manager's roles and responsibilities intended by the owner to be performed by the project manager firm			
3	Whether all the deliverables are listed in the RFP			
4	Whether all the information to enable bidders to submit the proposal is clear and unambiguous			
5	Whether documents are prepared taking into consideration the organization's Quality Management System (QMS)			
6	Whether regulatory/authority requirements are taken into consideration while preparing the RFP			
7	Whether the selection criteria and selection method for project manager firm selection is clearly specified in the organization's QMS			
8	Whether budget approval is obtained for project management services prior to the release of the RFP announcement			
9	Whether the RFP opening and closing dates are mentioned in the announcement			
10	Whether the schedule for required project management services is included in the RFP			
11	Whether the qualifications of personnel for design management services are requested in RFP			
12	Whether the qualifications of personnel for project management services are requested in RFP			
13	Whether project quality requirements are properly defined in the documents			
14	Whether rights and liabilities of all the parties are mentioned in the documents			
15	Whether change/variation clauses are included in the RFP			
16	Whether cancellation/termination clauses are included in the documents			
17	Whether bond and insurance clauses are included in the documents			
18	Whether the RFP documents are as per international standard documents used by the construction industry			
19	Whether evaluation and assessment criteria are well defined			
20	Whether review and analysis of RFP are included			

(Continued)

Table 6.71 (Continued)

Serial Number	Item To Be Assessed	Yes	No	Comments
A. Checklist				
21	Whether the agreement to be signed for project management services is included in the documents			
22	Whether all the records are properly maintained			
23	Whether the request to provide the project manager's contact details is included in the documents			

B. Interview Questionnaires

Serial Number	Description	Response	Comments
1	Whether any outside agency was involved in the preparation of the tender documents		
2	What guidelines were adopted to prepare the tender documents		
3	Whether any specific format/template was followed		
4	Whether all the requirements of RFP documents were coordinated and agreed by relevant stakeholders		
5	Whether risks of disputes and associated financial costs were considered while preparing the RFP		
6	Whether problems and challenges were identified while preparing RFP		
7	Who was responsible for authorizing bidding requirements		

Table 6.72 Audit methodology for the assessment of the bidding and tendering process for the selection of the project manager firm – Stage III (contract bid solicitation)

Serial Number	Item To Be Assessed	Yes	No	Comments
A. Checklist				
1	Whether RFP documents are properly organized			
2	Whether the RFP is distributed to all the prequalified/shortlisted/registered project manager firms to participate in the tender			
3	Whether tender is notified to all the prequalified project manager firms			

Serial Number	Item To Be Assessed	Yes	No	Comments
A. Checklist				
4	Whether all the prequalified/shortlisted/registered bidders participated in the tender			
5	Whether the attendance sheet for the meeting was signed by all the attendees			
6	Whether minutes of the meeting were circulated to all the attendees/participating bidders/project manager firms			
7	Whether sufficient time is provided to prepare and submit the proposal/quotation			
8	Whether the tender submission date is extended from the originally announced date			
9	Whether an addendum (if any) is issued/notified to all the participating bidders			
10	Whether there are any requests for an extension of submission date			
11	Whether the clarification meeting is attended by all participating bidders			
12	Whether any changes are made to the original RFP			
13	Whether the received bid envelopes are placed in safe custody			
14	Whether technical and financial envelopes are submitted at the same time			
15	Whether there are any delayed submissions			
16	Whether proposals are opened as per the announced date and time			
17	Whether the proposals are opened in presence of all the bidders			
18	Whether received tenders are acknowledged			
19	Whether received tenders are clearly identified and recorded			
20	Whether applicable fees/bid bond are submitted by all the bidders who participated in the tender			
21	Whether tenders are fairly evaluated as per the announced policy and procedure			
22	Whether the comparison of bids is tabulated			
23	Whether evaluation criteria, weightage and methodology are as specified			
24	Whether the bid cost is higher than the budgeted cost			
25	Whether any submitted tender is found to be incomplete			
26	Whether there are any late submissions			
27	Whether all the scheduled pages are signed by the bidder			

(*Continued*)

Table 6.72 (Continued)

Serial Number	Item To Be Assessed	Yes	No	Comments
A. Checklist				
28	Whether risk factors are considered while evaluating the bids			
29	Whether reasons for non-acceptance of the tender are conveyed to the unsuccessful bidders			

B. Interview Questionnaires

Serial Number	Description	Response	Comments
1	How the notification was announced		
2	How tenders were submitted: • Hand delivery • By post • Courier service		
3	Whether the tender was available for electronic distribution		
4	How was the response to participate in tender		
5	Whether tender boxes were placed at secured places and monitored through an electronic surveillance system		
6	How much time was allowed to open the bids after the notified tendering closing time		
7	Whether the tender submission date was extended and what was the reason for extension		
8	Whether all the tender envelopes/packages were sealed with marking of tenders and contain the original tender documents and required number of sets as per tender conditions		

Table 6.73 Audit methodology for the assessment of the bidding and tendering process for the selection of the project manager firm – Stage IV (contract award)

Serial Number	Item To Be Assessed	Yes	No	Comments
A. Checklist				
1	Whether the selection of the project manager firm satisfies all the conditions to be a successful bidder			
2	Whether the selected bidder (project manager) is a legal entity			

Serial Number	Item To Be Assessed	Yes	No	Comments

A. Checklist

3	Whether the selection was made by the selection committee			
4	Whether the selection of the project manager firm is approved by the relevant applicable authority, if mandated by the organization's policy			
5	Whether the selected project manager is capable of carrying out the specified work and has the necessary resources			
6	Whether the selected project manager has experience of selecting the designer (A/E), contractor, subcontractor			
7	Whether the selected project manager has the resources for design management and management of construction			
8	Whether risk factors are considered prior to the signing of the contract			
9	Whether the standard contract format is used for signing the contract			
10	Whether the contract terms and conditions are clearly written and unambiguous			
11	Whether an addendum and minutes of the meeting are included in the contract documents			
12	Whether clarifications to queries are included as part of the contract documents			
13	Whether the contract period/schedule is properly described			
14	Whether staff qualifications are properly defined			
15	Whether a performance bond is submitted by the consultant			
16	Whether variation/change management clauses are included in the contract agreement			
17	Whether dispute resolution and conflict resolution clauses are properly defined			
18	Whether contract documents are reviewed prior to signing			
19	Whether tender (proposal) documents and contract are properly archived			
20	Whether a letter of award is sent			
21	Whether final results are published and announced			

(*Continued*)

Table 6.73 (Continued)

B. Interview Questionnaires

Serial Number	Description	Response	Comments
1	On what basis the bidder (project manager) was selected: low bid, competitive bid or quality based system		
2	Was there any tie between two or more bidders		
3	Was there any dispute or objection raised by any of the bidders before the award of the contract to the successful bidder		
4	Whether the final price was negotiated or as per the bid (proposal) received		
5	Whether the contract amount is as per the budget		

6.4.2 Auditing process for the project manager's proposal submission process

Figure 6.88 illustrates a typical quality auditing process for the project manager's proposal submission process.

The main objectives of performing a quality audit of the project manager's proposal submission process are to:

1. Evaluate and assess process compliance with the organizational strategic policies and procedure
2. Assess conformance to QMS and that processes meet established quality system requirements
3. Review the processes used to prepare the proposal
4. Verify that the proposal preparation and submission comply with the organization's proposal submittal process
5. Validate that the response to the contents of the Request for Proposal is complete in all respects
6. Check that the availability of resources for project review and design management is considered
7. Check that the availability of resources for project management during the construction phase is considered
8. Check that the availability of resources for project management during the testing, commissioning and handover phase is considered
9. Review the process used to prepare the proposal
10. Validate the schedule for deployment of project staff
11. Validate the cost of the proposal
12. Review that the key risk factors are analyzed and managed
13. Check compliance with regulatory requirements

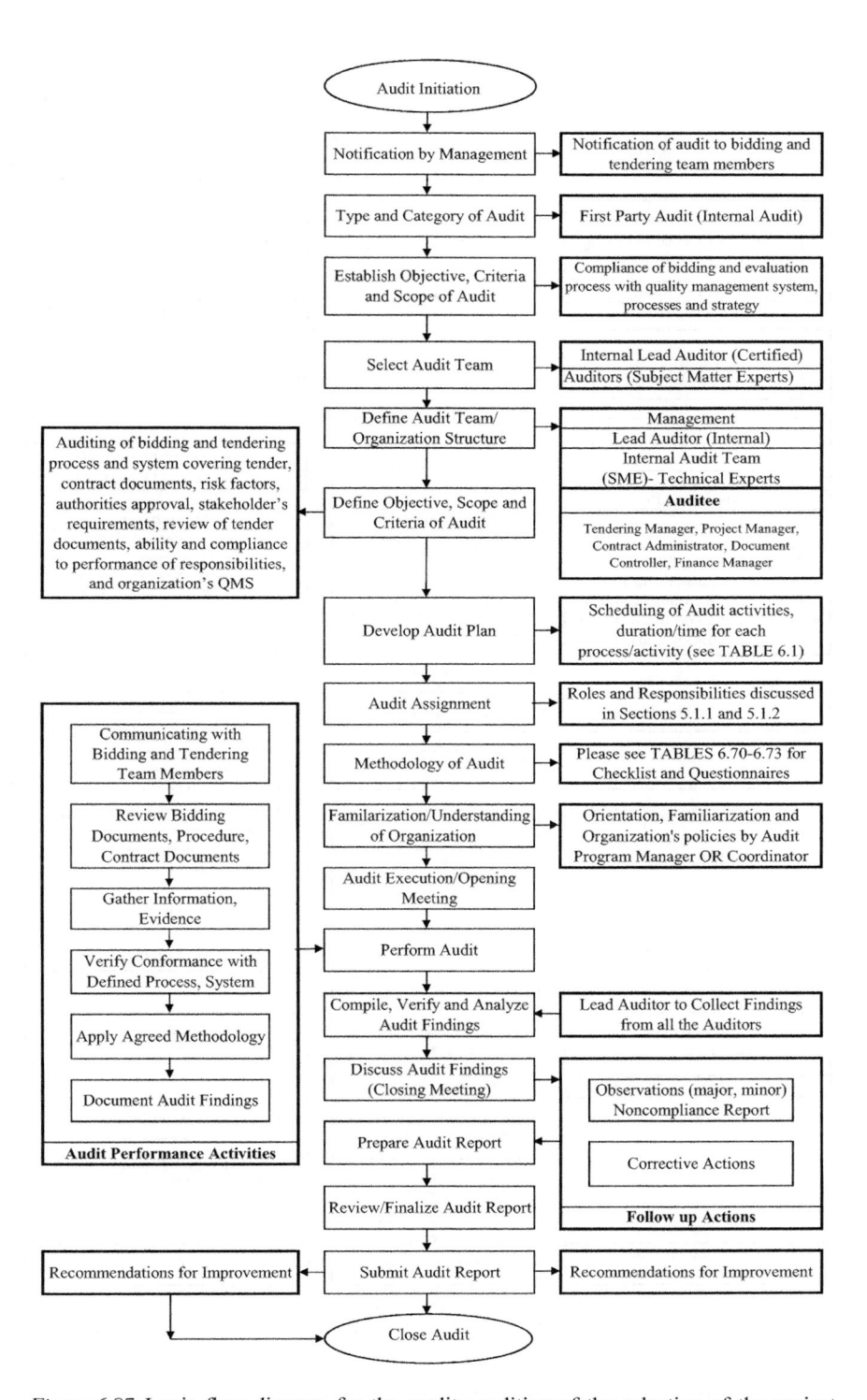

Figure 6.87 Logic flow diagram for the quality auditing of the selection of the project manager

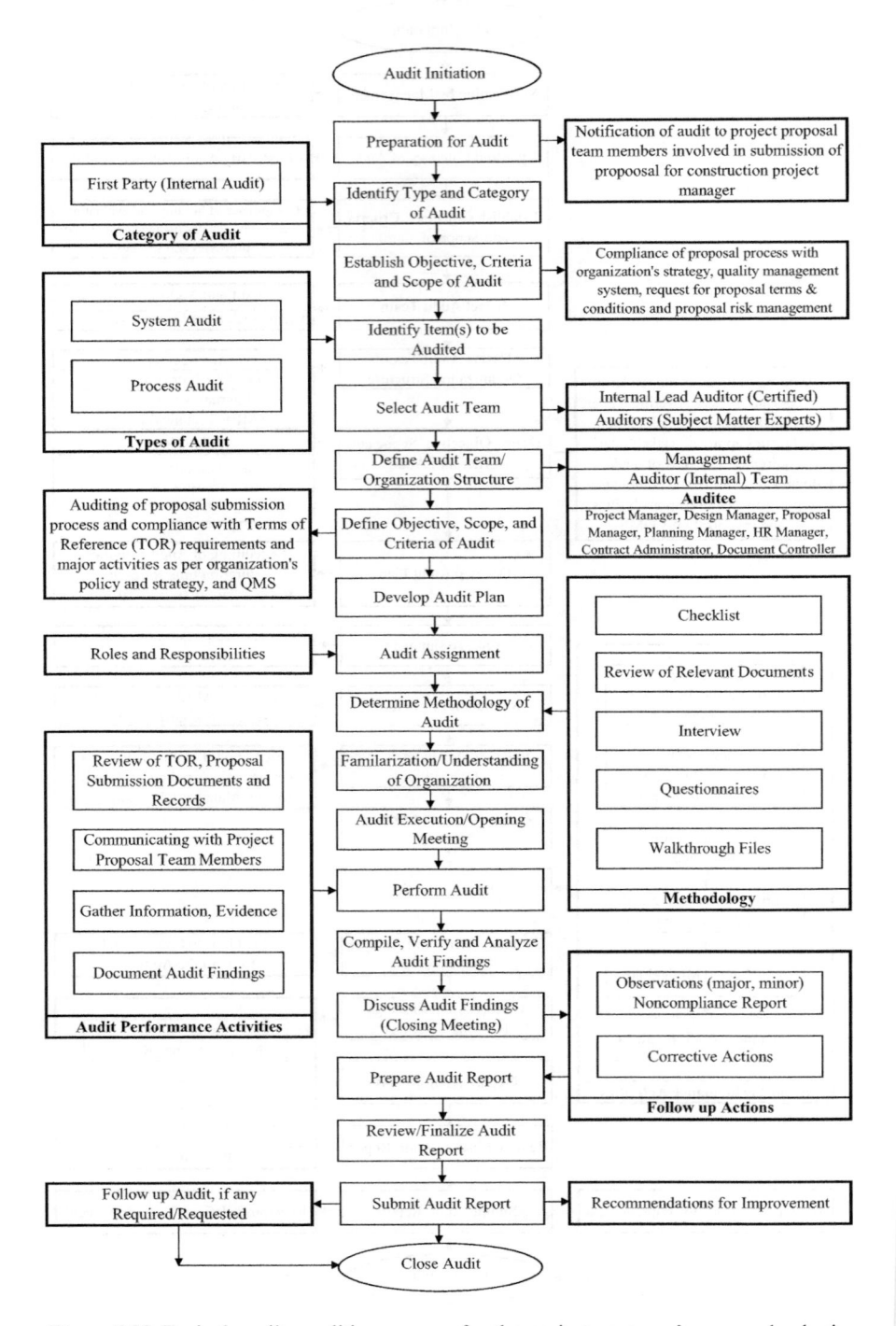

Figure 6.88 Typical quality auditing process for the project manager's proposal submission process

The following audit tools are used to perform a quality audit of the project manager's proposal submission process:

- Analytical tools
- Checklists
- Review of documents
- Interviews
- Questionnaires

The auditee team members are mainly from the following organizational groups:

- Project Manager
- Design Manager
- Proposal Manager
- Quality Manager
- Contract Administrator/Quantity Surveyor
- HR Manager
- Document Controller
- Financial Manager

Table 6.74 illustrates the audit methodology for the assessment of the project manager's proposal submission process and Figure 6.89 illustrates the logic flow diagram for the quality auditing of the project manager's proposal submission process.

Table 6.74 Audit methodology for the assessment of the project manager's bid (proposal) submission process

Serial Number	Item To Be Assessed	Yes	No	Comments
A. Checklist				
1	Whether the company is already registered as a project manager to participate in the tender			
2	Whether there are any fees to collect tender (RFP) documents			
3	Whether the scope of work is clearly defined in the RFP			
4	Whether the TOR (Terms of Reference) is clearly defining all the scope			
5	Whether there is any request for clarification raised			
6	Whether the response to the RFP is complete			
7	Whether sufficient time is provided to prepare and submit the completed RFP			

(*Continued*)

Table 6.74 (Continued)

Serial Number	Item To Be Assessed	Yes	No	Comments
A. Checklist				
8	Whether there are any requests for extension of the submission date			
9	Whether there are any delayed submissions			
10	Whether RFP is prepared taking into consideration/referring to the organization's quality management system			
10	Whether the resources and time required to select the designer (A/E), consultant and contractor were considered while preparing the schedule			
11	Whether the design review and management requirements are considered while preparing the proposal			
12	Whether the resources for project management during construction were considered while preparing the proposal			
13	Whether the proposal was reviewed prior to submission			
14	Whether the proposal was approved by the relevant committee/authorized person			
15	Whether the risk factors were considered and analyzed			

B. Interview Questionnaires

Serial Number	Description	Response	Comments
1	Whether the notification for proposal submission was announced in all technical newsletters and leading news media or it was only for registered bidders (project manager firm)		
2	How the proposal was submitted: • Hand delivery • By post • Courier service		
3	Whether the RFP was available for electronic distribution		
4	What tools and techniques were used to price the proposal		
5	Whether the resources for bidding and tendering and selection of the designer (A/E) and consultant were considered		
6	Whether the resources for bidding and tendering and selection of contractor and subcontractor were considered		

B. Interview Questionnaires

Serial Number	Description	Response	Comments
7	Whether the organization's design management and review plan were considered to evaluate the price schedule		
8	Whether the availability of qualified resources (technical) to review the project design was considered while preparing the price schedule		
9	Whether the availability of qualified resources (technical) to manage the construction phase, testing, commissioning and handover phase was considered while preparing the price schedule		
10	Whether the design review and management schedule were properly planned by taking into consideration the availability of qualified manpower		
11	Whether the project management schedule was properly planned by taking into consideration the availability of qualified manpower		
12	Whether the regulatory approval period was properly considered		
13	What risk factors were considered while evaluating the price schedule		

6.4.3 Auditing process for the project manager's activities

Normally the owner/client engages the project manager firm from the inception of the project through to the issuance of the substantial completion certificate to manage the entire construction project on behalf of the owner.

There are three main activities that are to be performed by the project manager. These are:

1 Selection of project teams

  • Selection of designer (A/E) and consultant
  • Selection of contractor and subcontractor

2 Design management

  • Review and approval of the project design (design phases)
  • Review and approval of construction documents

3 Construction supervision activities

  • Construction phase
  • Testing, commissioning and handover phase

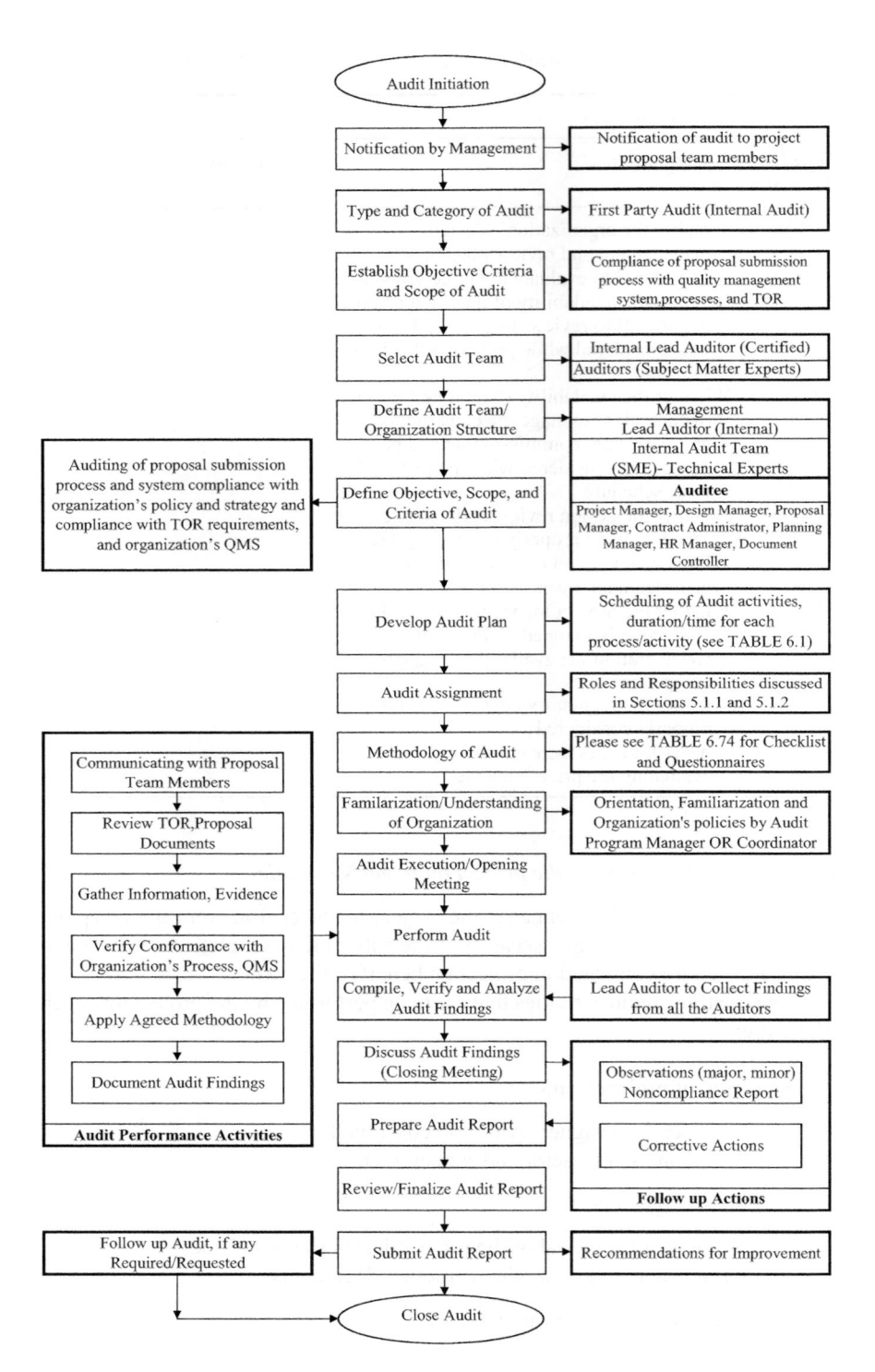

Figure 6.89 Logic flow diagram for the quality auditing process for the project manager's proposal submission process

6.4.3.1 Selection of project teams

The project manager is responsible for selecting the designer (A/E) and contractor. The project manager has to prepare the bidding and tendering documents as per the organizational strategy and follow the process to acquire all the related services, products/materials, systems, equipment, designer (A/E), contractor and subcontractor.

The quality auditing process that includes the aims, objectives and audit methodology for the assessment of the bidding and tendering process for the selection of the designer (A/E) discussed in section 6.2.1.2 can be used to perform the quality auditing process for the selection of the designer (A/E) by the project manager. Please refer to Figures 6.8–6.9 and Tables 6.6–6.10 discussed in Section 6.2.1.2.

Similarly, the quality auditing process that includes the aims, objectives and audit methodology for the assessment of the bidding and tendering process for the selection of the contractor discussed in section 6.2.3.1 can be used to perform the quality auditing process for the selection of the contractor by the project manager. Please refer to Figures 6.25–6.29 and Tables 6.24–6.29 discussed in Section 6.2.3.1.

6.4.3.2 Design management

The design management activities that are to be performed by the project manager mainly consist of:

- Review and approval of concept design
- Review and approval of schematic design
- Review and approval of design development
- Review and approval of construction documents

The project manager is responsible for ensuring that the project design and documents developed by the designer (A/E) meet the project goals and objectives to satisfy the owner's/end user's requirements.

6.4.3.2.1 AUDITING PROCESS FOR THE CONCEPT DESIGN PHASE

Figure 6.90 illustrates the typical quality auditing process for the concept design of the conceptual design phase.

The main objectives of performing a quality audit of the concept design of the conceptual design phase are to:

1. Evaluate and assess process compliance with the organizational strategic policies and procedure
2. Assess conformance to QMS and that processes meet established quality system requirements

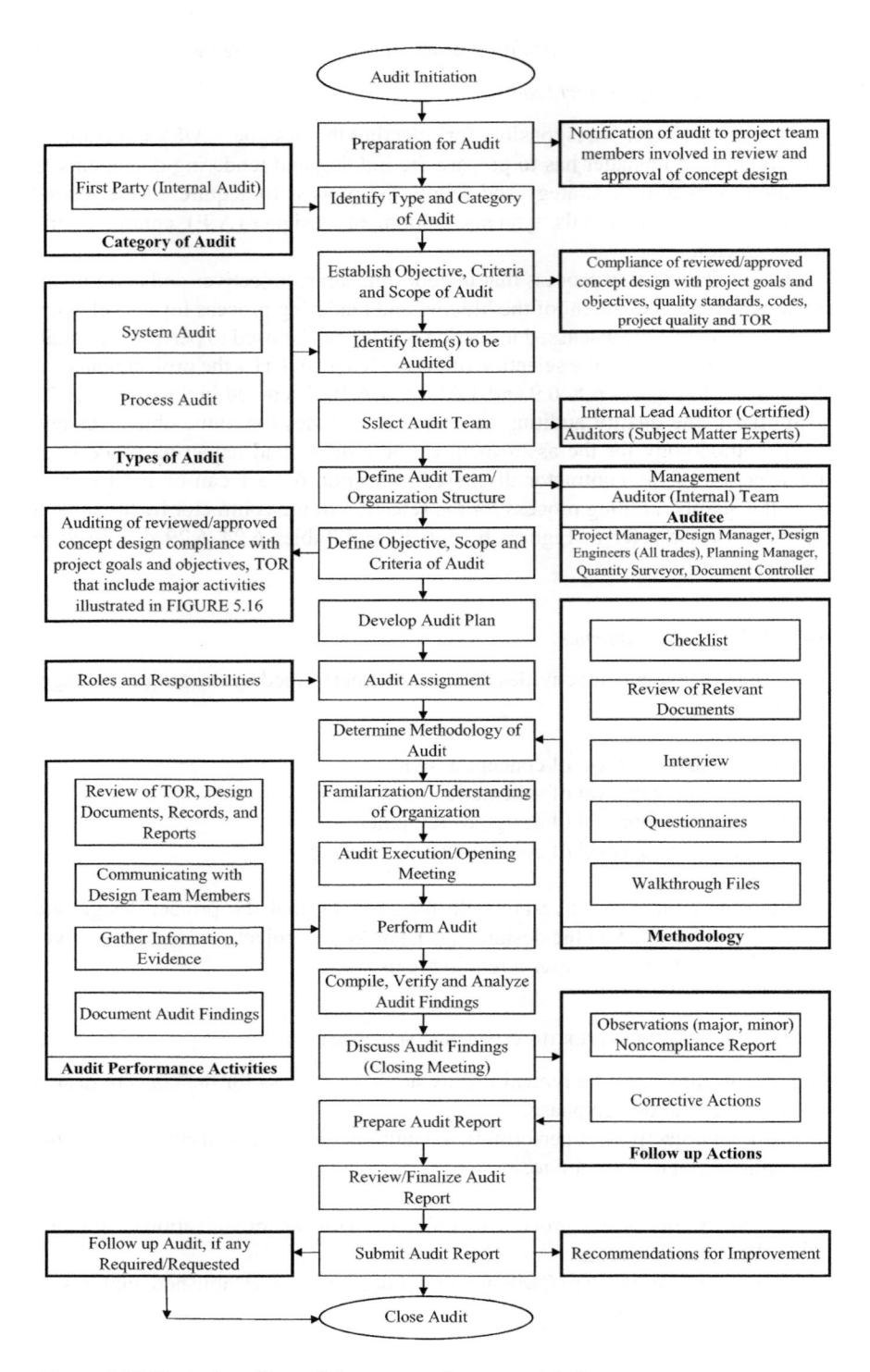

Figure 6.90 Typical quality auditing process for concept design

Auditing processes for project life cycle phases 483

3. Verify that the concept design prepared by the designer (A/E) is properly reviewed to comply with project goals and objectives and fully meet the owner's requirements
4. Check that the review comments on the design/drawings are specific or general
5. Check that design/drawings are reviewed for compliance with TOR requirements. (Please refer to Table 6.15 for typical contents of TOR for building construction projects)
6. Check that the submitted concept design report and model comply with contract requirements
7. Validate that the schedule and cost of the project estimated by the designer (A/E) meet the project schedule and budget
8. Validate that reviewed/approved design/drawings comply with specified quality standards and codes
9. Check that reviewed/approved design/drawings fully comply with regulatory requirements
10. Check that reviewed/approved design/drawings have sustainability considerations
11. Check that energy conservation requirements are considered in the reviewed/approved design/drawings
12. Check that facility management requirements are considered in the reviewed/approved design/drawings
13. Review that the key risk factors are considered by the designer (A/E)
14. Assess that all the major activities are adequately performed and managed
15. Assess the proper functioning of the project management system
16. Check the ratio of approved/not-approved drawings/design

The following audit tools are used to perform the quality audit of the concept design:

- Analytical tools
- Checklists
- Review of documents
- Interviews
- Questionnaires

The auditee team members are mainly from the following organizational groups:

- Project Manager
- Design Manager
- Design Engineers (All Trades)
- Planning Manager
- Contract Administrator/Quantity Surveyor
- Document Controller

Table 6.75 illustrates the audit methodology for the assessment of the concept design of the conceptual design phase and Figure 6.91 illustrates the logic flow diagram for the quality auditing of the concept design.

6.4.3.2.2 AUDITING PROCESS FOR THE SCHEMATIC DESIGN PHASE

Figure 6.92 illustrates the typical quality auditing process for the schematic design phase.

Table 6.75 Audit methodology for the assessment of the concept design

Serial Number	Item To Be Assessed	Yes	No	Comments
A. Checklist				
1	Whether the reviewed/approved concept design supports the owner's project goals and objectives			
2	Whether the owner's requirements are fully considered by the designer			
3	Whether the reviewed/approved concept design meets all the elements specified in the TOR			
4	Whether the design complies with organizational strategic policies			
5	Whether the reviewed/approved design is based on specified quality standards and codes			
6	Whether the reviewed/approved design meets all the performance requirements			
7	Whether regulatory/statutory requirements are considered			
8	Whether the project schedule estimated by the designer is achievable in practice			
9	Whether the designer has properly estimated the project cost			
10	Whether the designer has considered the availability of resources during the entire project life cycle			
11	Whether the designer has considered the project risks			
12	Whether health and safety requirements are considered in the design			
13	Whether the reviewed/approved design is based on environmental constraints			
14	Whether the designer has considered LEED requirements			
15	Whether the reviewed/approved design has considered cost-effectiveness over the entire project life cycle			

Serial Number	Item To Be Assessed	Yes	No	Comments
A. Checklist				
16	Whether the reviewed/approved reports is prepared to meet TOR requirements			
17	Whether energy conservation is considered in the reviewed/approved design			
18	Whether sustainability is considered in the reviewed/approved design			
19	Whether all reasonable alternative options/ systems are considered by the designer (A/E)			
20	Whether constructability has been considered			
21	Whether the reviewed/approved model meet the design objectives			
22	Whether the design supports proceeding to the next design stage			
23	Whether the reviewed/approved reports are complete and include adequate information about the project			
24	Whether the report is properly formatted and there is a table of contents for each report			

B. Interview Questionnaires

Serial Number	Description	Response	Comments
1	Whether there was any delay in submission of the design by the designer (A/E) for review and approval		
2	Whether the concept design was prepared within the agreed-upon schedule		
3	Was there any delay in getting authorities' approval		
4	Whether submitted numbers of sets were as per TOR		
5	Whether the concept design has any major comments		
6	Whether the concept design was not approved by the owner		
7	Whether the reasons (causes) for comments, non-approval, if any, were analyzed and corrective/preventive action taken		

The main objectives of performing a quality audit of the schematic design phase are to:

1. Evaluate and assess process compliance with the organizational strategic policies and procedure
2. Assess conformance to QMS and that processes meet established quality system requirements

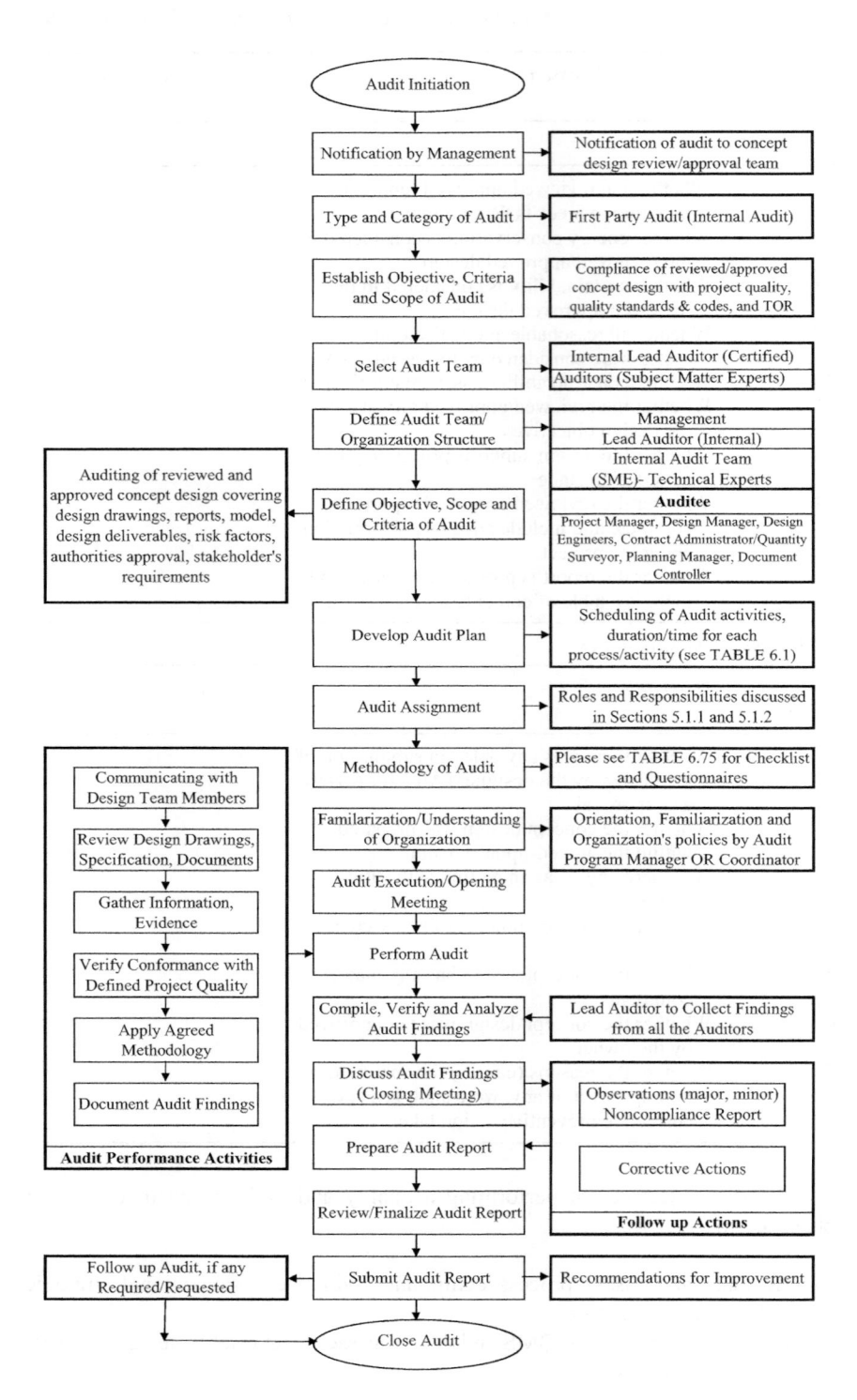

Figure 6.91 Logic flow diagram for the quality auditing of the concept design

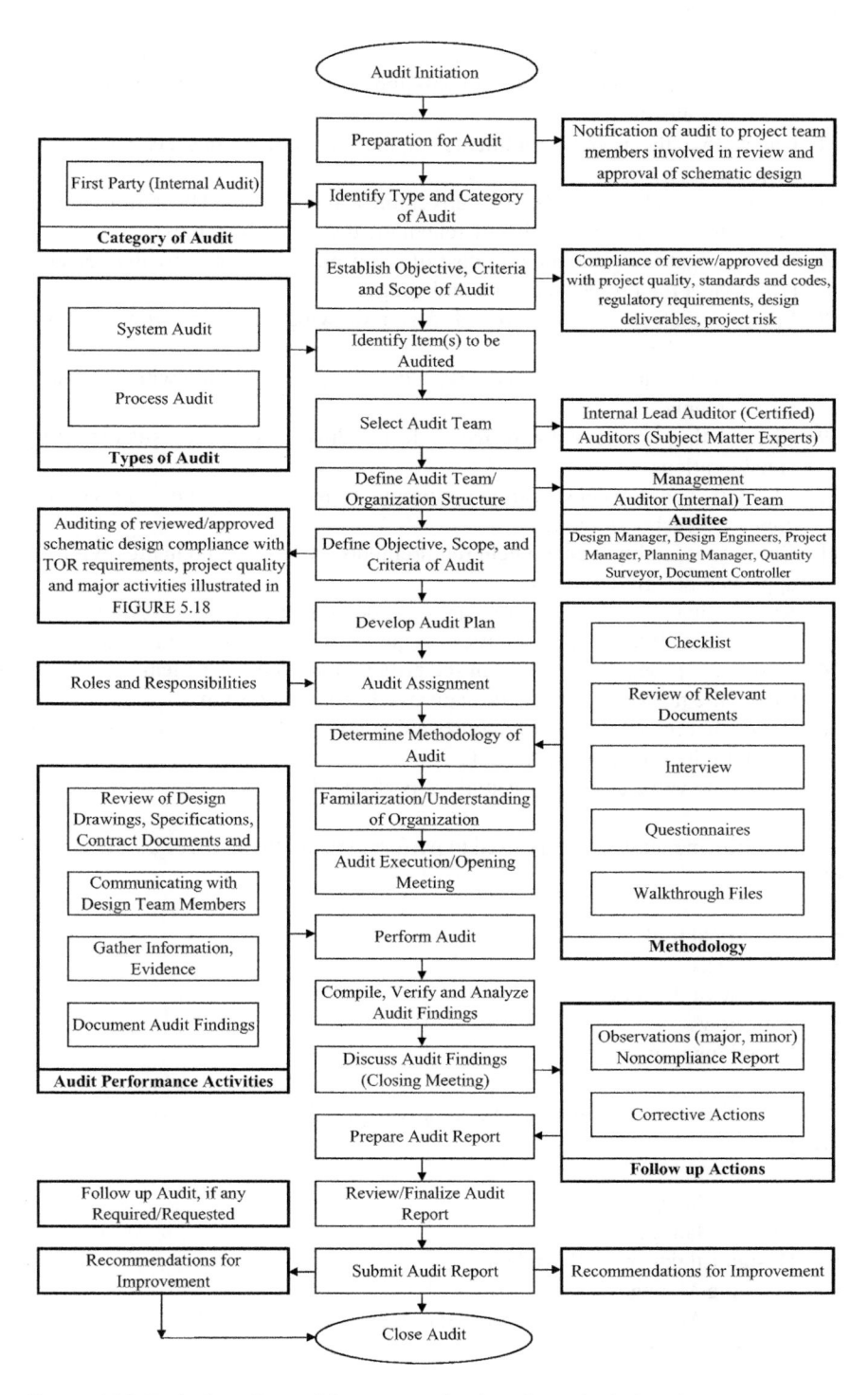

Figure 6.92 Typical quality auditing process for the schematic design phase

3. Verify that the schematic design prepared by the designer (A/E) is properly reviewed to comply with project goals and objectives and fully meet the owner's requirements
4. Check that the review comments on the design/drawings are specific or general
5. Verify that schematic design deliverables are as per TOR requirements. (Table 6.17 is an example list of schematic design deliverables)
6. Validate that the schedule and cost of the project estimated by the designer (A/E) meet the project schedule and budget
7. Validate that reviewed/approved design/drawings comply with specified quality standards and codes
8. Check that reviewed/approved design/drawings fully comply with regulatory requirements
9. Check that reviewed/approved design/drawings have sustainability considerations
10. Check that energy conservation requirements are considered in the reviewed/approved design/drawings
11. Check that facility management requirements are considered in the reviewed/approved design/drawings
12. Review that the key risk factors are considered by the designer (A/E)
13. Assess that all the major activities adequately performed and managed
14. Assess the proper functioning of the project management system
15. Check the ratio of approved/not-approved drawings/design
16. Check that reviewed preliminary specifications and contract documents comply with the project requirements
17. Check the value engineering study is performed

The following audit tools are used to perform the quality audit of the schematic phase:

- Analytical tools
- Checklists
- Review of documents
- Interviews
- Questionnaires

The auditee team members are mainly from the following organizational groups:

- Project Manager
- Design Manager
- Design Engineers (All Trades)
- Planning Manager
- Contract Administrator/Quantity Surveyor
- Document Controller

Auditing processes for project life cycle phases 489

Table 6.76 illustrates the audit methodology for the assessment of the schematic design and Figure 6.93 illustrates the logic flow diagram for the quality auditing of the schematic design.

6.4.3.2.3 AUDITING PROCESS FOR THE DESIGN DEVELOPMENT/DETAIL DESIGN PHASE

Figure 6.94 illustrates the typical quality auditing process for the design development phase.

Table 6.76 Audit methodology for the assessment of the schematic design

Serial Number	Item To Be Assessed	Yes	No	Comments
A. Checklist				
1	Whether the reviewed/approved schematic design supports the owner's project goals and objectives			
2	Whether the design complies with organizational strategic policies			
3	Whether the owner's requirements are fully considered by the designer (A/E)			
4	Whether comments on the concept design are taken into account by the designer while preparing the schematic design			
5	Whether the reviewed/approved schematic design meets all the elements specified in the TOR			
6	Whether regulatory/statutory requirements are considered			
7	Whether the design is based on specified quality standards and codes			
8	Whether the design meets all the performance requirements			
9	Whether the designer has considered project boundaries			
10	Whether the reviewed/approved design is constructible			
11	Whether the design was reviewed for technical and functional capability			
12	Whether sustainability is considered in the design			
13	Whether the design meets LEED requirements			
14	Whether energy conservation requirements are considered			
15	Whether the designer has considered facility management requirements			

(Continued)

Table 6.76 (Continued)

Serial Number	Item To Be Assessed	Yes	No	Comments
A. Checklist				
16	Whether the reviewed/approved design meets project objectives in respect of time, cost and quality			
17	Whether the reviewed/approved project schedule is achievable in practice			
18	Whether the reviewed/approved project cost is properly estimated			
19	Whether the reviewed/approved design is based on available resources during the entire project life cycle			
20	Whether design risks have been considered			
21	Whether the reviewed/approved design was fully coordinated with all the trades			
22	Whether health and safety requirements in the reviewed/approved design are considered			
23	Whether the reviewed/approved design conforms with fire and egress requirements			
24	Whether environmental constraints and requirements are considered			
25	Whether the reviewed/approved specification is prepared taking into consideration international construction contract documents			
26	Whether the designer performed a value engineering study and recommendations are considered			
27	Whether the reviewed/approved design has considered cost-effectiveness over the entire project life cycle			
28	Whether project boundaries are considered in the reviewed/approved design			
29	Whether the reviewed/approved drawings, specifications and documents meet TOR requirements			
30	Whether design deliverables conform as listed in Table 6.17			
31	Whether all the drawings are numbered.			
32	Whether the reports are complete and include adequate information about the project			
33	Whether the report is properly formatted and there is a table of contents for each report			
34	Whether the reviewed/approved design complies with the project quality requirements			
35	Whether the design supports proceeding to the next design development stage			

B. Interview Questionnaires

Serial Number	Description	Response	Comments
1	Whether the schematic design was submitted in time for review and approval		
2	Whether the schematic design was submitted in time for the owner's review		
3	Whether numbers of sets were prepared as per TOR and submitted		
4	Whether the schematic design has any major comments		
5	Whether the schematic design was not approved by the owner		
6	Whether the reasons (causes) for comments, non-approval, if any, were analyzed and corrective/preventive action taken		

The main objectives of performing a quality audit of the design development phase are to:

1. Evaluate and assess process compliance with the organizational strategic policies and procedure
2. Assess conformance to QMS and that processes meet established quality system requirements
3. Verify that the detail design prepared by the designer (A/E) is properly reviewed to comply with project goals and objectives and fully meet the owner's requirements
4. Check that the review comments on the design/drawings are specific or general
5. Verify that the design development deliverables are as per TOR requirements
6. Validate that the schedule and cost of the project estimated by the designer (A/E) meet the project schedule and budget
7. Validate that reviewed/approved design/drawings comply with specified quality standards and codes
8. Check that reviewed/approved design/drawings fully comply with regulatory requirements
9. Check that reviewed/approved design/drawings have sustainability considerations
10. Check that energy conservation requirements are considered in the reviewed/approved design/drawings
11. Check that facility management requirements are considered in the reviewed/approved design/drawings
12. Review that the key risk factors are considered by the designer (A/E)
13. Assess that the reviewed design has safety considerations
14. Check that the reviewed preliminary specifications and contract documents comply with the project requirements
15. Assess the proper functioning of the project management system

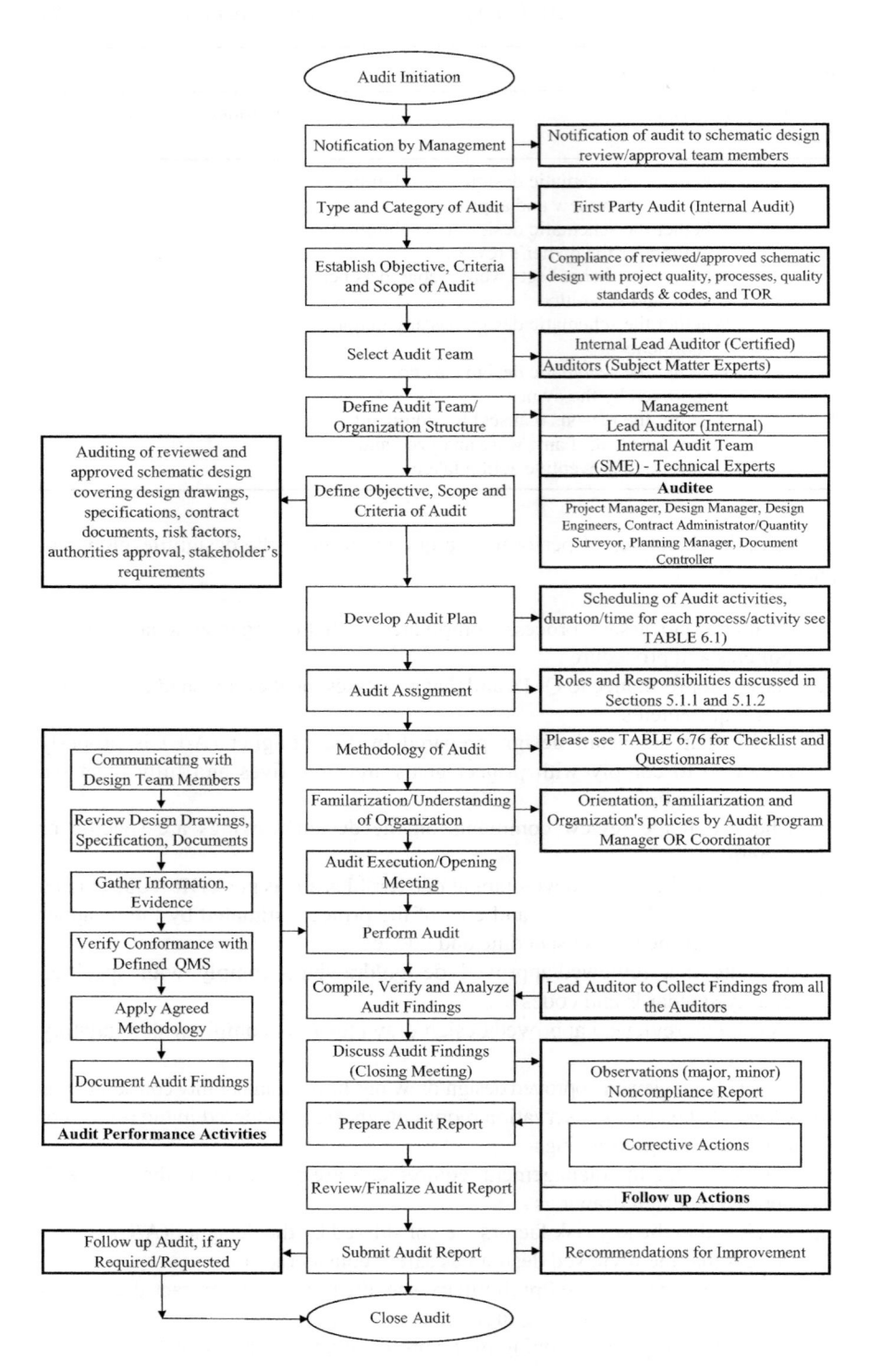

Figure 6.93 Logic flow diagram for the quality auditing of schematic design

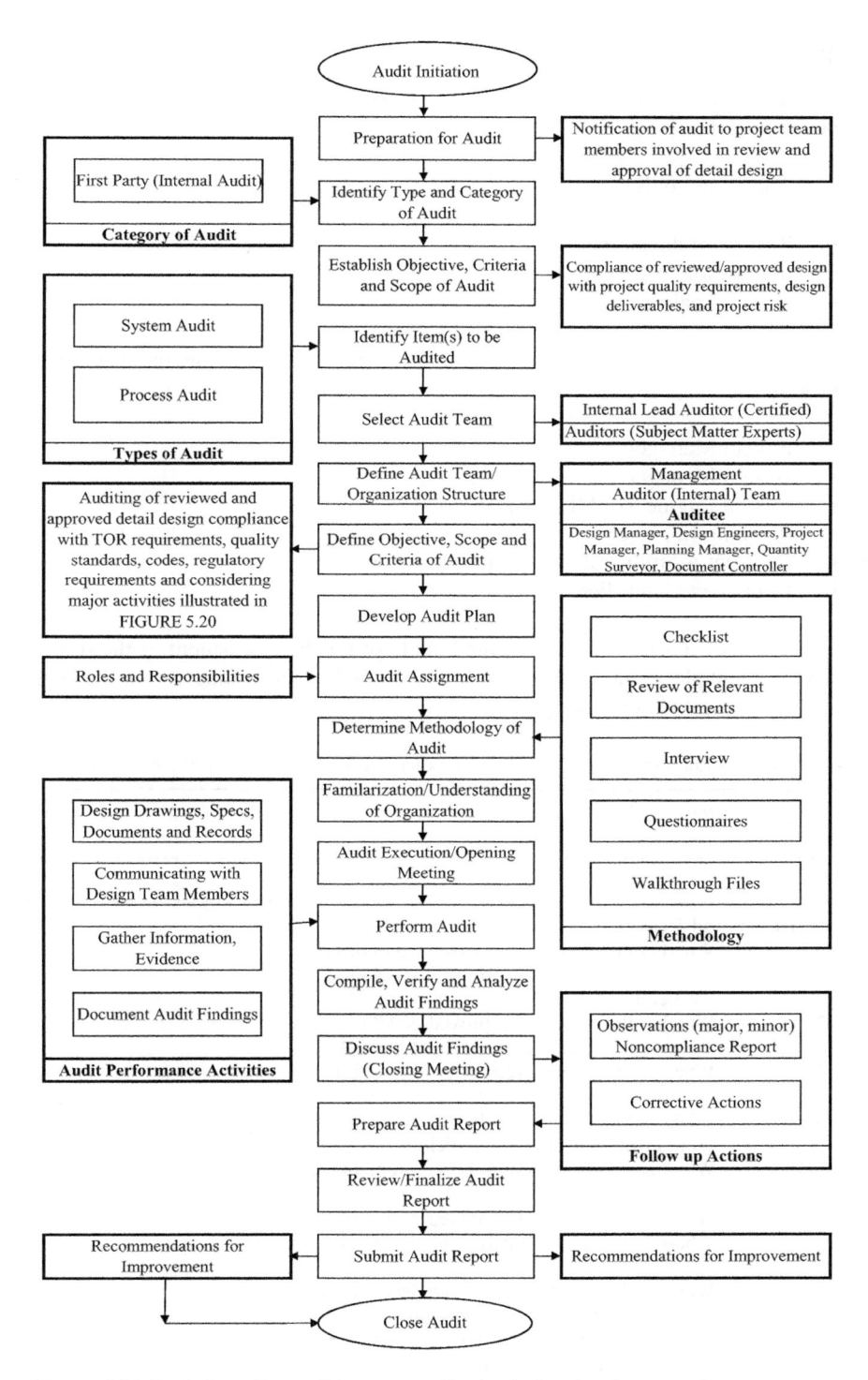

Figure 6.94 Typical quality auditing process for the design development phase

The following audit tools are used to perform the quality audit of the design development phase:

- Analytical tools
- Checklists
- Review of documents
- Interviews
- Questionnaires

The auditee team members are mainly from the following organizational groups:

- Project Manager
- Design Manager
- Design Engineers (All Trades)
- Planning Manager
- Contract Administrator/Quantity Surveyor
- Document Controller

Table 6.77 illustrates the audit methodology for the assessment of the design development phase and Figure 6.95 illustrates the logic flow diagram for the quality auditing of the detail design phase.

Table 6.77 Audit methodology for the assessment of the design development phase

Serial Number	Item To Be Assessed	Yes	No	Comments
A. Checklist				
A-1 Design and Drawings				
1	Whether the reviewed/approved detail design supports the owner's project goals and objectives			
2	Whether the reviewed/approved detail design meets all the elements specified in TOR			
3	Whether the owner's requirements are fully considered			
4	Whether the design complies with organizational strategic policies			
5	Whether regulatory/statutory requirements are considered			
6	Whether the reviewed/approved design is based on specified quality standards and codes			
7	Whether the reviewed/approved design meets all the performance requirements			
8	Whether constructability has been considered in the reviewed/approved design			

Serial Number	Item To Be Assessed	Yes	No	Comments

A. Checklist

9	Whether technical and functional capability is considered in the reviewed/approved design			
10	Whether sustainability is considered in the reviewed/approved design			
11	Whether the design meets LEED requirements in the reviewed/approved design			
12	Whether energy conservation is considered in the reviewed/approved design			
13	Whether the designer has considered facility management requirements			
14	Whether the designer has taken into consideration project boundaries and assumptions			
15	Whether value engineering study recommendations are considered in the reviewed/approved design			
16	Whether project objectives in respect of time, cost and quality are considered in the reviewed/approved design			
17	Whether the design supports proceeding to the next design development stage			

A-2 Specifications and Documents

1	Whether the reviewed/approved specifications are as per international construction contract documents			
2	Whether reviewed/approved specifications and documents prepared meet TOR requirements			
3	Whether the reviewed/approved reports are complete and include adequate information about the project			
4	Whether the reviewed/approved report is properly formatted and there is a table of contents for each report			
5	Whether division numbers in the specifications match with BOQ			
6	Whether numbers of submitted sets comply with TOR requirements			

A-3 Schedule

1	Whether the reviewed/approved project schedule is achievable in practice			
2	Whether the reviewed/approved schedule is prepared considering all the activities			
3	Whether the reviewed/approved schedule is developed considering specific internationally recommended schedule level practices			

(*Continued*)

Table 6.77 (Continued)

Serial Number	Item To Be Assessed	Yes	No	Comments
A. Checklist				
A-4 Cost				
1	Whether the reviewed/approved project cost is properly estimated			
2	Whether the reviewed/approved design has considered cost-effectiveness over the entire project life cycle			
3	Whether the reviewed/approved project cost is developed considering specific internationally recommended cost estimation level practices for construction projects			
A-5 Quality				
1	Whether the reviewed/approved design is fully coordinated for conflict between different trades			
2	Whether the reviewed/approved design has considered the availability of resources during the entire project life cycle			
3	Whether the reviewed/approved design complies with project quality requirements			
4	Whether all the drawings are numbered			
5	Whether the reviewed/approved design matches with property limits			
6	Whether legends match with layout			
7	Whether the reviewed/approved design drawings are properly numbered			
8	Whether the reviewed/approved design drawings have the owner logo and designer logo as per standard format			
9	Whether design calculation sheets are included in the set of documents			
10	Whether the project name and contract reference is shown on the drawings			
11	Whether the reviewed/approved BOQ matches with design drawings and specifications			
A-6 Risk				
1	Whether risk has been considered in the reviewed/approved design			
A-7 Safety				
1	Whether the reviewed/approved design has health and safety considerations			
2	Whether the reviewed/approved design has environmental constraints			
3	Whether the reviewed/approved design conforms with fire and egress requirements			
4	Whether the reviewed/approved design has environmental compatibility considerations			

Serial Number	Item To Be Assessed	Yes	No	Comments
A. Checklist					
A-8 General					
1	Whether design deliverables are as per TOR			
2	Whether regulatory approval is obtained and comments, if any, incorporated and all review comments responded to			
3	Whether the reviewed/approved design is approved by all the concerned stakeholders			

B. Interview Questionnaires

Serial Number	Description	Response	Comments
1	Whether the detail design was submitted in time for review and approval		
2	Was there any delay in getting authorities' approval		
3	Which international contract documents were followed to prepare the specification		
4	Whether numbers of submitted sets were as per TOR		
5	Whether the reviewed/approved detail design has any major comments		
6	Whether the detail design was not approved by the owner		
7	Whether the reasons (causes) for comments, non-approval, if any, were analyzed and corrective/preventive action taken		

6.4.3.2.4 AUDITING PROCESS FOR THE CONSTRUCTION DOCUMENTS PHASE

Figure 6.96 illustrate the typical quality auditing process for the construction documents phase.

The main objectives of performing a quality audit of the construction documents phase are to:

1. Evaluate and assess process compliance with the organizational strategic policies and procedure
2. Assess conformance to QMS and that processes meet established quality system requirements
3. Verify that the construction documents developed by the designer (A/E) are properly reviewed to comply with project goals and objectives and fully meet the owner's requirements
4. Check that the review comments on the construction documents are specific or general

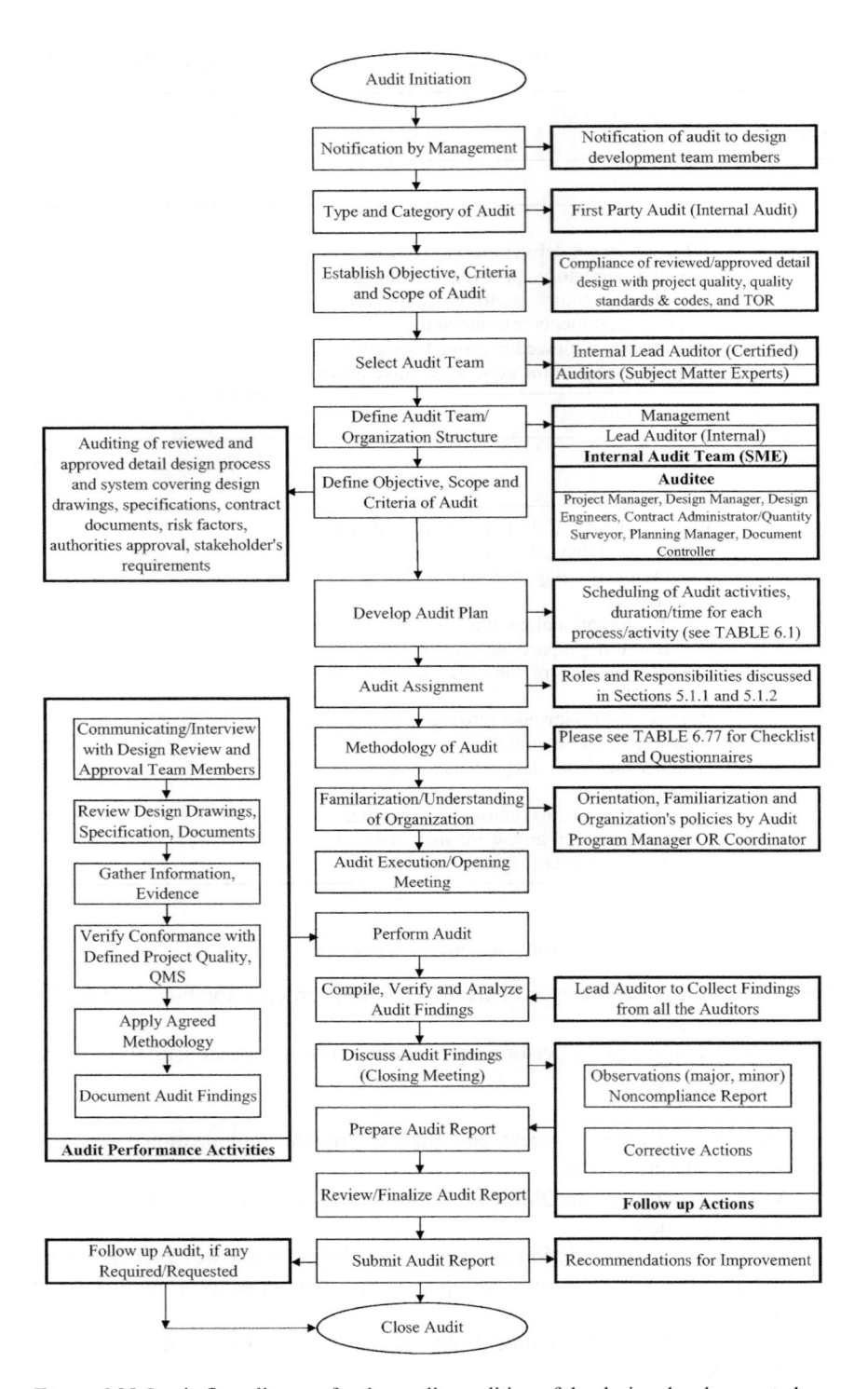

Figure 6.95 Logic flow diagram for the quality auditing of the design development phase

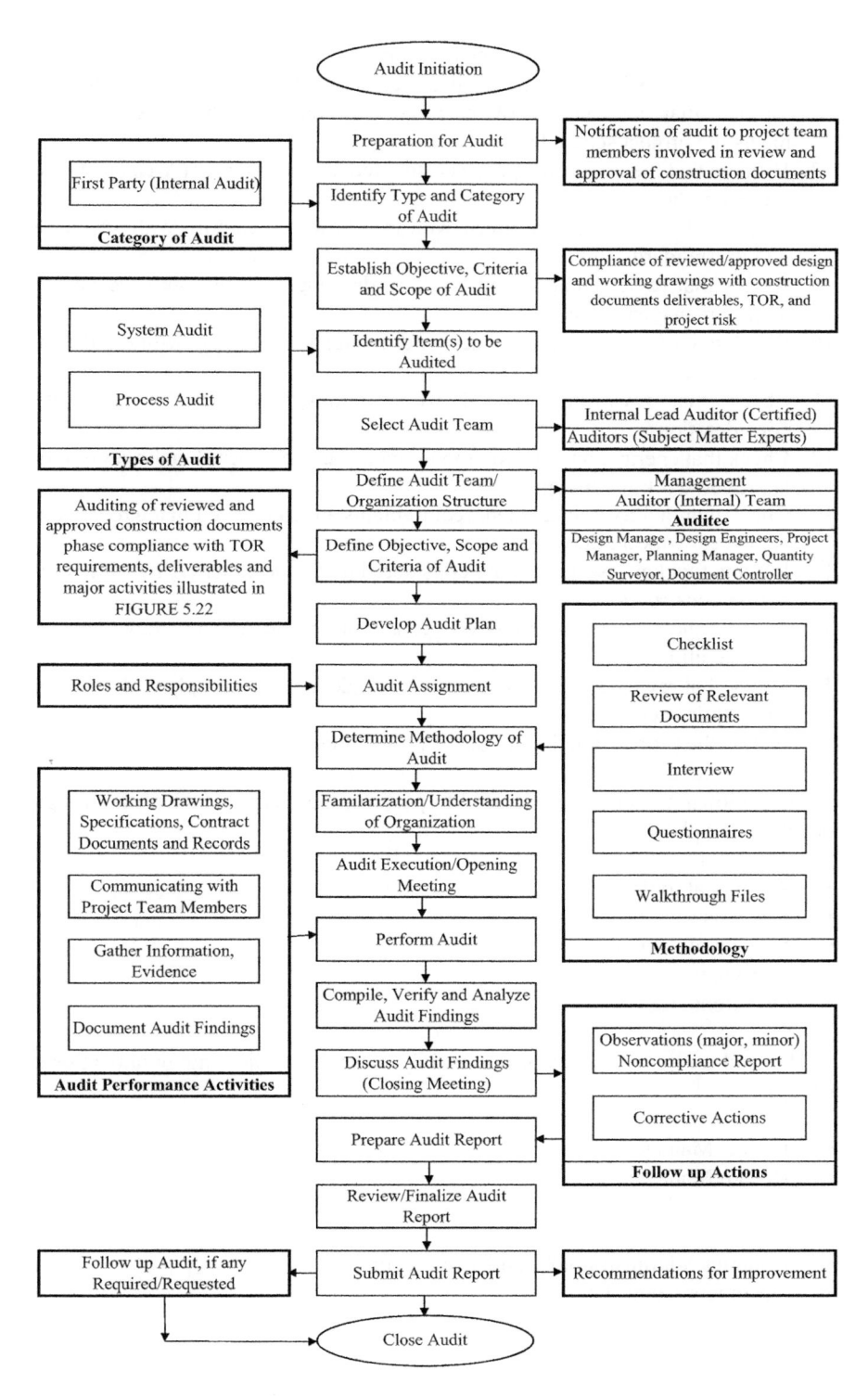

Figure 6.96 Typical quality auditing process for the construction documents phase

5 Verify that construction documents deliverables are as per TOR requirements. (Table 6.21 is an example list of construction documents deliverables)
6 Validate that the schedule of the project estimated by the designer (A/E) meets the project schedule
7 Validate that the cost of the project estimated by the designer (A/E) meets the project budget
8 Validate that reviewed/approved working drawings comply with specified quality standards and codes
9 Assess that reviewed/approved working drawings comply with the design quality management system
10 Check that reviewed/approved design/drawings fully comply with regulatory requirements
11 Check that reviewed/approved working drawings have sustainability considerations
12 Check that energy conservation requirements are considered in the reviewed/approved working drawings
13 Check that facility management requirements are considered in the reviewed/approved working drawings
14 Review that key risk factors are considered by the designer (A/E) while preparing construction documents
15 Assess that reviewed/approved construction documents have safety consideration
16 Check that reviewed/approved construction documents comply with the project requirements
17 Assess the proper functioning of the project management system

The following audit tools are used to perform the quality audit of the design development phase:

- Analytical tools
- Checklists
- Review of documents
- Interviews
- Questionnaires

The auditee team members are mainly from the following organizational groups:

- Project Manager
- Design Manager
- Design Engineers (All Trades)
- Planning Manager
- Contract Administrator/Quantity Surveyor
- Document Controller

Table 6.78 illustrates the audit methodology for the assessment of the construction documents phase and Figure 6.97 illustrates the logic flow diagram for the quality auditing of the construction documents phase.

6.4.3.3 Construction supervision activities

The project manager is the owner's representative and is directly responsible to the owner. The owner normally engages a professional firm with the expertise to supervise construction activities. The project manager has a contractual relationship with the construction supervisor. The project manager oversees the functioning of the construction supervisor as well as the contractor to ensure that construction works are carried out as per the contract requirements to ensure that the project/facility fully meets the project quality and satisfies the owner's goals and objectives.

Table 6.78 Audit methodology for the assessment of construction documents

Serial Number	Item To Be Assessed	Yes	No	Comments
A. Checklist				
1	Whether the reviewed/approved construction documents support the owner's project goals and objectives			
2	Whether the construction documents comply with organizational strategic policies			
3	Whether the reviewed/approved construction documents meet all the elements specified in TOR			
4	Whether regulatory/statutory requirements are considered			
5	Whether the reviewed/approved design has functional and technical compatibility			
6	Whether the reviewed/approved design is constructible			
7	Whether the reviewed/approved working drawings meet all the performance requirements			
8	Whether the reviewed/approved working drawings meet project objectives in respect of time, cost and quality			
9	Whether project boundaries are considered in the reviewed/approved working drawings			
10	Whether reviewed/approved project schedule is achievable in practice			
11	Whether reviewed/approved project cost are properly estimated			

(*Continued*)

Table 6.78 (Continued)

Serial Number	Item To Be Assessed	Yes	No	Comments
A. Checklist				
12	Whether reviewed/approved BOQ is coordinated with specifications and contract documents			
13	Whether reviewed/approved working drawings are based on specified quality standards and codes			
14	Whether the following information is provided in each of the trade drawings: • Working drawings produced at different scales and format • Plans • Sections • Elevations • Schedule • Drawing index			
15	Whether all the drawings are numbered			
16	Whether the following information is included on all the drawings: • Client name • Client logo • Location map • North orientation • Project name • Drawing title • Drawing number • Date of drawing • Revision number • Drawing scale • Contract reference number • Signature block • Signed by the designer for check and approval			
17	Whether the designer has considered the availability of resources during the construction and testing and commissioning phase			
18	Whether the recommended material meets the owner's objectives			
19	Whether construction documents conform with the project delivery system			
20	Whether reviewed/approved construction documents conform to to the adopted type of contract/pricing methodology			
21	Whether the following documents are included along with the working drawings: • Existing site conditions/site plan • Site surveys • Design calculations • Studies and reports			

Serial Number	Item To Be Assessed	Yes	No	Comments
A. Checklist				
22	Whether the reviewed/approved specification is prepared taking into consideration international construction specifications and division numbers			
23	Whether the reviewed/approved construction documents are prepared taking into consideration international construction contract documents			
24	Whether risks have been considered while preparing the working drawings			
25	Whether health and safety requirements are considered in the reviewed/approved working drawings			
26	Whether the reviewed/approved design conforms with fire and egress requirements			
27	Whether environmental constraints are considered			
28	Whether reviewed/approved working drawings, specifications and documents meet TOR requirements			
29	Whether reviewed/approved construction documents deliverables conform as listed in Table 6.21			
30	Whether reviewed/approved reports are complete and include adequate information about the project			
31	Whether the report is properly formatted and there is a table of contents for each report			
32	Whether the reviewed/approved working drawings support construction of the project/facility			
33	Whether the documents include a penalty for delay in completion of the project			
34	Whether the documents include a penalty for failing to provide the required resources (manpower) for the project			
35	Whether the documents include a penalty for delay and also violation of safety during execution of project			

Serial Number	Description	Response	Comments
B. Interview Questionnaires			
1	Whether construction documents were submitted in time for review and approval		
2	Whether there was any delay in getting authorities' approval		

(*Continued*)

Table 6.78 (Continued)

B. Interview Questionnaires

Serial Number	Description	Response	Comments
3	Whether the submitted numbers of sets were as per TOR		
4	Whether the construction documents have any major comments from the owner		
5	Whether the construction documents were not approved by the owner		
6	Whether the reasons (causes) for comments, non-approval, if any, were analyzed and corrective/preventive action taken		

6.4.3.3.1 CONSTRUCTION PHASE

Figure 6.98 illustrates the typical quality auditing process for the project manager's activities related to the construction phase.

The main objectives of performing a quality audit of the project manager's activities are to:

1. Evaluate and assess process compliance with the organizational strategic policies and procedure
2. Assess conformance to QMS and that processes meet project specific quality system requirements
3. Review the process used to perform the activities listed in the contract.
4. Validate the continued follow-up of supervision/inspection of the construction process activities of the construction supervisor
5. Validate the continued follow-up of supervision/inspection of the construction process activities of the construction contractor
6. Check that project performance as per schedule is properly monitored and controlled
7. Check that the project cost as per the budget/contracted value is properly monitored and monitored
8. Check if there are any changes in the contract documents submitted by the contractor and reasons for changes
9. Check that any variation order is properly evaluated and monitored and approved by the owner
10. Check if there was any delay in getting approval of the construction supervisor's project staff and any penalty clauses applied
11. Check if there was any delay in getting approval of the contractor's core staff and any penalty clauses applied

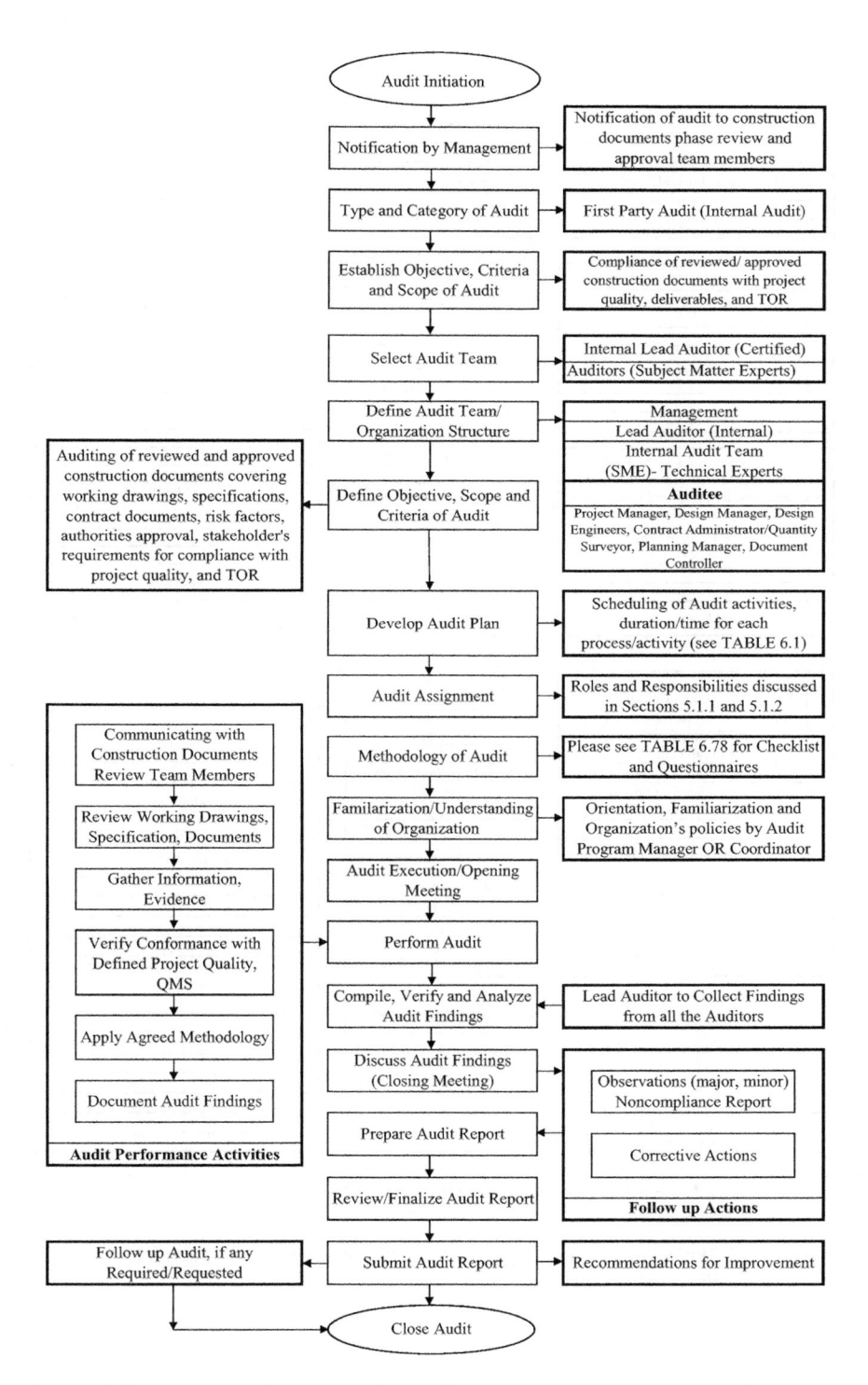

Figure 6.97 Logic flow diagram for the quality auditing of the construction documents phase

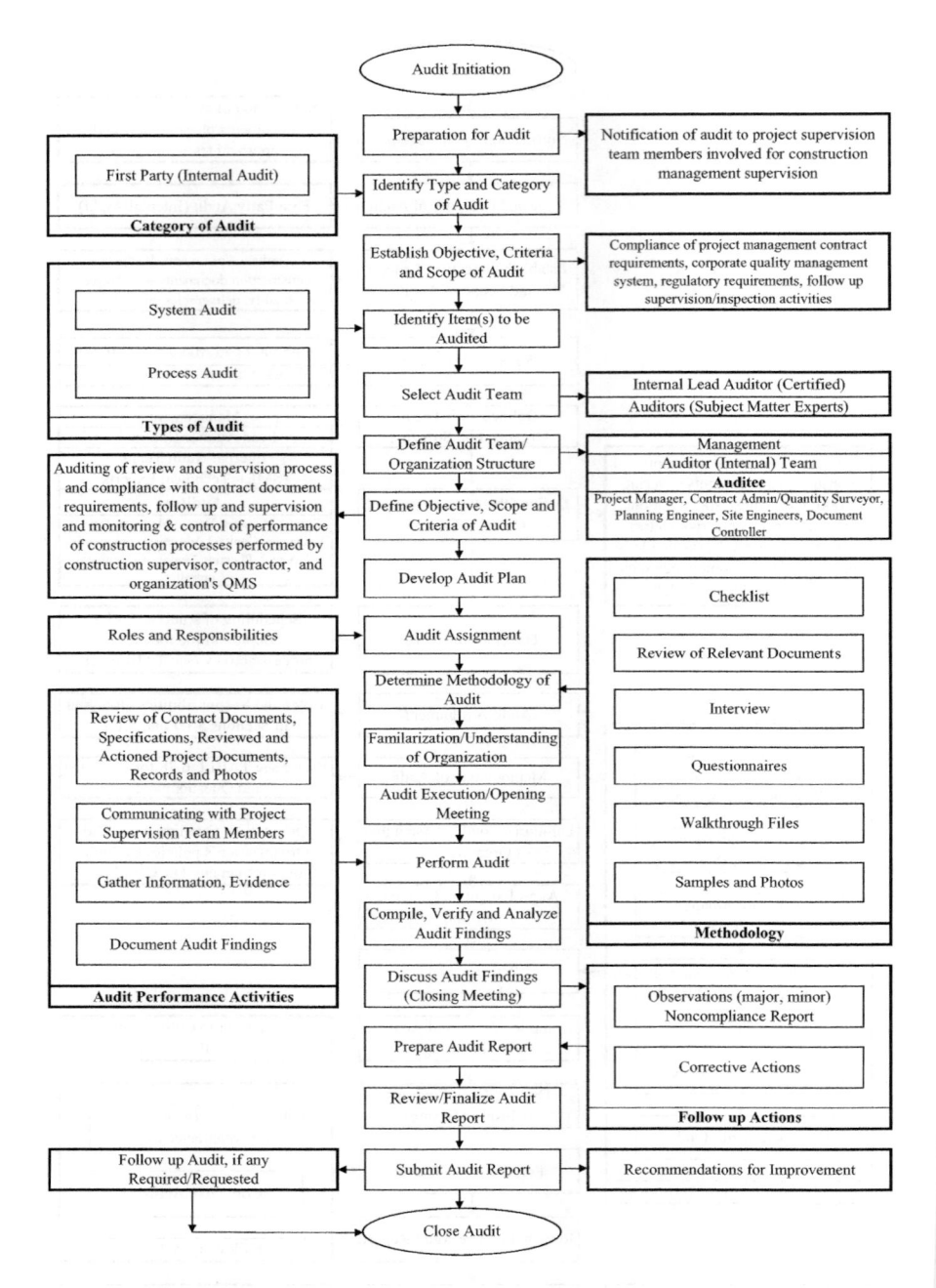

Figure 6.98 Typical quality auditing process for the project manager's activities related to the construction phase

Auditing processes for project life cycle phases 507

12 Check that the contractor submitted management plans as per schedule and the construction supervisor approved the same as per specified period for taking the action on submittals
13 Check all the project control documents were in place
14 Verify that the availability and presence of project team members on site as per schedule was being followed up regularly
15 Check if any conflicts were reported among team members and resolved
16 Check that all the contractor's submittals are as per schedule and reviewed and actioned/responded to by the construction supervisor as per specified period mentioned in the contract documents for taking action on the submittals
17 Validate that all the submittals are properly reviewed by the construction supervisor and actioned/responded to
18 Check the RFI and scope change request are actioned within the stipulated time
19 Review that the risk register was maintained and key risk factors are analyzed and managed
20 Check the payment applications are properly reviewed by the construction supervisor and submitted within the stipulated time
21 Verify if there is any claim by the contractor and if this is settled
22 Check that regulatory requirements are considered while approving the executed/installed works

The following audit tools are used to perform the quality audit of the project manager's activities related to the construction phase:

- Analytical tools
- Checklists
- Review of documents
- Interviews
- Questionnaires

The auditee team members are mainly from the following organizational groups:

- Project Manager
- Contract Administrator/Quantity Surveyor
- Planning/Scheduling Engineer
- Engineers from different trades such as architectural, structural, mechanical, HVAC, electrical, low voltage system, landscape, infrastructure
- Interior Designer
- Document Controller
- Office Secretary

Table 6.79 illustrates the audit methodology for the assessment of the project manager's activities related to the construction phase and Figure 6.99 illustrates

Table 6.79 Audit methodology for the assessment of the project manager's activities related to the construction phase

Serial Number	Item To Be Assessed	Yes	No	Comments

A. *Checklist*

A-1 Integration management
1. Whether contract documents are reviewed and project management requirements during construction phase are properly established
2. Whether all the management plans are developed by the contractor prior to the start of related activities and approved by the construction supervisor
3. Whether the necessary permits are obtained by the contractor prior to the start of construction works
4. Whether the supervision license of the construction supervisor is valid and has regulatory approval (if applicable)
5. Whether bonds and insurance were submitted by the contractor on time
6. Whether the project site was handed over to the contractor on time
7. Whether mobilization was started and completed on schedule
8. Whether mobilization plans were approved by the construction supervisor
9. Whether project control documents were in place

A-2 Stakeholders
1. Whether a responsibilities matrix of all the stakeholders was established by the construction supervisor and circulated
2. Whether the construction supervisor's staff were submitted and approved as per schedule and were available at the project site
3. Whether the contractor's core staff were submitted and approved as per schedule and were available at the project site
4. Whether the regulatory authorities were identified and the contractor was informed of relevant submittals as per schedule
5. Whether the concerned stakeholders were informed about the approved changes (if any) in the baseline (scope, schedule and cost)

A-3 Scope
1. Whether the scope baseline is correctly identified and regularly updated for any changes

Serial Number	Item To Be Assessed	Yes	No	Comments
A. Checklist				
2	Whether there is any change in the contracted scope of work and specifications			
3	Whether there was any Request for Information (RFI) raised and if this was properly actioned by the construction supervisor			
4	Whether there was any Request for Variation submitted by the contractor and the reason for variation was properly evaluated by the construction supervisor for further processing			
5	Whether there was any request for alternate material submitted by the contractor and approved			
6	Whether there was any site work instruction (SWI) issued by the construction supervisor			
7	Whether there was any Request for Proposal submitted by the contractor for additional/new works			
A-4 Schedule				
1	Whether the contractor's construction schedule was approved at first submission			
2	Whether the approved construction schedule was prepared by identifying the contract requirements			
3	Whether the approved construction schedule was prepared taking into consideration the availability of all the resources			
4	Whether the sequence of activities was properly coordinated and in an organized and structured manner			
5	Whether the performance baseline for all the activities was identified in the approved schedule			
6	Whether engineering categories and respective activities were properly listed in the schedule			
7	Whether the contractor's submittals were submitted as per approved schedule and actioned on time			
8	Whether the project progress status was monitored and updated to reflect actual progress versus planned progress			
9	Whether project forecasting was done to predict future progress and delay of activities to be performed			

(Continued)

Table 6.79 (Continued)

Serial Number	Item To Be Assessed	Yes	No	Comments
A. Checklist				
10	Whether the percentage of completion of each activity in the progress report was based on approved checklist			
11	Whether the contractor/construction supervisor updated the schedule based on approved changes and actual work performed/progress			
12	Whether the subcontractor's work progress was included in the progress report			
13	Whether resources (equipment and manpower) were included in the approved schedule			
14	Whether the cost loaded schedule was prepared by the contractor and submitted as per schedule for approval by the construction supervisor			
15	Whether work progress was properly monitored and the schedule updated			
16	Whether any delayed activities were noticed and remedial action was proposed by the contractor			
A-5 Cost				
1	Whether the project was completed as per approved construction budget			
2	Whether there was an approved S-Curve to monitor the expenses and progress payment			
3	Whether the project cost is exceeding the planned budget			
4	Whether cost loaded curve was prepared by the contractor for the project and approved by construction supervisor			
5	Whether there is any approved variation cost to the contracted project value			
6	Whether the S-Curve was regularly updated by the contractor/construction supervisor based on approved variation orders			
7	Whether the cost control process was in place			
8	Whether there was a need to change the cost baseline			
9	Whether there was cost overrun			
10	Whether all the changes were recorded and cash flow updated			
A-6 Quality				
1	Whether the contractor's quality control plan was approved as per schedule			

Serial Number	Item To Be Assessed	Yes	No	Comments
A. Checklist				
2	Whether the contractor's quality management system was properly monitored to comply with contract documents			
3	Whether builders' workshop drawings, coordinated drawings, composite drawings and shop drawings were approved before the start of construction activities			
4	Whether all the materials/systems were approved as per schedule			
5	Whether all incoming material was inspected upon receipt on site			
6	Whether there was any rejection to the material received on site			
7	Whether there were any factory inspections/ visits carried out for the material/system and what was the outcome			
8	Whether test certificates were received with materials supplied			
9	Whether laboratory tests were carried out for concrete works			
10	Whether quality meetings are conducted			
A-7 Resources				
1	Whether the construction supervisor's project staff were deployed as per project manpower plan			
2	Whether the contractor's core staff and project manpower were deployed as per the project manpower plan			
3	Whether the contractor had sufficient workforce available to meet the project execution plan			
4	Whether the contractor's project equipment was available on site as per schedule			
5	Whether all the equipment had a valid license and approval from the appropriate agencies			
6	Whether there was any conflict recorded among project team members			
7	Whether there was any delay in execution of works due to non-availability of resources			
8	Whether the subcontractor's workforce was available as per contract requirements			
9	Whether the staff attendance sheet was maintained			
10	Whether the owner-supplied material (if applicable) was available at the site as per schedule			

(Continued)

Table 6.79 (Continued)

Serial Number	Item To Be Assessed	Yes	No	Comments
A. Checklist				
11	Whether there was any delay in receipt of material, system, equipment that affected the project execution schedule			
A-8 Communication				
1	Whether the site administration matrix was followed properly			
2	Whether the correspondence procedure between all the stakeholders was established and approved			
3	Whether the communication method was established			
4	Whether submittals were distributed using the submittal transmittal form			
5	Whether logs are maintained			
6	Whether logs are updated on a regular basis			
7	Whether the project performance/status was regularly distributed to the stakeholders			
8	Whether meeting invitation and agenda were sent to all the stakeholders			
9	Whether progress meetings and coordination meetings are held as scheduled on a regular basis			
10	Whether daily, weekly and monthly reports were submitted by the contractor and circulated to the relevant stakeholders			
11	Whether project documents are kept in a safe and accessible location			
A-9 Risk				
1	Whether high risk items were identified and a response plan prepared			
2	Whether the risk register was maintained and updated			
3	Whether project risks were documented and recorded			
4	Whether financial risk was reported to the owner/client			
5	Whether occurrence of new risk was recorded to update project progress and performance			
6	Whether the risk management plan was maintained by the construction supervisor and followed			
A-10 Contract				
1	Whether the contract/project management process was established			
2	Whether there were major changes or additions to the signed contract			

Serial Number	Item To Be Assessed	Yes	No	Comments

A. Checklist

A-11 HSE

1	Whether the HSE management plan was approved			
2	Whether safety meetings were conducted			
3	Whether safety drills were conducted			
4	Whether a temporary fire fighting system was operative			
5	Whether any major accidents were recorded			
6	Whether the accident reporting system was followed properly			
7	Whether there were safety violation reports issued			
8	Whether safety disciplinary notices were issued			
9	Whether machinery, equipment and tools were regularly inspected for safe usage			
10	Whether the crane and hoist had valid third party certification			
11	Whether there was any report that sirens and bells were not in working condition			
12	Whether any insurance was claimed for accidents			
13	Whether there was any waste management plan operative			
14	Whether safety violations and accidents were recorded and documented			
15	Whether there was any accident investigation carried out and recorded			
16	Whether there was any non-compliance to HSE requirements			

A-12 Finance

1	Whether progress payment was being submitted by the contractor on time and actioned/responded to by the construction supervisor as per time allocated limit			
	Whether the cash flow is as per approved S-Curve			
2	Whether the project progress payment was made to the contractor(s) on time as per contract			
3	Whether there was regular deduction from the approved progress payment as per contract conditions			
4	Whether there was interim payment and advance payment for procurement of material			

(Continued)

Table 6.79 (Continued)

Serial Number	Item To Be Assessed	Yes	No	Comments
A. Checklist				
5	Whether the construction supervisor's staff salaries were paid on a regular basis			
6	Whether the contractor's staff/labors' salaries were paid on a regular basis			
7	Whether subcontractors are paid as per contract on a regular basis			
8	Whether there was any complaint for payment to the subcontractor not paid as per the agreed-upon terms			
9	Whether there was any complaint for payment to the suppliers not paid as per the agreed-upon terms			
10	Whether there was any complaint about salary to the project workforce not paid on time			
11	Whether finance records and logs are maintained properly			
A-13 Claim				
1	Whether any claim was submitted by the contractor and recorded by the construction supervisor for further approval by the owner			
2	Whether there was any dispute noticed related to contract documents			
3	Whether there was any claim that affected project progress			
4	Whether all the claims were resolved/settled			

B. *Interview Questionnaires*

Serial Number	Description	Response	Comments
1	Whether project management responsibilities are clearly defined in the contract		
2	Whether the project team members know the responsibilities		
3	Whether all the related data of each activity are collected and recorded for project monitoring and controlling purpose		
4	Whether there was any change request by the project owner		
5	Whether all the changes were promptly managed as they occur		
6	What was the frequency of progress status reporting		
7	Whether cost control structure and policy were established		

B. Interview Questionnaires

Serial Number	Description	Response	Comments
8	Whether there were any factory visits as per the contract		
8	Whether there was any problem for the approval of the construction supervisor's project staff and other project team members		
9	Whether there was any problem for the approval of the contractor's core staff and other project team members		
10	Whether any training and development program was conducted for project team members		
11	What methods were used to distribute correspondence to the stakeholders		
12	Whether all the documents and correspondences, transmittals, are saved electronically		
13	Whether identified risks were used to update the project schedule and cost		
14	Whether risk assessment was being done for each of the identified risks		
15	Whether all the risk factors were considered while selecting the project teams		
16	Whether all the risk factors were considered while selecting the construction supervisor's staff		
17	Whether all the risk factors were considered while selecting the contractor's core staff		
18	Whether safety awareness/training programs were conducted by the contractor		
19	Whether an emergency evacuation plan was approved and in place		
20	Whether evacuation routes were displayed on site		
21	Whether health surveillance was being carried out		
22	Whether there was any site visit by the regulatory authority		
23	Whether emergency evacuation plans were displayed at various locations		
24	Whether hazardous areas were properly monitored		
25	What method was followed to resolve/settle claims		
25	Whether any claim is pending to be settled		

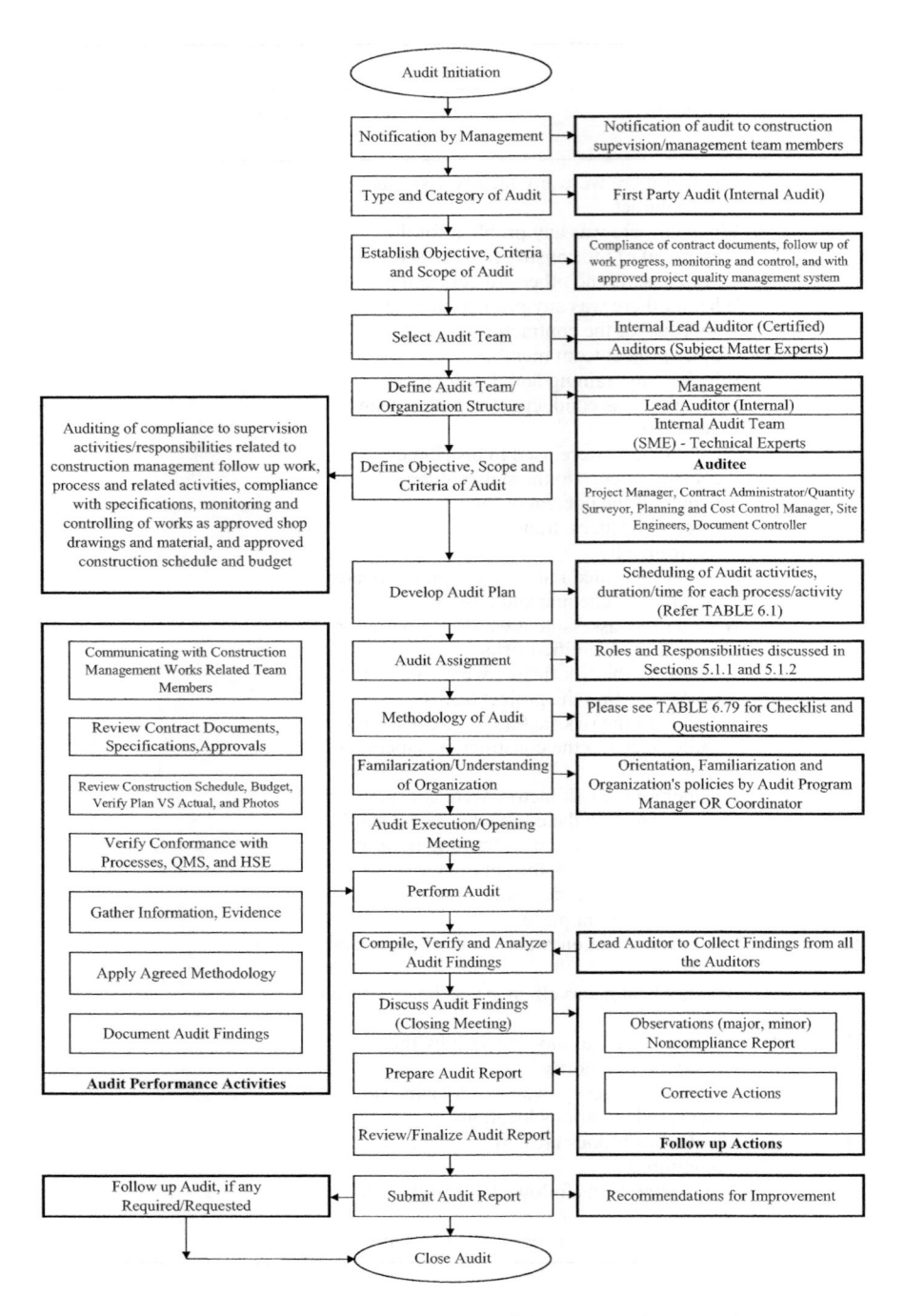

Figure 6.99 Logic flow diagram for the quality auditing process for the project manager's activities related to the construction phase

Auditing processes for project life cycle phases 517

the logic flow diagram for the quality auditing process for the project manager's activities related to the construction phase.

6.4.3.3.2 TESTING, COMMISSIONING AND HANDOVER

Figure 6.100 illustrates the typical quality auditing process for the project manager's activities related to the testing, commissioning and handover phase.

The main objectives of performing quality audit activities related to the testing, commissioning and handover phase are to:

1. Check that contract documents were reviewed and the scope of work for testing, commissioning and handover are properly identified
2. Assess conformance to QMS and that processes meet established quality system requirements
3. Validate that project team members were informed of the typical responsibilities as listed in Table 6.80
4. Verify that all the stakeholders and team members were identified and the responsibility matrix was established and circulated
5. Ascertain that all requirements of regulatory authorities were identified to ensure the smooth handing-over of the project/facility
6. Verify that the testing and commissioning schedule was submitted by the contractor and was approved
7. Check that a punch list/snag list was prepared by the contractor/construction supervisor and actioned
8. Verify that closeout documents as per Table 6.49 were submitted by the contractor and were reviewed, actioned and handed over to the client
9. Check that all the relevant documents were distributed by the construction supervisor to the concerned stakeholders
10. Check the risk management plan was developed by the construction supervisor
11. Check whether project completion was as per the approved schedule
12. Check that the handing-over certificate was signed by the concerned parties (Figure 6.66 illustrates a typical handing-over certificate)
13. Check that the contractor followed the HSE management plan and any accidents were recorded
14. Check that the payment to the contractor and subcontractor was approved and paid
15. Check that the payment to the construction supervisor was approved and paid
16. Check that all the claims were settled
17. Check if there is any "Lessons Learned" list prepared
18. Check that project records are archived as per the organization's policy

The following audit tools are used to perform the quality audit of the construction supervisor's activities related to the testing, commissioning and handover phase:

- Analytical tools
- Checklists

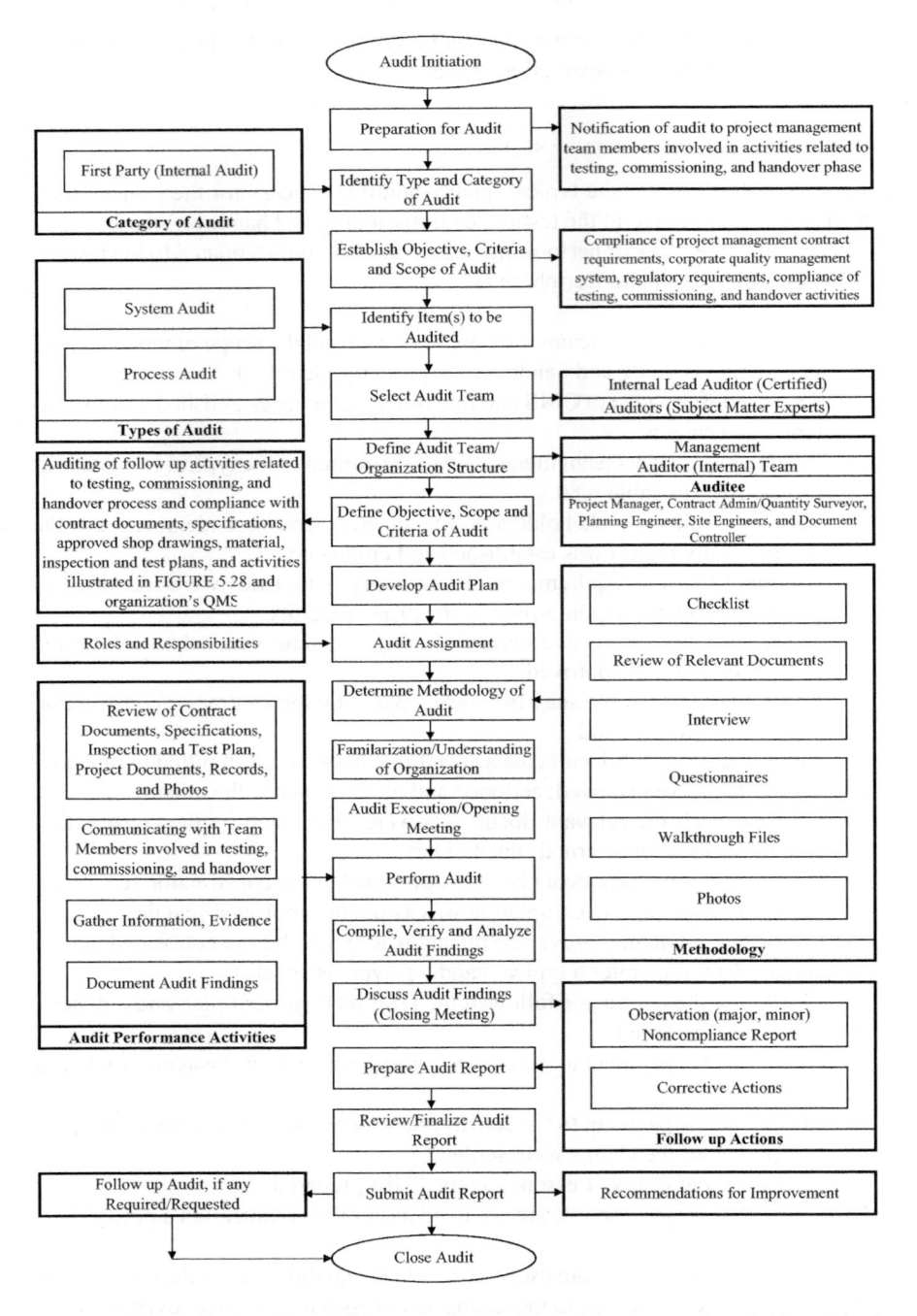

Figure 6.100 Typical quality auditing process for the project manager's activities related to the testing, commissioning and handover phase

Auditing processes for project life cycle phases 519

Table 6.80 Typical responsibilities of the project manager during the testing, commissioning and handover phase

Serial Number	Responsibilities
1	Ensure that the construction supervisor/contractor obtained the occupancy permit from the respective authorities
2	Ensure that all the systems are tested, functioning and operative
3	Ensure that the construction supervisor has closed any Job Site Instruction (JSI)and Non-Conformance Report (NCR)
4	Ensure that the contractor has cleaned the site and all the temporary facilities and utilities are removed
5	Ensure that master keys are handed over to the owner/end user
6	Ensure that the construction supervisor has handed over guarantees, warrantees and bonds to the client
7	Ensure that the construction supervisor has handed over operation and maintenance manuals to the client
8	Ensure that the construction supervisor has handed over test reports, test certificates and inspection reports to the client
9	Ensure as-built drawings are submitted by the contractor and actioned/responded to by the construction supervisor
10	Ensure as-built drawings are handed over to the client/end user
11	Ensure that spare parts are handed over to the client
12	Ensure that a snag list is prepared by the contractor/construction supervisor and handed over to the client
13	Ensure that training for client/end user personnel is completed
14	Ensure that all the dues of suppliers, subcontractors and contractor are paid
15	Ensure that retention money is released
16	Ensure that the substantial completion certificate issued and the maintenance period commissioned
17	Lessons learned are documented

- Review of documents
- Interviews
- Questionnaires

The auditee team members are mainly from the following organizational groups:

- Project Manager
- Contract Administrator/Quantity Surveyor
- Planning/Scheduling Engineer
- Engineers from different trades such as architectural, structural, mechanical, HVAC, electrical, low voltage system, landscape, infrastructure
- Interior Designer
- Document Controller
- Office Secretary

Table 6.81 illustrates the audit methodology for the assessment of the project manager's activities related to the testing, commissioning and handover phase and Figure 6.101 illustrates the logic flow diagram for the quality auditing process for the project manager's activities related to the testing, commissioning and handover phase.

Table 6.81 Audit methodology for the assessment of the project manager's activities related to the testing, commissioning and handover phase

Serial Number	Item To Be Assessed	Yes	No	Comments
A. Checklist				
1	Whether the scope of works was established taking into consideration contract documents and specifications			
2	Whether testing and startup requirements were identified as per the contract documents			
3	Whether the testing, commissioning and handover plan was developed taking into consideration all the executed/installed works/systems			
4	Whether all stakeholders were identified involved during the testing, commissioning and handover phase			
5	Whether contract closeout documents were identified as per the contract documents and specifications			
6	Whether the testing and commissioning works/systems were identified and listed and distributed to all the project team members			
7	Whether the testing schedule was developed and was approved			
8	Whether the commissioning schedule was developed and approved			
9	Whether the punch list/snag list was prepared			
10	Whether the startup risk management plan was established and managed			
11	Whether the project closeout documents as listed in Table 6.49 were submitted by the contractor, approved by the construction supervisor and accepted for the as-built record			
12	Whether the project handing-over schedule was developed and monitored			
13	Whether the "Lessons Learned" list was prepared			
14	Whether payments were made to the contractor and subcontractors			
15	Whether claims were resolved and settled			

B. Interview Questionnaires

Serial Number	Description	Response	Comments
1	Whether testing and commissioning requirements were identified and completed		
2	Whether project team members were notified about their responsibilities as listed in Table 6.80		
3	Whether authorities' approval was obtained as per the schedule		
4	Whether the handing-over certificate (substantial completion certificate) was issued		
5	Whether the project/facility was accepted/handed over		
6	Whether there are any incomplete works that are not executed/installed		
7	Whether the contractor's payment was certified and settled		
8	Whether supervision fees were paid to the construction supervisor		
9	Whether any claim was not resolved/settled		

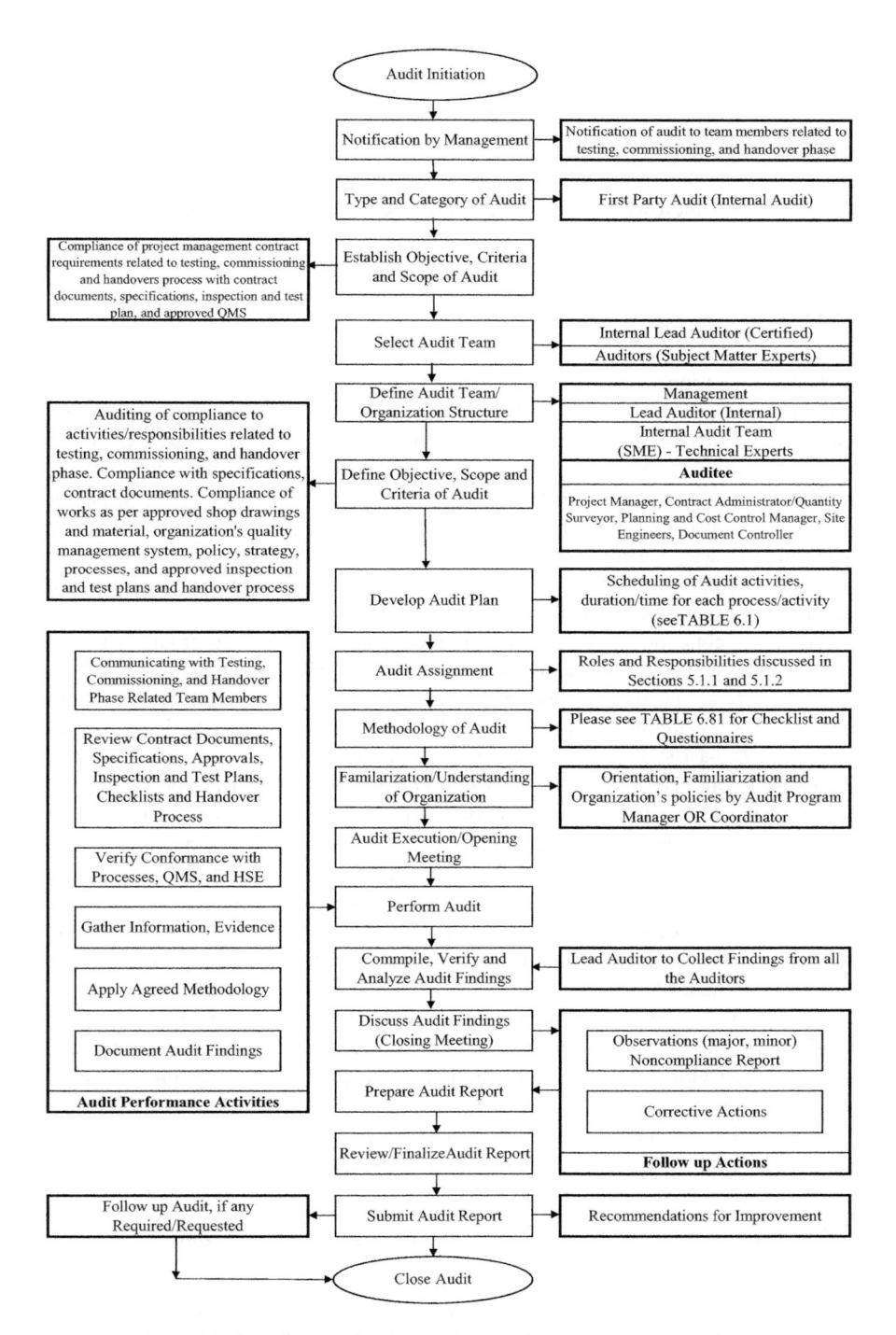

Figure 6.101 Logic flow diagram for the quality auditing process for the project manager's activities related to the testing, commissioning and handover phase

6.5 Auditing process for the construction management type of project delivery system

In this method, the owner contracts a construction management firm to coordinate the project for the owner and provide construction management services. There are two general forms of construction management. These are:

1 Agency Construction Management
2 Construction Management at Risk

The agency construction management type is a management process type of contract system having a four party arrangement involving the owner, designer, construction management firm and contractor. The construction manager provides advice to the owner regarding cost, time and safety, and about the quality of the materials/products/systems to be used on the project. The agency construction management firm performs no design or construction, but assists the owner in selecting the design firm(s), and contractor(s) to build the project.

Agency construction management could be implemented in conjunction with any type of project delivery system.

The basic concept of the construction management type of contract is that the firm is knowledgeable and capable of coordinating all aspects of the project to meet the intended use of the project by the owner. The agency construction manager acts as a principal agent to advise the owner/client, whereas the construction manager-at-risk is responsible for on-site performance and actually performs some of the project works. The construction management-at-risk type of contract has two stages. The first stage encompasses pre-construction services and during second stage the construction manager-at-risk is responsible for performing the construction work. The construction management-at-risk project delivery system is also known as construction manager/general contractor.

Figure 6.102 illustrates the contractual relationship for agency construction management and Figure 6.103 illustrates the contractual relationship for construction management-at-risk.

The agency construction management type of management system can be used with different types of project delivery system. Figure 6.104 illustrates the sequential activities of agency construction management in the Design-Bid-Build delivery system.

This section discusses the audit processes for the agency construction manager project delivery system for the Design-Bid-Build type of project.

6.5.1 Auditing process for the selection of the agency construction manager

Agency construction management is a management process where the construction manager acts an advisor to the owner. The owner may engage the construction manager for the entire life cycle of the construction project or during a specific phase of the construction project.

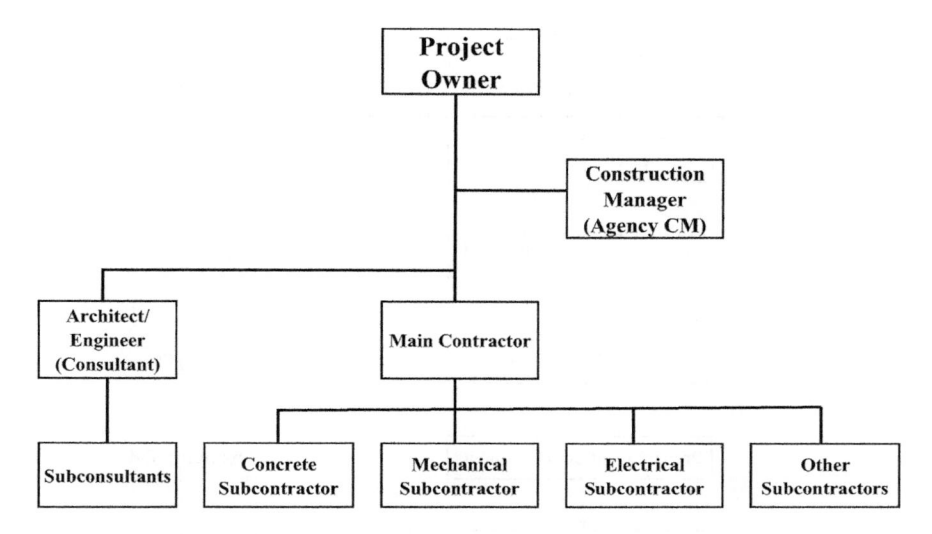

Figure 6.102 Construction manager contractual relationships (agency CM)

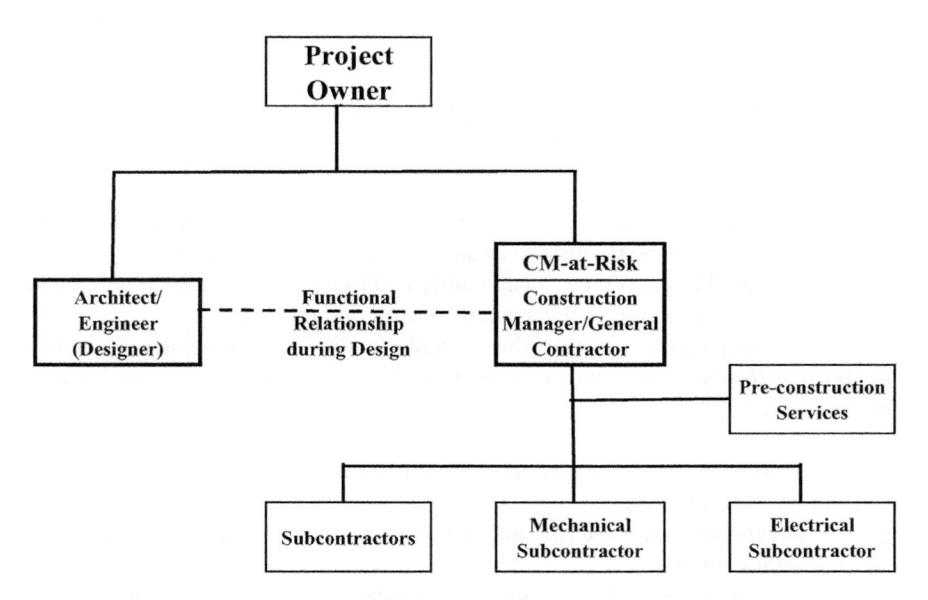

Figure 6.103 Construction manager contractual relationships (at risk CM)

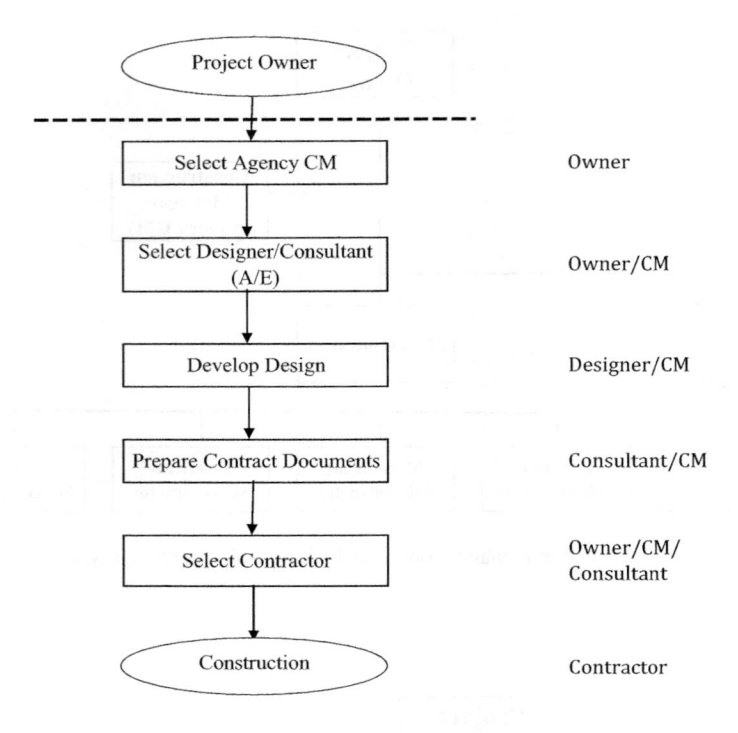

```
Project Owner
    │
----+-------------------
    ▼
Select Agency CM                    Owner
    │
    ▼
Select Designer/Consultant          Owner/CM
       (A/E)
    │
    ▼
Develop Design                      Designer/CM
    │
    ▼
Prepare Contract Documents          Consultant/CM
    │
    ▼
Select Contractor                   Owner/CM/
                                    Consultant
    │
    ▼
Construction                        Contractor
```

Figure 6.104 Sequential activities of agency CM – Design-Bid-Build

The roles and responsibilities of the construction manager in a project may vary substantially, and can be performed under a variety of contractual terms.

Figure 6.105 illustrates the typical quality auditing process for the bidding and tendering for the selection of the agency construction manager.

The main objectives of performing a quality audit for the bidding and tendering process to select the agency construction manager to manage the construction project are to:

1. Evaluate and assess process compliance with the organizational strategic policies and procedure
2. Assess conformance to QMS and that processes meet established quality system requirements
3. Verify that all the major activities are adequately performed and managed
4. Ensure that key risk factors are analyzed and managed to meet business requirements
5. Check the proper functioning and implementation of the project management system

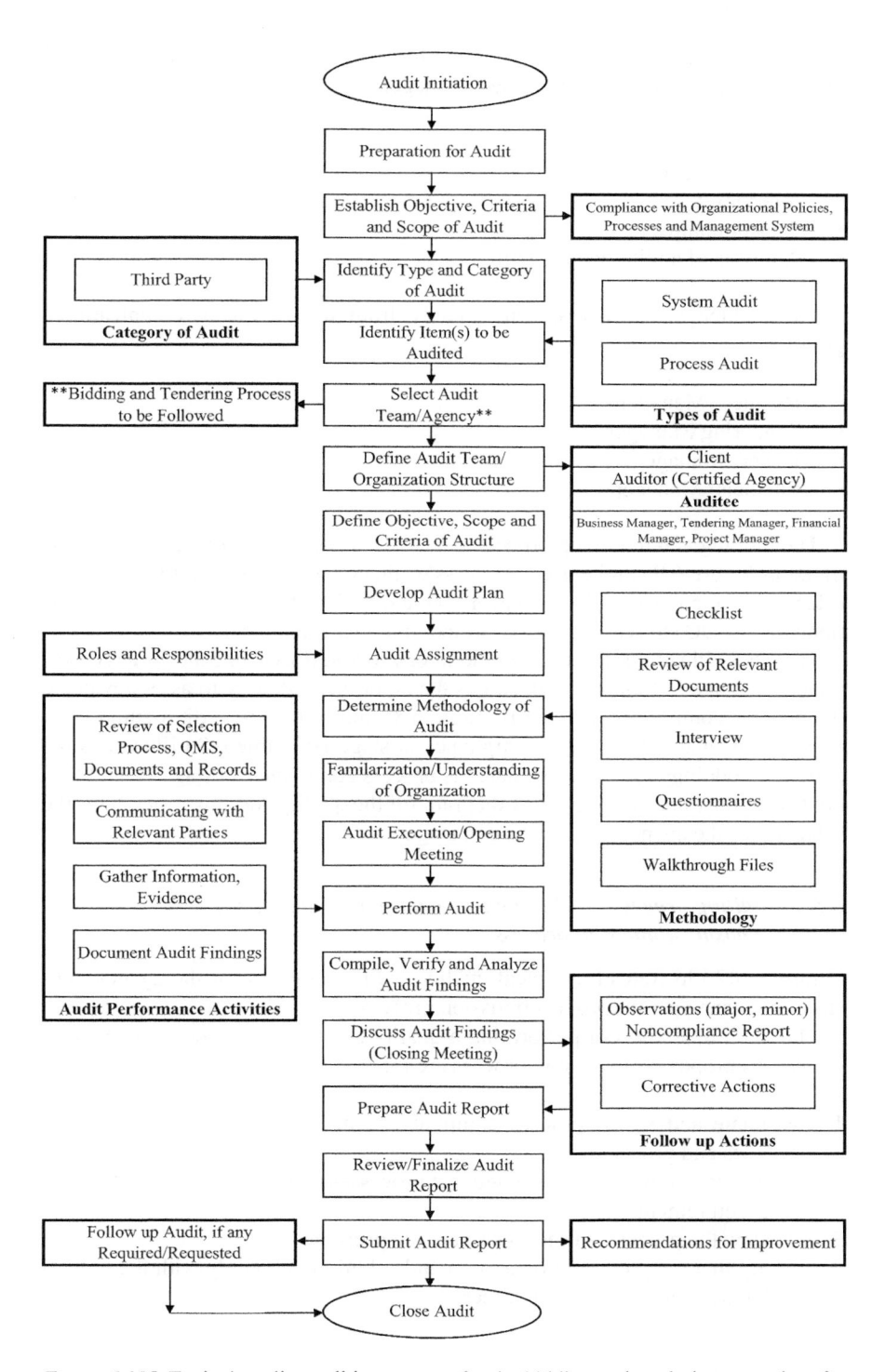

Figure 6.105 Typical quality auditing process for the bidding and tendering procedure for the selection of the agency construction manager

The following audit tools are used to perform a quality audit for the bidding and tendering process to select the project manager:

- Checklists
- Review of documents
- Interviews
- Questionnaires
- Analysis of audit documentation

The auditee team members are mainly from the following organizational groups:

- Business Manager
- Tendering Manager
- Project Manager
- Financial Manager

The bidding and tendering process stages discussed in Section 6.2.1.2 as illustrated in Figure 6.9 can be used to assess/evaluate the process compliance of the selection of the construction manager. Table 6.82 lists the prequalification questionnaires (PQQ) to register the construction management firm.

Table 6.83 illustrates the audit methodology to assess/evaluate the process compliance at stage I, Table 6.84 illustrates the audit methodology to assess/evaluate the process compliance at stage II, Table 6.85 illustrates the audit methodology to assess/evaluate the process compliance at stage III, Table 6.86 illustrates the audit methodology to assess/evaluate the process compliance at stage IV and Figure 6.106 illustrates the logic flow diagram for the quality auditing process for the selection of the agency construction manager.

6.5.2 Auditing process for the agency construction manager's proposal submission process

Figure 6.107 illustrates the typical quality auditing process for the agency construction manager's proposal submission process.

The main objectives of performing a quality audit of the agency construction manager's proposal submission process are to:

1. Evaluate and assess process compliance with the organizational strategic policies and procedure
2. Assess conformance to QMS and that processes meet established quality system requirements
3. Review the processes used to prepare the proposal
4. Verify that the proposal preparation and submission comply with the organization's proposal submittal process

Table 6.82 Prequalification questionnaires (PQQ) for selecting the construction manager

Serial Number	Question	Answer
1	Name of the organization and address	
2	Organization's registration and license number	
3	ISO certification	
4	Total experience (years)	
	4.1 Agency construction manager	
	4.2 Construction manager at risk	
5	Size of project (maximum amount single project)	
	5.1 Agency construction manager	
	5.2 Construction manager at risk	
6	List similar types (type to be mentioned) of projects completed	
	6.1 Agency construction manager	
	6.2 Construction manager at risk	
7	Type of services provided as agency construction manager/construction manager for above mentioned projects	
	7.1 During project study	
	7.2 During design	
	7.3 During bidding/tendering	
	7.4 During construction	
	7.5 During startup	
8	Total experience in green building construction	
9	Total management experience	
	9.1 Project scope	
	9.2 Project planning and scheduling	
	9.3 Project costs	
	9.4 Project quality	
	9.5 Technical and financial risk	
	9.6 HSE	
	9.7 Stakeholder management	
	9.8 Project monitoring & control	
	9.9 Conflict management	
	9.10 Negotiations	
	9.11 Information technology	
10	Experience in conducting value engineering	
11	Resources	
	11.1 Management	
	11.2 Engineering	
	11.3 Technician	
	11.4 Construction equipment	
12	Total turnover for last 5 years	
13	Submit audited financial reports for last 3 years	
14	Experience in training of owner's personnel	
15	Knowledge about regulatory procedures	
16	Litigation (dispute, claims) on earlier projects	

Table 6.83 Audit methodology for the assessment of the bidding and tender process for the selection of the agency construction manager firm – Stage I (shortlisting/registration of designers)

Serial Number	Item To Be Assessed	Yes	No	Comments
A. Checklist				
1	Whether the notification for registration was announced in all technical newsletters and leading news media as per the organization's policy			
2	Whether there are registered agency construction manager firms to participate in the tender			
3	Whether prequalification questionnaires (PQQ) are issued to all the intending bidders (please refer to Table 6.82 for guidelines)			
4	Whether tender documents clearly indicate the responsibilities of the agency construction manager for all relevant activities such as selection of designer (A/E), consultant, design management, selection of contractor and subcontractor and construction management			
5	Whether there are any requests for bid clarification			
6	Whether the meeting for bid clarification is attended by all the intending bidders			
7	Whether the attendance sheet for the meeting is signed by all the attendees			
8	Whether minutes of the meeting are circulated to all the attendees/participating bidders			
9	Whether sufficient time is provided to prepare and submit completed proposal			
10	Whether there are any requests for extension of submission date			
11	Whether all the intending bidders submitted the response to PQQ/tender			
12	Whether received bid envelopes are placed in safe custody			
13	Whether there are any delayed submissions			
14	Whether received responses are acknowledged			
15	Whether received responses are clearly identified and recorded			
16	Whether responses are fairly evaluated as per the organization's selection policy			
17	Whether evaluation criteria, weightage and methodology are specified for selection of project manager			
18	Whether any of the submitted responses are found to be incomplete			

Serial Number	Item To Be Assessed	Yes	No	Comments
A. Checklist				
19	Whether there are any late submissions			
20	Whether risk factors are considered while evaluating the response and registration of designers			
21	Whether reasons for non-registration are conveyed to unsuccessful participants			
22	Whether shortlisted construction firms are allotted a code and registration number			

Serial Number	Description	Response	Comments
B. Interview Questionnaires			
1	How responses were submitted: • Hand delivery • By post • Courier service		
2	Whether PQQ was available for electronic distribution		
3	How was the response for registration/ shortlisting		
4	Whether the quoted price was as per the allocated budget for construction management activities		

Table 6.84 Audit methodology for the assessment of the bidding and tendering process for the selection of the agency construction manager firm – Stage II (proposal documents)

Serial Number	Item To Be Assessed	Yes	No	Comments
A. Checklist				
1	Whether the Request for Proposal (RFP) includes all the relevant information required to select the agency construction manager firm			
2	Whether the RFP properly defines the construction manager's roles and responsibilities intended by the owner to be performed by the agency construction manager firm			
3	Whether all the deliverables are listed in the RFP			

(*Continued*)

Table 6.84 (Continued)

Serial Number	Item To Be Assessed	Yes	No	Comments
A. Checklist				
4	Whether all the information to enable bidders to submit the proposal is clear and unambiguous			
5	Whether documents are prepared taking into consideration the organization's Quality Management System (QMS)			
6	Whether regulatory/authority requirements are taken into consideration while preparing the RFP			
7	Whether the selection criteria and selection method for the project manager firm selection is clearly specified in the organization's QMS			
8	Whether budget approval is obtained for project management services prior to the release of the RFP announcement			
9	Whether the RFP opening and closing dates are mentioned in the announcement			
10	Whether the schedule for required project management services is included in the RFP			
11	Whether the qualifications of personnel for design management services are requested in RFP			
12	Whether the qualifications of personnel for project management services are requested in RFP			
13	Whether project quality requirements are properly defined in the documents			
14	Whether rights and liabilities of all the parties are mentioned in the documents			
15	Whether change/variation clauses are included in the RFP			
16	Whether cancellation/termination clauses are included in the documents			
17	Whether bond and insurance clauses are included in the documents			
18	Whether the RFP documents are as per international standard documents used by the construction industry			
19	Whether evaluation and assessment criteria are well defined			
20	Whether review and analysis of RFP are included			
21	Whether the agreement to be signed for construction management services is included in the documents			
22	Whether all the records are properly maintained			
23	Whether the request to provide the agency construction manager's contact details is included in the documents			

B. *Interview Questionnaires*

Serial Number	Description	Response	Comments
1	Whether any outside agency was involved in the preparation of the tender documents		
2	What guidelines were adopted to prepare the tender documents		
3	Whether any specific format/template was followed		
4	Whether all the requirements of RFP documents were coordinated and agreed by relevant stakeholders		
5	Whether risks of disputes and associated financial costs were considered while preparing the RFP		
6	Whether problems and challenges were identified while preparing RFP		
7	Who was responsible for authorizing bidding requirements		

Table 6.85 Audit methodology for the assessment of the bidding and tendering process for the selection of the agency construction manager firm – Stage III (contract bid solicitation)

Serial Number	Item To Be Assessed	Yes	No	Comments
A. *Checklist*				
1	Whether RFP are documents properly organized			
2	Whether the RFP is distributed to all the prequalified/shortlisted/registered agency construction manager firms to participate in the tender			
3	Whether the tender is notified to all the prequalified agency construction manager firms			
4	Whether all the prequalified/shortlisted/registered bidders participated in the tender			
5	Whether the attendance sheet for the meeting was signed by all the attendees			
6	Whether minutes of the meeting were circulated to all the attendees/participating bidders			
7	Whether sufficient time is provided to prepare and submit the proposal/quotation			
8	Whether the tender submission date is extended from the date originally announced			

(*Continued*)

Table 6.85 (Continued)

Serial Number	Item To Be Assessed	Yes	No	Comments
A. Checklist				
9	Whether an addendum (if any) is issued/notified to all the participating bidders			
10	Whether there are any requests for an extension of submission date			
11	Whether the clarification meeting is attended by all participating bidders			
12	Whether any changes are made to the originally RFP			
13	Whether the received bid envelopes are placed in safe custody			
14	Whether technical and financial envelopes are submitted at the same time			
15	Whether there are any delayed submissions			
16	Whether proposals are opened as per the announced date and time			
17	Whether proposals are opened in the presence of all the bidders			
18	Whether received tenders are acknowledged			
19	Whether received tenders are clearly identified and recorded			
20	Whether applicable fees/bid bond are submitted by all the bidders who participated in the tender			
21	Whether the tenders are fairly evaluated as per the announced policy and procedure			
22	Whether the comparison of bids is tabulated			
23	Whether evaluation criteria, weightage and methodology are as specified			
24	Whether the bid cost is higher than the budgeted cost			
25	Whether any submitted tender is found to be incomplete			
26	Whether there are any late submissions			
27	Whether all the scheduled pages are signed by the bidder			
28	Whether risk factors are considered while evaluating the bids			
29	Whether reasons for non-acceptance of the tender were conveyed to the unsuccessful bidders			

Serial Number	Description	Response	Comments
B. Interview Questionnaires			
1	How the notification was announced		
2	How tenders were submitted: • Hand delivery • By post • Courier service		

B. Interview Questionnaires

Serial Number	Description	Response	Comments
3	Whether the tender was available for electronic distribution		
4	How was the response to participate in tender		
5	Whether tender boxes were placed at secured places and monitored through an electronic surveillance system		
6	How much time was allowed to open the bids after the notified tendering closing time		
7	Whether the tender submission date was extended and what was the reason for extension		
8	Whether all the tender envelopes/packages were sealed with marking of the tenders and contain the original tender documents and required number of sets as per tender conditions		

Table 6.86 Audit methodology for the assessment of the bidding and tendering process for the selection of the agency construction manager firm – Stage IV (contract award)

Serial Number	Item To Be Assessed	Yes	No	Comments
A. Checklist				
1	Whether the selection of the agency construction manager firm satisfies all the conditions to be a successful bidder			
2	Whether the selected bidder (agency construction manager) is a legal entity			
3	Whether the selection was made by the selection committee			
4	Whether the selection of the agency construction manager is approved by relevant applicable authority, if mandated by the organization's policy			
5	Whether the selected agency construction manager is capable of carrying out the specified work and has the necessary resources			
6	Whether the selected agency construction manager has experience of selecting the designer (A/E), contractor, subcontractor			

(*Continued*)

Table 6.86 (Continued)

Serial Number	Item To Be Assessed	Yes	No	Comments
A. Checklist				
7	Whether the selected agency construction project manager has the resources for design review, design management and construction management services			
8	Whether risk factors are considered prior to the signing of the contract			
9	Whether the standard contract format is used for signing the contract			
10	Whether contract terms and conditions are clearly written and unambiguous			
11	Whether an addendum and minutes of the meeting are included in the contract documents			
12	Whether clarifications to queries are included as part of the contract documents			
13	Whether the contract period/schedule is properly described			
14	Whether staff qualifications are properly defined			
15	Whether a performance bond is submitted by the consultant			
16	Whether variation/change management clauses are included in the contract agreement			
17	Whether dispute resolution and conflict resolution clauses are properly defined			
18	Whether contract documents are reviewed prior to signing			
19	Whether tender (proposal) documents and contract are properly archived			
20	Whether a letter of award is sent			
21	Whether the final results are published and announced			

B. Interview Questionnaires

Serial Number	Description	Response	Comments
1	On what basis the bidder (agency construction manager) was selected: low bid, competitive bid or quality based system		
2	Was there any tie between two or more bidders		
3	Was there any dispute or objection raised by any of the bidders before the award of the contract to the successful bidder		
4	Whether the final price was negotiated or as per the bid (proposal) received		
5	Whether the contract amount is as per the budget		

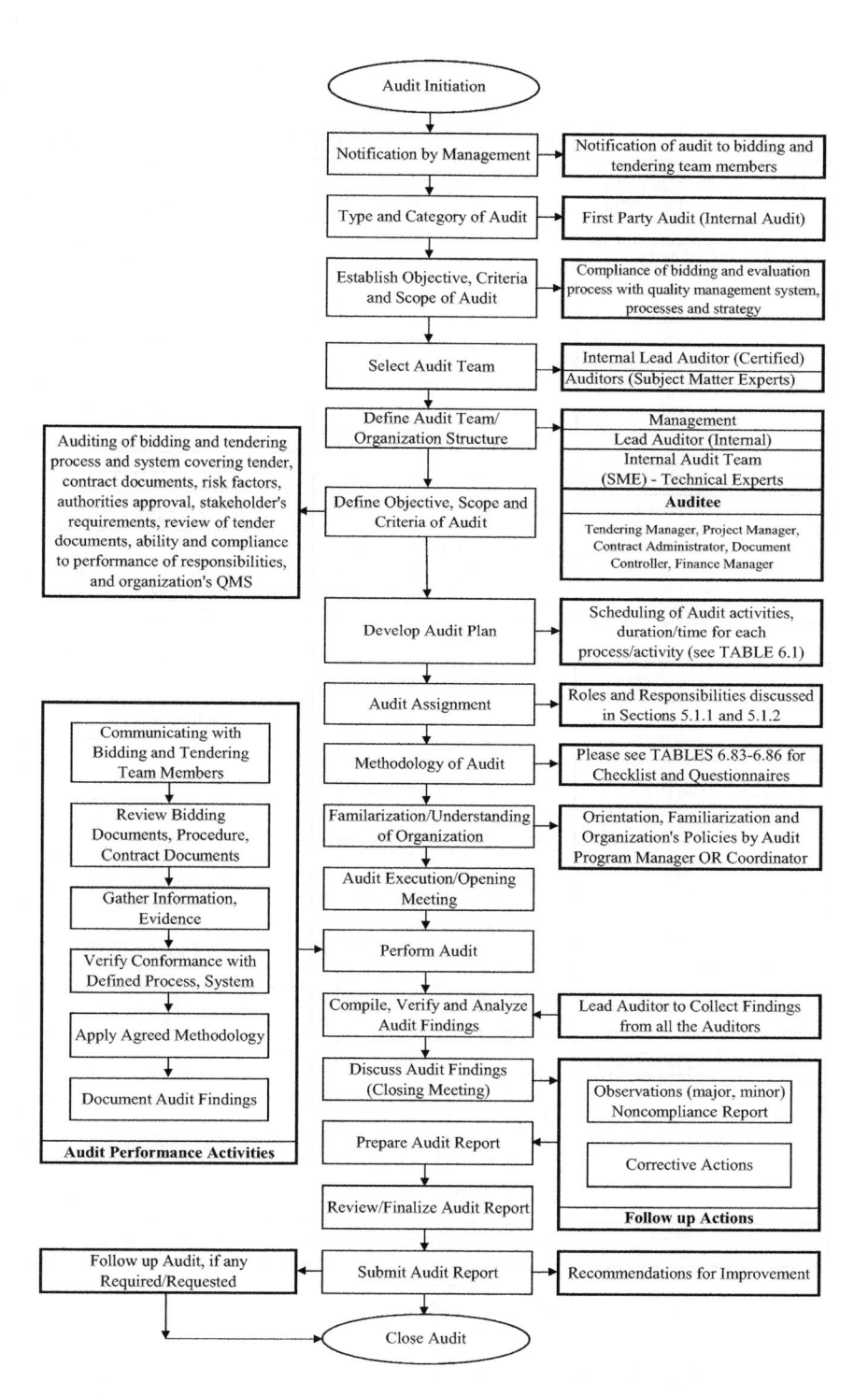

Figure 6.106 Logic flow diagram for the quality auditing of the selection of the agency construction manager

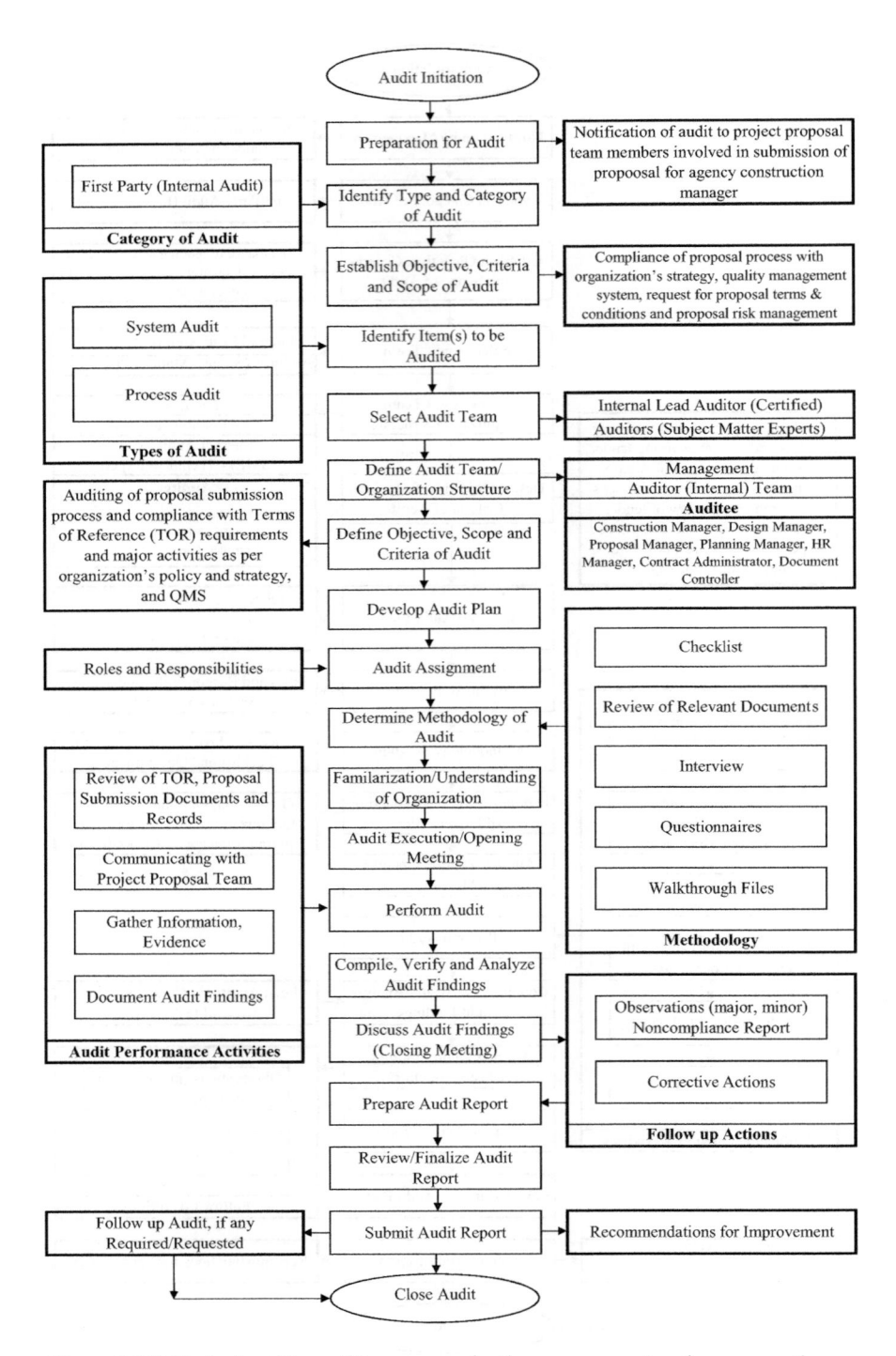

Figure 6.107 Typical quality auditing process for the agency construction manager's proposal submission process

5   Validate that the response to the contents of Request for Proposal is complete in all respect.
6   Check that the availability of resources for design input, design review and design management is considered
7   Check that the availability of resources for construction management during construction phase is considered
8   Check that the availability of resources for construction management during the testing, commissioning and handover phase is considered
9   Review the process used to prepare the proposal
10 Validate the schedule for deployment of construction management staff as per the contract
11 Validate the cost of the proposal for all the activities
12 Review that the key risk factors are analyzed and managed
13 Check compliance with regulatory requirements

The following audit tools are used to perform a quality audit of the agency constuction manager's proposal submission process:

- Analytical tools
- Checklists
- Review of documents
- Interviews
- Questionnaires

The auditee team members are mainly from the following organizational groups:

- Construction Manager
- Design Manager
- Proposal Manager
- Quality Manager
- Contract Administrator/Quantity Surveyor
- HR Manager
- Document Controller
- Financial Manager

Table 6.87 illustrates the audit methodology for the assessment of the agency construction manager's proposal submission process and Figure 6.108 illustrates the logic flow diagram for the quality auditing of the agency construction manager's proposal submission process.

6.5.3 Auditing process for construction management activities

The construction management delivery system is a project procurement/contract management process whereby the construction manager (firm or an individual with sound project manager skills) undertakes to manage the carrying out of the

Table 6.87 Audit methodology for the assessment of the agency construction manager's bid (proposal) submission process

Serial Number	Item To Be Assessed	Yes	No	Comments
A. Checklist				
1	Whether the company is already registered as an agency construction manager to participate in the tender			
2	Whether there are any fees to collect tender (RFP) documents			
3	Whether the scope of work is clearly defined in the RFP			
4	Whether the TOR (Terms of Reference) is clearly defining all the scope			
5	Whether there is any request for clarification raised			
6	Whether the response to the RFP is complete			
7	Whether sufficient time is provided to prepare and submit the completed RFP			
8	Whether there are any requests for extension of the submission date			
9	Whether there are any delayed submissions			
10	Whether the RFP is prepared taking into consideration/referring to the organization's quality management system			
10	Whether resources and time required to select the designer (A/E), consultant and contractor were considered while preparing the schedule			
11	Whether the design review and management requirements are considered while preparing the proposal			
12	Whether the resources for construction management during construction were considered while preparing the proposal			
13	Whether the proposal was reviewed prior to submission			
14	Whether the proposal was approved by the relevant committee/authorized person			
15	Whether the risk factors were considered and analyzed			

B. Interview Questionnaires

Serial Number	Description	Response	Comments
1	Whether the notification for the proposal submission was announced in all technical newsletters and leading news media or it was only for registered bidders (agency construction manager firm)		
2	How the proposal was submitted: • Hand delivery • By post • Courier service		
3	Whether the RFP was available for electronic distribution		
4	What tools and techniques were used to price the proposal		
5	Whether the resources for bidding and tendering and selection of the designer (A/E) and consultant were considered		
6	Whether the resources for bidding and tendering and selection of contractor and subcontractor were considered		
7	Whether the organization's design management and review plan were considered to evaluate the price schedule		
8	Whether the availability of qualified resources (technical) to review the project design was considered while preparing the price schedule		
9	Whether the availability of qualified resources (technical) to manage the construction phase, testing, commissioning and handover phase was considered while preparing the price schedule		
10	Whether the design review and management schedule were properly planned by taking into consideration the availability of qualified manpower		
11	Whether the construction management schedule was properly planned by taking into consideration the availability of qualified manpower		
12	Whether the regulatory approval period was properly considered		
13	What risk factors were considered while evaluating the price schedule		

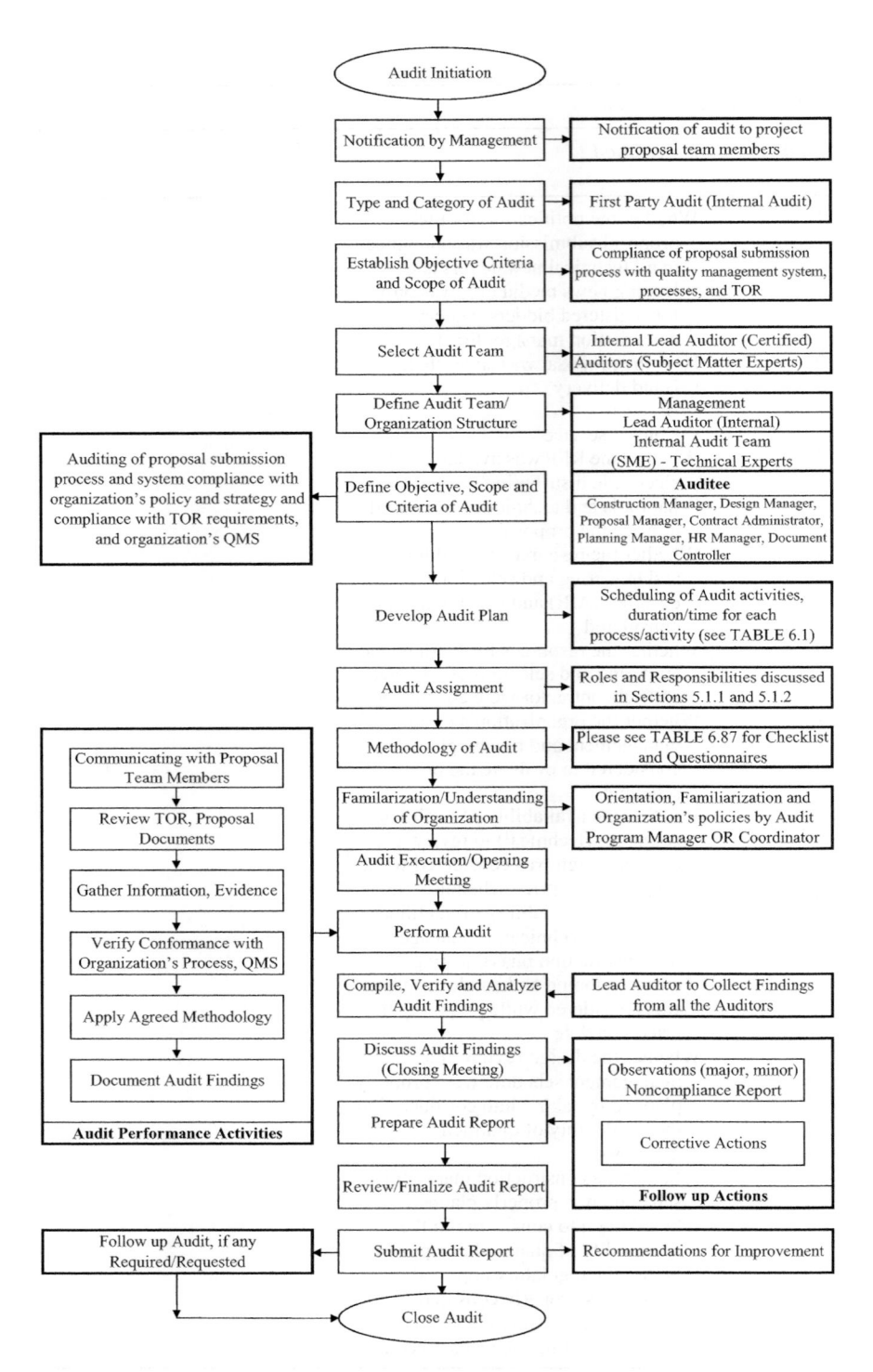

Figure 6.108 Logic flow diagram for the quality auditing process for the agency construction manager's proposal submission process

work through contractors, be they are general or trade specific, and oversee the performance throughout the project life cycle by systematic application of a set of management skills and principles.

The role of construction manager is to apply comprehensive management and control effort to the project beginning at the early project planning stages and continuing until project completion. The construction management process involves the application and integration of comprehensive project controls to the design and construction process to achieve successful project delivery/completion.

For the Design-Bid-Build type of construction projects, the agency construction manager provide following services during different stages of construction project:

1   Agency construction manager role during pre-design stage

- Defining overall performance requirements
- Defining overall project program
- Develop project scope of work (TOR)
- Development of project procedures and standards
- Establishing a management information system
- Development of project schedule
- Development of project budget
- Identify critical constituents of the project
- Identify authorities' approvals and permits required
- Selection of project delivery method
- Selection of designer (consultant)

2   Agency construction manager role during design stage

- Oversee and coordinate design
- Recommend alternative solutions
- Life cycle cost analysis
- Review design documents
- Constructability review
- Make suggestions to improve constructability
- Construction related recommendations
- Coordination with regulatory authorities to obtain permits and license
- Recommend selection of materials, building systems and equipment
- Value engineering suggestions
- Value enhancement suggestions
- Construction schedule
- Construction cost estimates
- Budget and schedule management
- Project risk management
- Help designer (A/E) develop multiple bid packages to expedite construction process

- Review of final drawings and specifications
- Preparing and revising project management plan

3. Agency construction manager role during bidding and tendering
   - Preparation of bid packages
   - Prequalification of bidders
   - Establish bidding schedule
   - Management of bidding documents
   - Pre-bid meetings to familiarize bidders with the bidding documents
   - Answering all queries related to bid
   - Addendum to bid documents
   - Receive bids
   - Bid review and evaluation
   - Bid analysis and comparison of bids
   - Contract negotiation
   - Prepare recommendation to owner to accept or reject the proposals
   - Participate in contractor selection
   - Conduct pre-award meeting(s) with the selected contractor
   - Incorporate addenda changes into contract documents
   - Prepare construction contract
   - Notice to proceed

4. Agency construction manager role during construction stage
   - Ensure that contractor submitted performance bond
   - Ensure that contractor submitted workers' insurance policy
   - Selecting and recommending contractor's core staff
   - Selecting and recommending subcontractor
   - Establishing and implementing procedures for processing and approval of shop drawings
   - Material approvals
   - Managing contractor's Request for Information (RFI)
   - Change order management
   - Construction schedule approval
   - Construction supervision
   - Quality management
   - Coordination of on-site, off-site inspection
   - Inspection of works
   - Construction contract administration
   - Conduction periodic progress meetings
   - Preparation of minutes of meeting and distribution as per agreed-upon matrix
   - Document control
   - Technical correspondence between contractor

- Managing submittals
- Monitor daily progress
- Monitoring contractor's performance and ensure that the work is performed as specified, as per approved shop drawings, and as per applicable codes
- Scope control
- Construction scheduling and monitoring
- Cost tracking and management
- Review, evaluation and documentation of claims
- Maintain project progress record
- Evaluation of payment request and recommending progress payments
- Monitor project risk
- Monitoring contractor's HSE plans
- Coordination of works by multiple contractors
- Coordination for delivery and storage of owner supplied materials and systems
- Testing of systems
- Punch list

5 Agency construction manager role during testing, commissioning and handover stage

- Testing and commissioning of systems
- Review of as-built drawings
- Review of record documents and manuals
- Warranties and guarantees
- Authorities' approval for occupancy
- Coordination of handover-take over procedure
- Move in plan
- Punch list
- Preparing list of lessons learned
- Substantial completion
- Archiving project documents
- Settlement of claims
- Project final account
- Project close-out

There are three activities that are to be performed by the agency construction manager. These are:

1 Selection of project teams

- Selection of designer (A/E) consultant
- Selection of contractor, subcontractor

2 Design management

- Review and approval of project design (design phases)
- Review and approval of construction documents

3 Construction management activities

- Construction phase
- Testing, commissioning and handover phase

6.5.3.1 Selection of project teams

The agency construction manager is responsible for selecting the designer (A/E) and contractor. The agency construction manager has to prepare the bidding and tendering documents as per the organizational strategy and follow the process to acquire all the related services, products/materials, systems, equipment, designer (A/E), contractor and subcontractor. The role of the agency construction manager during the bidding and tendering process is discussed above (please refer to item 3 in this section).

The quality auditing process that includes the aims, objectives and audit methodology for the assessment of the bidding and tendering process for the selection of the designer (A/E) discussed in section 6.2.1.2 can be used to perform the quality auditing process for the selection of the designer (A/E) by the agency construction manager. Please refer to Figures 6.8–6.9 and Tables 6.6–6.10 discussed in Section 6.2.1.2.

Similarly, the quality auditing process that includes the aims, objectives and audit methodology for the assessment of the bidding and tendering process for the selection of the contractor discussed in Section 6.2.3.1 can be used to perform the quality auditing process for the selection of the contractor by the agency construction manager. Please refer to Figures 6.25–6.29 and Tables 6.23–6.27 discussed in Section 6.2.3.1.

6.5.3.2 Design management

The design management activities that are to be performed by the agency construction manager already discussed above. (Please refer to item 2 role of agency construction manager during design stage.)

The agency construction manager is responsible for ensuring that the project design and documents developed by the designer (A/E) meet the project goals and objectives to satisfy the owner's/end user's requirements.

6.5.3.2.1 AUDITING PROCESS FOR THE CONCEPT DESIGN PHASE

The quality auditing process that includes the aims, objectives and audit methodology for the assessment of the auditing process for concept design discussed in Section 6.4.3.2.1 can be used to perform the quality auditing process for the

Auditing processes for project life cycle phases 545

concept design phase by the agency construction manager. Please refer to Figures 6.90–6.91 and Table 6.75 discussed in Section 6.4.3.2.1.

6.5.3.2.2 AUDITING PROCESS FOR THE SCHEMATIC DESIGN PHASE

The quality auditing process that includes the aims, objectives and audit methodology for the assessment of the auditing process for the schematic design discussed in Section 6.4.3.2.2 can be used to perform the quality auditing process for the schematic design phase by the agency construction manager. Please refer to Figures 6.92–6.93 and Table 6.76 discussed in Section 6.4.3.2.2.

6.5.3.2.3 AUDITING PROCESS FOR THE DESIGN DEVELOPMENT/DETAIL DESIGN PHASE

The quality auditing process that includes the aims, objectives and audit methodology for the assessment of the auditing process for the detail design discussed in Section 6.4.3.2.3 can be used to perform the quality auditing process for the detail design phase by the agency construction manager. Please refer to Figures 6.94–6.95 and Table 6.77 discussed in Section 6.4.3.2.3.

6.5.3.2.4 AUDITING PROCESS FOR THE CONSTRUCTION DOCUMENT PHASE

The quality auditing process that includes the aims, objectives and audit methodology for the assessment of the construction documents process discussed in Section 6.4.3.2.4 can be used to perform the quality auditing process for the construction documents phase by the agency construction manager. Please refer to Figures 6.96–6.97 and Table 6.78 discussed in Section 6.4.3.2.4.

6.5.3.3 Construction management activities

Construction management activities that are to be performed by the agency construction manager already discussed above. (Please refer items no. 4 and 5 the roles of agency construction manager during construction phase and testing, commissioning, and handover phase, respectively.)

The agency construction manager's role and responsibilities during the construction phase are similar to the construction supervisor's activities to supervise the construction phase.

The quality auditing process that includes the aims, objectives and audit methodology for the assessment of the auditing process for the construction phase (construction supervisor's activities) discussed in Section 6.2.4.2 can be used to perform the quality auditing process for the construction phase by the agency construction manager. Please refer to Figures 6.60–6.61 and Tables 6.45–6.48 discussed in Section 6.2.4.2.

Similarly the quality auditing process that includes the aims, objectives and audit methodology for the assessment of the auditing process for the testing, commissioning and handover phase (construction supervisor's activities) discussed in Section 6.2.5.2 can be used to perform the quality auditing process for the testing, commissioning and handover phase by the agency construction manager. Please refer to Figures 6.65–6.67 and Tables 6.51–6.52 discussed in Section 6.2.5.2.

7 Audit reporting

7.1 Audit report

Once the audit performance is completed, the auditor has to submit an audit report to the clients. The purpose of the audit report is to document and communicate the results of the audit to the client.

Table 7.1 illustrates a typical audit report format.

A typical audit process and the audit processes for various types of project delivery systems and at different stages are discussed in Chapter six. Normally the audit results are discussed during the closing meeting. Thereafter the report is reviewed and finalized and communicated to the client and the auditee as per the agreed-upon list of recipients. The audit report of construction project activities at different stages/phases provides conformation that:

- Activities are performed as per the corporate quality management system and conforms to the requirements of ISO 9001
- The contracted project requirements and project specific procedures have been verified for adherence to meet project quality
- All the processes have been implemented to ensure that the project quality is achieved
- All the design elements are considered to meet the owner's goals and objectives
- The works have been executed as per specification requirements
- The works have been executed as per specified codes and standards
- The project meets regulatory requirements
- This serves guidelines to reduce rejection, non-approval of executed works
- This serves as guidelines for subsequent improvement and prevention of mistakes, errors while developing new projects

Table 7.2 illustrates a typical audit finding report and Table 7.3 illustrates a typical nonconformance report.

Table 7.1 Typical contents of an audit report

	Name of Organization	
1	Audit Reference No.	
2	Date of Audit	
3	Project No.	
4	Phase/Stage	
4	Audit Location	
6	Name of Auditor	
7	Category of Audit	
8	Description:	
9	Objective of Audit	1
		2
		N
10	Scope of Audit	
11	Auditee	1
		2
		N
12	Audit Process Methodology	1
		2
		N
13	Audit Findings with evidence	
13.1	Nonconformance Report	1
		2
		N
13.2	Observations	1
	(Major /Minor)	2
		N
14	Recommendation	
15	Corrective Actions	1
		2
		N
16	Opportunities for Improvements	1
	(Areas for Improvements)	2
		N
17	Names of Report Recipients	1
		2
		N
18	Any Confidential Issues Addressed	Y/N

Table 7.2 Typical audit findings report

Audit Result

Organization Name
Project Name & No.
Phase/Stage
Period

Decision	*Location*	*ID*	*Process/Activity*	*Auditor*	*Completion Date*
Nonconformance					
Observations					
OK					
Area for Improvement					

Table 7.3 Typical nonconformance report

Internal Audits Nonconformance Report

Organization Name
Project Name & No.
Phase/Stage
Period

Location	ID	Process/Activity	Auditor	Summary	Completion Date	Corrective Action Requested

7.2 Corrective and preventive action

Upon completion of the audit program, and based on the findings of the audit, the auditor has to suggest improvements to be made in the processes, procedures and systems audited. The findings, nonconformance and observations, if any, must be acted upon and necessary action taken by the audited organization.

Table 7.4 illustrates a typical areas for improvement report.

The audited organization should review the audit findings and prepare corrective action plan. Table 7.5 illustrates a typical corrective action plan.

There are several types of tools, techniques and methods, in practice, which are used as quality improvement tools and have a variety of applications in the manufacturing and process industry. However not all of these tools are used in construction projects due to the nature of construction projects which are customized and non-repetitive. Some of these quality management tools that are most commonly used in the construction industry are listed under the following broader categories:

- Classic quality tools
- Management and planning tools
- Process analysis tools
- Process improvement tools
- Innovation and creative tools
- Lean tools
- Cost of quality
- Quality function deployment
- Six Sigma
- Triz (teirija rezhenijia izobretalenksh zadach)

Table 7.6 lists the process improvement tools that are normally used to improve the quality of the process and product.

Table 7.4 Typical areas for improvement report

Internal Audits
Areas for Improvement Report

Organization Name
Project Name & No.
Phase/Stage
Period

Location	ID	Process/ Activity	Auditor	Summary	Completion Date	Corrective Action Requested

Table 7.5 Typical corrective action plan

Non Conformance Report
Corrective Action Plan

Organization Name
Project Name & No.
Phase/Stage

Nonconformance Report Reference	ID	Process/ Activity	Audit Date	Possible Cause of Nonconformance	Proposed Corrective Action	Completion Date

Table 7.6 Process improvement tools

Sr. No.	Name of Quality Tool	Usage
Tool 1	Root cause analysis	To identify root causes that caused the problem to occur.
Tool 2	PDCA cycle	Used to plan for improvement followed by putting into action
Tool 3	SIPOC analysis	Used to identify supplier-input-process-output-customer realationship
Tool 4	Six Sigma DMAIC	Used as analytic tool for improvement
Tool 5	Failure Mode and Effects Analysis (FMEA)	To identify and classify failures according to effect and prevent or reduce failure
Tool 6	Statistical process control	Used to study how the process changes over a time.

Table 7.7 Typical corrective action taken report

Non Conformance Corrective Action Taken Report

Organization Name
Project Name & No.
Phase/Stage

Nonconformance Report Reference	ID	Process/ Activity	Audit Date	Corrective Action Details	Corrective Action Completion Date	Responsible Person for Action	Requested Verification Date

Report Prepared By   Date
Action Report Verified By   Date
Report Approved/ NCR Closed   YES ☐   NO ☐   Reason for Non Approval
Signature   Date

7.3 Project closeout audit

Follow-up audits may be required in some situations. The auditor may be engaged to verify that corrective actions have been taken.

Table 7.7 illustrates a typical action taken report.

If a follow-up audit or verification is not required then the audit may be declared closed and follow-up activities can be included with the next audit.

Bibliography

Albersmeier, F., Schulze, H., Jahn, G. and Spiller A. (2009). The reliability of third party certification in the food chain: From checklists to risk-oriented auditing. *Food Control*, 20(10), 927–935.

Arditi, D. and Gunaydin, H. M. (1997). Total quality management in the construction process. *International Journal of Project Management*, 15(4), 235–243.

ASQ/ANSI/ISO 19011:2018 (2018). *Guidelines for Auditing Management* Systems. Milwaukee, WI: Quality Press.

Mafi, S., Huber, M. and Shraim, M. (2016). ISO certification in the construction industry. In A. R. Razzak (Ed.), *Handbook of Construction Management Scope, Schedule, and Cost Control* (pp. 731–759). Boca Raton, FL: CRC Press (a Taylor & Francis Group Company).

Rumane, Abdul Razzak (2013). *Quality Tools for Managing Construction Projects*. Boca Raton, FL: CRC Press (a Taylor & Francis Group Company).

Rumane, Abdul Razzak (2016). *Handbook of Construction Management: Scope, Schedule, and Cost Control*. Boca Raton, FL: CRC Press (a Taylor & Francis Group Company).

Rumane, Abdul Razzak (2017). *Quality Management in Construction Projects*, Second Edition. Boca Raton, FL: CRC Press (a Taylor & Francis Group Company).

Sowards, D. (2013). Quality is key to lean's success. *Contractor Magazine*, (May), 44, 52.

Tang, S. L. (2005). *Construction Quality Management*. Hong Kong: Hong Kong University Press.

Index

Accelerated 205
Accident, Accidents 5, 123, 154, 193, 282, 350, 363, 367, 373–4, 376, 382, 418, 455, 513, 517
Addendum 186–7, 211, 213, 216, 218, 222, 228, 230, 281, 289, 291, 295, 302, 307, 309, 313, 372, 422, 424, 428, 436, 443, 445, 451, 471, 473, 532, 534, 542
Adequacy 2, 27, 47, 129, 135
Adequacy audit 203
Alternatives 10, 54, 59, 158, 163–4, 167–8, 209, 238, 242, 401–2
Analytical tool 234, 240, 247, 251, 261, 272, 283, 314, 319, 373, 387, 396, 408, 425, 437, 458, 477, 483, 488, 494, 500, 507, 517, 537
Analyze 17, 49, 58, 62, 64, 68, 74, 78, 80, 84, 93, 97, 108, 129, 146, 157, 162, 221, 232, 239, 244, 260, 269–70, 277, 279, 293–4, 298, 300, 311–2, 317, 322, 325, 330, 337, 353, 355, 357, 368, 370, 386, 388, 394–5, 409, 413, 426–7, 432, 435, 449–50, 452, 457, 462, 466, 475–6, 480, 482, 486–7, 492–3, 498–9, 505–6, 516, 518, 521, 525, 536–6, 540
ANSI 29, 116, 118
Appendix 48–9, 63, 80, 97, 271, 407
Approval, Approvals 77, 91–3, 106, 146, 156, 161, 164–5, 167, 174, 180, 184, 186, 189–90, 193, 195, 199, 214, 226, 233, 235, 238, 243, 249, 253, 256–9, 274, 278, 281, 287, 287, 299, 306, 318, 324, 326–7, 331, 335–6, 343, 346–7, 352, 353, 387, 390, 393, 399, 402, 411–3, 426, 431, 448–9, 461, 463, 475, 479, 486–7, 491, 493, 497–8, 502–5, 508, 535, 539, 541, 543
Arbitration 218

Architect/Engineer (A/E) 2–5, 159–160, 206, 218–23, 225–26, 228–29, 231, 262, 331, 405–406, 451, 464, 467, 473–474, 479, 481,483, 485, 488, 491, 497, 500, 524, 528, 533, 538–539, 541, 543–544
Architectural 17, 253, 369, 373, 396
Architectural design 246, 250, 259
Architectural plans 171
Architectural works 390
Archived 57, 73–4, 89, 105, 218, 230, 249, 256, 268, 275, 291, 310, 387, 396, 424, 445, 473, 517, 534
Arditi D. 123, 553
As Built 345
As Built drawings 10, 195, 197, 199, 375, 385, 390, 396, 463, 519, 543
ASHRAE
ASME 29
ASQ 29, 117–8, 120–1, 129, 553
ASTM 29,
Attendance sheet 215, 225, 228, 286, 289, 304, 307, 348, 361, 380, 419, 422, 439, 442, 467, 471, 511, 528, 531
Audit activities 116, 118, 142–4, 235, 257, 269, 277, 293, 298, 311, 317, 325, 330, 337, 353, 355, 368, 386–7, 393–4, 400, 413, 426, 432, 449, 452, 462, 475, 480, 486, 492, 498, 505, 516–7, 521, 535, 540
Audit report 109, 114, 115–6, 118, 128, 131, 142–6, 203–4, 221, 232, 236, 239, 244, 252, 257, 260, 262, 269–70, 277, 279, 293–4, 298, 300, 311–2, 317, 321–2, 325, 330, 337, 353, 355, 356, 368, 370, 386, 388, 394–5, 400, 409, 413, 415, 426–7, 423, 435, 449–50, 452, 457, 462, 466, 475–6, 480, 482, 486–7, 492–3, 498–9, 505–6, 516, 518, 521, 424, 535–6, 540, 547, 548

Auditee 111, 115–9, 124, 126–8, 134–8, 142–4, 204, 212, 220, 221, 232, 236, 239, 244, 252, 257, 260, 262, 269–70, 277, 279, 293–4, 298, 300, 311–2, 317, 321–2, 325, 330, 337, 353, 355, 357, 368, 370, 386, 388, 394–5, 400, 409, 413, 415, 426–7, 432, 435, 449–50, 452, 457, 462, 466, 475–6, 480, 482, 486–7, 492–3, 498–9, 505–6, 516, 518, 521, 525, 535–6, 540, 547–8
Auditee team 138–9, 140–1, 144–5, 234, 240, 247, 253, 261, 272, 280, 283, 301, 314, 320, 373, 389, 396, 410, 414, 429, 434, 448, 458, 465, 477, 483, 488, 494, 500, 507, 519, 526, 537
Auditor-in-training 137
Auditor selection 114, 167–8, 174, 180, 184, 189, 195, 199
Authority/Authorities' approval 10, 156, 158, 161, 164, 169, 172, 176–7, 182, 191, 197, 209, 259, 276–7, 293, 297, 311, 316, 324, 329, 335, 376, 385, 387, 390, 399, 402, 411–3, 426, 431, 448–9, 461, 463, 475, 479, 486–7, 491, 493, 497–8, 502–5, 508, 535, 539, 541, 543

Baseline 175, 358, 364, 377, 464, 508
  Cost baseline 360, 379, 510
  Measurement baseline 343
  Performance baseline 343, 346, 359, 378, 509
  Scope baseline 351, 358, 365, 377, 508
Bid bond 156, 161, 186, 216, 228, 281, 289, 308, 422, 443, 471, 532
Bidders 5, 9, 10, 153, 156, 159, 185–9, 213–7, 219, 222, 225, 228–9, 235, 271–2, 281, 286–9, 295–6, 302, 303–9, 313, 315, 407, 419, 422–5, 428, 430, 436, 439–40, 442–6, 451, 467, 469, 471–2, 474, 478, 528, 530–4, 539, 542
Bidding and tendering 6, 9, 21, 24, 37, 53, 59, 90–1, 105–6, 149, 151, 153–4, 156, 161, 179–80, 183–9, 208, 210–3, 215, 217, 219–22, 225, 228–9, 272, 278–81, 286, 288, 293–4, 299–304, 307, 309, 349, 362, 401, 404–5, 414–6, 422, 424, 426,433–6, 439–40, 465–8, 470, 472, 481, 524–9, 531, 533, 542, 544
BOQ 15, 23, 176, 261, 265–6, 268, 273, 283, 297, 331, 364, 372, 397, 410, 416, 431, 495–6, 502
BOOT 205
BOT 205

BSI 29
BTO 205
BREEAM xxxii
Budget 1, 10, 12, 13, 15, 20, 27, 109, 113, 154–5, 160, 170, 176, 183, 189–90, 201, 214, 216, 222, 226, 242, 245–6, 259, 281, 292, 302, 310, 314, 316, 318, 322, 326, 329, 357, 368, 386, 420, 425–6, 446, 455, 469, 474, 483, 488, 491, 500, 516, 530, 534, 541
  Allocated budget 468, 529
  Construction budget 192, 347, 360, 379, 510
  Controlling budget 21
  Planned budget 347, 360, 379, 510
  Preliminary budget 158
  Project budget 24, 155, 158, 167, 178, 182, 189
  Update budget 160, 181, 186, 281
Budget approval 287, 306, 441
Builder 36
Builder's drawings 193, 374
Builder's shop drawing 333
Builder's work 338
Builder's workshop drawing 347–8, 360, 379, 511
Building 1–3, 28, 155, 245, 250, 258–9, 345, 397, 401, 448, 483, 541
Building codes 123, 170–1
Building material 5, 27
Business manager 138, 212, 220–1, 239–40, 465–6, 525–6

Calculations 251, 259, 263
Design calculation 171, 177, 180, 182, 258, 266, 274, 411, 496, 502
Cash flow 176, 192
Casting 338, 350, 354, 363, 375–6, 382
Casting process xv, 336
Casting samples 336
Categories of;
  Audit 143, 203
  Auditing 121, 210, 401, 403
  Civil construction projects 3
  Engineering 359, 378, 509
  Project delivery system 205
  Risk 366, 384
CEN 29
CENLEC 29
Certification 47, 121
Certification ISO 285
Change order 299, 369, 434, 542
Change management 37, 42, 44–5, 54, 69, 85, 95, 102, 168, 174, 180, 191, 195,

556 Index

218, 230, 291, 310, 333, 346, 356, 358, 377, 424, 445, 473, 534
Checklist for audit preparation 116
Claim(s) 9, 95, 151, 166, 173, 179, 183, 188, 194–5, 198–9, 224, 283, 285, 351–2, 356, 364, 367, 383, 385, 387, 392–3, 396, 398–9, 406, 418, 465, 507, 524–5, 517, 520, 527, 543
Claim management 151, 166, 173, 179, 183, 188, 194–5, 198, 465
Client 2, 5, 17, 36, 41, 44, 46–7, 49, 61–2, 65, 74, 78–9, 82–3, 89, 92, 94, 96, 108, 110, 112–3, 212–2, 124, 127, 130, 132–5, 138, 142, 144–7, 151, 154–5, 167–9, 174–5, 184–5, 195, 200, 202–3, 208–9, 221, 231, 239, 241, 246, 259, 262, 264, 274, 292, 295, 313, 319, 323–4, 335, 341, 349, 351, 356, 362, 364, 367, 371, 381, 383, 396, 399, 401–2, 411, 428, 451, 456, 465–6, 479, 502, 517, 519, 522, 525, 547
Climate 246
Closing process group 150, 164, 171, 177, 182, 187, 191, 197
Code 74, 75, 225, 304, 345, 419, 439, 468, 529, 543
   Action code 331
   Building codes 123, 170–1
   Energy code 253
   Quality codes 165
Code of conduct 145
Code of ethics 112
Codes and standards please refer Standards and codes
Commercial 1–3
Commercial A/E projects 3
Commercial (mixed use) 223, 282, 405, 417
Committee 235, 296, 315, 447, 478, 538
   Selection Committee 217, 229, 290, 309, 424, 430, 444, 473, 533
   Tendering Committee 186, 278–80, 300–1, 305, 314, 414–5, 434–5, 440, 448
Communication 3, 18, 31, 33, 38, 41, 43, 46, 59, 72, 76, 88, 92, 104, 106, 110, 125, 146, 148, 190, 195, 202, 204
Communication internal and external 37, 40, 42, 45, 57, 72, 88, 104
Communication management 151, 165, 173, 179, 183, 188, 193, 198, 200
Communication matrix 179, 183, 372
Communication method 381, 512
Communication plan 9, 193, 318, 381, 455
Communication skills 118, 146–9

Communication technology 246, 299, 318, 321, 329, 348–9, 361, 369, 372, 381, 434, 454–5, 512
Competencies 113
Competencies and expertise 143, 146–9, 167–9, 180, 184–5, 189, 195, 199
Competitive 5, 6, 14, 18, 28, 56, 71, 87, 120, 123, 151, 154, 237, 451
Competitive bid (bidding) 208, 222, 230, 272, 281, 291, 302, 310, 425, 436, 445, 474, 534
Complaints, client 38, 414, 44, 46, 62, 79, 94, 108
Complaints, contractor 62
Compliance 4, 12–3, 15, 27–8, 30, 82, 96, 99, 109, 120, 122, 124, 128, 135, 146, 151, 156, 161, 163–4, 170–2, 177, 191, 198, 201, 212, 218, 221, 232, 272, 283, 314, 319, 336, 408, 425, 433, 458–9, 537
   Design compliance 165, 178, 183
   HSE compliance 179, 183
   Process compliance 220, 224, 238, 243, 251, 258, 268, 278, 280, 299, 301, 313, 319, 369, 408, 414, 416, 433, 437, 456, 465, 467, 474, 481, 485, 491, 497, 504, 524, 526
   Quality compliance 26, 248
   Regulatory compliance 25, 193
   Compliance audit 203
   Compliance with TOR 182, 184, 189, 245, 483
Concept Design 4, 7, 10–1, 23, 75, 95–6, 152, 162–5, 167–8, 170–1, 174, 209–10, 219, 238, 243–5, 247–50, 254, 266, 401–4, 408–10, 412–4, 420, 426, 431, 433, 456, 459, 481–6, 488–9, 491, 544–5
Conceptual Design 6, 7, 21–3, 25, 149, 162–6, 168–10, 237–40, 243, 247, 403, 406, 453, 481, 484
Confidentiality please see integrity, honesty, confidentiality pages 146–9
Conformance 25, 48, 93, 107, 110, 144, 165, 236, 257, 262, 269, 319, 321
Conformance to QMS 220, 224, 238, 245, 251, 258, 271, 278, 283, 299, 313, 323–4, 326, 332, 341, 356, 369, 387, 393, 408, 414, 416, 433, 437, 454, 456, 465, 474, 481, 485, 491, 497, 504, 517, 524, 526
Conservation of energy 245, 251, 258, 265, 271, 408
Cost management 150, 165, 172, 178, 182, 188, 192, 197

Index   557

Constraints 158, 237, 245, 254, 267, 460
   Environmental constraints 248, 255, 267, 275, 412, 460, 484, 490, 496, 503
   Financial constraints 241
   Project constraints 162, 167
   Triple constraints xiii, 201–2, 204
Construction manager 5, 38, 59, 61–2, 96, 124, 151, 205, 207, 278, 323, 522–31, 533–42, 544–6
Construction Project life cycle 5, 9, 10, 149, 154, 162, 175–6, 189, 195, 201, 204, 206, 208, 210, 237, 314, 385, 401, 403, 463
Construction documents 6, 10, 21, 23–5, 60–1, 68, 84, 90, 96, 176, 180–2, 184, 189, 195, 219, 237, 261–2, 268, 271–7, 323, 401, 404, 407–12, 414, 420, 453–4, 456, 458–61, 479, 481, 497, 500, 501–4, 544
Constructor 13, 27, 36
Consultant 4, 5, 13–5, 21, 26, 36, 39, 48, 58, 63, 73, 113, 133, 138–9, 151, 154–6, 159, 208, 212, 249, 278, 292, 299, 301, 367, 404, 433–4, 437, 464, 479
Continual improvement 17, 30–1, 33, 35, 38, 41, 44, 46, 52, 62, 66, 79, 82, 94, 100, 108
Contract award 212, 217, 220, 229, 280, 290, 301, 424, 434, 444, 472, 533
Contract bid solicitation 212, 215, 220, 228, 280, 288, 301, 307, 422, 434, 442, 470, 531
Contractor (Design-Bid-Build) 278–82, 286–7, 290
Contractor (Design-Build) 414, 418, 420, 422, 424, 428
Contractor (Supervisor) 305, 441
Contract documents 1, 4, 13, 15, 18, 21, 23–4, 25–7, 29, 82, 86–7, 96, 99, 149, 162, 189, 195, 201, 218, 223, 230, 233, 246, 251, 255, 258, 261–2, 265, 268, 272–4, 291–2, 299, 307, 309–10, 314, 319–20, 323–4, 326, 329, 335, 348, 351, 358, 361, 364, 367, 369, 371–2, 377, 380, 383, 387, 392–3, 408, 410–1, 460, 473, 488, 490–1, 496, 497, 502,-4, 507–8, 511, 517, 520, 534, 542
Contract management 11, 151, 165, 179, 183, 193, 195, 198, 318, 349, 356, 382, 455, 537
Contractor's construction schedule 9, 89, 192, 299, 326, 346, 359, 369, 378, 433, 509
Contractor's quality control plan 25, 327
Controlling budget 21

Controlling construction expenses 299, 369,433
Controlling construction time 299, 369, 433
Controlling project schedule 21
Control chart 21
Coordination 18, 29, 77, 165, 172–3, 180, 195, 414, 438, 447, 453, 541–2
Coordination meeting 73, 88, 165, 173, 256, 268, 349, 362, 381, 461, 512
Corrective action 31, 38, 41, 44, 46, 62, 78, 93, 107–8, 110, 115, 125, 130–2, 145, 191, 341, 547–52
Cost baseline 360, 379, 510
Cost loaded curve 379, 510
Cost loaded schedule 373, 378, 510
Cost of quality 549
Cost management 150, 165, 172, 178, 182, 188, 192, 197
Customer satisfaction 12, 14, 16–8, 20, 27–8, 30, 32, 35, 109, 121–3, 126

Daily Report 340, 344, 372, 376
Data Collection 164, 168, 170, 174, 180, 233, 264
Data gathering 127
Design approval 461
Design management 456, 459, 464, 467, 469, 473–4, 479, 481, 528, 530, 534, 537, 539, 544
Design management activities 237, 433
Design management experience 404
Design management stages 457–8, 461–2
Design Professional 219, 258,464
Designer please see A/E
Developer 27, 29, 36, 49, 54, 73
Disposition 110
Dispute(s) 5, 15, 154, 193–4, 215, 218–9, 224, 227, 230, 283, 285, 288, 291, 307, 310, 351, 364, 367, 383, 406, 418, 421, 424–5, 442, 445, 455, 470, 473–4, 514, 527, 531, 534

Economical 6, 14, 28, 71, 87, 158, 237, 241, 451
Electronic surveillance system 217, 229, 290, 308, 423, 444, 472, 533
End user 4, 6, 9–11, 14, 21, 155, 196, 201, 208, 242, 385, 396, 399, 463–4, 481, 519, 544
End user 9–11, 14, 21, 155, 196, 201, 208, 242, 385, 396, 464, 481, 519, 544
Energy conservation 172, 174, 246, 248, 254, 459, 483, 485, 488–9, 491, 495, 500

Index

Environmental 2–3, 18, 22–3, 28, 113, 158, 164, 173, 182–3, 242, 246, 248, 253, 267, 461, 496
Environmental constraints please see constraints
Environmental management system 18, 111
Environmental protection 157, 161, 167, 241, 341, 356, 375
Ethical 112, 145–9
EVM 347, 352, 365, 384
Execution process group 15, 164, 171, 177, 182, 187, 191, 197
Expertise please see competencies and expertise
External audit 122, 136–7, 210, 403
External auditor 114, 121, 142, 154, 201

Facility management 245, 251, 254, 261, 265, 271, 408, 459, 483, 488–9, 491, 495, 500
Feasibility study 154–5, 157, 167, 162, 164–5, 167, 238, 241
Financial 160, 162, 194, 198, 211, 213, 223, 238, 241, 246, 285, 302–3, 406, 438, 465
Financial management 151, 166, 173, 179, 183, 188, 194, 198
Financial manager 151, 166, 173, 179, 183, 188, 194, 212, 220, 234, 240, 477, 426, 537
First party audit/auditing 122, 136, 143

Gathering information 146–9, 204
Gathering point 273
Goals and objectives 1, 5, 23, 141, 151, 154–8, 163, 167–9, 174–5, 180, 189, 199, 201–2, 314, 226, 237, 242, 245, 247, 251, 254, 258, 264, 273, 314, 401, 410, 433, 458–9, 481, 483–4, 488–9, 491, 497, 501, 544, 547
Green building 19, 223, 233, 406, 418, 527
Gunaydin H. 123, 553

Hand delivery 217, 226, 229, 235, 287, 290, 297, 305, 308, 316, 420, 423, 430, 440, 443, 447, 468, 472, 478, 529, 532, 539
Health surveillance 367, 385, 515
Honesty please see integrity, honesty, confidentiality pages 146–9
House of Quality 247–8, 250
HSE 74, 86, 103, 168, 174, 179–80, 183, 190, 195, 199–200, 303, 321, 350, 363, 382, 418, 438, 513, 527
HSE Engineer 141, 320, 392

HSE management 151, 166, 173, 179, 183, 188, 193, 198, 326, 387, 396, 517
HSE meeting 349
HSE Plan 9, 329, 543
Huber M. 124, 553

Identification of alternatives 163
Identification of need 163
IEC 29
IEEE 29
Importance of standards 28
Importance of ISO standards 28
Improvement process 34,
Improvement report 549, 550
  Continuous improvement 13, 125
  Continual improvement 17, 30–1, 38, 41, 44, 46, 52, 62, 66, 79, 82, 94, 100, 108
  Organizational improvement 18
  Quality improvement 18, 120, 549
  Skills improvement 119
Independence 112, 114, 156, 161
Independent 112, 117, 120–2, 124, 128, 136, 150, 156, 161, 201
Infrastructure (QMS manual clause) 37, 40, 42, 45, 55, 70, 86, 102
Initiating audit 116
Initiating process 150, 164, 171, 173, 177–8, 182, 187, 191, 197
Inspection and test plan 92, 354, 387, 389
Integrated quality management 18
Integration management 358, 377, 508
Integrity, honesty, confidentiality 146–9
Internal audit (first party audit) 31, 33, 47, 62, 78, 93, 108–11, 114–5, 121–2, 145, 129–30, 134–6, 148, 210, 403, plus references in many figures
Internal auditor 43, 113, 119, 121, 137, 142, 147, 151, 154, 262, 321
ISO 12, 14, 16, 18, 27, 29–31, 36, 47, 109–14, 118–9, 122–5, 146–9
ISO 9001:2015 17, 32–4, 36, 48, 129
ISO audit procedure 167, 169, 174, 180, 185, 189, 195, 199, 327
ISO certification 160, 223, 282, 285

Job site instruction 340, 372, 374, 376, 395, 519

KESAA 147

Laboratory 26, 250, 324, 354, 360, 379, 511
Lead Auditor 122, 124, 129, 136–7, 142–5, 147, 262, 321 plus references in many figures

LEED 160, 168, 174, 223, 248, 254, 265, 405, 418, 459, 484, 489, 495
Litigation 224, 233, 283, 406, 418, 527
Lump sum 159, 276, 401

Mafi S. 120, 124, 553
Management plans 318, 321, 324, 326, 329–30, 335, 358, 375, 377, 455, 507
  Construction management plans 195
  Quality management plans 190, 199
Manpower 14–5, 21, 183, 193, 235, 275, 282–3, 292, 324, 329, 347–8, 359, 361, 378, 380, 412, 417, 425, 479, 503, 510–11, 539
Manufacturer 5, 26–7, 44, 91–2, 97, 297, 333, 342, 431
Manufacturing 3, 5, 12, 20, 26–7, 36, 45–6, 48, 101–6, 108, 110, 113,120, 123, 137, 459
Marcus Vitruvius Polo 1
Matrix:
  Administrative matrix 59, 326
  Authorization matrix 60, 77, 106
  Communication matrix 179, 183, 195, 372
  Responsibility(ies) matrix 22–3, 164, 191, 223, 233, 256, 267–8, 323, 326, 336, 346, 356, 358, 365, 370–1, 377, 393, 406, 455, 461, 508, 517
  Site administration matrix 193, 348, 361, 381, 521
  Stakeholder's matrix 177
Measurable 155, 241
Method statement 19, 26, 92, 101, 180, 193, 195, 327, 336, 345, 372, 374, 380
Middle ages 1
Mistake proofing 174, 180, 261, 264, 266
Mobilization 10, 190–1, 316, 318, 320–1, 323, 339, 358, 369, 371, 377, 455, 508
Model(s) 23, 32, 35–6, 133, 164, 168, 177, 245–6, 248, 253, 483, 485
Monitoring 17, 31, 33, 36–7, 39–41, 45, 49–50, 55, 63–4, 70, 78, 80–1, 90, 97–8, 116, 121, 146, 189–90, 192, 303, 316, 329, 341, 344–5, 351, 364–5, 372, 384, 401, 431, 464–5, 527, 543
Monitoring and controlling 10, 21, 150, 164, 171, 177, 182, 187, 191, 197, 292, 299, 303, 318, 341, 346, 367, 369, 433, 438, 455–6, 514
Monitoring and controlling process group 150, 164, 171, 177, 182, 187, 191, 197
Monthly report 344, 349, 362, 372, 381, 512
Move-in-plan 11, 197, 387, 543

Need analysis 152, 155, 157, 240
Need statement 155, 157, 164
Non conformance 13, 61, 79, 94, 110, 113, 115, 125, 127, 130, 131–2, 144, 297, 341, 372, 375, 395, 431, 519, 547–50
Non destructive 354
Normative references 17, 31, 36, 39, 41, 44, 49, 63, 80, 97, 111
NQA 123
NRA 121

Objectives please refer goals and objectives
Observation 118, 119, 124, 126, 144
Obstacle 143–4
Organizational knowledge 37, 40, 42, 45, 56, 71, 87, 104
Outsource 46, 106
Owner please refer client
Owner's representative 60, 191, 195, 385, 463, 501

Payment 15, 166, 173, 179, 183, 191–2, 194, 198–9, 273, 299, 341, 367, 369, 373,376, 385, 387, 392–3, 396, 398–9, 434, 465, 507, 514, 517, 520, 543
  Advance payment 364, 374, 383, 513
  Interim payment 351, 364, 383, 513
  Progress payment 195, 340, 350–1, 356, 363–4, 372, 375, 383, 510, 513, 543
PCDA 34, 114, 201–2, 550
Penalty clause for; 292, 310, 425, 446, 504
Penalty for delay 218, 246, 275, 412, 503
Planning process group 150, 164, 171, 177, 182, 187, 191, 197
Plan quality 20
PMBOK 150
Post-audit review 118
Pre-audit 123
Precedence 178
Preferred alternative 59, 158, 163, 238, 242
Preliminary/Schematic design 6, 10, 21, 169, 171, 175–6, 219, 249, 251, 433
Prevention 123, 193–4, 547
Preventive action 130, 191–2, 249, 256, 268, 276, 336, 340–1, 352, 365, 412, 485, 491, 497, 504, 549
Project charter 7, 58, 159, 168, 174, 180
Project Quality Plan 37, 39, 42, 52, 67, 83, 93
Punch list 197, 199, 340, 387, 390, 392–3, 398, 517, 520, 543
Pyramid 15–6, 30, 32

560 Index

Qualitative 6, 14, 28, 71, 87–8, 127, 154, 237, 451
Quality assurance 15, 20–4, 192
Quality based system (QBS) 59, 161, 211, 213, 219, 230, 291, 310, 366, 425, 445, 474, 534
Quality control 5, 9, 12, 15, 17, 19, 20–1, 24–6, 123, 190, 303, 318, 327, 341, 375, 438
Quality control plan 25, 319, 326–7, 339, 347, 360, 372, 375–6, 379, 455, 510
Quality goals 17, 292, 299, 367, 433, 456
Quality improvement 18, 120, 549
Quality in construction 12, 14–5, 27
Quality in manufacturing 12, 27
Quality manual 15–7, 26, 29, 30, 36, 39, 41, 44, 49, 63, 80, 83, 97, 127
Quality management 5, 10–2, 14–6, 20, 22, 24, 27, 48, 60, 64, 72, 74, 76, 80, 100, 150, 165, 172, 178, 183, 188, 192, 197, 223, 406, 442
Quality management certification 223, 303, 438
Quality management plan 21, 50, 162, 169, 174, 180, 190, 195, 199, 256, 266, 273, 297, 316, 326, 356, 410, 430, 445, 447
Quality management principles 30
Quality management policy 282, 418
Quality management procedure 237
Quality management process 249, 371
Quality management program 5, 64, 97
Quality management system 15–8, 20, 29–32, 34–6, 39, 42, 50–1, 65, 77, 81, 87–9, 109, 113–5, 121, 129, 133–4, 146, 151, 214, 226, 233, 235, 256, 258, 271, 296, 305, 315, 326, 341, 348, 361, 380, 408, 418, 420, 430, 440, 446, 469, 478, 500, 511, 530, 538, 547
Quality management tools 549
Quality planning 21, 22
Quality policy 5–6, 30, 33, 36–7, 39, 42, 45, 47, 51–2, 54, 66, 82–3, 99, 100, 144
Quality principles 14, 15, 32
Quality system 12, 14, 17, 20, 27, 29, 42, 45, 69, 73, 89, 105, 115, 120–3, 126, 129, 137, 144, 146, 220, 238, 245, 251, 258, 271, 278, 283, 299, 313, 319, 323–4, 326, 332, 341, 356, 369, 387, 393, 408, 414, 416, 433, 437, 454, 456, 465, 474, 481, 485, 497, 504, 517, 524, 526
Quantitative 127

Realistic 146–8, 241
Regulatory approval please see Authority approval
Resource management 9, 151, 178, 183, 190, 193, 195, 198, 200, 318, 329, 455, 465
Responsibilities matrix please see matrix
Review of submittal 231
Rework 14, 193, 338, 341, 352, 365, 384
RFI 85, 315, 346, 359, 372–3, 377, 447, 507, 509, 542
RFP 222, 224, 226–8, 231, 233–5, 468–71, 577–8, 529–32, 538–9
Risk management 9, 11, 19, 26, 42, 44–5, 53–4, 67–8, 84, 94, 101–2, 125, 151, 165, 173, 179, 183, 188, 193, 198, 326, 336, 349, 352, 356, 362, 366, 372, 381, 384, 387, 392–3, 398, 512, 517, 420, 541
Roles and responsibilities 138, 142–7, 165, 204, 363, 371, 382, 404, 469, 524, 529

S-Curve 9, 318, 326, 329, 333, 347, 350, 360, 363, 379, 383, 455, 510, 513
Safety management plan 193, 319, 372, 376
Safety officer 140
Samples 13, 112, 126, 128, 320, 336, 354, 390
Schedule control 192, 195
Schedule management 150, 165, 172, 178, 182, 187, 192, 197, 541
Schematic design please see Preliminary design
Scope management 150, 164, 172, 177, 182, 187, 191, 197
Scope of work (SOW) 9, 12–3, 15, 20, 22–3, 27, 161, 218–9, 226, 233–4, 245, 247, 254, 264, 273, 292, 296, 415, 319–20, 346, 351, 356, 358–9, 364, 371, 377, 392–3, 398, 401, 404, 407, 429, 446, 477, 509, 517, 520, 538, 541
Second party audit 111, 122, 136–7, 142–3, 147, 210, 403
Security 12, 250
Selection of auditor 114, 149, 151
Selection of certification body 47
Shop drawings approval 347, 360, 379
Shortlisting 156, 159, 220, 225–6, 280, 286–7, 301, 304–5, 418, 420, 434, 439–40, 467–8, 528–9
Shraim M. 109, 124, 553
Site investigation 174, 177, 180, 256, 267, 461
Site work instruction (SWI) 346, 359, 372, 378, 509
Site safety 11, 193, 340–1
Six sigma 549

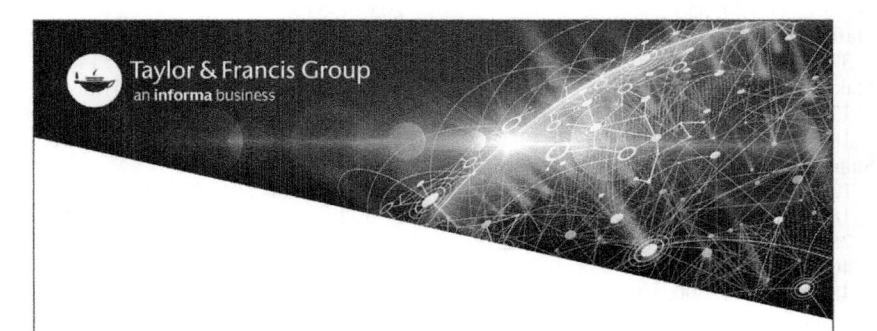

Sketches 168, 174, 253, 333
SMART 241
Sowards D. 124, 553
Staff approval 190, 318, 339, 372, 374, 376, 455
Stakeholder management 150, 164, 169, 171, 174, 177, 180, 182, 184, 187, 189–91, 195, 197, 199, 200, 527
Standards and codes 14, 21–3, 75, 91, 96, 120, 165, 168, 172, 174–5, 178, 180, 183, 195, 245, 247, 251, 254, 257–8, 264–5, 271, 273, 277, 286, 319, 372, 408, 410, 433, 454, 456, 458–9, 483–4, 489, 491, 494, 500, 502, 547
Statutory approval please see regulatory approval
Subcontractor approval 318, 374, 376, 455
Subject matter expert 137, 147, 149, 157, 162, 167–9, 175, 184–5, 190, 199, 262, 321
Submittal procedure 15, 19, 224, 327, 341
Submittal process 313, 326, 332, 437, 474, 526
Submittal plan 22–3
Submittal status log 326, 331, 375
Surveillance audit 123
Sustainability 158, 164, 242, 245, 248, 251, 254, 265, 271, 408, 458, 483, 485, 488–9, 491, 495, 500
Systems engineering 6, 32, 212, 237

Terms of Reference (TOR) 22, 154, 159, 161, 167, 234, 237–8, 243, 245, 262, 264, 477, 538
Test procedure please see inspection and test procedure

Third party 111, 122
Third party audit 122–3, 135–7, 143, 147, 203–4, 513
Third party certification 122–3, 350, 363, 382
Tools and techniques 14, 89, 129, 142, 146, 148–9, 235, 297, 316, 351, 365, 430, 447, 478, 539
Traditional construction project 4, 5, 36, 133
Triple constraints 201–2, 204
TUV 123

User's personnel 9, 195, 385, 463

Validate 9, 121, 224, 238, 245, 251, 258, 271, 283, 314, 319, 356, 358, 371, 373, 387, 393, 408, 425, 433, 437, 454, 456, 458, 474, 483, 488, 500, 504, 507, 517, 537
Value engineering 8, 11, 19–20, 169–70, 172, 174, 209, 223, 246, 251, 253, 255, 259, 265, 406, 459, 488, 490, 495, 527, 541
Variation order 191–2, 218, 299, 347, 360, 369, 371, 374, 376, 379, 434, 504, 510
Visual or non-destructive test 354

Weightage 160, 216, 225, 228, 285–6, 289, 419, 423, 468, 471, 528, 532
Work breakdown structure (WBS) 6–7, 149, 237, 258
Working drawings 23–4, 180, 182, 261, 271–5, 283, 500–03
Workmanship 12, 27–8, 120, 123
Workshop drawing 347, 360, 374, 379, 511